Lecture Notes in Computer Science 5250

Commenced Publication in 1973
Founding and Former Series Editors:
Gerhard Goos, Juris Hartmanis, and Jan van Leeuwen

Nadia Creignou Phokion G. Kolaitis
Heribert Vollmer (Eds.)

Complexity of Constraints

An Overview of Current Research Themes

Volume Editors

Nadia Creignou
LIF (CNRS UMR 6166)
Aix-Marseille Université
Marseille, France
E-mail: creignou@lif.univ-mrs.fr

Phokion G. Kolaitis
IBM Almaden Research Center
E-mail: kolaitis@almaden.ibm.com
and
Computer Science Department
University of California, Santa Cruz
E-mail: kolaitis@cs.ucsc.edu

Heribert Vollmer
Institut für Theoretische Informatik
Leibniz Universität Hannover
Hannover, Germany
E-mail: vollmer@thi.uni-hannover.de

Library of Congress Control Number: Applied for

CR Subject Classification (1998): F.2, E.1, G.2, I.2.8, I.3.5, G.1

LNCS Sublibrary: SL 1 – Theoretical Computer Science and General Issues

ISSN 0302-9743
ISBN-10 3-540-92799-9 Springer Berlin Heidelberg New York
ISBN-13 978-3-540-92799-0 Springer Berlin Heidelberg New York

springer.com

Printed in Germany

Typesetting: Camera-ready by author, data conversion by Scientific Publishing Services, Chennai, India
Printed on acid-free paper SPIN: 12592739 06/3180 5 4 3 2 1 0

Preface

In October 2006, the editors of this volume organized a Dagstuhl Seminar on "Complexity of Constraints" at the Schloss Dagstuhl Leibniz Center for Informatics in Wadern, Germany. This event consisted of both invited and contributed talks by some of the approximately 40 participants, as well as problem sessions and informal discussions. After the conclusion of the seminar, the organizers invited a number of speakers to write surveys presenting the state-of-the-art knowledge in their area of expertise. These contributions were peer-reviewed by experts in the field and revised before they were included in this volume. In addition, this volume contains a reprint of a survey by P.G. Kolaitis and M.Y. Vardi on the logical approach to constraint satisfaction that first appeared in "Finite Model Theory and Its Applications," (Springer 2007).

We thank the Directorate of Schloss Dagtuhl for its support, the speakers of the seminar for making it a successful event, and, above all, the contributors to this volume for their informative and well-written surveys. We also thank Arne Meier for technical assistance during the final compilation of this book, and Alfred Hofmann at Springer for his support and guidance.

July 1 (the birthday of Gottfried Wilhelm Leibniz) 2008

Nadia Creignou
Phokion G. Kolaitis
Heribert Vollmer

Table of Contents

Introduction

The first systematic complexity-theoretic study of constraints was carried out by T.J. Schaefer in 1978 with a paper on Boolean constraint satisfaction problems (CSPs). This volume opens with a survey of Boolean constraints by N. Creignou and H. Vollmer. Schaefer proved a *Dichotomy Theorem* about the satisfiability problem for Boolean CSPs, which asserts that each Boolean CSP is either NP-complete or in P (hence, assuming P $\neq$ NP, all infinitely many intermediate complexity degrees are avoided). Creignou and Vollmer present a modern algebraic proof of Schaefer's Dichotomy Theorem, a proof that makes use of Galois theory and of the structure of the lattice of Boolean clones, known as *Post's lattice*. In addition to the satisfiability problem, several other algorithmic problems, including counting, optimization, and circumscription, can be completely classified from a complexity point of view using Post's lattice. The first contribution to this volume surveys these results and poses the question: what is so special about certain algorithmic tasks that makes Post's lattice applicable to their classification.

The realm of Boolean universes is by now quite well understood. When moving to larger finite domains, the complexity-theoretic study of CSPs relies on Galois connections. The second contribution to this volume, by F. Börner, presents a systematic introduction to Galois theory, to the widely used Pol-Inv Galois connection, and also to variants of this Galois connection that only very recently have turned out to be important for the study of CSPs.

One complication that arises when moving from 2-element universes to larger ones is that the lattice of clones is well-understood only for the former case. As a matter of fact, this lattice becomes uncountable even for 3-element domains. Consequently, even though the algebraic underpinnings remain the same, the situation now becomes much more complicated in a fundamental way. In 1993, T. Feder and M.Y. Vardi articulated their *Dichotomy Conjecture* to the effect that the satisfiability problem will avoid all degrees between 0 (polynomial time) and 1 (NP-complete) in all finite domains. This conjecture was proved by A. Bulatov for 3-element universes in 2002, but it remains open to this day for all domains of cardinality bigger than 3. The survey by A. Bulatov and M. Valeriote describes recent algebraic attacks to the Feder-Vardi Dichotomy Conjecture, leading to some rather unexpected universal-algebraic formulations of this conjecture in the language of tame congruence theory.

The survey by A. Bulatov, A. Krokhin, and B. Larose focuses on *dualities* and their relationship to the Feder-Vardi Dichotomy Conjecture. In fact, the class of CSPs that exhibit the so called bounded treewidth duality is one of the largest classes with a tractable satisfiability problem. Additional large tractable classes can be identified using a logical approach, as described in the survey by Kolaitis and Vardi.

N. Creignou et al. (Eds.): Complexity of Constraints, LNCS 5250, pp. 1–2, 2008.

An important line of research identifies "islands of tractability" for the *uniform constraint satisfaction problem*, where not only the conjunctive query but also the algebraic structure of the constraints is part of the input. Interestingly, this field has a tight connection to database theory, and many algorithmic approaches from there lead to significant progress in the complexity of (uniform) CSPs. This is the topic of the survey by F. Scarcello, G. Gottlob, and S. Greco.

In another variant of CSPs, which has been studied only very recently, one considers the situation where the constraints are defined over an infinite domain. The survey by M. Bodirsky presents important examples of such problems, shows how the universal-algebraic approach finds applications here, and overviews both the main results and certain fundamental questions that remain open.

The contribution by H. Schnoor and I. Schnoor returns to question of determining for which algorithmic problems Post's lattice may help. While the usual Pol-Inv Galois connection is tailored towards the satisfiability problem (and for additional algorithmic tasks), Shnoor and Schnoor show that a similar Galois connection involving clones of partial functions is potentially applicable to a wide range of algorithmic problems. Unfortunately, even in the Boolean case, the lattice of partial clones is uncountable; nonetheless, Schnoor and Schnoor show that for such algorithmic problems as enumeration, equivalence, and many others, a countable spine of this lattice can be identified that is good enough for obtaining a complexity-theoretic classification.

The interesting issue of the complexity of approximation for optimization problems makes also sense in the context of CSPs. The survey by P. Jonsson and G. Nordh introduces an optimization variant of the constraint satisfaction problem, and presents complexity and approximability results. This variant, which associates a weight to every solution, captures many well-known combinatorial optimization problems. Thus, many different problems can be given a uniform treatment when it comes to solve them or to analyze their complexity.

Formulating algorithmic problems as propositional satisfiability (SAT) problems has become an important problem-solving technique that competes with the CSP approach. In particular, there is a considerable interest in transferring SAT techniques to the CSP context. In the last contribution to this volume, O. Kullmann presents an overview of current paradigms of SAT solving.

Boolean Constraint Satisfaction Problems: When Does Post's Lattice Help?

Nadia Creignou[1] and Heribert Vollmer[2]

[1] LIF (CNRS UMR 6166), Aix-Marseille Université, 163 avenue de Luminy, F-13288 Marseille, France
creignou@lif.univ-mrs.fr
[2] Institut für Theoretische Informatik, Leibniz Universität Hannover, Appelstr. 4, D-30167 Hannover, Germany
vollmer@thi.uni-hannover.de

1 Satisfiability Problems

The propositional satisfiability problem SAT, i.e., the problem to decide, given a propositional formula ϕ (without loss of generality in conjunctive normal form CNF), if there is an assignment to the variables in ϕ that satisfies ϕ, is the historically first and standard NP-complete problem [Coo71]. However, there are well-known syntactic restrictions for which satisfiability is efficiently decidable, for example if every clause in the CNF formula has at most two literals (2CNF formulas) or if every clause has at most one positive literal (Horn formulas) or at most one negative literal (dual Horn formulas), see [KL99]. To study this phenomenon more generally, we study formulas with "clauses" of arbitrary shapes, i.e., consisting of applying arbitrary relations $R \subseteq \{0,1\}^k$ to (not necessarily distinct) variables $x_1, \ldots, x_k$. A *constraint language* Γ is a finite set of such relations. In the rest of this chapter, Γ and Γ' will always denote Boolean constraint languages. A Γ*-formula* is a conjunction of clauses $R(x_1, \ldots, x_k)$ as above using only relations R from Γ. The for us central family of algorithmic problems, parameterized by a constraint language Γ, now is the problem to determine satisfiability of a given Γ-formula, denoted by $\mathrm{CSP}(\Gamma)$.

The NP-complete problem 3SAT, the satisfiability problem for CNF formulas with exactly three literals per clause, now is the problem $\mathrm{CSP}(\Gamma_{3\mathrm{SAT}})$, where $\Gamma_{3\mathrm{SAT}} = \{x \vee y \vee z, x \vee y \vee \neg z, x \vee \neg y \vee \neg z, \neg x \vee \neg y \vee \neg z\}$; here and in the sequel we do not distinguish between a formula ϕ and the logical relation R_ϕ it defines, i.e., the relation consisting of all satisfying assignments of ϕ. If every relation in Γ is definable by a Horn formula, then $\mathrm{CSP}(\Gamma)$ is polynomial-time decidable, also if every relation in Γ is definable by a 2-CNF formula. Hence we see that the family of problems $\mathrm{CSP}(\Gamma)$ has NP-complete members as well as easily solvable members.

A question attacked by Thomas Schaefer [Sch78] is the following: Can we determine for each constraint language Γ the complexity of $\mathrm{CSP}(\Gamma)$? Is there even a simple algorithm that, given Γ, determines the complexity of $\mathrm{CSP}(\Gamma)$? Are there more cases than NP-complete and polynomial-time solvable? In this chapter we will present a way to answer these questions that relies on notions

N. Creignou et al. (Eds.): Complexity of Constraints, LNCS 5250, pp. 3–37, 2008.

and results from universal algebra. A central rôle in our development will be an exploitation of the structure of *Post's lattice* of all Boolean clones, all classes of Boolean functions closed under superposition (composition). Post's lattice will turn out to be a very helpful tool to classify the complexity of CSP(Γ), but also of related algorithmic problems for Γ-formulas and generalizations thereof such as quantified Boolean formulas, for instance counting the number of satisfying assignments of quantified Boolean formulas, model checking for circumscription (minimal satisfiability), and many more.

In this chapter we will first present a full account of Schaefer's Theorem and related results for quantified formulas. Then we will survey complexity classifications obtained for many further computational problems for Boolean constraint satisfaction problems in the recent past, with a particular emphasis on the question when Post's lattice can be used in obtaining the classification and when not.

2 Background from Universal Algebra

A *logical relation* (or *constraint relation*) of arity k is a relation $R \subseteq \{0,1\}^k$. A *constraint* (or *constraint application*) is a formula $R(x_1, \ldots, k_k)$, where R is a logical relation of arity k and the $x_1, \ldots, x_k$ are (not necessarily distinct) variables. An assignment I of truth values to the variables *satisfies* the constraint if $\big(I(x_1), \ldots, I(x_k)\big) \in R$. A *constraint language* Γ is a finite set of logical relations. A *Γ-formula* is a conjunction of constraint applications using only logical relations from Γ. Such a formula ϕ is satisfied by an assignment I if I satisfies all constraints in ϕ simultaneously.

Problem:	CSP(Γ)
Input:	a Γ-formula ϕ
Question:	Is ϕ satisfiable, i.e., is there an assignment that satisfies ϕ?

When we want to determine the complexity of all CSP-problems, we will certainly need a way to compare the complexity of CSP(Γ) and CSP(Γ') for different constraint languages Γ and Γ'. For example, to show that some CSP(Γ) is NP-complete we might show that using Γ we can " simulate" or "implement" all relations in Γ_{3SAT}, and to show that CSP(Γ) is polynomial-time decidable we might implement all relations in Γ using Horn-formulas. As it turns out, a useful notion of implementation comes from universal algebra, from clone theory.

Definition 2.1. For a constraint language Γ, let $\langle\Gamma\rangle$, the *relational clone* (or *co-clone*) generated by Γ, be the smallest set of relations such that

- $\langle\Gamma\rangle$ contains the equality relation and all relations in Γ, and
- $\langle\Gamma\rangle$ is closed under primitive positive definitions, i.e., if ϕ is a $\langle\Gamma\rangle$-formula and $R(x_1, \ldots, x_n) \equiv \exists y_1 \ldots y_\ell\ \phi(x_1, \ldots, x_n, y_1, \ldots, y_\ell)$, then $R \in \langle\Gamma\rangle$.

Intuitively, $\langle\Gamma\rangle$ contains all relations that can be implemented by Γ and is thus called the *expressive power* of Γ, as justified by the following observation:

Proposition 2.2. *If* $\Gamma \subseteq \langle \Gamma' \rangle$ *then* $\text{CSP}(\Gamma) \leq_m^{\log} \text{CSP}(\Gamma')$.

Proof. Let ϕ be a Γ-formula. We construct a formula ϕ' by performing the following steps:

- Replace every constraint from Γ by its defining existentially quantified $(\Gamma' \cup \{=\})$-formula.
- Delete existential quantifiers.
- Delete equality clauses and replace all variables that are connected via a chain of equality constraints by a common new variable.

Then, obviously, ϕ' is a Γ'-formula, and moreover, ϕ is satisfiable if and only if ϕ' is satisfiable. The complexity of the above transformation is dominated by the last step, which is essentially an instance of the undirected graph reachability problem, which is solvable in logarithmic space [Rei05]. Hence we conclude that $\text{CSP}(\Gamma)$ is reducible to $\text{CSP}(\Gamma')$ under logspace reductions. □

In particular, we thus have shown that the complexity of $\text{CSP}(\Gamma)$ depends only on $\langle \Gamma \rangle$ in the following sense:

Proposition 2.3. *If* $\langle \Gamma \rangle = \langle \Gamma' \rangle$, *then* $\text{CSP}(\Gamma) \equiv_m^{\log} \text{CSP}(\Gamma')$,

Thus, we "only" have to study co-clones in order to obtain a full classification, and the question arises what co-clones there are. Astonishingly, all co-clones, each with a "simple" basis, are known. The key to obtain this list is to study closure properties of relations.

Definition 2.4. Let $f\colon \{0,1\}^m \to \{0,1\}$ and $R \subseteq \{0,1\}^n$. We say that f *preserves* R, $f \approx R$, if for all $x_1, \ldots, x_m \in R$, where $x_i = (x_i[1], x_i[2], \ldots, x_i[n])$, we have

$$\Big(f\big(x_1[1], \cdots, x_m[1]\big), f\big(x_1[2], \cdots, x_m[2]\big), \ldots, f\big(x_1[n], \cdots, x_m[n]\big)\Big) \in R.$$

In other words, $f \approx R$ if the coordinate-wise application of f to a sequence of m vectors in R always results in a vector that again is in R. Then we also say that R is *invariant* under f or that f is a *polymorphism* of R, and for a set of relations Γ we write $\text{Pol}(\Gamma)$ to denote the set of all polymorphisms of Γ, i.e., the set of all Boolean functions that preserve every relation in Γ. For technical reasons, we will exclude the empty relation (constraint) and nullary polymorphisms in the rest of this paper.

It is now straightforward to verify that for every Γ, $\text{Pol}(\Gamma)$ is a *clone*, i.e., a set of Boolean functions that contains all projections (all functions $\text{I}_k^n(x_1, \ldots, x_n) = x_k$ for $1 \leq k \leq n$) and is closed under composition; the smallest clone containing a set B of Boolean functions will be denoted by $[B]$ in the sequel (B is also called a *basis* for $[B]$). In fact, the connection between clones and relational clones is much tighter. For a set B of Boolean functions, let $\text{Inv}(B)$ denote the set of all *invariants* of B, i.e., the set of all Boolean relations that are preserved by every function in B. It can be observed that each $\text{Inv}(B)$ is a relational clone.

As shown first in [Gei68, BKKR69] (see also [Lau06, Sect. 2.9]), the operators Pol-Inv constitute a Galois correspondence between the lattice of sets of Boolean relations and the lattice of sets of Boolean functions. In particular, for every set Γ of Boolean relations and every set B be of Boolean functions,

Proposition 2.5. – $\mathrm{Inv}\big(\mathrm{Pol}(\Gamma)\big) = \langle\Gamma\rangle$,
– $\mathrm{Pol}\big(\mathrm{Inv}(B)\big) = [B]$.

Hence, Proposition 2.2 can equivalently be stated as follows:

Proposition 2.6. *If* $\mathrm{Pol}(\Gamma) \supseteq \mathrm{Pol}(\Gamma')$ *then* $\textsc{Csp}(\Gamma) \leq_m^{\log} \textsc{Csp}(\Gamma')$.

Thus, there is a one-to-one correspondence between clones and co-clones and we may compile a full list of relational clones from the list of clones obtained by Emil Post in [Pos20, Pos41]. In these papers, Post presented a complete list of Boolean clones, the inclusion structure among them (see Fig. 1), and a finite basis for each of them (Fig. 2). We do not have enough space here to give a full account of *Post's lattice*, as the structure became known, but we refer the interested reader to [Pip97] for a gentle introduction to clones, co-clones, the Galois connection, and Post's results. A rigorous comprehensive study is given in [Lau06]. Complexity-theoretic applications of Post's lattice in the constraint context but also the Boolean circuit context are surveyed in [BCRV03, BCRV04]. A compilation of all co-clones with simple bases is given in [BRSV05].

For the purpose of this paper, we define the clones by simply giving a basis for each of them, see Fig. 2, i.e., the third column of the table gives for each clone its defining basis. One function appearing in the bases that is maybe not so familiar is the threshold function T_k^n, where $\mathrm{T}_k^n(x_1,\ldots,x_n) = 1 \iff \sum_{i=1}^n x_i \geq k$. Also, for a function f, dual(f), the dual function of f, is given by $\mathrm{dual}\big(f(a_1,\ldots,a_n)\big) = \overline{f(\overline{a_1},\ldots,\overline{a_n})}$.

Let us turn in a little bit more detail to that part of the lattice that will be important here. First, let us give an additional definition. Assuming a canonical order on variables, one can regard assignments as tuples. Thus, with each quantifier free propositional formula ϕ one can associate the relation R_ϕ of all satisfying assignments of ϕ. In the following, we say that a relation R is *defined by* a formula ϕ if $R = R_\phi$. The clone generated by the logical AND function is denoted by E_2. A relation is preserved by AND if and only if it is Horn, that is, definable by a Horn-formula, i.e., $\mathrm{Inv}(\mathrm{E}_2)$ is the set of all Horn relations. Similarly, $\mathrm{V}_2 = [\{\mathrm{OR}\}]$, and $\mathrm{Inv}(\mathrm{V}_2)$ is the set of all dual Horn relations. Relations definable by 2CNF formulas, the so-called *bijunctive* relations, are exactly those in $\mathrm{Inv}(\mathrm{D}_2)$, where D_2 is the clone generated by the 3-ary majority function. Finally, the clone L_2 is generated by the 3-ary exclusive-or $x \oplus y \oplus z$ (the 3-ary addition in GF[2]), and $\mathrm{Inv}(\mathrm{L}_2)$ is the set of all *affine* formulas, i.e., conjunctions of XOR-clauses (consisting of an XOR of some variables plus maybe the constant 1)—these formulas may also be seen as systems of linear equations over GF[2].

Let us say that a constraint language is *Schaefer*, if it belongs to one of the above four types, i.e., Γ is Horn (i.e., every relation in Γ is Horn), dual Horn, bijunctive, or affine. If Γ is Schaefer then $\textsc{Csp}(\Gamma)$ is polynomial-time

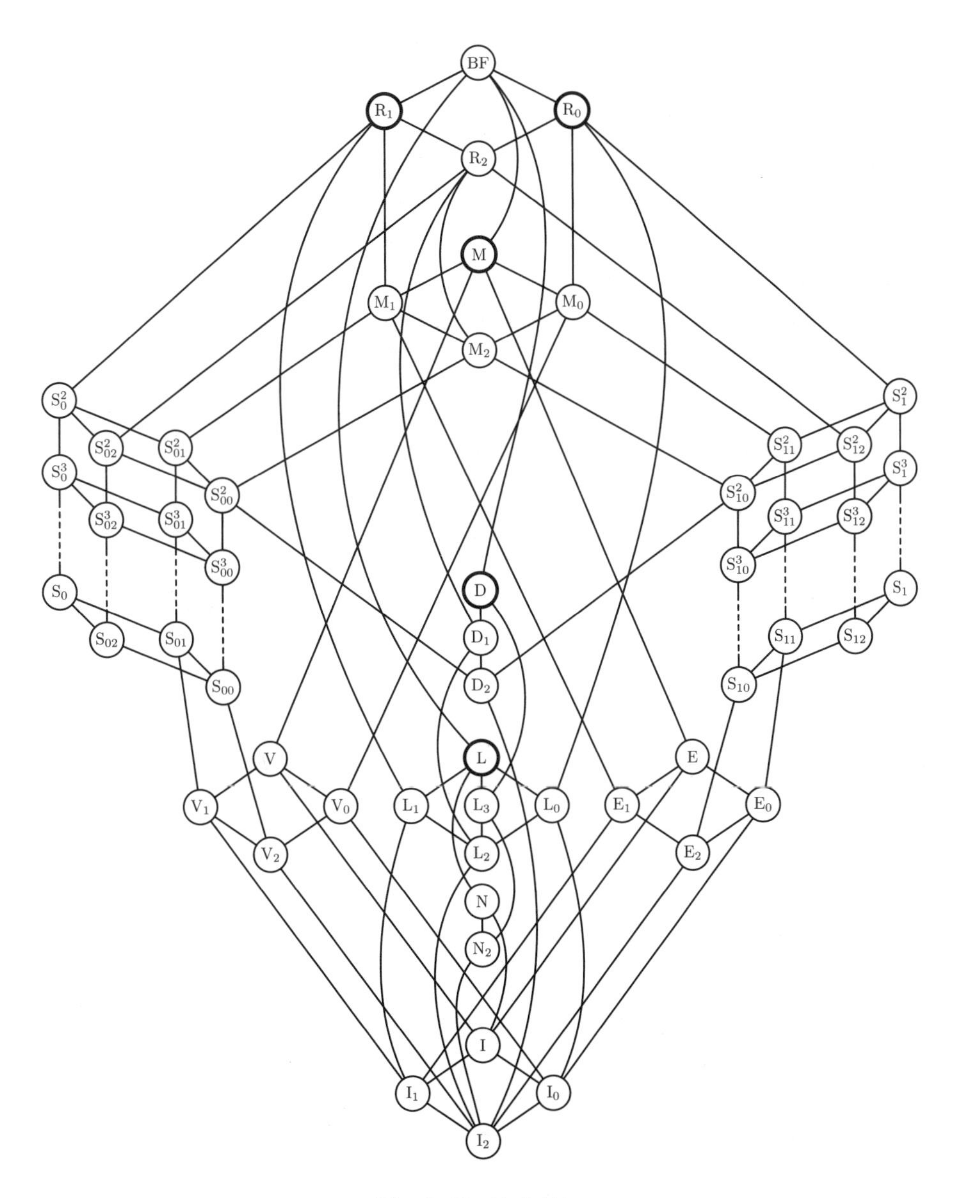

Fig. 1. Post's lattice

solvable, as already noted above for the cases Horn, dual Horn, and bijunctive; for the remaining case of affine relations we remark that we use the interpretation as equations over GF[2] and thus may check satisfiability efficiently using the Gaussian algorithm. (A detailed exposition can be found in [CKS01].)

There is a unique minimal relational clone that is not Schaefer: this is the co-clone Inv(N), where the clone N is generated by the negation function NOT plus the Boolean constants 0, 1. This relational clone consists of all relations

Class	Description	Base
BF	all Boolean functions	$\{\wedge, \neg\}$
R_0	0-reproducing functions	$\{\wedge, \oplus\}$
R_1	1-reproducing functions	$\{\vee, x \oplus y \oplus 1\}$
R_2	$\mathrm{R}_1 \cap \mathrm{R}_0$	$\{\vee, x \wedge (y \oplus z \oplus 1)\}$
M	monotone functions	$\{\wedge, \vee, 0, 1\}$
M_1	$\mathrm{M} \cap \mathrm{R}_1$	$\{\wedge, \vee, 1\}$
M_0	$\mathrm{M} \cap \mathrm{R}_0$	$\{\wedge, \vee, 0\}$
M_2	$\mathrm{M} \cap \mathrm{R}_2$	$\{\wedge, \vee\}$
S_0^n	functions that are 0-separating of degree n	$\{\rightarrow, \mathrm{dual}(\mathrm{T}_n^{n+1})\}$
S_0	0-separating functions	$\{\rightarrow\}$
S_1^n	functions that are 1-separating of degree n	$\{x \wedge \overline{y},\ \mathrm{T}_n^{n+1}\}$
S_1	1-separating functions	$\{x \wedge \overline{y}\}$
S_{02}^n	$\mathrm{S}_0^n \cap \mathrm{R}_2$	$\{x \vee (y \wedge \overline{z}), \mathrm{dual}(\mathrm{T}_n^{n+1})\}$
S_{02}	$\mathrm{S}_0 \cap \mathrm{R}_2$	$\{x \vee (y \wedge \overline{z})\}$
S_{01}^n	$\mathrm{S}_0^n \cap \mathrm{M}$	$\{\mathrm{dual}(\mathrm{T}_n^{n+1}), 1\}$
S_{01}	$\mathrm{S}_0 \cap \mathrm{M}$	$\{x \vee (y \wedge z), 1\}$
S_{00}^n	$\mathrm{S}_0^n \cap \mathrm{R}_2 \cap \mathrm{M}$	$\{x \vee (y \wedge z), \mathrm{dual}(\mathrm{T}_n^{n+1})\}$
S_{00}	$\mathrm{S}_0 \cap \mathrm{R}_2 \cap \mathrm{M}$	$\{x \vee (y \wedge z)\}$
S_{12}^n	$\mathrm{S}_1^n \cap \mathrm{R}_2$	$\{x \wedge (y \vee \overline{z}), \mathrm{T}_n^{n+1}\}$
S_{12}	$\mathrm{S}_1 \cap \mathrm{R}_2$	$\{x \wedge (y \vee \overline{z})\}$
S_{11}^n	$\mathrm{S}_1^n \cap \mathrm{M}$	$\{\mathrm{T}_n^{n+1}, 0\}$
S_{11}	$\mathrm{S}_1 \cap \mathrm{M}$	$\{x \wedge (y \vee z), 0\}$
S_{10}^n	$\mathrm{S}_1^n \cap \mathrm{R}_2 \cap \mathrm{M}$	$\{x \wedge (y \vee z), \mathrm{T}_n^{n+1}\}$
S_{10}	$\mathrm{S}_1 \cap \mathrm{R}_2 \cap \mathrm{M}$	$\{x \wedge (y \vee z)\}$
D	self-dual functions	$\{x\overline{y} \vee x\overline{z} \vee \overline{yz}\}$
D_1	$\mathrm{D} \cap \mathrm{R}_2$	$\{xy \vee x\overline{z} \vee y\overline{z}\}$
D_2	$\mathrm{D} \cap \mathrm{M}$	$\{\mathrm{T}_2^3\}$
L	linear functions	$\{\oplus, 1\}$
L_0	$\mathrm{L} \cap \mathrm{R}_0$	$\{\oplus\}$
L_1	$\mathrm{L} \cap \mathrm{R}_1$	$\{\equiv\}$
L_2	$\mathrm{L} \cap \mathrm{R}_2$	$\{x \oplus y \oplus z\}$
L_3	$\mathrm{L} \cap \mathrm{D}$	$\{x \oplus y \oplus z \oplus 1\}$
V	$\vee$-functions plus constant functions	$\{\vee, 0, 1\}$
V_0	$[\{\vee\}] \cup [\{0\}]$	$\{\vee, 0\}$
V_1	$[\{\vee\}] \cup [\{1\}]$	$\{\vee, 1\}$
V_2	$[\{\vee\}]$	$\{\vee\}$
E	$\wedge$-functions plus constant functions	$\{\wedge, 0, 1\}$
E_0	$[\{\wedge\}] \cup [\{0\}]$	$\{\wedge, 0\}$
E_1	$[\{\wedge\}] \cup [\{1\}]$	$\{\wedge, 1\}$
E_2	$[\{\wedge\}]$	$\{\wedge\}$
N	$[\{\neg\}] \cup [\{0\}] \cup [\{1\}]$	$\{\neg, 1\}$, $\{\neg, 0\}$
N_2	$[\{\neg\}]$	$\{\neg\}$
I	$\mathrm{I}_2 \cup [\{1\}] \cup [\{0\}]$	$\{0, 1\}$
I_0	$\mathrm{I}_2 \cup [\{0\}]$	$\{0\}$
I_1	$\mathrm{I}_2 \cup [\{1\}]$	$\{1\}$
I_2	all projections	$\emptyset$

Fig. 2. List of all Boolean clones with their defining bases

$\text{Pol}(R) \supseteq \text{V}_2$	$\Leftrightarrow$	R is dual Horn	$\text{Pol}(R) \supseteq \text{E}_2$	$\Leftrightarrow$	R is Horn
$\text{Pol}(R) \supseteq \text{V}_0$	$\Leftrightarrow$	R is definite dual Horn	$\text{Pol}(R) \supseteq \text{E}_1$	$\Leftrightarrow$	R is definite Horn
$\text{Pol}(R) \supseteq \text{S}_{00}$	$\Leftrightarrow$	R is IHSB+	$\text{Pol}(R) \supseteq \text{S}_{10}$	$\Leftrightarrow$	R is IHSB−
$\text{Pol}(R) \supseteq \text{L}_2$	$\Leftrightarrow$	R is affine	$\text{Pol}(R) \supseteq \text{D}_2$	$\Leftrightarrow$	R is bijunctive
$\text{Pol}(R) \supseteq \text{D}_1$	$\Leftrightarrow$	R is affine with width 2	$\text{Pol}(R) \supseteq \text{N}_2$	$\Leftrightarrow$	R is complementive
$\text{Pol}(R) \supseteq \text{I}_1$	$\Leftrightarrow$	R is 1-valid	$\text{Pol}(R) \supseteq \text{I}_0$	$\Leftrightarrow$	R is 0-valid
$\text{Pol}(R) \supseteq \text{I}_2$	$\Leftrightarrow$	R is any relation	$\text{Pol}(R) \supseteq \text{N}$	$\Leftrightarrow$	R is compl., 0- and 1-valid

Fig. 3. Characterizations of some classes of relations

that are at the same time *complementive* (negating all entries of a tuple in the relations leads again to a tuple in the relation), *1-valid* (the all-1 tuple is in the relation), and *0-valid* (the all-0 tuple is in the relation). Because of these latter two properties, satisfiability for CSPs build using only relations from Inv(N) is again efficiently decidable (in fact, they are all satisfiable). If we drop the requirement 1-valid and 0-valid we arrive at the relational clone $\text{Inv}(\text{N}_2)$ consisting of all complementive relations ($\text{N}_2 = [\{\text{NOT}\}]$). Obviously, $\text{Inv}(\text{N}) \subseteq \text{Inv}(\text{N}_2)$, and from Post's lattice it can be seen that there is no relational clone in between. The only super-co-clone of $\text{Inv}(\text{N}_2)$ is the co-clone $\text{Inv}(\text{I}_2)$ of all relations ($\text{I}_2 = [\emptyset]$). These remarks are summarized in Fig. 3.

For some of the classifications we give below, we want to introduce other classes of relations. Let us first introduce classes of formulas which form subclasses of and are less expressive than the class of Horn and dual Horn formulas, namely definite Horn and IHSB (for implicative hitting set bounded); for more background the reader is asked to consult [CKS01]. A *definite Horn* (resp. *definite dual Horn*) formula is a CNF formula having exactly one positive (resp., negative) literal in each clause. A clause is said to be IHSB− if it is of one of the following types: (x_i), $(\neg x_{i_1} \lor x_{i_2})$ or $(\neg x_{i_1} \lor \ldots \lor \neg x_{i_k})$ for some $k \geq 1$. Dually, a clause is said to be IHSB+ if it is of one of the following types: $(\neg x_i)$, $(\neg x_{i_1} \lor x_{i_2})$ or $(x_{i_1} \lor \ldots \lor x_{i_k})$ for some $k \geq 1$. Finally, a formula is said to be IHSB− (resp. IHSB+) if all its clauses are IHSB− (resp. IHSB+).

As usual a Boolean relation R is said to be IHSB− (resp. IHSB+, definite Horn, definite dual Horn) if R can be defined by a CNF formula which is IHSB− (resp. IHSB+, , definite Horn, definite dual Horn). A relation is *affine with width 2* if it is definable by a conjunction of clauses, each of which being either a unary clause or a 2-XOR-clause (consisting of an XOR of 2 variables plus maybe the constant 1)— such a conjunctive formula may also be seen as a system of linear equations over GF[2] with at most two variables per equation. It can be proved that a relation is affine with width 2 if and only if it is both affine and bijunctive. Finally, a constraint language Γ is said to be affine with width 2 (resp. IHSB−, IHSB+, definite Horn, definite dual Horn) if every relation in Γ is affine with width 2 (resp. IHSB−, IHSB+, definite Horn, definite dual Horn).

As for the above introduced classes, all these subclasses of affine and (dual) Horn relations can be characterized by their polymorphisms, see again Fig. 3.

3 Complexity of Satisfiability for Γ-Formulas and Quantified Γ-Formulas

We have seen that if Γ is Schaefer or 0-valid or 1-valid then $\text{Csp}(\Gamma)$ is decidable in polynomial time. If Γ is not of this form, then we have seen that that $\langle\Gamma\rangle \supseteq \text{Inv}(\text{N}_2)$, the co-clone of all complementive relations. A particular example here is the relation

$$\text{R}_{\text{NAE}} = \{(0,0,1),(0,1,0),(0,1,1),(1,0,0),(1,0,1),(1,1,0)\}$$

("NAE" here stands for "not all equal"). The language $\text{Csp}(\{\text{R}_{\text{NAE}}\})$ thus consists of 3CNF formulas with only positive literals where we require that in every clause not all literals obtain the same truth value. This is the so-called NOT-ALL-EQUAL problem, known to be NP-complete (see, e.g., [Pap94]). Thus, we have proved *Schaefer's Theorem*:

Theorem 3.1. *If* $\langle\Gamma\rangle \supseteq \text{Inv}(N_2)$ *then* $\text{Csp}(\Gamma)$ *is* NP*-complete, in all other cases,* $\text{Csp}(\Gamma)$ *is polynomial-time decidable.*

In making use of Post's lattice (Fig. 1) this theorem can be reformulated as follows: If $\text{Pol}(\Gamma) \supseteq \text{E}_2$ or $\text{Pol}(\Gamma) \supseteq \text{V}_2$ or $\text{Pol}(\Gamma) \supseteq \text{D}_2$ or $\text{Pol}(\Gamma) \supseteq \text{L}_2$, or if $\text{Pol}(\Gamma) \supseteq \text{I}_0$ or $\text{Pol}(\Gamma) \supseteq \text{I}_1$ then $\text{Csp}(\Gamma)$ is polynomial-time decidable, otherwise $\text{Csp}(\Gamma)$ is NP-complete. Finally, in using the characterizations summarized in Fig. 3, Schaefer's theorem can be stated in more familiar terms: If Γ is Schaefer or 0-valid or 1-valid then $\text{Csp}(\Gamma)$ is polynomial-time decidable, otherwise Γ can express all complementive relations and $\text{Csp}(\Gamma)$ is NP-complete. In the list of complexity results that we will present in Sect. 6 below we will most of the time prefer the formulation as in Theorem 3.1; only in a few particular cases we will additionally present the classification using classical terms.

Because each member of the infinite family of the CSP-problems falls in two complexity cases and avoids the (under the assumption $\text{P} \neq \text{NP}$) infinitely many intermediate degrees, this theorem is also known as Schaefer's *Dichotomy Theorem.*

Recently there has been growing interest in quantified constraints, and we want to survey some of the developments here. The $\text{Csp}(\Gamma)$ problem is equivalent to asking if a Γ-formula with all variables existentially quantified evaluates to true. In the quantified CSP problem one allows also universal quantifiers.

Let us first go one step back and look at usual propositional formulas again. The problem QBF of deciding, whether a given closed quantified propositional formula is true, is PSPACE-complete [SM73], even if the formula is restricted to 3CNF. If the number of quantifier alternations is bounded, the problem is complete in the *polynomial-time hierarchy*, which was defined by Meyer and Stockmeyer [MS72]. Following the notation of [Pap94], $\Sigma_0\text{P} = \Pi_0\text{P} = \text{P}$ and for all $i \geq 0$, $\Sigma_{i+1}\text{P} = \text{NP}^{\Sigma_i\text{P}}$ and $\Pi_{i+1}\text{P} = \text{coNP}^{\Sigma_i\text{P}}$. The set QBF_k of all closed, true quantified Boolean formulas with $k-1$ quantifier alternations starting with an $\exists$-quantifier, is complete for $\Sigma_k\text{P}$ for all $k \geq 1$ [SM73]. This problem remains $\Sigma_k\text{P}$-complete if we restrict the Boolean formula to be 3CNF for k odd, and 3DNF for k even [Wra77]. Since disjunctive normal forms cannot be naturally modelled in a constraint satisfaction context, in order to generalize QBF_k to arbitrary set of constraints Γ in the same way we generalized SAT to $\mathrm{CSP}(\Gamma)$, we consider the unsatisfiability problem for these cases and we adopt the following definition for $\mathrm{QCSP}_k(\Gamma)$ from [Hem04].

Let Γ be a constraint language and $k \geq 1$. For k odd, a $\mathrm{QCSP}_k(\Gamma)$ formula is a closed formula of the form $\phi = \exists X_1 \forall X_2 \ldots \exists X_k \psi$, and for k even, a $\mathrm{QCSP}_k(\Gamma)$ formula is a closed formula of the form $\phi = \forall X_1 \exists X_2 \ldots \exists X_k \psi$, where the X_j, $j = 1, \ldots, k$, are disjoint sets of variables and ψ is a quantifier-free Γ-formula with variables from $\bigcup_j X_j$.

Problem:	$\mathrm{QCSP}_k(\Gamma)$
Input:	a $\mathrm{QCSP}_k(\Gamma)$-formula ϕ
Question:	If k is odd: Is ϕ true?
	If k is even: Is ϕ false?

As in the case of simple CSPs above, we note that the Galois connection still helps to study the complexity of QCSP:

Proposition 3.2. *If* $\Gamma \subseteq \langle \Gamma' \rangle$ *then* $\mathrm{QCSP}_k(\Gamma) \leq_m^{\log} \mathrm{QCSP}_k(\Gamma')$ *for all* $k \geq 1$.

Proof. The proof of this is very similar to the one for Proposition 2.2: Given a Γ-formula ϕ, we construct a formula ϕ' by replacing every constraint from Γ by its defining existentially quantified $(\Gamma' \cup \{=\})$-formula. The newly introduced quantified variables will be quantified in the final quantifier block which is by definition of $\mathrm{QCSP}_k(\Gamma)$-formulas always existential. All that remains to do now is to delete equality clauses as above. □

Certainly, for every constraint language Γ and every $k \geq 1$, $\mathrm{QCSP}_k(\Gamma) \in \Sigma_k\mathrm{P}$ and $\mathrm{QCSP}_k(\Gamma_{3\mathrm{SAT}})$ is $\Sigma_k\mathrm{P}$-complete. In fact, even for the single constraint relation

$$\mathrm{R}_{1\text{-IN-}3} = \{(1,0,0),(0,1,0),(0,0,1)\}$$

we have that $\mathrm{QCSP}_k(\{\mathrm{R}_{1\text{-IN-}3}\})$ is $\Sigma_k\mathrm{P}$-complete. This follows since $\mathrm{R}_{1\text{-IN-}3}$ is only closed under projections and, thus, $\mathrm{Pol}(\mathrm{R}_{1\text{-IN-}3})$ is the minimal clone I_2 in Post's lattice and $\langle \mathrm{R}_{1\text{-IN-}3} \rangle$ is the co-clone $\mathrm{Inv}(\mathrm{I}_2)$ of all Boolean relations. Thus, from Proposition 3.2 we conclude $\mathrm{QCSP}_k(\Gamma_{3\mathrm{SAT}}) \leq_m^{\log} \mathrm{QCSP}_k(\{\mathrm{R}_{1\text{-IN-}3}\})$.

In the case of Schaefer's theorem for CSP, already the constraint language consisting of the relation $\mathrm{R}_{\mathrm{NAE}}$ is hard. We want to show an analogous result for QCSP next. To show this, we will reduce $\mathrm{QCSP}_k(\{\mathrm{R}_{1\text{-IN-}3}\})$ to $\mathrm{QCSP}_k(\{\mathrm{R}_{\mathrm{NAE}}\})$:

Let ϕ be a $\mathrm{QCSP}_k(\{\mathrm{R}_{1\text{-IN-}3}\})$-formula,

$$\phi = Q_1 X_1 \ldots \exists X_k \bigwedge_{j=1}^{p} \mathrm{R}_{1\text{-IN-}3}(x_{j_1}, x_{j_2}, x_{j_3}),$$

where Q is existential if k is odd and universal if k is even. We now replace each constraint $\mathrm{R}_{1\text{-IN-}3}(x_{j_1}, x_{j_2}, x_{j_3})$ by the following conjunction:

$$\bigwedge_{j \neq k \in \{j_1, j_2, j_3\}} \mathrm{R}_{\mathrm{NAE}}(x_j, x_k, t) \wedge \mathrm{R}_{\mathrm{NAE}}(x_{j_1}, x_{j_2}, x_{j_3}).$$

It can be checked that this conjunction is true if and only if exactly two of the four variables $x_{j_1}, x_{j_2}, x_{j_3}, t$ are true, hence we will abbreviate the above formula by $\mathrm{R}_{2\text{-IN-}4}(x_{j_1}, x_{j_2}, x_{j_3}, t)$. Now let $\phi' = Q_1 X_1 \ldots \exists X_k \bigwedge_{j=1}^{p} \mathrm{R}_{2\text{-IN-}4}(x_{j_1}, x_{j_2}, x_{j_3}, t)$. Since $\mathrm{R}_{1\text{-IN-}3}(x, y, z) = \mathrm{R}_{2\text{-IN-}4}(x, y, z, 1)$, the formula $\phi'[t = 1]$ (every occurrence of t in ϕ is replaced by 1) is true if and only if ϕ is true. Since $\mathrm{R}_{1\text{-IN-}3}(\bar{x}, \bar{y}, \bar{z}) = \mathrm{R}_{2\text{-IN-}4}(x, y, z, 0)$, the formula $\phi'[t/0]$ is true if and only if $\mathrm{Ren}(\phi)$ is true, where $\mathrm{Ren}(\phi)$ is obtained from ϕ by renaming all variables x by their negation $\bar{x}$. Finally, since $\mathrm{Ren}(\phi)$ is true if and only if ϕ is true, we proved that ϕ is true if and only if ϕ' is true. Thus, $\mathrm{QCSP}_k(\{\mathrm{R}_{1\text{-IN-}3}\}) \leq_m^{\log} \mathrm{QCSP}_k(\{\mathrm{R}_{\mathrm{NAE}}\})$.

Hence we now know that if $\langle \Gamma \rangle \supseteq \mathrm{Inv}(\mathrm{N}_2)$, then $\mathrm{QCSP}_k(\Gamma)$ is complete for $\Sigma_k\mathrm{P}$ for every $k \geq 1$. What about the next lower relational clone Inv(N)? In the case of $\mathrm{CSP}(\Gamma)$ (i.e., $\mathrm{QCSP}_1(\Gamma)$), satisfiability is trivial for all $\Gamma \subseteq \mathrm{Inv}(\mathrm{N})$, since every formula is satisfied by the constant-0 or constant-1 assignment. However, this tells us nothing about $\mathrm{QCSP}_k(\Gamma)$ for $k \geq 2$. Let us look at the relation

$$R_0 = \left\{ (u, v, x_1, x_2, x_3) \,\middle|\, u = v \text{ or } \mathrm{R}_{\mathrm{NAE}}(x_1, x_2, x_3) \right\}.$$

It is easy to see that R_0 is complementive, 0-valid, and 1-valid. We will show that $\mathrm{QCSP}_k(\{\mathrm{R}_{\mathrm{NAE}}\})$ reduces to $\mathrm{QCSP}_k(\{R_0\})$. Let

$$\phi = Q_1 X_1 \ldots \exists X_k \bigwedge_{j=1}^{p} \mathrm{R}_{\mathrm{NAE}}(x_{j_1}, x_{j_2}, x_{j_3}),$$

where Q_1 is existential if k is odd and universal if k is even, be an instance of $\mathrm{QCSP}_k(\{\mathrm{R}_{\mathrm{NAE}}\})$. We define

$$\phi' = Q_1 X_1 \ldots \forall X_{k-1} \forall u \forall v \exists X_k \bigwedge_{j=1}^{p} R_0(u, v, x_{j_1}, x_{j_2}, x_{j_3}).$$

Clearly ϕ is true if and only if ϕ' is true, thus $\mathrm{QCSP}_k(\{\mathrm{R}_{\mathrm{NAE}}\}) \leq_m^{\log} \mathrm{QCSP}_k(\{R_0\})$ for all $k \geq 2$.

We conclude that if $\langle \Gamma \rangle \supseteq \mathrm{Inv}(\mathrm{N})$, then $\mathrm{QCSP}_k(\Gamma)$ is complete for $\Sigma_k\mathrm{P}$ for every $k \geq 2$. If we drop the bound on the number of quantifier alternations and denote the resulting problem by $\mathrm{QCSP}(\Gamma)$, we know from [SM73] that $\mathrm{QCSP}(\Gamma_{3\mathrm{SAT}})$ is PSPACE-complete. The just given reductions thus also show that if $\langle \Gamma \rangle \supseteq \mathrm{Inv}(\mathrm{N})$, then $\mathrm{QCSP}(\Gamma)$ is PSPACE-complete.

If Γ does not include Inv(N), we know from the structure of Post's lattice that it must be Schaefer. However, it is known that in all four cases (Horn, dual Horn, bijunctive, and affine), the evaluation of quantified formulas is computable in polynomial time (the algorithms for the first three cases rely on Q-resolution, a variant of resolution for quantified propositional formulas, see [KL99]; the algorithm for the affine case is a refinement of the Gaussian algorithm, see [CKS01]). Thus we have proved the following classification:

Theorem 3.3. *If Γ is Schaefer then* $\mathrm{QCSP}(\Gamma)$ *is polynomial-time decidable, in all other cases,* $\mathrm{QCSP}(\Gamma)$ *is* PSPACE*-complete.*

This result was stated without proof and only for constraint languages that include the constants in Schaefer's paper [Sch78]. In its full form it was stated and proven for the first time in [Dal97] and later published in [CKS01].

Looking at QCSPs with bounded quantifier alternations we obtain with the same proof as above *Hemaspaandra's Theorem* [Hem04]. For all $k \geq 2$ (the case $k = 1$ is given by Schaefer's Theorem) the following holds:

Theorem 3.4. *If Γ is Schaefer then* $\text{QCSP}_k(\Gamma)$ *is polynomial-time decidable, in all other cases,* $\text{QCSP}_k(\Gamma)$ *is* $\Sigma_k\text{P}$*-complete.*

Theorems 3.1, 3.3, and 3.4 were originally proven in a different much more involved way in [Sch78, Hem04]. The above simple proofs using Galois theory appeared later. The proof of Theorem 3.1 is implicit in [JCG97, Dal00]. The proofs of Theorems 3.3 and 3.4 are from [BBC$^+$07]. Yet a different proof is given in [Che06].

4 When Does Post's Lattice Help?

Many further results, classifying the computational complexity of different algorithmic tasks for Boolean CSPs have been obtained in the past decades. Some of these rely on the algebraic approach explained above, for others this approach does not seem to be useful. To make this a little bit more precise, let $\Pi(\Gamma)$ be any computational problem defined for Γ-formulas or quantified Γ-formulas. If a result as

$$\text{If } \Gamma \subseteq \langle \Gamma' \rangle \text{then } \Pi(\Gamma) \leq_m^{\log} \Pi(\Gamma') \tag{1}$$

can be proven and then be used to obtain a complexity classification of Π, then we will say that the Galois connection holds *a priori* for Π. For the problems studied in the previous section, the Galois connection holds *a priori.*

For many problems that we will address below, a classification cannot be obtained with the help of a result as (1). Instead, the classification was obtained in sometimes very involved and technically complicated ways making use of different types of implementing one constraint relation by another. However, once the full classification is obtained, it sometimes happens incidentally that it obeys the borders among co-clones, that is, (1) holds but it can only be read from the obtained classification and not be used to obtain the classification. In such a case, we will say that the Galois connection holds *a posteriori* for the problem Π under consideration.

Also, there are some problems where the Galois connection simply does not hold, i.e., an implication as (1) is not true and the known complexity classification does not follow Post's lattice; there might, e.g., exist constraint languages with different complexities that nevertheless give rise to the same co-clone.

The distinction between *a priori* and *a posteriori* should of course not be taken as a mathematical definition—after all, in both cases (1) holds. The distinction should better be regarded as a historical notion.

In Sect. 6 we will survey complexity classifications obtained for many computational problems in the past decades. For each of them, we will pay particular attention to the question if Post's lattice could be used. Before we do so, however, we would like to address an important point concerning the type of reduction that is obtained from the Galois connection.

5 Reducibilities

Schaefer's Dichotomy theorem states that $\mathrm{CSP}(\Gamma)$ is either NP-complete or polynomial-time decidable. This means that under polynomial-time many-one reductions, there are only two possible degrees of complexity for this decision problem.

However, the statement of the Galois connection, Proposition 2.2, speaks of logspace many-one reductions. Hence, together with Post's lattice this can be used to determine all degrees of $\mathrm{CSP}(\Gamma)$ with respect to logspace m-reductions. First results in this direction appear already in Schaefer's paper [Sch78, Theorem 5.1] and in [CKS01, Theorem 6.5]. A thorough examination has been undertaken by Allender, Bauland, Immerman, Schnoor and Vollmer [ABI+05], and it was shown there that the complexity of $\mathrm{CSP}(\Gamma)$ falls into one of five logspace m-degrees: NP-complete, P-complete, NL-complete, $\oplus$L-complete, or decidable in logspace. Here, a logspace m-degree is a class of the form $[A]_{\equiv_m^{\log}} = \{B \mid B \equiv_m^{\log} A\}$ for some language A, where $B \equiv_m^{\log} A$ denotes that $A \leq_m^{\log} B$ and $B \leq_m^{\log} A$. The name stems from the fact that we are talking about logspace m-reductions $\leq_m^{\log}$.

In fact, Allender *et al.* even make a further step by looking at still stricter reductions, namely the so-called AC^0 many-one-reductions $\leq_m^{\mathrm{AC}^0}$. The class AC^0 consists of all languages/functions computable by uniform families of Boolean circuits of polynomial size and constant depth; for an exact definition and a thorough discussion of the type of uniformity involved we refer the reader to [Vol99]. Now $\leq_m^{\mathrm{AC}^0}$ reductions are just many-one reductions where the reduction function is computable by AC^0 circuits. These reductions are also known as FO-reductions, since the reduction function can be defined by first-order formulas, see [Imm99].

Example 5.1. Let $\Gamma_1 = \{\overline{x}, x\}$. An easy calculation, using Post's lattice, shows that $\mathrm{Pol}(\Gamma_1) = \mathrm{R}_2$, the class of all Boolean functions f that are at the same time 0-reproducing and 1-reproducing, i.e., $f(0,\ldots,0) = 0$ and $f(1,\ldots,1) = 1$. Now, define $\Gamma_2 = \Gamma_1 \cup \{=\}$, then obviously $\mathrm{Pol}(\Gamma_1) = \mathrm{Pol}(\Gamma_2)$.

Formulas over Γ_1 only contain clauses of the form x or $\overline{x}$ for some variables x, such a formula is unsatisfiable if and only if for some variable x, both x and $\overline{x}$ are clauses. This is easily decidable by AC^0 circuits, and $\mathrm{CSP}(\Gamma_1) \in \mathrm{AC}^0$.

In Γ_2 we additionally have the binary equality predicate, and we will now show that $\mathrm{CSP}(\Gamma_2)$ is complete for L under $\leq_m^{\mathrm{AC}^0}$ reductions: The complement of the graph accessibility problem (GAP) for undirected graphs, which is known to be complete for L [Rei05], can be reduced to $\mathrm{CSP}(\Gamma_2)$ as follows: Given a finite, undirected graph $G = (V, E)$ and vertices s, t in V, we build, for every edge $(v_1, v_2) \in E$, a constraint $v_1 = v_2$. Also we build the two clauses $\overline{s}$ and t. Finally, let ϕ_G be the conjunction of all these constraints. It is easy to see that there exists a path in G from s to t if and only if ϕ_G is not satisfiable. Since, moreover, $\text{CSP}(\Gamma_2) \in \text{L}$, we conclude that $\text{CSP}(\Gamma_2)$ is complete for L under $\leq_m^{\text{AC}^0}$ reductions.

This example shows that the Galois connection Proposition 2.2 does *not* hold for $\leq_m^{\text{AC}^0}$-recutions, since we constructed two constraint languages with the same expressive power but provably different complexity (note that $\text{AC}^0 \neq \text{L}$). In other words, $\Gamma_2 \subseteq \langle \Gamma_1 \rangle$ yet $\text{CSP}(\Gamma_2) \not\leq_m^{\text{AC}^0} \text{CSP}(\Gamma_1)$. All we can state is the following proposition:

Proposition 5.2. *If* $\Gamma \subseteq \langle \Gamma' \rangle$ *then* $\text{CSP}(\Gamma) \leq_m^{\text{AC}^0} \text{CSP}(\Gamma' \cup \{=\}) \leq_m^{\log} \text{CSP}(\Gamma')$.

Proof. The proof follows the one for Proposition 2.2: The first reduction $\text{CSP}(\Gamma) \leq_m^{\text{AC}^0} \text{CSP}(\Gamma' \cup \{=\})$ just consists of the local replacement of relations from Γ by their defining primitive positive definitions using relations in $\Gamma' \cup \{=\}$ and removing the existential quantifiers; this can be computed in AC^0. For the second reduction $\text{CSP}(\Gamma' \cup \{=\}) \leq_m^{\log} \text{CSP}(\Gamma')$ we delete all equality clauses and identify variables forced to be equal, and, as argued before, this can be computed in logspace. □

Some constraint languages, however, can express equality in the following sense:

Recall from Definition 2.1 that $R \in \langle \Gamma \rangle$ if and only if R can be defined by a primitive positive definition using relations in $\Gamma \cup \{=\}$. Now we explicitly do not allow equality constraints to define new relations. More precisely, we say that a constraint language Γ *can express a relation* R, if R can be defined with a primitive positive definition using only relations in Γ. Obviously, if Γ can express equality then $\text{CSP}(\Gamma \cup \{=\}) \leq_m^{\text{AC}^0} \text{CSP}(\Gamma)$, using local replacement of all equality constraints by defining Γ-formulas.

Proposition 5.3. *If* $\Gamma \subseteq \langle \Gamma' \rangle$ *and* Γ' *can express equality then*

$$\text{CSP}(\Gamma) \leq_m^{\text{AC}^0} \text{CSP}(\Gamma').$$

Allender *et al.* [ABI+05] present a simple algorithm to determine if a constraint language can express equality.

With these tools at hand, Allender *et al.* obtained the following classification:

Theorem 5.4.
- *If* $\langle \Gamma \rangle \supseteq \text{Inv}(\text{N}_2)$ *then* $\text{CSP}(\Gamma)$ *is* $\leq_m^{\text{AC}^0}$*-complete for* NP.
- *If* $\langle \Gamma \rangle = \text{Inv}(\text{V}_2)$ *or* $\langle \Gamma \rangle = \text{Inv}(\text{E}_2)$ *then* $\text{CSP}(\Gamma)$ *is* $\leq_m^{\text{AC}^0}$*-complete for* P.
- *If* $\text{Inv}(\text{L}_3) \subseteq \langle \Gamma \rangle \subseteq \text{Inv}(\text{L}_2)$ *then* $\text{CSP}(\Gamma)$ *is* $\leq_m^{\text{AC}^0}$*-complete for* $\oplus$L.
- *If* $\text{Inv}(\text{S}_{00}^2) \subseteq \langle \Gamma \rangle \subseteq \text{Inv}(\text{S}_{00})$ *or* $\text{Inv}(\text{S}_{10}^2) \subseteq \langle \Gamma \rangle \subseteq \text{Inv}(\text{S}_{10})$ *or* $\langle \Gamma \rangle = \text{Inv}(\text{D}_2)$ *or* $\langle \Gamma \rangle = \text{Inv}(\text{M}_2)$ *then* $\text{CSP}(\Gamma)$ *is* $\leq_m^{\text{AC}^0}$*-complete for* NL.
- *If* $\text{Inv}(\text{D}) \subseteq \langle \Gamma \rangle \subseteq \text{Inv}(\text{D}_1)$ *then* $\text{CSP}(\Gamma)$ *is* $\leq_m^{\text{AC}^0}$*-complete for* L.
- *If* $\text{Inv}(\text{R}_2) \subseteq \langle \Gamma \rangle \subseteq \text{Inv}(\text{S}_{02})$ *or* $\text{Inv}(\text{R}_2) \subseteq \langle \Gamma \rangle \subseteq \text{Inv}(\text{S}_{12})$ *then either* $\text{CSP}(\Gamma)$ *is in* AC^0, *or* $\text{CSP}(\Gamma)$ *is complete for* L *under* $\leq_m^{\text{AC}^0}$. *There is an algorithm deciding which case occurs.*

– *If* $\Gamma \subseteq \mathrm{Inv}(\mathrm{I}_0)$ *or* $\Gamma \subseteq \mathrm{Inv}(\mathrm{I}_1)$ *then every constraint formula over* Γ *is satisfiable, and therefore* $\mathrm{CSP}(\Gamma)$ *is trivial.*

Hence we see that the problem $\mathrm{CSP}(\Gamma)$ can assume only one of six different AC^0 m-degrees (assuming all relevant classes are different). Using Agrawal's First-Order Isomorphism Theorem [Agr01] an even stronger statement can be made: For any set of relations Γ, $\mathrm{CSP}(\Gamma)$ is AC^0-isomorphic either to $0\Sigma^*$ or to the standard complete set for one of the following complexity classes: NP, P, $\oplus$L, NL, L. In a sense there is not an infinite family of CSP-problems but in fact there are only 6 different such problems.

The question which type of reduction one obtains from the Galois connection will come up again in a different context in the next section. For counting problems for quantified constraints, the Galois connection yields parsimonious reductions (Proposition 6.11). Parsiminious reductions (to be defined in Sect. 6.3), however, are often too strict—not many completeness results for counting problems under parsimonious reductions are known—and one has to look for suitable coarser reductions.

6 Complexity Classifications

This section is intended to give a survey of complexity classifications obtained up to now in the framework of Boolean CSPs. Such results are so numerous that it would be too ambitious to make an exhaustive list of them. However we will review a large selection of results, organized according to the kind of computational task they address. In each case we will try to highlight the features of the problems for which the Galois holds *a priori* or at least can help. As we have seen in Theorem 3.1 and in the discussion that followed, a classification that matches Post's lattice can be stated in different ways, in any case it is easy to go from one statement to the other in using Fig. 1 and Fig. 3.

6.1 Nonmonotonic Reasoning

Circumscription is an important and well-studied formalism in the realm of nonmonotonic reasoning. The model checking and inference problem for propositional circumscription have been studied from the viewpoint of computational complexity.

Given a formula ϕ, let $\mathrm{Var}(\phi)$ denote its set of variables. If $|\mathrm{Var}(\phi)| = n$, then any assignment of truth values to the variables of ϕ can be seen as a word in $\{0,1\}^n$. Circumscription is defined in using a partial order $\leq_{(P,Q)}$ on $\{0,1\}^n$, where P, Q are two disjoint subsets of $\{1,\ldots,n\}$. Let $\beta = (b_1,\ldots,b_n)$ and $\alpha = (a_1,\ldots,a_n)$ be two truth assignments. We write $\beta \leq_{(P,Q)} \alpha$ to denote that $b_i \leq a_i$ for all $i \in P$ and $b_j = a_j$ for all $j \in Q$. Let $\beta <_{(P,Q)} \alpha$ denote that $\beta \leq_{(P,Q)} \alpha$ and there exists $i \in P$ such that $b_i \neq a_i$. Given a formula ϕ, we say that α is a minimal model of ϕ with respect to the partial order $\leq_{(P,Q)}$ if α satisfies ϕ and and there is no β satisfying ϕ such that $\beta <_{(P,Q)} \alpha$.

The model checking for propositional circumscription is defined as follows,

Problem: MIN-CSP(Γ)
Input: a Γ-formula ϕ, two disjoint subsets P, Q of $\mathrm{Var}(\phi)$ and a truth assignment I to $\mathrm{Var}(\phi)$
Question: Is I a model of ϕ minimal with respect to the partial order $\leq_{(P,Q)}$?

while the inference problem is

Problem: MIN-INF-CSP(Γ)
Input: a Γ-formula ϕ, two disjoint subsets P, Q of $\mathrm{Var}(\phi)$ and a clause c
Question: Is c satisfied in every model of ϕ which is minimal with respect to the partial order $\leq_{(P,Q)}$?

It was proved in [NJ04] that the Galois connection holds *a priori* for MIN-CSP and MIN-INF-CSP.

Proposition 6.1. *If $\Gamma \subseteq \langle \Gamma' \rangle$, then*

- MIN-CSP(Γ) $\leq_m^{\log}$ MIN-CSP(Γ'), *and*
- MIN-INF-CSP(Γ) $\leq_m^{\log}$ MIN-INF-CSP(Γ').

Proof. The proof is very similar to the one for Proposition 2.2: given an input (ϕ, P, Q, I) of MIN-CSP(Γ), we construct an input (ϕ', P', Q', I') of MIN-CSP(Γ). The formula ϕ' is obtained by first replacing every constraint from Γ by its defining existentially quantified $(\Gamma' \cup \{=\})$-formula, and second deleting the existential quantifiers. The newly introduced quantified variables are in $\mathrm{Var}(\phi')$. At this step the sets P' and Q' are unchanged, i.e., $P' = P$ and $Q' = Q$. It remains to delete equality clauses and replace all variables that are connected via a chain of equality constraints by a common new variable. If one of the variables in the chain is a variable from Q, then the new variable is added to Q', otherwise if one of the variables in the chain is a variable from P, then the new variable is added to P'. Finally we remove from P' and Q' the variables that do not occur anymore in ϕ'. It is then easy to check that I is a model of ϕ minimal with respect to the partial order $\leq_{(P,Q)}$ if and only if I' is a model of ϕ' minimal with respect to the partial order $\leq_{(P',Q')}$. A similar proof applies to the problem MIN-INF-CSP. □

Using this algebraic connection Nordh and Jonsson [NJ04] proved the following dichotomy theorem.

Theorem 6.2. [Model checking for propositional circumscription]

- *If Γ is Schaefer then* MIN-CSP(Γ) *is polynomial-time decidable.*
- *In all other cases,* MIN-CSP(Γ) *is* coNP*-complete.*

For the inference problem Nordh [Nor05] obtained a trichotomy classification.

Theorem 6.3. [Inference for propositional circumscription]

- *If* $\langle\Gamma\rangle \supseteq \mathrm{Inv}(\mathrm{N})$ *then* MIN-INF-CSP(Γ) *is* Π_2P*-complete.*
- *Otherwise if* $\langle\Gamma\rangle \supseteq \mathrm{Inv}(\mathrm{S}_{11}^2)$ *or* $\langle\Gamma\rangle \supseteq \mathrm{Inv}(\mathrm{S}_0^2)$ *or* $\langle\Gamma\rangle \supseteq \mathrm{Inv}(\mathrm{E})$ *or* $\langle\Gamma\rangle \supseteq \mathrm{Inv}(\mathrm{V})$ *or* $\langle\Gamma\rangle \supseteq \mathrm{Inv}(\mathrm{L})$*, then* MIN-INF-CSP($\Gamma$) *is* coNP*-complete.*
- *In all other cases,* MIN-INF-CSP(Γ) *is polynomial-time decidable.*

Kirousis and Kolaitis studied basic circumscription, where $Q = \emptyset$ and $P = \mathrm{Var}(\phi)$, which means that the partial order taken into account is the usual coordinate-wise order. Actually this special case is much harder to study. Indeed, there is no way to restrict P and Q in order to hide the newly introduced existential variables as in the proof of Theorem 6.1. Thus the Galois connection cannot be applied a priori and the results are obtained in a much more involved way. For the MIN-CSP problem in which $Q = \emptyset$ and $P = \mathrm{Var}(\phi)$ (denoted by MIN-SAT in the original paper [KK03]) Kirousis and Kolaitis obtained a dichotomy classification, P/coNP-complete, from which Theorem 6.2 can be derived. As for the inference problems (denoted by INF-CIRC in [KK04]) they obtained only a dichotomy classification (in coNP/Π_2P-complete), the possible trichotomy remains an open problem. It is worth noticing that these classifications follow Post's lattice, thus showing that for these problems the Galois connection holds *a posteriori*.

Abduction is another fundamental nonmonotonic process, which consists in searching for an explanation of a given observed manifestation with respect to some background knowledge. We define below the abduction problem we are interested in, the query to be explained consists of a single positive literal (whence PQ below for positive query). Given a set of variables A, $\mathrm{Lit}(A)$ denotes the set of literals x and $\neg x$ one can build upon the variables x from A.

Problem:	PQ-ABDUCTION(Γ)
Input:	a Γ-formula ϕ, a set of variables $A \subseteq \mathrm{Var}(\phi)$ and a variable $q \in \mathrm{Var}(\phi) \setminus A$.
Question:	Is there a set $E \subset \mathrm{Lit}(A)$ such that $\phi \wedge \bigwedge E$ is satisfiable but $\phi \wedge \bigwedge E \wedge (\neg q)$ is not? (If one exists such a set E is called a solution of the abduction problem, or sometimes an "explanation" of q.)

The Galois connection helps to study the complexity of PQ-ABDUCTION.

Proposition 6.4. *If* $\Gamma \subseteq \langle\Gamma'\rangle$ *then*

$$\text{PQ-ABDUCTION}(\Gamma) \leq_m^{\log} \text{PQ-ABDUCTION}(\Gamma').$$

Proof. Given (ϕ, A, q) an instance of PQ-ABDUCTION(Γ), we procede as in the proof of Proposition 2.2 and construct ϕ' by replacing every constraint from Γ by its defining existentially quantified $(\Gamma' \cup \{=\})$-formula, and then by deleting the equality clauses. Then, it can be shown that (ϕ, A, q) and (ϕ', A, q) have the same solutions for the PQ-ABDUCTION problem (see [CZ06, Corollary 18]). □

From this proposition one can obtain the following classification.

Theorem 6.5. [Abduction]

- *If* $\langle \Gamma \rangle \supseteq \mathrm{Inv}(\mathrm{N})$ *then* PQ-ABDUCTION(Γ) *is* $\Sigma_2\mathrm{P}$*-complete.*
- *Otherwise if* $\langle \Gamma \rangle \supseteq \mathrm{Inv}(\mathrm{V})$ *or* $\langle \Gamma \rangle \supseteq \mathrm{Inv}(\mathrm{E}_0)$, *then* PQ-ABDUCTION($\Gamma$) *is* NP*-complete.*
- *In all other cases,* PQ-ABDUCTION(Γ) *is polynomial-time decidable.*

This classification was originally obtained in [CZ06] without making use of the Galois connection. Maybe here, it is interesting to state the classification in classical more familiar terms: Polynomial-time decidability holds for constraint languages that are affine, bijunctive, definite Horn, IHSB− or IHSB+; the NP-case arises for constraint languages that are none of the above but Horn or dual-Horn; and the remaining hard case is the case of constraint languages that can express all relations that are simultaneously complementive, 0-valid, and 1-valid.

Nordh and Zanuttini studied such an abduction problem in the case where hypotheses as well as manifestations are sets of literals. In using the Galois connection they also obtained a trichotomy classification (see [NZ05]).

As a third example for nonmonotonic reasoning, we would like to mention that Ilka Schnoor [Sch07b] (building on previous work in [CHS07]) classified the complexity of some problems for default logic where the formulas are Γ-formulas for some constraint language Γ.

6.2 Equivalence, Implication and Isomorphism

Two formulas ϕ_1 and ϕ_2 are *equivalent* if every truth assignment to the variables of ϕ_1 and ϕ_2 satisfies ϕ_1 if and only if it satisfies ϕ_2. As usual we write this as $\phi_1 \equiv \phi_2$. Thus we can define the following equivalence problem.

Problem:	EQUIV(Γ)
Input:	two Γ-formulas ϕ_1 and ϕ_2
Question:	Does $\phi_1 \equiv \phi_2$ hold?

This problem also makes sense in the context of quantified formulas. Given two quantified formulas ϕ_1 and ϕ_2, we say that ϕ_1 is equivalent to ϕ_2, and we still write $\phi_1 \equiv \phi_2$, if every truth assignment to the free variables of ϕ_1 and ϕ_2 is a solution for ϕ_1 (i.e., makes ϕ_1 true) if and only if it is a solution for ϕ_2. Thus we can define the equivalence problem for QCSP$_k(\Gamma)$ formulas.

Problem:	QEQUIV$_k(\Gamma)$
Input:	two QCSP$_k(\Gamma)$-formulas ϕ_1 and ϕ_2
Question:	Does $\phi_1 \equiv \phi_2$ hold?

The algebraic framework will still be of use for this problem. However we have to be careful with the equality constraints resulting from the application of the Galois connection. Indeed, the removal of the equality constraints by identifying variables may not preserve the equivalence of the formulas. However, if our constraint language can express equality (in the sense of the definition given in Sect. 5), then this problem does not arise.

Proposition 6.6. *If $\Gamma \subseteq \langle \Gamma' \rangle$ and Γ' can express equality, then*

$$\textsc{QEquiv}_k(\Gamma) \leq_m^{\log} \textsc{QEquiv}_k(\Gamma').$$

Proof. Given a formula ϕ we construct a formula ϕ' by replacing every constraint from Γ by its defining existentially quantified $(\Gamma' \cup \{=\})$-formula. Any occurring equality constraint can be removed in using the Γ'-implementation of the equality relation, which exists due to the prerequisites. All the newly introduced variables will be quantified in the final (existential) quantifier block. So, we obtain a formula ϕ' which is equivalent to ϕ. Thus, applying this transformation to the pair of formulas given as instance for the problem $\textsc{QEquiv}(\Gamma)$ provides a proof of the above claim. □

With this tool the following classification was obtained in [BBC+07, Sch07a].

Theorem 6.7. [Equivalence for quantified formulas]

- *If $\langle \Gamma \rangle \supseteq \mathrm{Inv(N)}$ then $\textsc{QEquiv}_k(\Gamma)$ is $\Pi_{k+1}\mathrm{P}$-complete.*
- *Otherwise if $\langle \Gamma \rangle \supseteq \mathrm{Inv(V)}$ or $\langle \Gamma \rangle \supseteq \mathrm{Inv(E)}$, then $\textsc{QEquiv}_k(\Gamma)$ is* coNP-*complete.*
- *In all other cases, $\textsc{QEquiv}_k(\Gamma)$ is polynomial-time decidable.*

Here again it is interesting to state the classification in classical terms: Polynomial-time decidability holds for constraint languages that are affine, bijunctive, IHSB− or IHSB+; the coNP-case arises for constraint languages that are none of the above but Horn or dual-Horn; and the remaining hard case is the case of constraint languages can express all relations that are simultaneously complementive, 0-valid, and 1-valid.

When considering unquantified formulas there is no way to hide the new existential variables resulting from the application of the Galois connection. Therefore the Galois connection does not hold *a priori* to study the complexity of the equivalence problem $\textsc{Equiv}(\Gamma)$. However, as proved by Böhler, Hemaspaandra, Reith and Vollmer in [BHRV02], it holds *a posteriori* since the complexity classification for this problem follows the structure of Post's lattice.

Theorem 6.8. [Equivalence of formulas]

- *If Γ is Schaefer then $\textsc{Equiv}(\Gamma)$ is polynomial-time decidable.*
- *In all other cases, $\textsc{Equiv}(\Gamma)$ is* coNP-*complete.*

Besides equivalence another problem of interest is the one of isomorphism. We say that two formulas ϕ_1 and ϕ_2 over a set of variables X are *isomorphic*, and we write $\phi_1 \cong \phi_2$, if there exists a permutation π of X such that $\pi(\phi_1) \equiv \phi_2$. Thus one can define the following isomorphism problem.

Problem:	$\textsc{Iso}(\Gamma)$
Input:	two Γ-formulas ϕ_1 and ϕ_2
Question:	Does $\phi_1 \cong \phi_2$ hold?

The complexity of this problem is related to the one of the Graph Isomorphism problem (GI for short). The exact complexity of GI is not known. It is only known that it is in NP; it is not known to be in P and it is unlikely to be NP-complete, cf. [KST93]. The following classification result was obtained by Böhler, Hemaspaandra, Reith and Vollmer in [BHRV04].

Theorem 6.9. [Isomorphism]

- *If* Γ *is affine with width 2 then* $\textsc{Iso}(\Gamma)$ *is polynomial-time decidable.*
- *Otherwise, if* Γ *is Schaefer, then* $\textsc{Iso}(\Gamma)$ *is polynomial-time many-one equivalent to GI.*
- *In all other cases,* $\textsc{Iso}(\Gamma)$ *is* coNP*-hard and GI-hard.*

Whereas the Galois connection does not apply *a priori* for this problem, the complexity classification follows the structure of Post's lattice since it can be reformulated in the following way. This follows from the observation that a relation R is 2-affine if and only if it is affine and bijunctive, i.e., $\mathrm{Pol}(R) \supseteq \mathrm{D}_1$.

Theorem 6.10. [Isomorphism classification with co-clones]

- *If* $\langle\Gamma\rangle \supseteq \mathrm{Inv(N)}$ *then* $\textsc{Iso}(\Gamma)$ *is* coNP*-hard and GI-hard.*
- *Otherwise if* $\langle\Gamma\rangle \supseteq \mathrm{Inv}(\mathrm{S}_0^2)$ *or* $\langle\Gamma\rangle \supseteq \mathrm{Inv}(\mathrm{S}_1^2)$ *or* $\langle\Gamma\rangle \supseteq \mathrm{Inv(M)}$ *or* $\langle\Gamma\rangle \supseteq \mathrm{Inv(L)}$*, then* $\textsc{Iso}(\Gamma)$ *is polynomial-time many-one equivalent to GI.*
- *In all other cases,* $\textsc{Iso}(\Gamma)$ *is polynomial-time decidable.*

Additional results were obtained by Bauland and Hemaspaandra [BH05] for the *implication* and *isomorphic implication problems.* These problems are defined in a similar way as the ones above in weakening the equivalence requirement to an implication. For these problems again, while the Galois connection does not apply *a priori*, the complexity classifications they obtain follow Post's lattice.

6.3 Counting and Enumeration

The problems considered up to now have been decision problems. We now turn to counting problems where typically we do not ask whether a formula has a satisfying assignment but want to determine the number of satisfying assignments of a formula. Let us recall the definition of some complexity classes and reducibility notions relevant for such computational problems (see [Val79a], [Val79b], [Tod91a], [Tod91b], [Vol94] and [HV95]; a somewhat more detailed exposition of the material below can be found in [BBC+07]). Let Σ and Δ be alphabets and let $R \subseteq \Sigma^* \times \Delta^*$ be a binary relation between strings such that, for each $x \in \Sigma^*$, the set $R(x) = \{y \in \Delta^* \mid R(x, y)\}$ is finite. We write $\#R$ to denote the following counting problem: Given a string $x \in \Sigma^*$, find the cardinality $|R(x)|$, of the set $R(x)$ associated with x. The members of this set $R(x)$ are called *witnesses for* x and the relation R is called the *witness relation* of the counting problem. If $\mathcal{C}$ is a complexity class of decision problems, then $\#{\cdot}\mathcal{C}$ is the class of all counting problems whose witness relation R satisfies the following conditions:

1. There is a polynomial $p(n)$ such that for every x and every y with $R(x,y)$ we have $|y| \leq p(|x|)$.
2. The witness recognition problem "*given x and y, does $R(x,y)$ hold?*" is in $\mathcal{C}$.

The following inclusion chain is well-known [Tod91a, Vol94]: $\#\cdot\Sigma_k\mathrm{P} \subseteq \#\cdot\Pi_k\mathrm{P} = \#\mathrm{P}^{\Sigma_k\mathrm{P}} \subseteq \#\cdot\Sigma_{k+1}\mathrm{P}$ for each k.

Several notions of *reducibilities* among counting problems have been defined. We say that $\#A$ reduces to $\#B$ by a *Turing reduction* (see [Val79b]) if $\#A$ can be computed in polynomial-time by an oracle Turing machine with oracle $\#B$. Zankó [Zan91] introduced counting reductions (essentially truth-table-reduction with one oracle query) i.e., $\#A$ reduces to $\#B$ by a *counting reduction* if there exist polynomial-time computable functions f and g such that $\#A(x) = f(\#B(g(x)))$. The strongest notion of reduction is the one of *parsimonious reduction*, which exactly preserves the number of solutions and thus which is a special case of counting reduction where the function f is simply the identity. Unlike parsimonious reductions, counting reductions do not have the property that the classes of the hierarchy $\#\cdot\Sigma_k\mathrm{P}$ are closed under these reductions, unless this hierarchy collapses (see [TW92]).

This is the reason why researchers have looked for a reduction stricter than Turing reductions but not as strict as parsimonious ones. Building on a previous notion of *subtractive* reductions introduced in [DHK05], so called *complementive* reductions were introduced in [BCC+05]. For this, let $\Sigma_1, \Sigma_2, \Delta_1, \Delta_2$ be alphabets and let $\#A$ and $\#B$ be two counting problems determined by the binary relations $A \subseteq \Sigma_1^* \times \Delta_1^*$ and $B \subseteq \Sigma_2^* \times \Delta_2^*$.

We say that $\#A$ reduces to $\#B$ via a *strong complementive reduction*, if:

- for every string $x \in \Sigma_2^*$, $B(x)$ is complementive, i.e., there is a permutation π on Δ_2 such that for all words x, y we have that $y \in B(x) \iff \pi(y) \in B(x)$,
- there exist polynomial-time computable functions $f, g\colon \Sigma_1^* \to \Sigma_2^*$ such that for every string $x \in \Sigma_1^*$:
  - $B(g(x)) \subseteq B(f(x))$,
  - $2 \cdot |A(x)| = |B(f(x))| - |B(g(x))|$.

A *complementive reduction* $\#A \leq^{\mathrm{P}}_{compl} \#B$ is a sequence of strong complementive reductions. It is known that the classes $\#\cdot\Pi_k\mathrm{P}$ are closed under complementive reductions, and the closure of $\#\cdot\Sigma_k\mathrm{P}$ under these reductions is $\#\cdot\Pi_k\mathrm{P}$ [BCC+05].

Let us now examine the counting problems associated with quantified formulas. Given a formula with free variables $\phi(Y) = Q_1X_1 \ldots \exists X_k\psi(Y, X_1, \ldots, X_k)$ where ψ is quantifier free and the $X_1, \ldots, X_k, Y$ are vectors of variables, we are interested in the number of assignments for Y such that $\phi(Y)$ holds. We denote this number by $\#\mathrm{sat}(\phi)$ (and by $\#\mathrm{unsat}(\phi)$ we denote the number of assignments for Y such that $\phi(Y)$ does not hold). The following problems are the counting versions of the decision problems studied in Sect. 3.

Problem:	$\#\text{QCSP}_k(\Gamma)$
Input:	a $\text{QCSP}_k(\Gamma)$-formula ϕ with free variables
Output:	If k is odd: $\#\text{sat}(\phi)$
	If k is even: $\#\text{unsat}(\phi)$

It is tempting to conclude that (by a proof essentially identical to the one of Proposition 3.2) the Galois connection holds for these counting problems, i.e., if $\Gamma \subseteq \langle\Gamma'\rangle$ then there is a parsimonious reduction from $\#\text{QCSP}_k(\Gamma)$ to $\#\text{QCSP}_k(\Gamma')$. An easy counterexample shows, however, that this is not the case: Take $\Gamma = \{x\}$ and $\Gamma' = \{x, =\}$. These two constraint sets certainly have the same polymorphisms and obviously, every Γ-formula will have exactly one solution, since every appearing variable must be set to true, but with Γ', we can build formulas with exactly 2^k solutions using the formula $(x_1 = x_1) \wedge \cdots \wedge (x_k = x_k)$. However, it turns out that the only problematic case is that of constraint sets Γ consisting only of relations with exactly one tuple. Hence one obtains (see [BBC+07]):

Proposition 6.11. *If $\Gamma \subseteq \langle\Gamma'\rangle$ and not every relation in Γ is equivalent to a conjunction of literals, then there is a parsimonious reduction from $\#\text{QCSP}_k(\Gamma)$ to $\#\text{QCSP}_k(\Gamma')$.*

Using this algebraic result Bauland, Böhler, Creignou, Reith, Schnoor, and Vollmer [BBC+07], building on previous results from [BCC+05] for the case $k = 1$, obtained the following classification.

Theorem 6.12. [Counting for quantified formulas]

- *If $\langle\Gamma\rangle \supseteq \text{Inv}(\text{N})$ then $\#\text{QCSP}_k(\Gamma)$ is $\#\cdot\Sigma_k\text{P}$-complete under complementive reductions.*
- *Otherwise, if $\langle\Gamma\rangle \supseteq \text{Inv}(\text{M})$ or $\langle\Gamma\rangle \supseteq \text{Inv}(\text{S}_0^2)$ or $\langle\Gamma\rangle \supseteq \text{Inv}(\text{S}_1^2)$ then $\#\text{QCSP}_k(\Gamma)$ is $\#\cdot\text{P}$-complete under Turing reductions.*
- *In all other cases $\#\text{QCSP}_k(\Gamma)$ can be solved in polynomial time.*

The use of different reducibility notions in the statements of the previous theorem may be confusing, let us mention, however, that (under reasonable complexity-theoretic assumptions) the theorem places the counting problem for quantified Γ-formulas optimally in the $\Sigma_k\text{P}$-classes, since the following holds [BBC+07]:

Corollary 6.13. *1. If $\langle\Gamma\rangle \supseteq \text{Inv}(\text{N})$ then for any k, $\#\text{QCSP}_k(\Gamma) \in \#\cdot\Sigma_k\text{P}$ but $\#\text{QCSP}_k(\Gamma) \notin \#\cdot\Sigma_{k-1}\text{P}$ unless $\#\cdot\Sigma_k\text{P} = \#\Pi_{k-1}\text{P}$ (implying that $\Sigma_k\text{P}$-computations can be made "unambiguous"), and $\#\text{QCSP}_k(\Gamma)$ is not polynomial-time solvable unless $\text{P} = \text{NP}$.*
2. Otherwise, if $\langle\Gamma\rangle \supseteq \text{Inv}(\text{M})$ or $\langle\Gamma\rangle \supseteq \text{Inv}(\text{S}_0^2)$ or $\langle\Gamma\rangle \supseteq \text{Inv}(\text{S}_1^2)$ then for any k, $\#\text{QCSP}_k(\Gamma) \in \#\cdot\text{P}$ but $\#\text{QCSP}_k(\Gamma)$ is not polynomial-time solvable unless $\text{P} = \text{NP}$.

Let us now examine the counting problem for unquantified formulas.

Problem:	$\#\text{CSP}(\Gamma)$

Input: a Γ-formula ϕ
Output: number of satisfying assignments of ϕ

When dealing with unquantified formulas the problem is more complicated since one cannot take advantage of the existentially quantified variables in order to hide the new variables resulting from the application of the Galois connection. However, in [BD03] Bulatov and Dalmau obtained in a more involved way the following result, which shows that the Galois connection applies for counting problems under Turing reductions.

Proposition 6.14. *If $\Gamma \subseteq \langle \Gamma' \rangle$ then there is a Turing reduction from $\#\mathrm{CSP}(\Gamma)$ to $\#\mathrm{CSP}(\Gamma')$.*

Proof. Given a Γ-formula ϕ, one constructs a formula ϕ' by replacing every constraint from Γ by its defining existentially quantified $(\Gamma' \cup \{=\})$-formula, and then omitting the existential quantifiers. The difficulty is that if an individual constraint from Γ can be implemented by:

$$R_i(x_1, \ldots, x_{r_i}) = \exists y_1 \ldots \exists y_{q_i} \phi'_i(x_1, \ldots, x_{r_i}, y_1, \ldots, y_{q_i}),$$

where ϕ'_i is a Γ'-formula, then for each tuple $\mathbf{a_j} = (a_1, \ldots, a_{r_i}) \in R_i$ there are several tuples $(b_1, \ldots, b_{q_i})$ such that $\phi'_i(a_1, \ldots, a_{r_i}, b_1, \ldots, b_{q_i})$ holds (we denote by u_j the number of such extensions of $\mathbf{a_j}$). Therefore, the number of satisfying assignments of ϕ cannot be directly computed from the number of satisfying assignments of ϕ'. The trick is not to compute the formula ϕ' but p such formulas $\phi'^{(l)}$ for $l = 1, \ldots, p$, for a well chosen polynomial p in the input size, such that for each l, $\phi'^{(l)}$ is an expanded copy of ϕ', in which each Γ'-constraint is replicated l times. Let N_l denote the number of solutions of $\phi'^{(l)}$. It turns out that the number of solutions of the original formula ϕ can be expressed as the sum of p terms, which verify a linear system whose coefficients involve the u_j's and the N_l's defined above. This linear system has the good property of being Vandermonde, and therefore inversible. Thus these terms, and hence the number of solutions of ϕ can be computed in polynomial time from the N_l's (and the u_j's). Therefore, there is a Turing reduction from $\#\mathrm{CSP}(\Gamma)$ to $\#\mathrm{CSP}(\Gamma')$ with p queries to the oracle. □

Remark 6.15. For most of the above computational problems where the Galois connection helps *a priori*, the definition of the problem somehow allows to "hide" the new existentially quantified variables that are introduced by the application of the Galois connection. For the above proposition, however, it is the power of the reduction that helps to remove these variables.

In [CH96] Creignou and Hermann proved the following classification.

Theorem 6.16. [Counting]

- *If Γ is affine then $\#\mathrm{CSP}(\Gamma)$ can be solved in polynomial-time.*
- *Otherwise, $\#\mathrm{CSP}(\Gamma)$ is $\#\cdot\mathrm{P}$-complete under Turing reductions.*

Related to the counting problem is the unique satisfiability problem. Thus, UNIQUE-SAT(Γ) is the problem of deciding whether a given Γ-formula has a unique model. This problem is in the complexity class DP, the class of languages equal to an intersection of two languages, one from NP and the other from coNP. Unless the polynomial hierarchy collapses, DP is a strict superclass of NP and coNP. In fact, following [BG82], let us define the class US as the closure of UNIQUE-SAT (i.e., without restrictions on the clauses) under usual polynomial-time many-one reductions. Then US is a (supposedly proper) subclass of DP, showing that while UNIQUE-SAT $\in$ DP it is most likely not complete in this class.

In order to reduce a problem UNIQUE-SAT(Γ) to another one we need a parsimonious reduction that exactly preserves the number of solutions. Whereas as noticed above the Galois connection cannot help *a priori* for such a reduction, it holds *a posteriori*: a classification following Post's lattice was obtained by Juban [Jub99].

Theorem 6.17. [Unique satisfiability]

- *If Γ is Schaefer or Γ is both* 0*-valid and* 1*-valid or Γ is complementive, then* UNIQUE-SAT(Γ) *is polynomial-time decidable.*
- *Otherwise,* UNIQUE-SAT(Γ) *is* coNP*-hard.*

This classification theorem can indeed be reformulated as follows.

Theorem 6.18. [Unique satisfiability classification with co-clones]

- *If* $\langle\Gamma\rangle \supseteq \mathrm{Inv}(\mathrm{I}_0)$ *or* $\langle\Gamma\rangle \supseteq \mathrm{Inv}(\mathrm{I}_1)$*, then* UNIQUE-SAT($\Gamma$) *is* coNP*-hard.*
- *Otherwise* UNIQUE-SAT(Γ) *is polynomial-time decidable.*

Looking at the proof given by Juban it is easy to see that in the cases where UNIQUE-SAT(Γ) is coNP-hard, it is in fact complete for the class US.

Besides counting another computational goal of interest is to enumerate all the solutions. *Polynomial-delay algorithms* (see [JPY88]) are efficient enumerating algorithms that generate all the solutions, one after the other (i.e., without repetitions), in such a way that the delay until the first is output, and thereafter, the delay between any two consecutive solutions (and between the last solution and the halting) is bounded by a polynomial in the input size. Creignou and Hébrard [CH97] obtained a complexity classification for the enumeration problem associated with Γ-formulas which shows that in the Boolean case the property of having an efficient enumeration algorithm only depends on the polymorphisms of the set of relations Γ.

Theorem 6.19. [Enumeration]

- *If Γ is Schaefer, then there is a polynomial-delay algorithm that generates all satisfying assignments of a Γ-formula.*
- *Otherwise such an algorithm does not exist unless* P $=$ NP.

The fact that the Galois connection holds *a posteriori* for the enumeration problem in the Boolean case is rather unexpected. Indeed in [SS07] Henning and Ilka

Schnoor showed that the Galois connection does not work for the enumeration problems associated with CSPs over (non Boolean) finite domains; namely they showed that there exist relations R and R' (over a non Boolean domain) that give rise to the same co-colone, i.e., $\langle R\rangle = \langle R'\rangle$, and such that $\text{CSP}(R)$ has an efficient enumerating algorithm whereas $\text{CSP}(R')$ has none unless P = NP.

6.4 Optimisation Problems

In contrast to the decision and counting worlds, where there is essentially a unique way to obtain constraint satisfaction versions of the corresponding class, the case of optimization is somewhat more general.

The study of optimization problems in computational complexity started with the work of Krentel [Kre88, Kre92]; he defined the class OptP as follows. We say that a function h is in MinP if there are a function f in FP (i.e., is computable in polynomial-time) and a polynomial p such that $h(x) = \min_{|y|=p(|x|)} f(x,y)$, where minimization is taken with respect to the lexicographical order. The class MaxP is defined by taking the maximum of these values. Finally OptP = MinP$\cup$ MaxP. Krentel considered the following reducibility in connection with these classes. A function f is *metric reducible* to a function h, $f \leq^{\mathrm{P}}_{met} h$, if there exist two functions $g_1, g_2 \in \mathrm{FP}$ such that for all x, $f(x) = g_1(h(g_2(x)), x)$. The class OptP is a subclass of $\mathrm{FP}^{\mathrm{NP}}$ and it is well known that the closure of all three classes MinP, MaxP and OptP under metric reductions coincide with the class $\mathrm{FP}^{\mathrm{NP}}$, which means that showing completeness for a problem in MinP generally implies hardness for the same problem for MaxP and completeness for OptP. In the context of Boolean constraint satisfaction the typical problem for OptP is defined as follows.

Problem:	$\text{Lex-Min-Sat}(\Gamma)$
Input:	a Γ-formula ϕ
Output:	the lexicographically smallest satisfying assignment of ϕ

The algebraic framework will still be of use for this problem. However once again we have to be careful with the equality constraints resulting from the application of the Galois connection. Indeed, the removal of the equality constraints by identifying variables may make some variables disappear. However, if our constraint language can implement the equality relation, then this problem does not arise.

Proposition 6.20. *If $\Gamma \subseteq \langle\Gamma'\rangle$ and Γ' can express equality, then*

$$\text{Lex-Min-Sat}(\Gamma) \leq^{\mathrm{P}}_{met} \text{Lex-Min-Sat}(\Gamma').$$

Proof. Let ϕ be a Γ-formula with $\mathrm{Var}(\phi) = \{x_1, \dots, x_n\}$ ordered by their index, i.e., $x_1 < \dots < x_n$. We construct a formula by replacing in ϕ every constraint from Γ by its defining existentially quantified $(\Gamma' \cup \{=\})$-formula. Any occurring equality constraint can be removed in using the Γ'-implementation of the equality relation, which exists due to the prerequisites. Finally we delete all the existential quantifiers. The Γ'-formula ϕ' so obtained is over a set of variables $\{x_1, \dots, x_n\} \cup \{y_1, \dots, y_m\}$. This process corresponds to the function g_2 in the definition of metric reductions. The variables are ordered by their index and their alphabet, i.e., $x_1 < x_2 < \ldots x_n < y_1 < y_2 < \cdots < y_m$. The ordering of the variables ensures that in the minimal satisfying assignment of ϕ' the variables in $\{x_1, \ldots, x_n\}$ will be minimal with respect to the satisfaction of ϕ. Now the function g_1 shortens the assignment and removes all bits belonging to the variables y_j. Hence g_1 applied to the minimal satisfying assignment of ϕ' produces the minimal satisfying assignment of ϕ, thus concluding the proof. □

It is worth noticing that, here again as in the case of counting problems with Turing reductions (see Remark 6.15), it is the power of the reduction that makes the Galois connection help *a priori.*

Though not stated explicitly, this was the main tool to obtain the following classification in [RV03] by Reith and Vollmer.

Theorem 6.21. [Lexicographically minimal satisfying assignment]

- *If Γ is Schaefer or* 0*-valid then* LEX-MIN-SAT(Γ) *is polynomial-time computable.*
- *Otherwise,* LEX-MIN-SAT(Γ) *is* OptP*-complete under metric reductions.*

Also it was shown that to determine the last bit of the lexicographically minimal satisfying assignment of Γ-formulas is either complete for the class P^{NP} or in P, and which case occurs depends on Γ in exactly the same way as for LEX-MIN-SAT(Γ). Similar results (in switching 0-valid to 1-valid) hold for the corresponding maximization problems.

In the last decades, a lot of work in complexity theory has been devoted to classify the approximability of NP optimization problems (NPO). We propose to review here some of the results obtained on this subject for Boolean CSPs in an informal way. For a more detailed exposition we refer the reader to [CKST99] and [CKS01]. Recall that a generic NPO problem is specified by a four-tuple $(\mathcal{I}, S, m, opt)$ where $\mathcal{I}$ describes the set of instances, S the set of feasible solutions, m defines the value of a given solution and *opt* describes the goal, max or min, to achieve. We will focus here on maximization problems, i.e., in which $opt = max$. We say that an approximation algorithm has a *performance ratio* $\mathcal{R}(n)$ if, given an instance $x \in \mathcal{I}$ such that $|x| = n$ and the maximum value of its feasible solutions is $opt(x)$, it computes a solution $y \in S(x)$ whose value, $m(x, y)$ satisfies $m(x, y) \leq \mathcal{R}(n) \cdot opt(x)$. An NPO problem Π is in the class APX if there exists a polynomial-time approximation algorithm for Π whose performance ratio is bounded by a constant, and it is in poly-APX if there exists a polynomial-time approximation algorithm for Π whose performance ratio is bounded by a polynomial factor in the size of the input. Completeness in approximation classes is defined via appropriate approximation preserving reducibilities. We refer here to *AP*-reducibility $\leq_{AP}$, whose precise definition can be found in [CKST99] or in [CKS01, Definition 2.44]. In particular this notion of reducibility preserves membership in poly-APX as well as in APX.

We review results concerning two different optimisation problems MAX-ONES and MAX-SAT. In these problems the instances are still the same: Γ-formulas.

These problems differ in their definition of feasible solutions and the manner in which the value of a feasible solution is computed. First let us examine the MAX-ONES problem, where a Γ-formula has as its feasible solution space the set of satisfying assignments. The objective function is then simply set to that of maximizing the number of variables set to 1.

Problem:	MAX-ONES(Γ)
Input:	a Γ-formula ϕ
Output:	A satisfying assignment of ϕ that maximizes the number of variables set to true

In the study of optimization problems it is natural to consider weighted problems.

Problem:	weighted-MAX-ONES(Γ)
Input:	a Γ-formula ϕ over n variables and n non-negative integers $w_1, \ldots, w_n$
Output:	A satisfying assignment of ϕ that maximizes the sum of the weights of the variables set to true

When dealing with weighted problems the Galois connection works in the approximation framework.

Proposition 6.22. *If* $\Gamma \subseteq \langle \Gamma' \rangle$ *then*

$$\text{weighted-MAX-ONES}(\Gamma) \leq_{AP} \text{weighted-MAX-ONES}(\Gamma').$$

Proof. Given ϕ a Γ-formula with weighted variables, we construct a formula ϕ' by performing the following steps:

- Replace every constraint from Γ by its defining existentially quantified $(\Gamma' \cup \{=\})$-formula.
- Delete existential quantifiers.
- All the newly introduced variables get the weight 0.
- Delete equality clauses and replace all variables that are connected via a chain of equality constraints by a common new variable. This new variable gets as weight the sum of the weights of the variables in the chain.

Then, obviously, ϕ' is a Γ'-formula. Moreover, for any t, ϕ is satisfied by an assignment of weight t if and only ϕ' is satisfied by an assignment of weight t (an assignment has weight t if the sum of the weights of the variables it sets to true is t). Therefore, this provides a reduction that preserves approximability. □

It turns out that when the set Γ is Schaefer or 1-valid, then there is no essential difference in the approximability of the weighted and unweighted problems, and otherwise the task of finding either any feasible solutions or a solution of positive value is already NP-hard. For this reason the Galois connection can help in the framework of approximability of MAX-ONES problems and thus Khanna, Sudan and Williamson [KSW97] obtained the classification below, which follows Post's lattice.

Theorem 6.23. [Maximum ones satisfiability]

- *If Γ is* 1*-valid or dual Horn or affine with width* 2*, then* MAX-ONES(Γ) *is in* PO.
- *Else if Γ is affine, then* MAX-ONES(Γ) *is* APX*-complete.*
- *Else if Γ is Horn or bijunctive, then* MAX-ONES(Γ) *is poly-*APX*-complete.*
- *Else if Γ is* 0*-valid, then finding a feasible solution to* MAX-ONES(Γ) *is in* P *but finding a solution of positive value is* NP*-hard.*
- *Else the task of finding any feasible solution to* MAX-ONES(Γ) *is* NP*-hard.*

Now let us examine the problem MAX-SAT in which a given Γ-formula ϕ has as feasible solutions space the set of all truth assignments. The objective is to maximize the number of satisfied constraints in ϕ.

Problem:	MAX-SAT(Γ)
Input:	a Γ-formula ϕ
Output:	A truth assignment that maximizes the number of constraints satisfied in ϕ

Creignou [Cre95], and Khanna and Sudan [KS96] obtained a dichotomy classification for the approximability of MAX-SAT(Γ). Before stating the corresponding theorem we recall that a relation R is said to be 2-*monotone* if it is expressible as a DNF-formula either of the form $(x_1 \wedge \ldots \wedge x_p)$ or $(\neg y_1 \wedge \ldots \wedge \neg y_q)$ or $(x_1 \wedge \ldots \wedge x_p) \vee (\neg y_1 \wedge \ldots \wedge \neg y_q)$. A set Γ of relations is said to be 2-monotone if every relation R from Γ is 2-monotone.

Theorem 6.24. [Maximum satisfiability]

- *If Γ is* 0*-valid or* 1*-valid or* 2*-monotone, then* MAX-SAT(Γ) *is in* PO.
- *Otherwise* MAX-SAT(Γ) *is* APX*-complete.*

It is worth pointing out that this classification does not follow Post's lattice. Indeed it is easy to check that 2-monotone relations do not constitute a co-clone. (The usual implication is obviously 2-monotone, however the conjunction $x \to y \wedge u \to v$ is not.) Therefore the Galois connection is of no help in order to get the complexity classification of MAX-SAT(Γ). A similar dichotomy (P/PLS-complete) was obtained by Chapdelaine and Creignou [CC05] for the local search problems, in which the objective is to find an assignment that is locally maximum, that is to say such that flipping any bit does not increase the number of satisfied constraints. Note that the minimization problems, whose approximability properties might be different from the ones of the maximization problems, were also studied in [KSTW01].

6.5 Further Problems

We examine here further problems that do not fall in the previous categories. Let us start with the Inverse satisfiability problem.

Problem:	INVERSE-SAT(Γ)
Input:	a set $M \subseteq \{0,1\}^n$
Question:	Does there exist a Γ-formula over n variables that has M as its set of models?

As noticed in [KS98] the fact that $\Gamma \subseteq \langle \Gamma' \rangle$ is not sufficient to ensure that INVERSE-SAT(Γ) is reducible to INVERSE-SAT(Γ'). Indeed one needs a stronger notion of implementation, which from a Γ-formula ϕ allows to build a Γ'-formula ϕ' such that not only ϕ is satisfiable if and only if ϕ' is satisfiable, but there is also a one-to-one onto correspondence between their respective set of models. However Kavvadias and Sideri [KS98] obtained a classification theorem that follows Post's lattice.

Theorem 6.25. [Inverse satisfiability] *Let Γ be a constraint language that contains the constants* 0 *and* 1*.*

- *If Γ is Schaefer then* INVERSE-SAT(Γ) *is in* P.
- *Otherwise* INVERSE-SAT(Γ) *is* coNP*-complete.*

Observe that a complexity classification for sets Γ that do not necessarily contain the constants is still an open question.

The parameterized complexity of constraint satisfaction problems is also of interest. In parameterized complexity we are dealing with problems where each instance has a distinguished part called the *parameter*. A parameterized problem is *fixed-parameter-tractable* (FPT) if it can be solved in polynomial-time for every fixed value of the problem parameter k, and moreover, the degree of the polynomial in the time bound does not depend on k. By showing that a problem is NP-complete one gives strong evidence that it does not have a polynomial-time algorithm. There is a similar completeness program in parameterized complexity that allows to show that certain problems are unlikely to be in FPT. In particular the class W[1] contains the parameterized problems that can be reduced to the problem *"Does the given nondeterministic Turing machine accept input x in at most k steps?"*. (The parameter for this problem is k.) It is believed that W[1]-complete problems are not fixed-parameter-tractable. For more background on parameterized complexity the reader is asked to consult the monograph [FG06].

In [Mar05] Marx investigated the parameterized complexity of p-SAT(Γ) defined as follows.

Problem:	p-SAT(Γ)
Input:	a Γ-formula ϕ where each variable can occur at most once in each constraint, and an integer k
Parameter:	k
Question:	Is there a truth assignment setting exactly k variables to true that satisfies ϕ?

He introduced a new property, weak separability, that plays a crucial role in the parameterized complexity of problems.

Definition 6.26. A relation R is *weakly separable* if

1. whenever x_1 and x_2 are in R, if $x_1 \wedge x_2$ is in R, then so is $x_1 \vee x_2$,
2. whenever $x_1 < x_2 < x_3$ are in R (where $<$ refers to the coordinate-wise order) then so is $x_1 \oplus x_2 \oplus x_3$.

All operations are taken coordinate-wise.

Marx got the following classification result.

Theorem 6.27. [Parameterized complexity]

- *If every relation in Γ is weakly separable then p-*SAT(Γ) *is in* FPT.
- *Otherwise p-*SAT(Γ) *is* W[1]*-complete.*

This classification does not follow Post's lattice. Indeed it is easy to check that weakly separable relations do not constitute a co-clone (for instance the relation $R = \{000, 101, 110\}$ is weakly separable, however $R'(y, z) = \exists x R(x, y, z)$, $R' = \{00, 01, 10\}$ is not).

To conclude let us mention other complexity results for Boolean constraint satisfaction problems that are related to geometric properties of the solution space. A first result in this vein (even if it was not originally advertised in this way) was obtained by Jonsson and Krokhin [JK04], who studied the frozen variables problem, actively used e.g. in the study of phase transition phenomena, but also closely related to a problem from the database context (of "auditing variables" [KPR03]):

Problem:	FROZEN-VAR(Γ)
Input:	a Γ-formula ϕ and V' a nonempty subset of Var(ϕ)
Question:	Is every variable x in V' frozen? (a variable x is said to be *frozen* if $\lvert\{I(x) \mid I \text{ is a truth assignment satisfying } \phi\}\rvert = 1$)

The problem to recognize frozen variables is a generalization of the UNIQUE-SAT problem in which the uniqueness requirement is applied to all variables. As proved in [JK04] the Galois connection applies *a priori* for this problem and leads to the following classification.

Theorem 6.28. [Frozen variables]

- *If $\langle\Gamma\rangle \supseteq$* Inv$(\mathrm{N}_2)$*, then* FROZEN-VAR$(\Gamma)$ *is* DP*-complete.*
- *Otherwise, if $\langle\Gamma\rangle \supseteq$* Inv$(\mathrm{I}_0)$ *or $\langle\Gamma\rangle \supseteq$* Inv$(\mathrm{I}_1)$*, then* FROZEN-VAR$(\Gamma)$ *is* coNP*-complete.*
- *In all other cases* FROZEN-VAR(Γ) *is in* P.

Connectivity properties of the solution space were investigated by Gopalan, Kolatitis, Maneva and Papadimitriou in [GKMP06]. In particular they studied the st-connectivity problem defined as follows.

Problem:	ST-CONN(Γ)
Input:	a Γ-formula ϕ, and two satisfying assignments s and t of ϕ
Question:	Is there a path in $G(\phi)$, the subgraph of the n-dimensional hypercube induced by the solutions of ϕ, from s to t?

Gopalan *et al.* [GKMP06] introduced the notion of a *tight* set of Boolean relations, defined as follows: A relation R is tight if it is componentwise bijunctive (that is, every connected component in the hypercube of all tuples in R is bijunctive), OR-free (that is, the binary disjunction is not definable from R by setting all but two of the variables to constants), or NAND-free (that is, the binary NAND is not definable from R by setting all but two of the variables to constants); and as usual a set of relations is tight if every relation in it is tight. The class of tight constraint languages properly includes the class of Schaefer constraint languages. Gopalan *et al.* showed that if Γ is tight then $\textsc{st-Conn}(\Gamma)$ is in P, otherwise it is PSPACE-complete. This classification does not follow Post's lattice, therefore one can deduce that the Galois connection does not help in studying the connectivity properties of the solution space of constraint satisfaction problems. Gopalan *et al.* also investigated the more general problem $\textsc{Conn}(\Gamma)$, in which the question is whether $G(\phi)$ is connected. They proved that if Γ is tight then $\textsc{Conn}(\Gamma)$ is in coNP, otherwise it is PSPACE-complete. The only tight cases for which $\textsc{Conn}(\Gamma)$ is not known to be coNP-complete or in P are Horn and dual Horn. While these cases were conjectured in [GKMP06] to be polynomial, Makino, Tamaki and Yamamoto recently exhibited in [MTY07] a Horn set Γ such that $\textsc{Conn}(\Gamma)$ is coNP-complete. So, a complete classification for the connectivity problem is still an open problem.

7 Conclusion

We have seen that Post's lattice helps if the definition of the considered computational problem allows to hide the existential quantifiers that are obtained from the usual local-replacement reduction, and to remove the equality clauses by identifying variables. The article by Ilka and Henning Schnoor in this volume addresses similar but different Galois connections that apply if one or both of these conditions are not given. These connections do not involve Post's lattice but finer structures. Nevertheless it turns out that also here, Post's lattice is often the important spine.

While the complexity of Boolean constraint satisfaction problems has been extensively studied, there still remain some interesting open questions as we have seen along this survey. For instance, a complete complexity classification for the above mentionned connectivity problem appears as a challenging problem. Another problem that has raised considerable attention is a complexity classification of the inference problem for *basic circumscription*, where "basic" here means that a model is defined to be minimal if it is minimal in the componentwise order involving all variables (in the notation of the definition given in Sect. 6.1 on p. 16 we thus require that P is the set of all variables); the Galois connection is not known to hold for this case.

The interest of the complexity of Boolean CSPs lies in its strong connection to Post's lattice. The latter being well-known, we can take advantage of it in order to study the complexity of Boolean CSPs, which in turn can help to develop new tools and strategies towards the more ambitious goal of the complexity of constraint satisfaction problems over arbitrary finite domains. To name only one example, the recent detailed complexity study of CSPs solvable in logarithmic space published in [ELT07] gained many of its initial ideas from a corresponding fine classification of Boolean CSPs (presented in Sect. 5 of this survey). Moreover, the dual connection between Boolean CSPs and Post's lattice works also the other way round. Indeed, complexity studies have renewed the interest in Post lattice and motivated some advanced studies in the description of the lattice itself (see for instance [BRSV05] and [CKZ08] where different bases of the co-clones were proposed) as well as in the description of finers structures underlying Post lattice [Sch07b, SS].

Boolean satisfiability problems and related counting and optimization problems form the nucleus of most complexity classes since they usually provide the first and canonical complete problems. Their study, thus, is a hardly disguised study of complexity classes, their inclusion structure, and their properties. We think we have made this very clear in the course of this chapter by talking about the power of reductions, the structure of complete sets, the complexity of isomorphism problems—these address typical complexity-theoretic questions, and this provides a main motivation for a further pursue of the study of Boolean constraint satisfaction problems.

Acknowledgement

We are grateful to Henning Schnoor for many helpful comments on a previous version of this paper.

References

[ABI+05] Allender, E., Bauland, M., Immerman, N., Schnoor, H., Vollmer, H.: The complexity of satisfiability problems: Refining Schaefer's theorem. In: Jedrzejowicz, J., Szepietowski, A. (eds.) MFCS 2005. LNCS, vol. 3618, pp. 71–82. Springer, Heidelberg (2005)

[Agr01] Agrawal, M.: The first-order isomorphism theorem. In: Hariharan, R., Mukund, M., Vinay, V. (eds.) FSTTCS 2001. LNCS, vol. 2245, pp. 70–82. Springer, Heidelberg (2001)

[BBC+07] Bauland, M., Böhler, E., Creignou, N., Reith, S., Schnoor, H., Vollmer, H.: The complexity of problems for quantified constraints. Technical Report 07-023, Electronic Colloquium on Computational Complexity (2007)

[BCC+05] Bauland, M., Chapdelaine, P., Creignou, N., Hermann, M., Vollmer, H.: An algebraic approach to the complexity of generalized conjunctive queries. In: Hoos, H.H., Mitchell, D.G. (eds.) SAT 2004. LNCS, vol. 3542, pp. 30–45. Springer, Heidelberg (2005)

[BCRV03] Böhler, E., Creignou, N., Reith, S., Vollmer, H.: Playing with Boolean blocks, part I: Post's lattice with applications to complexity theory. ACM-SIGACT Newsletter 34(4), 38–52 (2003)

[BCRV04] Böhler, E., Creignou, N., Reith, S., Vollmer, H.: Playing with Boolean blocks, part II: Constraint satisfaction problems. ACM-SIGACT Newsletter 35(1), 22–35 (2004)

[BD03] Bulatov, A., Dalmau, V.: Towards a dichotomy theorem for the counting constraint satisfaction problem. In: Proceedings Foundations of Computer Science, pp. 562–572. ACM Press, New York (2003)

[BG82] Blass, A., Gurevich, Y.: On the unique satisfiability problem. Information and Control 82, 80–88 (1982)

[BH05] Bauland, M., Hemaspaandra, E.: Isomorphic implication. In: Jedrzejowicz, J., Szepietowski, A. (eds.) MFCS 2005. LNCS, vol. 3618, pp. 119–130. Springer, Heidelberg (2005)

[BHRV02] Böhler, E., Hemaspaandra, E., Reith, S., Vollmer, H.: Equivalence and isomorphism for Boolean constraint satisfaction. In: Bradfield, J.C. (ed.) CSL 2002 and EACSL 2002. LNCS, vol. 2471, pp. 412–426. Springer, Heidelberg (2002)

[BHRV04] Böhler, E., Hemaspaandra, E., Reith, S., Vollmer, H.: The complexity of Boolean constraint isomorphism. In: Diekert, V., Habib, M. (eds.) STACS 2004. LNCS, vol. 2996, pp. 164–175. Springer, Heidelberg (2004)

[BKKR69] Bodnarchuk, V.G., Kalužnin, L.A., Kotov, V.N., Romov, B.A.: Galois theory for Post algebras I, II. Cybernetics 5, :243–252, 531–539 (1969)

[BRSV05] Böhler, E., Reith, S., Schnoor, H., Vollmer, H.: Bases for Boolean co-clones. Information Processing Letters 96, 59–66 (2005)

[CC05] Chapdelaine, P., Creignou, N.: The complexity of boolean constraint satisfaction local search problems. Annals of Mathematics and Artificial Intelligence 43(1-4), 51–63 (2005)

[CH96] Creignou, N., Hermann, M.: Complexity of generalized satisfiability counting problems. Information and Computation 125, 1–12 (1996)

[CH97] Creignou, N., Hébrard, J.-J.: On generating all solutions of generalized satisfiability problems. Informatique Théorique et Applications/Theoretical Informatics and Applications 31(6), 499–511 (1997)

[Che06] Chen, H.: A rendezvous of logic, complexity, and algebra. ACM-SIGACT Newsletter 37(4), 85–114 (2006)

[CHS07] Chapdelaine, P., Hermann, M., Schnoor, I.: Complexity of default logic on generalized conjunctive queries. In: Baral, C., Brewka, G., Schlipf, J. (eds.) LPNMR 2007. LNCS, vol. 4483, pp. 58–70. Springer, Heidelberg (2007)

[CKS01] Creignou, N., Khanna, S., Sudan, M.: Complexity Classifications of Boolean Constraint Satisfaction Problems. Monographs on Discrete Applied Mathematics. SIAM, Philadelphia (2001)

[CKST99] Crescenzi, P., Kann, V., Silvestri, R., Trevisan, L.: Structure in approximation classes. SIAM Journal on Computing 28(5), 1759–1782 (1999)

[CKZ08] Creignou, N., Kolaitis, Ph., Zanuttini, B.: Structure Identification for Boolean Relations and Plain Bases for co-Clones. Journal of Computer and System Sciences 74, 1103–1115 (2008)

[Coo71] Cook, S.A.: The complexity of theorem proving procedures. In: Proceedings 3rd Symposium on Theory of Computing, pp. 151–158. ACM Press, New York (1971)

[Cre95] Creignou, N.: A dichotomy theorem for maximum generalized satisfiability problems. Journal of Computer and System Sciences 51, 511–522 (1995)

[CZ06] Creignou, N., Zanuttini, B.: A complete classification of the complexity of propositional abduction. SIAM Journal on Computing 36(1), 207–229 (2006)

[Dal97] Dalmau, V.: Some dichotomy theorems on constant-free quantified boolean formulas. Technical Report LSI-97-43-R, Department de Llenguatges i Sistemes Informàtica, Universitat Politécnica de Catalunya (1997)

[Dal00] Dalmau, V.: Computational complexity of problems over generalized formulas. Ph.D thesis, Department de Llenguatges i Sistemes Informàtica, Universitat Politécnica de Catalunya (2000)

[DHK05] Durand, A., Hermann, M., Kolaitis, P.G.: Subtractive reductions and complete problems for counting complexity classes. Theoretical Computer Science 340(3), 496–513 (2005)

[ELT07] Egri, L., Larose, B., Tesson, P.: Symmetric datalog and constraint satisfaction problems in logspace. In: 22nd IEEE Symposium on Logic in Computer Science, pp. 193–202. IEEE Computer Society, Los Alamitos (2007)

[FG06] Flum, J., Grohe, M.: Parameterized complexity theory. Springer, Heidelberg (2006)

[Gei68] Geiger, D.: Closed systems of functions and predicates. Pac. J. Math. 27(2), 228–250 (1968)

[GKMP06] Gopalan, P., Kolaitis, P.G., Maneva, E.N., Papadimitriou, C.H.: The connectivity of boolean satisfiability: Computational and structural dichotomies. In: Bugliesi, M., Preneel, B., Sassone, V., Wegener, I. (eds.) ICALP 2006. LNCS, vol. 4051, pp. 346–357. Springer, Heidelberg (2006)

[Hem04] Hemaspaandra, E.: Dichotomy theorems for alternation-bounded quantified boolean formulas. CoRR, cs.CC/0406006 (2004)

[HV95] Hemaspaandra, L., Vollmer, H.: The satanic notations: counting classes beyond #P and other definitional adventures. Complexity Theory Column 8, ACM-SIGACT News 26(1), 2–13 (1995)

[Imm99] Immerman, N.: Descriptive Complexity. Graduate Texts in Computer Science. Springer, New York (1999)

[JCG97] Jeavons, P.G., Cohen, D.A., Gyssens, M.: Closure properties of constraints. Journal of the ACM 44(4), 527–548 (1997)

[JK04] Jonsson, P., Krokhin, A.: Recognizing frozen variables in constraint satisfaction problems. Theoretical Computer Science 329(1-3), 93–113 (2004)

[JPY88] Johnson, D.S., Papadimitriou, C.H., Yannakakis, M.: On generating all maximal independent sets. Information Processessing Letters 27(3), 119–123 (1988)

[Jub99] Juban, L.: Dichotomy theorem for generalized unique satisfiability problem. In: Ciobanu, G., Păun, G. (eds.) FCT 1999. LNCS, vol. 1684, pp. 327–337. Springer, Heidelberg (1999)

[KK03] Kirousis, L.M., Kolaitis, P.G.: The complexity of minimal satisfiability problems. Information and Computation 187(1), 20–39 (2003)

[KK04] Kirousis, L.M., Kolaitis, P.G.: A dichotomy in the complexity of propositional circumscription. Theory of Computing Systems 37, 695–715 (2004)

[KL99] Kleine Büning, H., Lettmann, T.: Propositional Logic: Deduction and Algorithms. Cambridge Tracts in Theoretical Computer Science. Cambridge University Press, Cambridge (1999)

[KPR03] Kleinberg, J., Papadimitriou, C., Raghavan, P.: Auditing Boolean attributes. Journal of Computer and System Sciences 66(1), 244–253 (2003)

[Kre88] Krentel, M.W.: The complexity of optimization functions. Journal of Computer and System Sciences 36, 490–509 (1988)

[Kre92] Krentel, M.W.: Generalizations of OptP to the polynomial hierarchy. Theoretical Computer Science 97, 183–198 (1992)

[KS96] Khanna, S., Sudan, M.: The optimization complexity of constraint satisfaction problems. Technical Report STAN-CS-TN-96-29, Stanford University (1996)

[KS98] Kavvadias, D., Sideri, M.: The inverse satisfiability problem. SIAM Journal of Computing 28(1), 152–163 (1998)

[KST93] Köbler, J., Schöning, U., Torán, J.: The Graph Isomorphism Problem: its Structural Complexity. Progress in Theoretical Computer Science. Birkhäuser, Basel (1993)

[KSTW01] Khanna, S., Sudan, M., Trevisan, L., Williamson, D.P.: The approximability of constraint satisfaction problems. SIAM Journal on Computing 30, 1863–1920 (2001)

[KSW97] Khanna, S., Sudan, M., Williamson, D.: A complete classification of the approximability of maximization problems derived from Boolean constraint satisfaction. In: Proceedings 29th Symposium on Theory of Computing, pp. 11–20. ACM Press, New York (1997)

[Lau06] Lau, D.: Function Algebras on Finite Sets. Monographs in Mathematics. Springer, Heidelberg (2006)

[Mar05] Marx, D.: Parameterized complexity of constraint satisfaction problems. Computational Complexity 14(2), 153–183 (2005)

[MS72] Meyer, A.R., Stockmeyer, L.J.: The equivalence problem for regular expressions with squaring requires exponential time. In: Proceedings 13th Symposium on Switching and Automata Theory, pp. 125–129. IEEE Computer Society Press, Los Alamitos (1972)

[MTY07] Makino, K., Tamaki, S., Yamamoto, M.: On the boolean connectivity problem for horn relations. In: Marques-Silva, J., Sakallah, K.A. (eds.) SAT 2007. LNCS, vol. 4501, pp. 187–200. Springer, Heidelberg (2007)

[NJ04] Nordh, G., Jonsson, P.: An algebraic approach to the complexity of propositional circumscription. In: 19th Symposium on Logic in Computer Science, pp. 367–376. IEEE Computer Society, Los Alamitos (2004)

[Nor05] Nordh, G.: A trichotomy in the complexity of propositional circumscription. In: Baader, F., Voronkov, A. (eds.) LPAR 2004. LNCS, vol. 3452, pp. 257–269. Springer, Heidelberg (2005)

[NZ05] Nordh, G., Zanuttini, B.: Propositional abduction is almost always hard. In: Proceedings International Joint Conference on Artificial Intelligence, pp. 534–539 (2005)

[Pap94] Papadimitriou, C.H.: Computational Complexity. Addison-Wesley, Reading (1994)

[Pip97] Pippenger, N.: Theories of Computability. Cambridge University Press, Cambridge (1997)

[Pos20] Post, E.L.: Determination of all closed systems of truth tables. Bulletin of the AMS 26, 437 (1920)

[Pos41] Post, E.L.: The two-valued iterative systems of mathematical logic. Annals of Mathematical Studies 5, 1–122 (1941)

[Rei05] Reingold, O.: Undirected st-connectivity in log-space. In: Proceedings of the 37th Symposium on Theory of Computing, pp. 376–385. ACM Press, New York (2005)

[RV03] Reith, S., Vollmer, H.: Optimal satisfiability for propositional calculi and constraint satisfaction problems. Information and Computation 186(1), 1–19 (2003)

[Sch78] Schaefer, T.J.: The complexity of satisfiability problems. In: Proccedings 10th Symposium on Theory of Computing, pp. 216–226. ACM Press, New York (1978)

[Sch07a] Schnoor, H.: Algebraic Techniques for Satisfiability Problems. PhD thesis, Leibniz Universität Hannover, Fakultät für Elektrotechnik und Informatik (2007)

[Sch07b] Schnoor, I.: The Weak Base Method for Constraint Satisfaction. Ph.D thesis, Leibniz Universität Hannover, Fakultät für Elektrotechnik und Informatik (2007)

[SM73] Stockmeyer, L.J., Meyer, A.R.: Word problems requiring exponential time. In: Proceedings 5th ACM Symposium on the Theory of Computing, pp. 1–9. ACM Press, New York (1973)

[SS] Schnoor, H., Schnoor, I.: Partial polymorphisms and constraint satisfaction problems. In: Creignou, N., Kolaitis, P.G., Vollmer, H. (eds.) Complexity of Constraints. LNCS, vol. 5250. Springer, Heidelberg (2008)

[SS07] Schnoor, H., Schnoor, I.: Enumerating all solutions for constraint satisfaction problems. In: 24th Symposium on Theoretical Aspects of Computer Science. LNCS, pp. 694–705. Springer, Heidelberg (2007)

[Tod91a] Toda, S.: Computational Complexity of Counting Complexity Classes. Ph.D thesis, Tokyo Institute of Technology, Department of Computer Science, Tokyo (1991)

[Tod91b] Toda, S.: PP is as hard as the polynomial time hierarchy. SIAM Journal on Computing 20, 865–877 (1991)

[TW92] Toda, S., Watanabe, O.: Polynomial time 1-Turing reductions from #PH to #P. Theoretical Computer Science 100, 205–221 (1992)

[Val79a] Valiant, L.G.: The complexity of computing the permanent. Theoretical Computer Science 8, 189–201 (1979)

[Val79b] Valiant, L.G.: The complexity of enumeration and reliability problems. SIAM Journal of Computing 8(3), 411–421 (1979)

[Vol94] Vollmer, H.: Komplexitätsklassen von Funktionen. Ph.D thesis, Universität Würzburg, Fakultät für Mathematik und für Informatik (1994)

[Vol99] Vollmer, H.: Introduction to Circuit Complexity – A Uniform Approach. Texts in Theoretical Computer Science. Springer, Heidelberg (1999)

[Vol07] Vollmer, H.: Computational complexity of constraint satisfaction. In: Cooper, S.B., Löwe, B., Sorbi, A. (eds.) CiE 2007. LNCS, vol. 4497, pp. 748–757. Springer, Heidelberg (2007)

[Wra77] Wrathall, C.: Complete sets and the polynomial-time hierarchy. Theoretical Computer Science 3, 23–33 (1977)

[Zan91] Zankó, V.: #P-completeness via many-one reductions. International Journal of Foundations of Computer Science 2, 77–82 (1991)

Basics of Galois Connections

Ferdinand Börner

Universität Potsdam, Institut für Informatik,
PF 90 03 27, D–14 439 Potsdam
fboerner@rz.uni-potsdam.de

Abstract. We give an overview of basic properties of Galois connections between sets of relations and sets of functions or generalized functions. First we focus on the Galois connections Inv–Pol and Inv–mPol. Then we use these results to provide some tools for the representation of several closure operators on relations as closure operators of some Galois connections.

Introduction

Galois connections appear very frequently in all parts of Mathematics. Often they are of importance, they simplify things, they relate different objects to each other, and they are easy to define. We only need two sets $\mathcal{X}$ and $\mathcal{Y}$ and a defining relation $\Xi \subseteq \mathcal{X} \times \mathcal{Y}$. Then the operators

$$\alpha : \mathcal{X} \supseteq X \mapsto \alpha X := \{y \in \mathcal{Y} \mid (\forall x \in X)\Xi(x,y)\} \subseteq \mathcal{Y}$$
$$\beta : \mathcal{Y} \supseteq Y \mapsto \beta Y := \{x \in \mathcal{X} \mid (\forall y \in Y) \in \Xi(x,y)\} \subseteq \mathcal{X}$$

form a Galois connection, and all Galois connections can be defined in this way. In Section 1 we give a short introduction to the general properties of such abstract Galois connections. (This is not the most general way to consider Galois connections. For a study in the framework of lattice theory we refer to [1].)

The "Galois closed subsets" of $\mathcal{X}$ and $\mathcal{Y}$ are the sets $X \subseteq \mathcal{X}$ and $Y \subseteq \mathcal{Y}$ with $X = \beta\alpha X$ and $Y = \alpha\beta Y$. The closed subsets of $\mathcal{X}$ and $\mathcal{Y}$ are in one-one correspondence, "small" subsets of $\mathcal{X}$ determine "large" closed subsets of $\mathcal{Y}$ and vice versa. To "characterize" a Galois connection means to describe the Galois closed subsets with other tools.

Of course, the name "Galois connection" reminds of E. Galois' famous connection between elements a of a field E and its automorphisms π, based on the fixpoint relation $\pi(a) = a$. Marc Krasner was impressed by this connection, and, with the intention to generalize the notion of a field, he extended it to a "Galois connection" between permutations π of a set D and relations ϱ on D ([16, 17]). The defining relation was given by "$\pi(\varrho) \subseteq \varrho$", where $\pi(\varrho) := \{\pi(\boldsymbol{a}) \mid \boldsymbol{a} \in \varrho\}$. If D is finite, then the Galois closed permutation sets are just the permutation groups, and they correspond to those sets of relations that are closed under all operations that can be defined by first-order formulas.

This was the starting point for an investigation of Galois connections between sets of relations and sets of functions or generalized functions. Krasner's approach

N. Creignou et al. (Eds.): Complexity of Constraints, LNCS 5250, pp. 38–67, 2008.

was extended to unary functions, arbitrary functions, partial functions and so on ([2, 3, 26, 29, 31, 32, 27]). In particular, L.A. Kalužnin and his collaborators simplified proofs, generalized the notation and found applications in various fields of algebra. Corresponding investigations in the english literature can be found in the paper [13] by D. Geiger.

In the center of this approach stands the Galois connection Inv–Pol between sets of functions and sets of relations on a basic set D. (In this article we always assume that D is finite.) Inv–Pol is based on the "invariance relation" $f(\varrho) \subseteq \varrho$, where f is an n-ary function, ϱ an m-ary relation and $f(\varrho) = \{f(\boldsymbol{a}_1, \ldots, \boldsymbol{a}_n) \mid \boldsymbol{a}_1, \ldots, \boldsymbol{a}_n \in \varrho\}$. The closed sets of functions are clones (i.e. sets of functions, closed under superposition and containing all projections), and the closed sets of relations, called relational clones, are sets of relations that are closed under definitions with first-order primitive positive formulas. This connection became one of the main tools for the investigation of clones and clone lattices with applications from Universal Algebra to Technical Computer Science. We refer to the monographs [20] and [18].

Then it was a nice surprise when P. Jeavons et.al. discovered a very successful application of this Galois connection to the investigation of the algorithmic complexity of Constraint Satisfaction Problems (CSPs) ([14, 15, 10]). In Section 3 we shortly introduce the starting idea.

Motivated by these results, we want to provide tools for the description of more closure operators on sets of relations with the help of Galois connections. For this aim, in Section 4, we first introduce the general notion of a "multifunction" (also called multioperation, partial hyperfunction, correspondence, ...), and characterize the corresponding Galois connection Inv–mPol ([32, 4, 27]).

In Section 5 we start with this general connection and obtain new closure operators by restricting the set of multifunctions to various subsets $\mathcal{M}$. If a closure operator $\mathcal{H}$ on sets of relations is given, then we want to find a set $\mathcal{M}$ of multifunctions such that $\mathcal{H} = \mathsf{Inv}(\mathcal{M} \cap \mathsf{mPol})$. If this is possible, then $\mathcal{H}$ can be described with the help of a Galois connection. Theorem 5.3 shows under which conditions this is really possible. Then we give a short overview of some known connections that can be obtained in this way. (Here we restrict ourselves to the case of a finite basic set D; for a more general survey we refer to [22].)

Finally, in Section 6, we investigate a special closure operator which is connected with the algorithmic Quantified Constraint Satisfaction Problem (QCSP), a generalization of the CSP. Using the results of the previous Sections, we show that the set of surjective functions can serve as a suitable base for the representation of this closure operator via a new Galois connection. This result leads to a similar application to the QCSP as before for the CSP.

1 Basic Properties of Galois Connections

Galois connections can be considered as tools for the description of closure operators. In this article, closure operators on sets of relations (or on sets of functions or generalized or specialized functions) will play an important role. In this

Section we collect some general properties of closure operators and Galois connections. Almost everything here is standard and well known (see e.g. [18, 19, 20]). In the last Subsection we investigate the problem of how to approximate a closure operator $\mathcal{H}$ with the help of a given Galois connection α–β.

The proofs in this Section are mainly straight forward, and we usually skip them.

Closure Operators

If $\mathcal{Z}$ is a set, then $\mathcal{P}\,\mathcal{Z} := \{Z \mid Z \subseteq \mathcal{Z}\}$ is the set of all subsets of $\mathcal{Z}$. An operator $\mathcal{H} : \mathcal{P}\,\mathcal{Z} \to \mathcal{P}\,\mathcal{Z}$ is a *closure operator on* $\mathcal{Z}$ if for all $Z_0, Z_1 \subseteq \mathcal{Z}$ the following hold:

1. $Z_0 \subseteq \mathcal{H}Z_0$ ($\mathcal{H}$ is extensive.)
2. $Z_0 \subseteq Z_1 \Longrightarrow \mathcal{H}Z_0 \subseteq \mathcal{H}Z_1$ ($\mathcal{H}$ is monotone.)
3. $\mathcal{H}\mathcal{H}Z_0 = \mathcal{H}Z_0$ ($\mathcal{H}$ is idempotent.)

A set $Z \subseteq \mathcal{Z}$ is *closed under the closure operator* $\mathcal{H}$ if $Z = \mathcal{H}Z$. The set of all closed subsets of $\mathcal{Z}$ is the *closure system of* $\mathcal{H}$ and denoted by

$$\mathcal{C}^{\mathcal{H}}_{\mathcal{Z}} := \{Z \subseteq \mathcal{Z} \mid Z = \mathcal{H}Z\}.$$

If $\mathcal{C} = \mathcal{C}^{\mathcal{H}}_{\mathcal{Z}}$ for some closure operator $\mathcal{H}$, then

1. $\mathcal{Z} \in \mathcal{C}$ and
2. if $Z_i \in \mathcal{C}$ for all $i \in I$, I arbitrary index set, then $\bigcap_{i\in I} Z_i \in \mathcal{C}$.

Vice versa, every set $\mathcal{C} \subseteq \mathcal{P}\,\mathcal{Z}$ with (1.) and (2.) is called a *closure system.* Such a closure system $\mathcal{C}$ induces a closure operator $\mathcal{H}_{\mathcal{C}}$ by

$$\mathcal{H}_{\mathcal{C}}(Z_0) := \bigcap\{Z \in \mathcal{C} \mid Z_0 \subseteq Z\}.$$

Closure systems $\mathcal{C}$ and closure operators $\mathcal{H}$ on $\mathcal{Z}$ are in one-one connection; we always have $\mathcal{C}^{\mathcal{H}_{\mathcal{C}}}_{\mathcal{Z}} = \mathcal{C}$ and $\mathcal{H}_{\mathcal{C}_{\mathcal{H}}} = \mathcal{H}$.

Closure systems are complete lattices.

Theorem 1.1. *Let* $\mathcal{H}$ *be a closure operator on* $\mathcal{Z}$ *and* $\mathcal{C}^{\mathcal{H}}_{\mathcal{Z}}$ *the corresponding closure system. Let* $Z_1, Z_2, Z_i \in \mathcal{C}$, *(*$i \in I$, I *arbitrary index set) and consider the operations*

$$Z_1 \wedge Z_2 = Z_1 \cap Z_2, \; \bigwedge_{i\in I} Z_i = \bigcap_{i\in I} Z_i$$

$$Z_1 \vee Z_2 = \mathcal{H}(Z_1 \cup Z_2), \; \bigvee_{i\in I} Z_i = \mathcal{H}\left(\bigcup_{i\in I} Z_i\right).$$

Then $\mathcal{C}^{\mathcal{H}}_{\mathcal{Z}}$ *with the operations* $\wedge$ *and* $\vee$ *is a complete lattice. The order of this lattice is just the set theoretical inclusion.* □

A closure operator $\mathcal{H}$ and the system $\mathcal{C}^{\mathcal{H}}_{\mathcal{Z}}$ are called *algebraic* if $\mathcal{H}Z = \bigcup\{\mathcal{H}Z_0 \mid Z_0 \subseteq Z \text{ and } Z_0 \text{ is finite}\}$ for all $Z \subseteq \mathcal{Z}$. In this case, $\langle \mathcal{C}^{\mathcal{H}}_{\mathcal{Z}}; \wedge, \vee \rangle$ is an algebraic lattice. The set of all subalgebras of a universal algebra is an example of an algebraic closure system.

For closure operators $\mathcal{H}_1$ and $\mathcal{H}_2$ on $\mathcal{Z}$ we write $\mathcal{H}_1 \leq \mathcal{H}_2$ if $\mathcal{H}_1 Z_0 \subseteq \mathcal{H}_2 Z_0$ for all $Z_0 \subseteq \mathcal{Z}$.

Lemma 1.2. *The following are equivalent for two closure operators $\mathcal{H}_1$, $\mathcal{H}_2$.*

1. $\mathcal{H}_1 \leq \mathcal{H}_2$
2. $\mathcal{H}_1\mathcal{H}_2 = \mathcal{H}_2$
3. $\mathcal{C}^{\mathcal{H}_2}_{\mathcal{Z}} \subseteq \mathcal{C}^{\mathcal{H}_1}_{\mathcal{Z}}$ □

So every $\mathcal{H}_2$–closed subset $Z \subseteq \mathcal{Z}$ is also $\mathcal{H}_1$–closed. But note that $\mathcal{C}^{\mathcal{H}_2}_{\mathcal{Z}}$ is in general not a sublattice of $\mathcal{C}^{\mathcal{H}_1}_{\mathcal{Z}}$

Galois connections

We start with a general definition of "Galois connection".

Definition 1.3. *Let $\mathcal{X}$ and $\mathcal{Y}$ be nonempty sets. A pair α–β of operators $\alpha : \mathcal{P}\mathcal{X} \to \mathcal{P}\mathcal{Y}$ and $\beta : \mathcal{P}\mathcal{Y} \to \mathcal{P}\mathcal{X}$ is called a* Galois connection (G.C.) *between $\mathcal{X}$ and $\mathcal{Y}$ if the following hold for all $X, X_1, X_2 \subseteq \mathcal{X}$ and all $Y, Y_1, Y_2 \subseteq \mathcal{Y}$.*

1. $X_1 \subseteq X_2 \Longrightarrow \alpha X_2 \subseteq \alpha X_1$.
2. $Y_1 \subseteq Y_2 \Longrightarrow \beta Y_2 \subseteq \beta Y_1$.
3. $X \subseteq \beta\alpha X$ *and* $Y \subseteq \alpha\beta Y$. □

Throughout this section, let α–β denote a Galois connection between $\mathcal{X}$ and $\mathcal{Y}$. The following useful facts are immediate consequences of Def. 1.3.

Corollary 1.4. *The following hold for all $x \in \mathcal{X}$, $X, X_i \subseteq \mathcal{X}$ and all $y \in \mathcal{Y}$, $Y, Y_i \subseteq \mathcal{Y}$ ($i \in I$, arbitrary index set).*

1. $\alpha\emptyset = \mathcal{Y}$ *and* $\beta\emptyset = \mathcal{X}$.
2. $X \subseteq \beta Y \iff Y \subseteq \alpha X$ *and* $x \in \beta\{y\} \iff y \in \alpha\{x\}$.
3. $\beta\alpha\beta Y = \beta Y$ *and* $\alpha\beta\alpha X = \alpha X$.
4. $\alpha \bigcup_{i\in I} X_i = \bigcap_{i\in I} \alpha X_i$ *and* $\beta \bigcup_{i\in I} Y_i = \bigcap_{i\in I} \beta Y_i$.
5. *The operators $\beta\alpha : \mathcal{P}\mathcal{X} \to \mathcal{P}\mathcal{X}$ and $\alpha\beta : \mathcal{P}\mathcal{Y} \to \mathcal{P}\mathcal{Y}$ are closure operators on $\mathcal{X}$ and $\mathcal{Y}$ respectively.* □

So every G.C. generates two closure operators $\alpha\beta$ and $\beta\alpha$. Next we consider a general definition scheme for Galois connections.

Theorem 1.5. *1. Let $\Xi \subseteq \mathcal{X} \times \mathcal{Y}$ be a relation between $\mathcal{X}$ and $\mathcal{Y}$ and define operators $\alpha_\Xi : \mathcal{P}\mathcal{X} \to \mathcal{P}\mathcal{Y}$ and $\beta_\Xi : \mathcal{P}\mathcal{Y} \to \mathcal{P}\mathcal{X}$ by*

$$\alpha_\Xi X := \{y \in \mathcal{Y} \mid (\forall x \in X)\Xi(x,y)\}$$
$$\beta_\Xi Y := \{x \in \mathcal{X} \mid (\forall y \in Y)\Xi(x,y)\}.$$

Then the pair α_Ξ– β_Ξ is a Galois connection between $\mathcal{P}\mathcal{X}$ and $\mathcal{P}\mathcal{Y}$.

2. *Let α–β be a G.C. between $\mathcal{P}\mathcal{X}$ and $\mathcal{P}\mathcal{Y}$ and define a relation $\Xi \subseteq \mathcal{X} \times \mathcal{Y}$ by*

$$\Xi := \{(x,y) \in \mathcal{X} \times \mathcal{Y} \mid x \in \beta\{y\}\} \ (= \{(x,y) \in \mathcal{X} \times \mathcal{Y} \mid y \in \alpha\{x\}\} \ ).$$

Then $\alpha_\Xi = \alpha$ and $\beta_\Xi = \beta$. □

Consequently every G.C. is determined by a relation $\Xi \subseteq \mathcal{X} \times \mathcal{Y}$ between $\mathcal{X}$ and $\mathcal{Y}$, and each such relation defines a Galois connection. If α–β is a G.C., then we call the corresponding relation Ξ the *defining relation* for this connection, or we say that α–β is *based on* Ξ.

Usually the subsets of $\mathcal{X}$ and $\mathcal{Y}$ which are closed under the closure operators $\beta\alpha$ and $\alpha\beta$ play an important role.

Definition 1.6. *The sets $X \subseteq \mathcal{X}$ and $Y \subseteq \mathcal{Y}$ that are closed under $\beta\alpha$ or $\alpha\beta$, i.e. $X = \beta\alpha X$ and $Y = \alpha\beta Y$ resp., are called* Galois closed *with respect to the G.C. α–β.*

Consequently the closure systems $\mathcal{C}_{\mathcal{X}}^{\beta\alpha}$ and $\mathcal{C}_{\mathcal{Y}}^{\alpha\beta}$ for $\beta\alpha$ and $\alpha\beta$ consist of the Galois closed sets. If the G.C. α–β is fixed, then we simply write $\mathcal{C}_{\mathcal{X}}$ and $\mathcal{C}_{\mathcal{Y}}$. □

According to (1.1), $\mathcal{C}_{\mathcal{X}}^{\beta\alpha}$ and $\mathcal{C}_{\mathcal{Y}}^{\alpha\beta}$ are complete lattices. These lattices are interconnected.

Theorem 1.7. *The lattices $\left\langle \mathcal{C}_{\mathcal{X}}^{\beta\alpha}; \wedge, \vee \right\rangle$ and $\left\langle \mathcal{C}_{\mathcal{Y}}^{\alpha\beta}; \wedge, \vee \right\rangle$ are dually isomorphic. The dual isomorphisms are the operators $\alpha : \mathcal{C}_{\mathcal{X}} \to \mathcal{C}_{\mathcal{Y}}$ and $\beta : \mathcal{C}_{\mathcal{Y}} \to \mathcal{C}_{\mathcal{X}}$.* □

This shows a great advantage of Galois connections. The "large" elements of $\mathcal{C}_{\mathcal{X}}$ are determined by the "small" elements of $\mathcal{C}_{\mathcal{Y}}$ and vice versa. Information on one side of the G.C. often can be carried over to information on the other side.

Derived Galois Connections

There is an easy way to derive new Galois connections from α–β. Let $\mathcal{M} \subseteq \mathcal{X}$ be a so called *restriction set* and consider the new relation

$$\Xi_{\mathcal{M}} := \Xi \cap (\mathcal{M} \times \mathcal{Y}) \subseteq \mathcal{M} \times \mathcal{Y}.$$

Then this relation $\Xi_{\mathcal{M}}$ defines a new Galois connection.

Theorem 1.8. *If $\mathcal{M} \subseteq \mathcal{X}$ then the operator pair α - $\mathcal{M}$-β,*

$$\begin{aligned} \alpha &: \mathcal{P}\mathcal{M} \to \mathcal{P}\mathcal{Y}, \quad X \mapsto \alpha X \\ \mathcal{M}\text{-}\beta &: \mathcal{P}\mathcal{Y} \to \mathcal{P}\mathcal{M}, \quad Y \mapsto \mathcal{M} \cap \beta Y, \end{aligned}$$

forms a Galois connection between $\mathcal{M}$ and $\mathcal{Y}$, based on the relation $\Xi_{\mathcal{M}}$.

A subset $N \subseteq \mathcal{M}$ is Galois closed under $\mathcal{M}$-$\beta\,\alpha$ if and only if $N = \mathcal{M} \cap X$ for some $X \in \mathcal{C}_{\mathcal{X}}^{\beta\alpha}$, i.e. if N is the restriction of some $\beta\alpha$—closed set X to $\mathcal{M}$.

For the closure operator $\alpha\,\mathcal{M}$-β always holds $\alpha\beta \leq \alpha\,\mathcal{M}$-$\beta$. Therefore (1.2) $\mathcal{C}_{\mathcal{Y}}^{\alpha\,\mathcal{M}\text{-}\beta} \subseteq \mathcal{C}_{\mathcal{Y}}^{\alpha\beta}$ and $\alpha\beta\alpha\,\mathcal{M}$-$\beta = \alpha\,\mathcal{M}$-$\beta$. □

(The operator α in the new G.C. is not exactly the α of the old G.C. $\alpha - \beta$, it is the restriction of α to $\mathcal{P}\mathcal{M}$. But we will use the old symbol α further on, instead of writing $\alpha|_{\mathcal{P}\mathcal{M}}$.)

Now we consider, how $\mathcal{C}_{\mathcal{M}}^{\mathcal{M}\text{-}\beta\,\alpha}$ and $\mathcal{C}_{\mathcal{X}}^{\beta\alpha}$ are related. From (1.8) we already know that $X \in \mathcal{C}_{\mathcal{X}}^{\beta\alpha}$ implies $\mathcal{M} \cap X \in \mathcal{C}_{\mathcal{M}}^{\mathcal{M}\text{-}\beta\,\alpha}$. Conversely, if $N \in \mathcal{C}_{\mathcal{M}}^{\mathcal{M}\text{-}\beta\,\alpha}$, then $\beta\alpha N \in \mathcal{C}_{\mathcal{X}}^{\beta\alpha}$. This mapping is injective.

Theorem 1.9. *Let γ denote the operator $\gamma : \mathcal{C}_{\mathcal{M}}^{\mathcal{M}\text{-}\beta\,\alpha} \to \mathcal{C}_{\mathcal{X}}^{\beta\alpha}, \quad N \mapsto \beta\alpha N$.*

Then γ is injective, and for all $N \in \mathcal{C}_{\mathcal{M}}^{\mathcal{M}\text{-}\beta\,\alpha}$ holds $N = \mathcal{M} \cap \gamma(N)$. Moreover, $\alpha\gamma\,\mathcal{M}\text{-}\beta = \alpha\,\mathcal{M}\text{-}\beta$.

If $N_1, N_2 \in \mathcal{C}_{\mathcal{M}}^{\mathcal{M}\text{-}\beta\,\alpha}$, then $\gamma(N_1 \vee N_2) = \gamma(N_1) \vee \gamma(N_2)$ and $\gamma(N_1 \wedge N_2) \subseteq \gamma(N_1) \wedge \gamma(N_2)$.

If $\mathcal{M} \subseteq \mathcal{X}$ is closed under $\beta\alpha$, then $\mathcal{C}_{\mathcal{M}}^{\mathcal{M}\text{-}\beta\,\alpha}$ is a sublattice of $\mathcal{C}_{\mathcal{X}}^{\beta\alpha}$ (and γ is the identity). □

In general, γ is not a lattice embedding. But in all cases, γ gives the opportunity to represent the elements of $\mathcal{C}_{\mathcal{M}}^{\mathcal{M}\text{-}\beta\,\alpha}$ with the help of some elements in $\mathcal{C}_{\mathcal{X}}^{\beta\alpha}$.

Approximation of Closure Operators

Let $\mathcal{H}$ be a closure operator on $\mathcal{Y}$ with $\alpha\beta \leq \mathcal{H}$. We want to find a subset $\mathcal{M} \subseteq \mathcal{X}$ with $\mathcal{H} \leq \alpha\,\mathcal{M}\text{-}\beta$ and such that $\alpha\,\mathcal{M}\text{-}\beta$ is a good "approximation" for $\mathcal{H}$.

$\mathcal{H} \leq \alpha\,\mathcal{M}\text{-}\beta$ implies $\mathcal{C}_{\mathcal{Y}}^{\alpha\,\mathcal{M}\text{-}\beta} \subseteq \mathcal{C}_{\mathcal{Y}}^{\mathcal{H}}$ (1.2). Because of (1.4(3)) the sets $\alpha\{x\}$ with $x \in \mathcal{M}$ are closed under $\alpha\,\mathcal{M}\text{-}\beta$. So, if $\mathcal{H} \leq \alpha\,\mathcal{M}\text{-}\beta$ then $\alpha\{x\}$ must be closed under $\mathcal{H}$ whenever $x \in \mathcal{M}$. This motivates the next Definition.

Definition 1.10. *The* approximation set of $\mathcal{H}$ *is the set*

$$\mathbf{App}(\mathcal{H}) := \{x \in \mathcal{X} \mid \alpha\{x\} = \mathcal{H}\alpha\{x\}\}. \qquad \square$$

Theorem 1.11. *Let $\mathcal{M}_1, \mathcal{M}_2, \mathcal{M} \subseteq \mathcal{X}$. Then the following hold.*

1. $\mathcal{M}_1 \subseteq \mathcal{M}_2 \implies \alpha\mathcal{M}_2\text{-}\beta \leq \alpha\mathcal{M}_1\text{-}\beta$
2. $\mathcal{H} \leq \alpha\,\mathcal{M}\text{-}\beta \iff \mathcal{M} \subseteq \mathbf{App}(\mathcal{H})$

Consequently, $\alpha\mathbf{App}(\mathcal{H})\text{-}\beta$ is the least closure operator $\alpha\,\mathcal{M}\text{-}\beta$ with $\mathcal{H} \leq \alpha\,\mathcal{M}\text{-}\beta$.

Proof. 1. If $\mathcal{M}_1 \subseteq \mathcal{M}_2$ then for all $Y \subseteq \mathcal{Y}$ holds $\mathcal{M}_1\text{-}\beta Y = \mathcal{M}_1 \cap \beta Y \subseteq \mathcal{M}_2 \cap \beta Y = \mathcal{M}_2\text{-}\beta Y$, and therefore $\alpha\mathcal{M}_2\text{-}\beta Y \subseteq \alpha\mathcal{M}_1\text{-}\beta Y$.

2. "⇒" If $\mathcal{H} \leq \alpha\,\mathcal{M}\text{-}\beta$, then by (1.2) every $\alpha\,\mathcal{M}\text{-}\beta$–closed set is also closed under $\mathcal{H}$. If $x \in \mathcal{M}$, then $\alpha\{x\}$ is closed under $\alpha\,\mathcal{M}\text{-}\beta$, therefore it must be closed under $\mathcal{H}$, therefore $x \in \mathbf{App}(\mathcal{H})$.
"⇐" If $\mathcal{M} \subseteq \mathbf{App}(\mathcal{H})$ and $Y \subseteq \mathcal{Y}$, then

$$\mathcal{M}\text{-}\beta Y = \mathcal{M} \cap \beta Y = \bigcup\{\{x\} \mid x \in \mathcal{M} \text{ and } x \in \beta Y\}.$$

Therefore $\alpha\,\mathcal{M}\text{-}\beta Y = \bigcap\{\alpha\{x\} \mid x \in \mathcal{M} \text{ and } x \in \beta Y\}$ (1.4(4)). Now $x \in \beta Y$ implies $Y \subseteq \alpha\beta Y \subseteq \alpha\{x\}$ and $x \in \mathbf{App}(\mathcal{H})$ implies $\mathcal{H}Y \subseteq \mathcal{H}\alpha\{x\} = \alpha\{x\}$. Consequently $\mathcal{H}Y \subseteq \bigcap\{\alpha\{x\} \mid x \in \mathcal{M} \text{ and } x \in \beta Y\} = \alpha\,\mathcal{M}\text{-}\beta Y$. □

The question, whether we can reach $\mathcal{H} = \alpha\mathcal{M}\text{-}\beta$ can only be answered if the special properties of the basic Galois connection $\alpha - \beta$ are known.

Remark. Every closure operator $\mathcal{H}$ on $\mathcal{Y}$ can be defined in a trivial way by some Galois connection. (Choose $\mathcal{X} = \mathcal{C}^{\mathcal{H}}_{\mathcal{Y}}$ and $\Xi = \{(Y, y) \in \mathcal{X} \times \mathcal{Y} \mid y \in Y\}$.) But this trivial Galois connection provides no new information.

2 The Galois Connection Inv – Pol

In the following we consider Galois connections between sets of functions and sets of relations. They all are based on certain "preservation relations". During the last ten years it became clear that these G.C.s also are of importance for the description of the complexity of Constraint Satisfaction Problems. We will focus on this interconnection in the next section.

Now we consider the G.C. Inv – Pol between sets of functions and sets of relations on a common basic set D. This Galois connection is well investigated ([20], [18]), and can serve as a model for all other connections that we will consider later. Therefore we study this special case in greater detail.

Notions and Notations

First we introduce the sets that correspond to $\mathcal{X}$ and $\mathcal{Y}$ in the last section. $\mathbf{N}^+$ denotes the set of all positive natural numbers. Let D be a nonempty finite *basic set* with at least two elements. For $n \in \mathbf{N}^+$, the set of all *n–ary functions on D* is $\mathbf{O}_D^{(n)} := \{f \mid f : D^n \to D\}$, and the *set of all finitary functions on D* is $\mathbf{O}_D := \bigcup_{n \in \mathbf{N}^+} \mathbf{O}_D^{(n)}$. If $F \subseteq \mathbf{O}_D$, then $F^{(n)} := F \cap \mathbf{O}_D^{(n)}$ denotes the set of all n-ary functions in F.

An m–ary relation on D is a subset $\varrho \subseteq D^m$. The elements of ϱ are m–tuples $\boldsymbol{a} = (a_1, \ldots, a_m) \in D^m$. We write $\boldsymbol{a}(i)$ to denote the ith entry a_i of $\boldsymbol{a}$. The set of all m–ary relations on D is $\mathbf{R}_D^{(m)} := \{\varrho \mid \varrho \subseteq D^m\}$, and the set of all finitary relations on D is $\mathbf{R}_D := \bigcup_{m \in \mathbf{N}^+} \mathbf{R}_D^{(m)}$. For a relation set $\Gamma \subseteq \mathbf{R}_D$ we put $\Gamma^{(m)} := \Gamma \cap \mathbf{R}_D^{(m)}$.

We do not distinguish sharply between relations and predicates. So instead of $\boldsymbol{a} \in \varrho$ we also write $\varrho(\boldsymbol{a})$.

The arity of a function f or a relation ϱ is denoted by $\mathsf{ar}(f)$ and $\mathsf{ar}(\varrho)$.

Let $f \in \mathbf{O}_D^{(n)}$, $f : D^n \to D$, $m \in \mathbf{N}^+$. Then we can assign to f also a function $f : (D^m)^n \to D^m$. For $i = 1 \ldots n$ let $\boldsymbol{a}_i \in D^m$. Then we put

$$f(\boldsymbol{a}_1, \ldots, \boldsymbol{a}_n) := (f(\boldsymbol{a}_1(1), \ldots, \boldsymbol{a}_n(1)), \ldots, f(\boldsymbol{a}_1(m), \ldots, a_n(m))) \in D^m.$$

If $\varrho \in \mathbf{R}_D^{(m)}$, then we define $f(\varrho) := \{f(\boldsymbol{a}_1, \ldots, \boldsymbol{a}_n) \mid \boldsymbol{a}_1, \ldots, \boldsymbol{a}_n \in \varrho\} \subseteq A^m$, which is again an m–ary relation.

The sets $\mathbf{O}_D$ and $\mathbf{R}_D$ will play the role of $\mathcal{X}$ and $\mathcal{Y}$ in Section 1. Now we describe the defining relation Ξ for our Galois connection.

Definition 2.1. *Let* $f \in \mathbf{O}_D^{(n)}$ *and* $\varrho \in \mathbf{R}_D^{(m)}$. *We say that* f preserves ϱ, *or that* ϱ *is an* invariant relation for f, *or that* f *is a* polymorphism of ϱ, *if for all* $\boldsymbol{a}_1, \ldots, \boldsymbol{a}_n \in D^m$ *holds*

$$\boldsymbol{a}_1, \ldots, \boldsymbol{a}_n \in \varrho \Longrightarrow f(\boldsymbol{a}_1, \ldots, \boldsymbol{a}_n) \in \varrho,$$

or equivalently, if $f(\varrho) \subseteq \varrho$. *With* **pres** *we denote the defining relation*

$$\textbf{pres} := \{(f, \varrho) \in \mathbf{O}_D \times \mathbf{R}_D \mid f \textit{ preserves } \varrho\}.$$

Based on this relation **pres**, *we define a Galois connection* $\mathsf{Inv}_D - \mathsf{Pol}_D$ *by*

$$\mathsf{Inv}_D : \mathcal{P}\mathbf{O}_D \to \mathcal{P}\mathbf{R}_D, \quad F \mapsto \{\varrho \in \mathbf{R}_D \mid (\forall f \in F)\ f \,\textbf{pres}\, \varrho\}$$
$$\mathsf{Pol}_D : \mathcal{P}\mathbf{R}_D \to \mathcal{P}\mathbf{O}_D, \quad \Gamma_0 \mapsto \{f \in \mathbf{O}_D \mid (\forall \varrho \in \Gamma_0)\ f \,\textbf{pres}\, \varrho\}.$$

As long as the basic set D *is fixed, we suppress the index* D. □

Examples

- Let $D = \{0, 1\}$. Then the relation $\leq$ $(= \{(0,0), (0,1), (1,1)\})$ is invariant for f iff f is monotone, $\mathsf{Pol}\,\{\leq\}$ is the set of all monotone Boolean functions.
- Let $\Delta_D = \{(a, a) \mid a \in D\}$ denote the *equality relation on* D. Every function $f \in \mathbf{O}_D$ preserves Δ_D, $\mathsf{Pol}\,\{\Delta_D\} = \mathbf{O}_D$.
- Let $\mathsf{Co}\,\Delta_D = \{(a, b) \in D^2 \mid a \neq b\}$ be the *inequality relation on* D. If $D = \{0, 1\}$, then $\mathsf{Pol}\,\{\mathsf{Co}\,\Delta_D\}$ is the set of all *selfdual Boolean functions*, i.e. functions satisfying $\overline{f(\overline{x}_1, \ldots, \overline{x}_n)} = f(x_1, \ldots, x_n)$ for all $(x_1, \ldots, x_n) \in D^n$. An example of such a function is the ternary function $x_1 \oplus x_2 \oplus x_3$. (Here $\oplus$ denotes addition modulo 2.)
  But if $|D| \geq 3$, then $\mathsf{Pol}_D\{\mathsf{Co}\,\Delta_D\}$ consists only of functions f of the form $f(x_1, \ldots, x_n) = \pi(x_i)$, where $1 \leq i \leq n$ and π is a permutation.

Now we are going to *characterize the Galois connection*, i.e. we want to describe the Galois closed sets of functions and relations. We have to introduce certain operations on functions and on relations.

Clones of Functions on D

The *elementary functions* or *projections* on D are the functions

$$e_{i,D}^n \;:\; D^n \to D, \quad (a_1, a_2, \ldots, a_n) \mapsto a_i \;\; (\text{with } 1 \leq i \leq n).$$

The set of all elementary functions on D is denoted by

$$\mathbf{J}_D := \{e_{i,D}^n \mid i, n \in \mathbf{N}^+ \text{ and } 1 \leq i \leq n\}.$$

If D is fixed, then we write e_i^n instead of $e_{i,D}^n$.

Let $f \in \mathbf{O}_D^{(n)}$ and $g_1, \ldots, g_n \in \mathbf{O}_D^{(m)}$ be functions on D. Then the *superposition of* f *and* $g_1, \ldots, g_n$ is the m-ary function

$$f[g_1, \ldots, g_n] \;:\; D^m \to D,$$
$$(a_1, \ldots, a_m) \mapsto f(g_1(a_1, \ldots, a_m), \ldots, g_n(a_1, \ldots, a_m)).$$

Definition 2.2. *A set* $C \subseteq \mathbf{O}_D$ *is called a* clone of functions on D, *if* C *contains all elementary functions* e_i^n *and if* C *is closed under superposition, i.e. if* $f \in C^{(n)}$, $g_1, \ldots, g_n \in C^{(m)}$ *then always* $f[g_1, \ldots, g_n] \in C$. *The set of all clones of functions on* D *is denoted by* $\mathcal{L}_{\mathbf{O}_D}$.

If $F \subseteq \mathbf{O}_D$ *is a set of functions, then*

$$\mathsf{clone}\, F := \bigcap \{C \subseteq \mathbf{O}_D \mid C \textit{ is a clone and } F \subseteq C\}$$

denotes the clone, generated by F. *(This is the least clone that contains* F.*)* □

Examples of clones are the set $\mathbf{O}_D$ of all functions on D, the set $\mathbf{J}_D$ of all elementary functions, or the set of all monotone Boolean functions on the two-element set $D = \{0, 1\}$.

We mention that clone is an algebraic closure operator on $\mathbf{O}_D$, and $\mathcal{L}_{\mathbf{O}_D}$ is the corresponding algebraic lattice. We can obtain $\mathsf{clone}\, F$ also in the following way: Let $C_0 := F \cup \mathbf{J}_D$ and

$$C_{k+1} := C_k \cup \{f[g_1, \ldots, g_n] \mid (\exists n, m \in \mathbf{N}^+) f \in F^{(n)}, g_1, \ldots, g_n \in C_k^{(m)}\},$$

then $\mathsf{clone}\, F = \bigcup_{k=0}^{\infty} C_k$.

In the Boolean case $|D| = 2$ there are countably many clones and the structure of $\mathcal{L}_{\mathbf{O}_D}$ is completely known by the work of E.L. Post ([23, 24, 25]). Figure 1 on page 47 shows the structure of this lattice.

But for $|D| \geq 3$ there are continuum many clones on D, and the structure of $\mathcal{L}_{\mathbf{O}_D}$ is largely unknown.

The first interconnection with our G.C. Inv–Pol comes from the facts that obviously the elementary functions preserve every relation, and that the superposition $f[g_1, \ldots, g_n]$ preserves a relation ϱ whenever f and $g_1, \ldots, g_n$ preserve ϱ. This immediately yields:

Corollary 2.3. *1. If* $\Gamma_0 \subseteq \mathbf{R}_D$, *then* Pol Γ_0 *is a clone.*
2. For all $F \subseteq \mathbf{O}_D$, $\mathsf{clone}\, F \subseteq \mathsf{Pol\ Inv}\, F$; *i.e.* $\mathsf{clone} \leq \mathsf{Pol\ Inv}$.

Proof. The preceding remarks show that Pol Γ_0 is a clone. And $F \subseteq \mathsf{Pol\ Inv}\, F$ implies $\mathsf{clone}\, F \subseteq \mathsf{clone\, Pol\ Inv}\, F = \mathsf{Pol\ Inv}\, F$. □

Operations on Relations

We turn to the relational side. In order to define operations on the relations in $\mathbf{R}_D$, we use a first–order predicate calculus with logical symbols from the set

$$\text{Symb} := \{\exists, \forall, \wedge, \vee, \neg, =, \neq, \mathsf{f}, \mathsf{t}\}. \tag{1}$$

The symbols f and t are included for technical reasons; f is a formula that is always false and t stands for the always true formula.

Let $\mathcal{S}$ be a list of logical symbols from Symb. With $\Phi(\mathcal{S})$ we denote the *set of all first–order formulas that contain (besides variables, predicate symbols and brackets) only logical symbols from the list* $\mathcal{S}$.

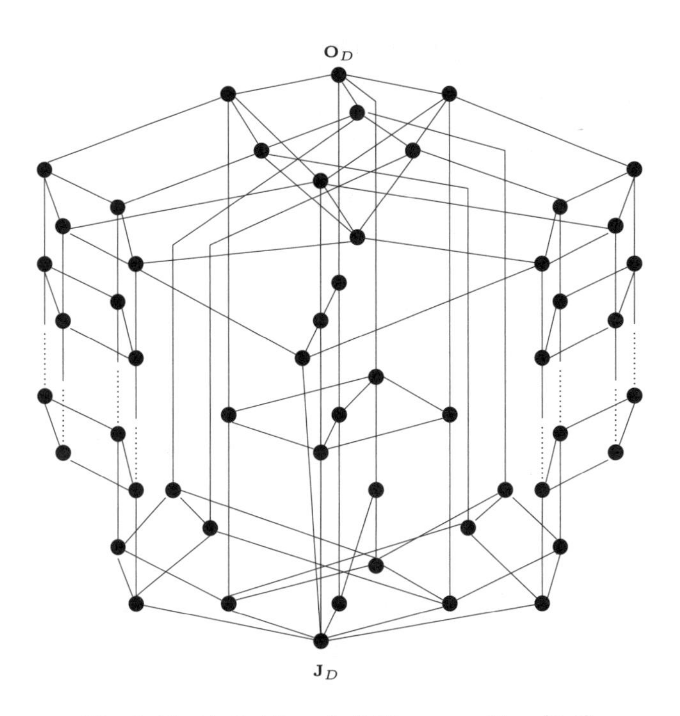

Fig. 1. Post's Lattice of all Clones on $D = \{0, 1\}$

Now let $\varphi = \varphi(P_1, \ldots, P_k; x_1, \ldots, x_m)$ be a first–order formula with predicate symbols P_i of arity m_i, and with free variables among $x_1, \ldots, x_m$. (We always assume that the number m of free variables assigned to φ is known. It is not necessary that all variables explicitly occur in φ.) If φ holds in the model $\langle D; \sigma_1, \ldots, \sigma_k; a_1, \ldots, a_m\rangle$, then we denote this by $\models \varphi_D(\sigma_1, \ldots, \sigma_k; a_1, \ldots, a_m)$.

In the following we will use instead of the predicate symbols P_i mostly the symbols σ_i or ϱ_i for the relations (predicates) itself in our formulas φ. Moreover, we also allow y, z, y_j, ... to occur as variable symbols in our formulas.

To φ we assign a *logical operation* L_φ $(= L_{\varphi,m})$ by

$$\begin{aligned} L_\varphi : \mathbf{R}_D^{(m_1)} \times \cdots \times \mathbf{R}_D^{(m_k)} &\to \mathbf{R}_D^{(m)} \\ (\sigma_1, \ldots, \sigma_k) &\mapsto \{(a_1, \ldots, a_m) \in A^m \mid \models \varphi_D(\sigma_1, \ldots, \sigma_k; a_1, \ldots, a_m)\} \end{aligned}$$

Some Examples and Definitions

- $\Phi(\emptyset)$ contains all formulas of the form $\sigma(x_{i_1}, \ldots, x_{i_n})$.
- $\Phi(\exists, \wedge, =)$ consists of all *primitive positive formulas.*
- If $\varphi \equiv \sigma_1(x_1, \ldots, x_m) \wedge \sigma_2(x_1, \ldots, x_m)$, then L_φ describes the intersection of two m-ary relations, $L_\varphi(\sigma_1, \sigma_2) = \sigma_1 \cap \sigma_2$.
- For the formula f (considered with m free variables) we obtain as L_f the (m-ary) *empty relation* $L_\mathsf{f} = \emptyset^{(m)} \in \mathbf{R}_D^{(m)}$.
  Similarly, the formula t (considered with m free variables) describes the *full relation* $L_\mathsf{t} = D^m \in \mathbf{R}_D^{(m)}$.

- If $\varphi \equiv (x_1 = x_2)$, then L_φ is the binary *equality relation* Δ_D on D.
- If $\varphi \equiv (\exists x_{m+1})\sigma(x_1, \ldots, x_m, x_{m+1})$, then the logical operation L_φ describes the *projection* of an $(m+1)$-ary relation σ to its first m components.

$$L_\varphi = \mathrm{Pr}^{(m)} : \mathbf{R}_D^{(m+1)} \to \mathbf{R}_D^{(m)}, \ \sigma \mapsto \{\boldsymbol{a} \in D^m \mid (\exists a_{m+1})\ \sigma(\boldsymbol{a}, a_{m+1})\}$$

- The formula $\varphi \equiv \neg\sigma(x_1, \ldots, x_m)$ defines the *complement* of an m-ary relation,

$$L_\varphi(\sigma) = \mathsf{Co}\,\sigma := D^m \setminus \sigma = \{\boldsymbol{a} \in D^m \mid \neg\sigma(\boldsymbol{a})\}.$$

Each set of logical operations defines an algebraic closure operator on $\mathbf{R}_D$.

Definition 2.4. *Let $\mathcal{S}$ be a list of logical symbols from the set* Symb *(see Eq. (1)) and let $\Gamma \subseteq \mathbf{R}_D$. We say that Γ is $\mathcal{S}$*–closed, *if Γ is closed under all logical operations L_φ with $\varphi \in \Phi(\mathcal{S})$.*

If $\Gamma_0 \subseteq \mathbf{R}_D$, then the $\mathcal{S}$–closed set of relations, generated by Γ_0 *is denoted with*

$$\langle \Gamma_0 \rangle_{\mathcal{S}} := \bigcap \{\Gamma \subseteq \mathbf{R}_D \mid \Gamma_0 \subseteq \Gamma \text{ and } \Gamma \text{ is } \mathcal{S}\text{–closed}\}.$$

A set $\Gamma \subseteq \mathbf{R}_D$ is called a relational clone *if Γ is $\exists, \wedge, =, \mathsf{f}$–closed.*

$$\Gamma \text{ relational clone } :\Longleftrightarrow \langle \Gamma \rangle_{\exists,\wedge,=,\mathsf{f}} = \Gamma$$

The set of all relational clones ist denoted by $\mathcal{L}_{\mathbf{R}_D}$.

For $\Gamma_0 \subseteq \mathbf{R}_D$ we call $\langle \Gamma_0 \rangle_{\exists,\wedge,=,\mathsf{f}}$ the relational clone, generated by Γ_0. □

Clearly, the operator $\langle\ \rangle_{\mathcal{S}} : \Gamma_0 \mapsto \langle \Gamma_0 \rangle_{\mathcal{S}}$ is an algebraic closure operator and if $\mathcal{S} = \exists, \wedge, =, \mathsf{f}$, then $\mathcal{L}_{\mathbf{R}_D}$ is the corresponding algebraic lattice.

By Definition 2.4, a set Γ is a relational clone iff

- Γ contains the empty relations $\emptyset^{(m)}$ and the equality relation Δ_D.
- Γ is closed under intersections $\bigcap$ of m–ary relations.
- Γ is closed under projections $\mathrm{Pr}^{(m)}$.
- Γ is closed under rearrangements of arguments, i.e. under operations L_φ : $\mathbf{R}_D^{(m)} \to \mathbf{R}_D^{(k)}$ where $\varphi \equiv \sigma(x_{s(1)}, \ldots, x_{s(m)})$ for some function $s : \{1, \ldots, m\} \to \{1, \ldots, k\}$.

In particular this implies that the full relations D^m also belong to every relational clone on D.

It is not difficult to see that the equality relation Δ_D, the empty relations $\emptyset^{(m)}$ and the full relations D^m are invariant for all functions in $\mathbf{O}_D$. If $\varrho_1, \varrho_2 \in \mathbf{R}_D^{(m)}$ are both invariant for a function $f \in \mathbf{O}_D$, then the intersection $\varrho_1 \cap \varrho_2$ is also invariant for f. Moreover, if ϱ is an $(m+1)$-ary invariant relation for f, then the projection $\mathrm{Pr}^{(m)}\varrho$ is also invariant for f.

Corollary 2.5. *1. If $F \subseteq \mathbf{O}_D$, then* $\mathsf{Inv}\ F$ *is a relational clone.*
2. For all $\Gamma_0 \subseteq \mathbf{R}_D$, $\langle \Gamma_0 \rangle_{\exists,\wedge,=,\mathsf{f}} \subseteq \mathsf{Inv\ Pol}\ \Gamma_0$, i.e. $\langle\ \rangle_{\exists,\wedge,=,\mathsf{f}} \leq \mathsf{Inv\ Pol}$.

Proof. The preceding remarks imply the first statement. And $\Gamma_0 \subseteq \mathsf{Inv\ Pol}\ \Gamma_0$ implies $\langle \Gamma_0 \rangle_{\exists,\wedge,=,\mathsf{f}} \subseteq \langle \mathsf{Inv\ Pol}\ \Gamma_0 \rangle_{\exists,\wedge,=,\mathsf{f}} = \mathsf{Inv\ Pol}\ \Gamma_0$. □

Now we want to prove that (for finite D) the clones of functions and the relational clones are always Galois closed.

The Characterization of Inv – Pol

First we define an important technical tool.

Definition 2.6. *Let $\Gamma \subseteq \mathbf{R}_D$ and $\varrho \in \mathbf{R}_D^{(m)}$. We define*

$$\mu_\Gamma(\varrho) := \bigcap\{\sigma \in \Gamma^{(m)} \mid \varrho \subseteq \sigma\}.$$

(If the set of relations on the right side is empty, then we put $\mu_\Gamma(\varrho) = D^m$.)

If $\boldsymbol{a}_1, \boldsymbol{a}_2, \ldots, \boldsymbol{a}_k \in D^m$ then we write $\mu_\Gamma(\boldsymbol{a}_1, \boldsymbol{a}_2, \ldots, \boldsymbol{a}_k)$ instead of $\mu_\Gamma(\{\boldsymbol{a}_1, \boldsymbol{a}_2, \ldots, \boldsymbol{a}_k\})$. □

Lemma 2.7. *1. Let $\Gamma \subseteq \mathbf{R}_D$ such that for all $m \in \mathbf{N}^+$ the set $\Gamma^{(m)}$ is closed under the intersection of m–ary relations and $D^m \in \Gamma^{(m)}$. Then $\varrho \subseteq \mu_\Gamma(\varrho) \in \Gamma$ for all $\varrho \in \mathbf{R}_D$ and $\varrho \in \Gamma \iff \varrho = \mu_\Gamma(\varrho)$.*

2. Let $F \subseteq \mathbf{O}_D$ and $\boldsymbol{a}_1, \ldots, \boldsymbol{a}_n \in \mathbf{O}_D^{(m)}$. Then

$$\mu_{\mathsf{Inv}\ F}(\boldsymbol{a}_1, \ldots, \boldsymbol{a}_n) = \{f(\boldsymbol{a}_1, \ldots, \boldsymbol{a}_n) \mid f \in \mathsf{clone}^{(n)} F\}.$$

Proof. (1.) is obvious. In order to see (2.) note that $\Gamma = \mathsf{Inv}\ F$ satisfies the assumptions of (1.) (Lemma 2.5) and therefore $\mu_{\mathsf{Inv}\ F}(\boldsymbol{a}_1, \ldots, \boldsymbol{a}_n)$ is the smallest m-ary relation in $\mathsf{Inv}\ F$ that contains $\boldsymbol{a}_1, \ldots, \boldsymbol{a}_n$. Such a relation must contain all $f(\boldsymbol{a}_1, \ldots, \boldsymbol{a}_n)$ with $f \in \mathsf{clone}^{(n)} F$. It is easy to see that the relation on the right side is also invariant for the functions in F. Therefore on both sides of the equation stands the least relation which contains $\boldsymbol{a}_1, \boldsymbol{a}_2, \ldots, \boldsymbol{a}_n$ and is invariant for all functions from F. □

We still need a special relation for the characterization.

Definition 2.8. *Let $n \in \mathbf{N}^+$. We define a relation $\chi_{n,D}$ on D with $\mathsf{ar}(\chi_{n,D}) = |D|^n$ and $|\chi_{n,D}| = n$. Let $k = |D|^n$ and let $\boldsymbol{c}_1, \ldots, \boldsymbol{c}_n \in D^k$ be k-tuples such that for $i = 1 \ldots k$ the n-tuple $(\boldsymbol{c}_1(i), \ldots, \boldsymbol{c}_n(i))$ runs through all k elements of D^n in a lexicographic order. Then we put $\chi_{n,D} := \{\boldsymbol{c}_1, \ldots, \boldsymbol{c}_n\}$.* □

Now we are able to give a characterization for the closure operator Pol Inv.

Theorem 2.9. *Let D be a finite basic set.*

1. *A subset $C \subseteq \mathbf{O}_D$ is Galois closed for the G.C. Inv –Pol, (i.e. $C = \mathsf{Pol\ Inv}\ C$) if and only if C is a clone of functions.*
2. *For all $F \subseteq \mathbf{O}_D$, $\mathsf{clone}\,F = \mathsf{Pol\ Inv}\ F$.*

Proof. It suffices to prove (2.). Because of 2.3 we only have to prove $\mathsf{Pol\ Inv}\ F \subseteq \mathsf{clone}\,F$. Let $g \in \mathsf{Pol}^{(n)}\ \mathsf{Inv}\ F$. Then g has to preserve the relation $\mu_{\mathsf{Inv}\ F}(\chi_{n,D}) \in \mathsf{Inv}\ F$. Because of $\chi_{n,D} = \{\boldsymbol{c}_1, \ldots, \boldsymbol{c}_n\}$ it follows from 2.7 that

$$g(\boldsymbol{c}_1, \ldots, \boldsymbol{c}_n) \in \mu_{\mathsf{Inv}\ F}(\boldsymbol{c}_1, \ldots, \boldsymbol{c}_n) = \{f(\boldsymbol{c}_1, \ldots, \boldsymbol{c}_n) \mid f \in \mathsf{clone}^{(n)} F\}.$$

Consequently $g(\boldsymbol{c}_1, \ldots, \boldsymbol{c}_n) = f(\boldsymbol{c}_1, \ldots, \boldsymbol{c}_n)$ for some $f \in \mathsf{clone}^{(n)} F$. Therefore $g(\boldsymbol{c}_1(i), \ldots, \boldsymbol{c}_n(i)) = f(\boldsymbol{c}_1(i), \ldots, \boldsymbol{c}_n(i))$ for all $i = 1 \ldots |D|^n$. But the n-tuples $(\boldsymbol{c}_1(i), \ldots, \boldsymbol{c}_n(i))$ run through all possible n-tuples in D^n, so $g = f \in \mathsf{clone}\,F$. □

The characterization of the Galois closed relation sets is a bit more difficult. The crucial point in the proof is the following:

Lemma 2.10. *Let $\Gamma \subseteq \mathbf{R}_D$ be a relational clone and let $\boldsymbol{a}_1, \ldots, \boldsymbol{a}_n \in D^m$. Then*

$$\mu_\Gamma(\boldsymbol{a}_1, \ldots, \boldsymbol{a}_n) = \{f(\boldsymbol{a}_1, \ldots, \boldsymbol{a}_n) \mid f \in \mathsf{Pol}^{(n)}\Gamma\}.$$

Proof. The inclusion $\supseteq$ is clear, because Γ is closed under intersection and contains the relations D^m, therefore by 2.7 $\mu_\Gamma(\boldsymbol{a}_1, \ldots, \boldsymbol{a}_n) \in \Gamma$ and every $f \in \mathsf{Pol}\,\Gamma$ has to preserve $\mu_\Gamma(\boldsymbol{a}_1, \ldots, \boldsymbol{a}_n)$.

It remains to prove the other inclusion $\subseteq$. For each $\boldsymbol{b} \in \mu_\Gamma(\boldsymbol{a}_1, \ldots, \boldsymbol{a}_n)$ we must find an $f \in \mathsf{Pol}\,\Gamma$ with $\boldsymbol{b} = f(\boldsymbol{a}_1, \ldots, \boldsymbol{a}_n)$.

First we prove that this is true for the elements $\boldsymbol{c}_1, \ldots, \boldsymbol{c}_n$ of $\chi_{n,D}$ with $\mathsf{ar}(\chi_{n,D}) = k := |D|^n$. Let $\boldsymbol{d} \in \mu_\Gamma(\boldsymbol{c}_1, \ldots, \boldsymbol{c}_n)$. We define a function $g_{\boldsymbol{d}}$ by $g_{\boldsymbol{d}}(\boldsymbol{c}_1(i), \ldots, \boldsymbol{c}_n(i)) = \boldsymbol{d}(i)$ for all i with $1 \le i \le k$. Then we have $\boldsymbol{d} = g_{\boldsymbol{d}}(\boldsymbol{c}_1, \ldots, \boldsymbol{c}_n)$ and we want to show that $g_{\boldsymbol{d}} \in \mathsf{Pol}\,\Gamma$. Assume that this is not true. Then there exist $r \in \mathbf{N}^+$, $\sigma \in \Gamma^{(r)}$ and $\boldsymbol{a}'_1, \ldots, \boldsymbol{a}'_n \in \sigma$ with $g_{\boldsymbol{d}}(\boldsymbol{a}'_1, \ldots, \boldsymbol{a}'_n) \notin \sigma$. Now for all j with $1 \le j \le r$ there exists $s(j) \in \{1, \ldots, k\}$ with $(\boldsymbol{a}'_1(j), \ldots, \boldsymbol{a}'_n(j)) = (\boldsymbol{c}_1(s(j)), \ldots, \boldsymbol{c}_n(s(j)))$. Let γ be the relation

$$\gamma := \{(c'_1, \ldots, c'_k) \in D^k \mid \sigma(c'_{s(1)}, \ldots, c'_{s(r)})\}.$$

The relation γ is defined from σ by the formula $\sigma(x_{s(1)}, \ldots, x_{s(r)}) \in \Phi(\exists, \wedge, =, \mathsf{f})$, therefore $\gamma \in \Gamma$. Moreover, $\boldsymbol{c}_1, \ldots, \boldsymbol{c}_n \in \gamma$, consequently $\mu_\Gamma(\boldsymbol{c}_1, \ldots, \boldsymbol{c}_n) \subseteq \gamma$. But $\boldsymbol{d} \notin \gamma$, and this contradicts $\boldsymbol{d} \in \mu_\Gamma(\boldsymbol{c}_1, \ldots, \boldsymbol{c}_n)$. Consequently $g_{\boldsymbol{d}} \in \mathsf{Pol}\,\Gamma$.

Now we turn to the general case $\boldsymbol{b} \in \mu_\Gamma(\boldsymbol{a}_1, \ldots, \boldsymbol{a}_n)$. Again, for all j with $1 \le j \le m$ there exists $s(j)$ with $1 \le s(j) \le k$ such that $(\boldsymbol{a}_1(j), \ldots, \boldsymbol{a}_n(j)) = (\boldsymbol{c}_1(s(j)), \ldots, \boldsymbol{c}_n(s(j)))$. Consider the formula

$$\varphi(\mu; x_1, \ldots, x_m) :\Longleftrightarrow (\exists y_1) \cdots (\exists y_k)(\ \mu(y_1, \ldots, y_k) \wedge \bigwedge_{j=1}^{m} x_j = y_{s(j)}\ ).$$

Then $\varphi \in \Phi(\exists, \wedge, =, \mathsf{f})$ and therefore $L_\varphi(\mu_\Gamma(\boldsymbol{c}_1, \ldots, \boldsymbol{c}_n)) \in \Gamma$. Moreover, $\boldsymbol{a}_1, \ldots, \boldsymbol{a}_n \in L_\varphi(\mu_\Gamma(\boldsymbol{c}_1, \ldots, \boldsymbol{c}_n))$ and so $\mu_\Gamma(\boldsymbol{a}_1, \ldots, \boldsymbol{a}_n) \subseteq L_\varphi(\mu_\Gamma(\boldsymbol{c}_1, \ldots, \boldsymbol{c}_n))$. Then $\boldsymbol{b} \in \mu_\Gamma(\boldsymbol{a}_1, \ldots, \boldsymbol{a}_n)$ implies $\boldsymbol{b} \in L_\varphi(\mu_\Gamma(\boldsymbol{c}_1, \ldots, \boldsymbol{c}_n))$, hence there exists $\boldsymbol{d} \in \mu_\Gamma(\boldsymbol{c}_1, \ldots, \boldsymbol{c}_n)$ with $\boldsymbol{b}(j) = \boldsymbol{d}(s(j))$ for all $j = 1 \ldots m$. But then $\boldsymbol{b} = g_{\boldsymbol{d}}(\boldsymbol{a}_1, \ldots, \boldsymbol{a}_n)$, and because of $g_{\boldsymbol{d}} \in \mathsf{Pol}\,\Gamma$ the proof is finished. □

Now the characterization Theorem follows from 2.5, 2.10 and the general Theorem 5.3 that we prove a bit later.

Theorem 2.11. *Let D be a finite basic set.*

1. *A subset $\Gamma \subseteq \mathbf{R}_D$ is Galois closed for the G.C.* Inv –Pol *, (i.e. $\Gamma = \mathsf{Inv}\,\mathsf{Pol}\,\Gamma$) if and only if Γ is a relational clone.*
2. *For all $\Gamma_0 \subseteq \mathbf{R}_D$, $\langle \Gamma_0 \rangle_{\exists,\wedge,=,\mathsf{f}} = \mathsf{Inv}\,\mathsf{Pol}\,\Gamma_0$; i.e. $\langle\ \rangle_{\exists,\wedge,=,\mathsf{f}} = \mathsf{Inv}\,\mathsf{Pol}$.* □

An immediate consequence of 1.7, 2.9 and 2.11 is the following

Corollary 2.12. *The lattice* $\mathcal{C}_{\mathbf{O}_D}^{\mathsf{Pol\ Inv}}$ *of all Galois closed function sets coincides with the lattice* $\mathcal{L}_{\mathbf{O}_D}$ *of all clones on* D*. The lattice* $\mathcal{C}_{\mathbf{R}_D}^{\mathsf{Inv\ Pol}}$ *of all Galois closed relation sets coincides with the lattice* $\mathcal{L}_{\mathbf{R}_D}$ *of all relational clones on* D*.*

These lattices are dually isomorphic, the dual isomorphisms are the operators Inv *and* Pol. □

Figure 2 on page 51 illustrates this interconnection. ($\mathbf{W}_D$ denotes the least relational clone on D, $\mathbf{W}_D = \langle\emptyset\rangle_{\exists,\wedge,=,\mathsf{f}} = \mathsf{Inv}\ \mathbf{O}_D$.) Consequently, Post's lattice of all clones on $D = \{0,1\}$ in Fig. 1 on page 47 also describes the structure of the lattice of all relational clones on D, if we turn it around.

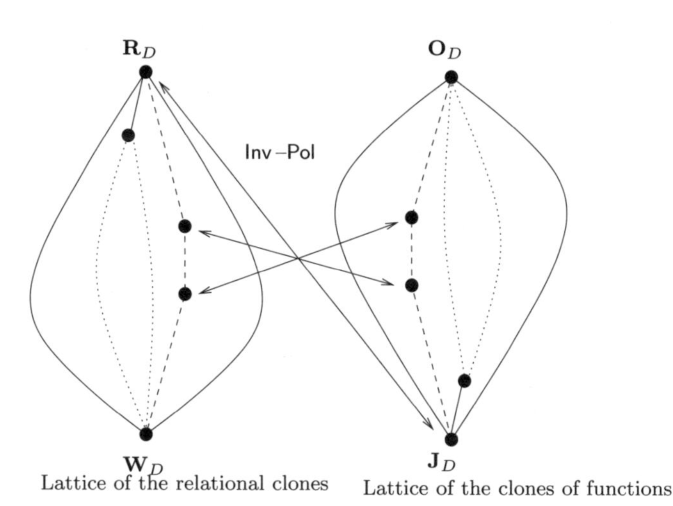

Fig. 2. Interconnection between clones of functions and relational clones

3 An Application to the CSP

Constraint Satisfaction Problems have a wide range of applications and so the study of their algorithmic complexity is of some importance. In the last decade, P. Jeavons and others discovered an interconnection between these complexity theoretic questions and the theory of clones ([11, 14, 15]). The link between these fields is the Galois connection Inv – Pol. In this section we want to give a short insight into the starting idea. We start with a (slightly simplified) definition of the Constraint Satisfaction Problems.

Definition 3.1. *Let* D *be a finite basic set and let* $\Gamma_0 \subseteq \mathbf{R}_D$ *be a finite set of relations. Then the* Constraint Satisfaction Problem on Γ_0, denoted by CSP(Γ_0), *is the following decision problem:*

Input: *A sentence of the form*

$$(\exists x_1)\ldots(\exists x_n)\,(\varrho_1(x_{i_{1,1}},\ldots)\wedge\ldots\wedge\varrho_m(x_{i_{m,1}},\ldots)),$$

where the ϱ_j are (symbols for the) relations $\varrho_j \in \Gamma_0$ and the indices $i_{s,t}$ are in $\{1, \ldots, n\}$.
Problem: *Is this sentence true?*

(The relations $\varrho_j \in \Gamma_0$ are also called the constraints *of the problem.)* □

Let $\Gamma_1 = \{\sigma_1, \ldots \sigma_k\} \subseteq \mathbf{R}_D$ and $\Gamma_2 = \{\varrho_1, \ldots, \varrho_n\} \subseteq \mathbf{R}_D$ be two finite sets of relations with $\Gamma_1 \subseteq \langle \Gamma_2 \rangle_{\exists,\wedge,=,\mathsf{f}}$. Then for each nonempty relation $\sigma_j \in \Gamma_1$ there exists a formula $\varphi_j \in \Phi(\exists, \wedge, =)$ such that $\sigma_j = L_\varphi(\varrho_1, \ldots, \varrho_n)$. If an instance of CSP($\Gamma_1$) is given, then the relation symbols σ_j in this input sentence can be replaced by the formulas $\varphi_j(\varrho_1, \ldots, \varrho_n)$. Then the equality signs can be removed, the new sentence can be converted into prenex normal form, and we obtain an instance of CSP(Γ_2). It was shown in [15] that this reduction can be done in polynomial time. Consequently, CSP(Γ_1) is polynomially time reducible to CSP(Γ_2).

Now we can apply our Galois connection: Because of Theorem 2.11 we have $\langle \Gamma_2 \rangle_{\exists,\wedge,=,\mathsf{f}} = \mathsf{Inv}\ \mathsf{Pol}\ \Gamma_2$ and therefore $\Gamma_1 \subseteq \langle \Gamma_2 \rangle_{\exists,\wedge,=,\mathsf{f}}$ is equivalent with $\Gamma_1 \subseteq \mathsf{Inv}\ \mathsf{Pol}\ \Gamma_2$, and this is equivalent with $\mathsf{Pol}\ \Gamma_2 \subseteq \mathsf{Pol}\ \Gamma_1$ (1.4). This yields the following Theorem.

Theorem 3.2. *([15]) Let $\Gamma_1, \Gamma_2 \subseteq \mathbf{R}_D$ be finite sets of relations. If $\mathsf{Pol}\ \Gamma_2 \subseteq \mathsf{Pol}\ \Gamma_1$, then* CSP($\Gamma_1$) *is polynomially time reducible to* CSP(Γ_2)*. If $\mathsf{Pol}\ \Gamma_2 = \mathsf{Pol}\ \Gamma_1$, then* CSP($\Gamma_1$) *and* CSP($\Gamma_2$)*, are polynomially time equivalent.* □

Consequently, the complexity of CSP(Γ_0) is (modulo polynomial time equivalence) in some sense determined by the clone $\mathsf{Pol}\,(\Gamma_0)$.

The interconnection between CSP(Γ_0) and $\mathsf{Pol}\ \Gamma_0$ has further consequences. It is known that CSP(Γ_0) is **NP**–complete if $\langle \Gamma_0 \rangle_{\exists,\wedge,=,\mathsf{f}} = \mathbf{R}_D$. This is the case if $\mathsf{Pol}\ \Gamma_0 = \mathsf{Pol}\ \mathbf{R}_D = \mathbf{J}_D$, the least clone in $\mathcal{L}_{\mathbf{O}_D}$. But if $\langle \Gamma_0 \rangle_{\exists,\wedge,=,\mathsf{f}} \neq \mathbf{R}_D$, then $\mathsf{Pol}\ \Gamma_0 \neq \mathbf{J}_D$. Now it is well known that the clone lattice $\mathcal{L}_{\mathbf{O}_D}$ is atomic with finitely many atoms – the so called *minimal clones*. These minimal clones must be generated by a single function. At present there is no exact general classification of the minimal clones, except for some small values of $|D|$. But a beautiful Theorem of I.G. Rosenberg ([30]) gives a list of functions such that all minimal clones are generated by a function from this list.

If $\mathsf{Pol}\ \Gamma_0 \neq \mathbf{J}_D$, then $\mathsf{Pol}\ \Gamma_0$ must contain a minimal clone and therefore also one of the functions f from Rosenberg's list. But then f is a polymorphism for all relations in Γ_0. Very often this knowledge helps to find algorithms to solve the CSP(Γ_0). To give a trivial example: If $\mathsf{Pol}\ \Gamma_0$ contains a constant function with the constant value c, then every relation $\varrho \in \Gamma_0$ is preserved by this function, and for $\varrho \neq \emptyset$ this means $(c, c, \ldots, c) \in \varrho$. But then all instances of CSP(Γ_0) are trivially true, provided that $\emptyset \notin \Gamma_0$.

So the lattice structure of $\mathcal{L}_{\mathbf{O}_D}$ is of some interest for the CSP. In the Boolean case $D = \{0, 1\}$ there are 7 minimal clones (see Fig. 1), generated by the functions 0, 1, $\wedge$, $\vee$, $x \oplus y \oplus z$, $xy \vee xz \vee yz$ and the negation $\overline{x}$. It can be shown (see [15]) that CSP(Γ_0) is in **PTIME** whenever $\mathsf{Pol}\ \Gamma_0$ contains one of the first 6 functions. Post's lattice shows that this is true for all clones on D, with the exceptions $\mathbf{J}_D$ and $\mathsf{clone}\{\overline{x}\}$. If $\mathsf{Pol}\,\Gamma_0$ is one of these 2 clones, then $\mathrm{CSP}(\Gamma_0)$ is **NP**-complete. (This "dichotomy result" was already obtained by Schaefer in [33] with other combinatorial methods. But these methods are difficult to generalize to the case $|D| \geq 3$.)

The "algebraic" interconnection between clones and the CSP has provided a lot of strong complexity theoretical results, also in the general case, we refer to [14, 15, 10].

4 The Galois Connection Inv–mPol

We want to describe more closure operators on sets of relations with the help of Galois connections. Therefore we first generalize our concept of functions, in order to have a general starting point for the next investigations.

Notions and Notations

An *n-ary multifunction on the basic set D* is a function $f : D^n \to \mathcal{P}D$. The set of all n-ary multifunctions on D is denoted with $\mathbf{M}_D^{(n)}$ and the set of all finitary multifunctions is $\mathbf{M}_D := \bigcup_{n=1}^{\infty} \mathbf{M}_D^{(n)}$. For a subset $F \subseteq \mathbf{M}_D$ we put $F^{(n)} := F \cap \mathbf{M}_D^{(n)}$.

The *domain* $\operatorname{dom} f$ of a multifunction is the set

$$\operatorname{dom} f := \{\boldsymbol{a} \in D^n \mid f(\boldsymbol{a}) \neq \emptyset\}.$$

If $\operatorname{dom} f = D^n$, then f is called *total.* The *set of all total multifunctions* is denoted by $\mathbf{tM}_D$.

Remark. These multifunctions appear under various names, e.g. multioperations, correspondences etc. In [32, 27], they are called *partial hyperfunctions,* and the total multifunctions are called *hyperfunctions.*

If $g, f \in \mathbf{M}_D^{(n)}$ such that $g(\boldsymbol{a}) \subseteq f(\boldsymbol{a})$ for all $\boldsymbol{a} \in D^n$, then g is called a *submultifunction* of f and we write $g \subseteq f$. If $F \subseteq \mathbf{M}_D$, then $\Downarrow F$ ("down F") denotes the set of all submultifunctions of multifunctions in F.

$$\Downarrow F := \{g \in \mathbf{M}_D \mid (\exists f \in F) g \subseteq f\}$$

Clearly, $\Downarrow$ is a closure operator. A set F with $\Downarrow F = F$ is called *down closed* or *strong.* If $\mathcal{M} \subseteq \mathbf{M}_D$, then we call a subset $F \subseteq \mathcal{M}$ *$\mathcal{M}$-strong* if $F = \mathcal{M} \cap \Downarrow F$.

We can assign to each "ordinary" function $g \in \mathbf{O}_D^{(n)}$ a total multifunction

$$\widetilde{g} : D^n \to \mathcal{P}D, \ \boldsymbol{a} \mapsto \{g(\boldsymbol{a})\}.$$

For $G \subseteq \mathbf{O}_D$ we put $\widetilde{G} := \{\widetilde{g} \mid g \in G\}$. In this sense we sometimes identify $\mathbf{O}_D$ with the subset $\widetilde{\mathbf{O}_D}$ of $\mathbf{tM}_D$.

Similar as for functions, we can extend a multifunction $f : D^n \to \mathcal{P}\,D$ to a multifunction on D^m, $f : (D^m)^n \to \mathcal{P}\,D^m$ by

$$f(\boldsymbol{a}_1,\ldots \boldsymbol{a}_n) := f(\boldsymbol{a}_1(1),\ldots,\boldsymbol{a}_n(1)) \times \cdots \times f(\boldsymbol{a}_1(m),\ldots,\boldsymbol{a}_n(m)).$$

(If $(\boldsymbol{a}_1(i),\ldots,\boldsymbol{a}_n(i)) \notin \operatorname{dom} f$ for some $i \in \{1,\ldots,m\}$, then $f(\boldsymbol{a}_1,\ldots \boldsymbol{a}_n) = \emptyset$.) For $\varrho \in \mathbf{R}_D$ we put $f(\varrho) := \bigcup\{f(\boldsymbol{a}_1,\ldots \boldsymbol{a}_n) \mid \boldsymbol{a}_1,\ldots,\boldsymbol{a}_n \in \varrho\}$.

Now we extend our defining relation **pres** $\subseteq \mathbf{O}_D \times \mathbf{R}_D$ to a relation **mpres** $\subseteq \mathbf{M}_D \times \mathbf{R}_D$ between multifunctions and relations. Then this extended relation again defines a Galois connection. Here the sets $\mathbf{M}_D$ and $\mathbf{R}_D$ play the role of $\mathcal{X}$ and $\mathcal{Y}$ in Section 1.

Definition 4.1. *Let $f \in \mathbf{M}_D^{(n)}$ and $\varrho \in \mathbf{R}_D^{(m)}$. We say that f* preserves *ϱ, or that ϱ is an* invariant relation *for f, or that f is a* multi-polymorphism *of ϱ, if for all $\boldsymbol{a}_1,\ldots,\boldsymbol{a}_n$ inD^m holds*

$$\boldsymbol{a}_1,\ldots,\boldsymbol{a}_n \in \varrho \Longrightarrow f(\boldsymbol{a}_1,\ldots,\boldsymbol{a}_n) \subseteq \varrho,$$

or equivalently, if $f(\varrho) \subseteq \varrho$. With **mpres** *we denote the defining relation*

$$\mathbf{mpres} := \{(f,\varrho) \in \mathbf{M}_D \times \mathbf{R}_D \mid f \textit{ preserves } \varrho\}.$$

Based on this relation **mpres**, *we define a Galois connection* Inv_D – mPol_D *by*

$$\begin{aligned} \mathsf{Inv}_D : \mathcal{P}\,\mathbf{M}_D \to \mathcal{P}\,\mathbf{R}_D, \quad & F \mapsto \{\varrho \in \mathbf{R}_D \mid (\forall f \in F)\ f\,\mathbf{mpres}\,\varrho\} \\ \mathsf{mPol}_D : \mathcal{P}\,\mathbf{R}_D \to \mathcal{P}\,\mathbf{M}_D, \quad & \Gamma_0 \mapsto \{f \in \mathbf{M}_D \mid (\forall \varrho \in \Gamma_0)\ f\,\mathbf{mpres}\,\varrho\}. \end{aligned}$$

(As long as the basic set D is fixed, we suppress the index D.) □

For the characterization of this G.C. we mainly follow the lines of the proofs in Section 2. First we need the notion of a clone of multifunctions.

Clones of Multifunctions on *D*

The *elementary multifunctions* are the functions

$$\widetilde{e}_i^n : D^n \to D,\ (a_1,\ldots,a_n) \mapsto \{a_i\}.$$

$\widetilde{\mathbf{J}}_D$ is the set of all elementary multifunctions.

The *superposition* of multifunctions $f \in \mathbf{M}_D^{(n)}$, $g_1,\ldots,g_n \in \mathbf{M}_D^{(m)}$ is the multifunction

$$\begin{aligned} f[g_1,\ldots,g_n] : \ & D^m \to \mathcal{P}\,D \\ & \boldsymbol{a} \mapsto \bigcup\{f(b_1,\ldots,b_n) \mid b_i \in g_i(\boldsymbol{a}) \text{ for } i = 1\ldots n\}. \end{aligned}$$

This notion is a generalization of the superposition of ordinary functions. If $f, g_1,\ldots,g_n$ are in $\mathbf{O}_D$, then $\widetilde{f}[\widetilde{g}_1,\ldots\widetilde{g}_n] = \widetilde{f[g_1,\ldots,g_n]}$.

Definition 4.2. *A set $C \subseteq \mathbf{M}_D$ is called a* clone of multifunctions on D, *if C contains all elementary multifunctions $\widetilde{e}_i^n$ and if C is closed under superposition, i.e. if $f \in C^{(n)}$, $g_1, \ldots, g_n \in C^{(m)}$, then also $f[g_1, \ldots, g_n] \in C$.*

If $F \subseteq \mathbf{M}_D$, then

$$\mathsf{clone}\, F := \bigcap \{C \subseteq \mathbf{M}_D \mid C \text{ is a clone and } F \subseteq C\}$$

denotes the clone, generated by F. *This is the least clone of multifunctions on D that contains F.* □

The sets $\widetilde{\mathbf{J}_D}$, $\widetilde{\mathbf{O}_D}$, $\mathbf{M}_D$, $\mathbf{tM}_D$ are examples of clones of multifunctions. Moreover, $C \subseteq \mathbf{O}_D$ is a clone of functions iff $\widetilde{C}$ is a clone of multifunctions. The lattice of all clones of multifunctions is uncountable, even in the case $|D| = 2$. The same holds for the sublattice of all clones of total multifunctions.

The superposition of multifunctions is monotone with respect to $\subseteq$. If $f' \subseteq f$ and $g_i' \subseteq g_i$, then $f'[g_1', \ldots, g_n'] \subseteq f[g_1, \ldots, g_n]$. Therefore, if C is a clone of multifunctions, then $\Downarrow C$ is also a clone. A clone C of multifunctions with $\Downarrow C = C$ is called a *strong clone* or a *down closed clone.*

The *set of all strong clones on D* also forms an algebraic lattice. We denote this lattice with $\mathcal{L}_{\mathbf{M}_D}$. The meet in this lattice is just the set theoretical intersection while the join of two strong clones is given by $C_1 \vee C_2 = \Downarrow \mathsf{clone}(C_1 \cup C_2)$.

The elementary multifunctions preserve every relation. If $f, g_1, \ldots, g_n$ preserve a relation ϱ, then ϱ is also invariant under $f[g_1, \ldots, g_n]$. Therefore $\mathsf{mPol}\, \Gamma_0$ is always a clone. Moreover, if f preserves ϱ and $g \subseteq f$, then g also preserves ϱ. We obtain:

Corollary 4.3. *If $\Gamma_0 \subseteq \mathbf{R}_D$, then $\mathsf{mPol}\, \Gamma_0$ is a $\Downarrow$-closed clone of multifunctions, $\Downarrow \mathsf{clone}\, \mathsf{mPol}\, \Gamma_0 = \mathsf{mPol}\, \Gamma_0$. Therefore $\Downarrow \mathsf{clone}\, F \subseteq \mathsf{mPol}\, \mathsf{Inv}\, F$ for all $F \subseteq \mathbf{M}_D$, i.e. $\Downarrow \mathsf{clone} \leq \mathsf{mPol}\, \mathsf{Inv}$.* □

Weak Systems of Relations

The sets $\mathsf{Inv}\, F$ with $F \subseteq \mathbf{M}_D$ are in general not closed under projections, and do not contain the equality relation. But it is easy to see that $\mathsf{Inv}\, F$ is closed under intersection of m-ary relations and contains the empty and the full relations.

Definition 4.4. *A set $\Gamma \subseteq \mathbf{R}_D$ is called a* weak system with zero, *if Γ is $\wedge, \mathsf{f}, \mathsf{t}$-closed, i.e. if $\Gamma = \langle \Gamma \rangle_{\wedge,\mathsf{t},\mathsf{f}}$. If $\Gamma_0 \subseteq \mathbf{R}_D$, then $\langle \Gamma_0 \rangle_{\wedge,\mathsf{t},\mathsf{f}}$ is the* weak system with zero, generated by Γ_0. □

(A *weak system* is a set $\Gamma \subseteq \mathbf{R}_D$ with $\Gamma = \langle \Gamma \rangle_{\wedge,\mathsf{t}}$.)

Corollary 4.5. *If $F \subseteq \mathbf{M}_D$ then $\mathsf{Inv}\, F$ is a weak system with zero, $\langle \mathsf{Inv}\, F \rangle_{\wedge,\mathsf{t},\mathsf{f}} = \mathsf{Inv}\, F$. Therefore $\langle \Gamma_0 \rangle_{\wedge,\mathsf{t},\mathsf{f}} \subseteq \mathsf{Inv}\, \mathsf{mPol}\, \Gamma_0$ for all $\Gamma_0 \subseteq \mathbf{R}_D$, i.e. $\langle\ \rangle_{\wedge,\mathsf{t},\mathsf{f}} \leq \mathsf{Inv}\, \mathsf{mPol}$.* □

The Characterization of Inv–mPol

The following statements correspond to 2.7 and 2.10.

Lemma 4.6. *Let* $\boldsymbol{a}_1, \ldots, \boldsymbol{a}_n \in D^m$.

1. *For all* $F \subseteq \mathbf{M}_D$ *holds*

$$\mu_{\mathsf{Inv}\ F}(\boldsymbol{a}_1, \ldots, \boldsymbol{a}_n) = \bigcup\{f(\boldsymbol{a}_1, \ldots, \boldsymbol{a}_n) \mid f \in \mathsf{clone}^{(n)} F\}.$$

2. *Let* $\Gamma_0 \subseteq \mathbf{R}_D$ *and* $\Gamma = \langle \Gamma_0 \rangle_{\wedge,\mathsf{t},\mathsf{f}}$. *Then*

$$\mu_\Gamma(\boldsymbol{a}_1, \ldots, \boldsymbol{a}_n) = \bigcup\{f(\boldsymbol{a}_1, \ldots, \boldsymbol{a}_n) \mid f \in \mathsf{mPol}^{(n)} \Gamma\}.$$

Proof. 1. The proof is analogous to the proof of 2.7(2.). Both sides of the equation describe the least relation containing $\boldsymbol{a}_1, \ldots, \boldsymbol{a}_n$ which is invariant for all functions in F.

2. "$\supseteq$" is clear: Γ is closed under intersection, therefore $\mu_\Gamma(\boldsymbol{a}_1, \ldots, \boldsymbol{a}_n) \in \Gamma$ and so the multifunctions in $\mathsf{mPol}\ \Gamma$ preserve $\mu_\Gamma(\boldsymbol{a}_1, \ldots, \boldsymbol{a}_n)$.
"$\subseteq$": Let $\boldsymbol{b} \in \mu_\Gamma(\boldsymbol{a}_1, \ldots, \boldsymbol{a}_n)$. We claim $\boldsymbol{b} \in f(\boldsymbol{a}_1, \ldots, \boldsymbol{a}_n)$ for some $f \in \mathsf{mPol}^{(n)} \Gamma$. We define a multifunction $g_{\boldsymbol{b}} : D^n \to \mathcal{P}\, D$ by

$$g_{\boldsymbol{b}}(a_1, \ldots, a_n) := \{\boldsymbol{b}(i) \mid 1 \le i \le m \text{ and } (\boldsymbol{a}_1(i), \ldots, \boldsymbol{a}_n(i)) = (a_1, \ldots, a_n)\}.$$

Then $\boldsymbol{b} \in g_{\boldsymbol{b}}(\boldsymbol{a}_1, \ldots, \boldsymbol{a}_n)$. Assume $g_{\boldsymbol{b}} \notin \mathsf{mPol}^{(n)} \Gamma$. Then there exists k, $\sigma \in \Gamma^{(k)}$, $\boldsymbol{c}_1, \ldots, \boldsymbol{c}_n \in \sigma$ and $\boldsymbol{d} \in g_{\boldsymbol{b}}(\boldsymbol{c}_1, \ldots, \boldsymbol{c}_n)$ with $\boldsymbol{d} \notin \sigma$. Then for each j with $1 \le j \le k$ there is a number $s(j)$ with $1 \le s(j) \le m$ such that

$$(\boldsymbol{c}_1(j), \ldots, \boldsymbol{c}_n(j), \boldsymbol{d}(j)) = (\boldsymbol{a}_1(s(j)), \ldots, \boldsymbol{a}_n(s(j)), \boldsymbol{b}(s(j))).$$

Now consider the formula $\varphi(x_1, \ldots, m; \sigma) :\iff \sigma(x_{s(1)}, \ldots, x_{s(k)})$. This formula is in $\Phi(\wedge, \mathsf{f}, \mathsf{t})$, therefore $L_\varphi(\sigma) \in \Gamma$. Now $\boldsymbol{a}_1, \ldots, \boldsymbol{a}_n \in L_\varphi(\sigma)$, therefore $\mu_\Gamma(\boldsymbol{a}_1, \ldots, \boldsymbol{a}_n) \subseteq L_\varphi(\sigma)$. But $\boldsymbol{d} \notin \sigma$ implies $\boldsymbol{b} \notin L_\varphi(\sigma)$ – a contradiction. This shows $g_{\boldsymbol{b}} \in \mathsf{mPol}\ \Gamma$ and finishes the proof. □

Now we can give the characterizations for $\mathsf{mPol\ Inv}$ and $\mathsf{Inv\ mPol}$. The proof of 4.7 is similar to the proof 2.9 and we skip it here. Theorem 4.8 follows from 4.6 and Theorem 5.3.

Theorem 4.7. *Let* D *be a finite basic set.*

1. *A subset* $C \subseteq \mathbf{M}_D$ *is Galois closed for the G.C.* Inv–mPol, *(i.e.* $C = \mathsf{mPol\ Inv}\ C$*) if and only if* C *is a* $\Downarrow$*-closed clone of multifunctions.*
2. *For all* $F \subseteq \mathbf{M}_D$, $\Downarrow \mathsf{clone}\, F = \mathsf{mPol\ Inv}\ F$. □

Theorem 4.8. *Let* D *be a finite basic set.*

1. *A subset* $\Gamma \subseteq \mathbf{R}_D$ *is Galois closed for the G.C.* Inv–mPol, *(i.e.* $\Gamma = \mathsf{Inv\ mPol}\ \Gamma$*) if and only if* Γ *is a weak system with zero.*
2. *For all* $\Gamma_0 \subseteq \mathbf{R}_D$, $\langle \Gamma_0 \rangle_{\wedge,\mathsf{t},\mathsf{f}} = \mathsf{Inv\ mPol}\ \Gamma_0$. □

5 Derived Galois Connections

In this section we want to generate more Galois connections in order to describe other closure operators on sets of relations. We start with the general connection $\mathsf{Inv} - \mathsf{mPol}$ and then we restrict the set $\mathbf{M}_D$ of multifunctions to various subsets $\mathcal{M}$. Theorem 1.8 ensures that we obtain new Galois connections.

We choose a *restriction set* $\mathcal{M} \subseteq \mathbf{M}_D$ and consider the operators

$$\begin{aligned} \mathsf{Inv} &: \mathcal{P}\mathcal{M} \to \mathcal{P}\mathbf{R}_D, \quad F \mapsto \mathsf{Inv}\ F \quad \text{and} \\ \mathcal{M}\text{-}\mathsf{mPol} &: \mathcal{P}\mathbf{R}_D \to \mathcal{P}\mathcal{M}, \quad \Gamma_0 \mapsto \mathcal{M} \cap \mathsf{mPol}\ \Gamma_0. \end{aligned}$$

The defining relation for this G.C. is $\Xi_{\mathcal{M}} = \{(f, \varrho) \in \mathcal{M} \times \mathbf{R}_D \mid f\, \mathbf{mpres}\, \varrho\}$. From Theorems 1.8 and 4.7 we already obtain the description of the Galois closed subsets of $\mathcal{M}$.

Corollary 5.1. *If $F \subseteq \mathcal{M}$, then $\mathcal{M}$-$\mathsf{mPol}\ \mathsf{Inv}\ F = \mathcal{M} \cap \Downarrow \mathsf{clone}\, F$.*

A subset $C \subseteq \mathcal{M}$ is Galois closed for $\mathsf{Inv} - \mathcal{M}$-$\mathsf{mPol}$ (i.e. $C = \mathcal{M}$-$\mathsf{mPol}\ \mathsf{Inv}\ C$) if and only if C is an $\mathcal{M}$-strong clone of multifunctions, i.e. $C = \mathcal{M} \cap \Downarrow \mathsf{clone}\, C$. □

At this point we want to agree on a small inexactness of our notation: If $f \in \mathbf{O}_D$ is an ordinary function, then we identify f with the corresponding multifunction $\widetilde{f}$ even if we do not write the tilde. In this sense we have the inclusions $\mathbf{O}_D \subseteq \mathbf{tM}_D \subseteq \mathbf{M}_D$. We also use this agreement for the following sets:

$$\begin{aligned} \mathbf{P}_D^{(n)} &:= \{f \mid f \ :\ D^n \supseteq \operatorname{dom} f \to D\} \\ \mathbf{P}_D &:= \bigcup_{n \in \mathbf{N}^+} \mathbf{P}_D^{(n)} \quad \text{(set of all } \textit{partial functions}\text{)} \\ \mathbf{S}_D &:= \{\pi \mid \pi \ :\ D \to D \text{ is a permutation }\} \end{aligned}$$

So an n-ary partial function $f \in \mathbf{P}_D^{(n)}$ is identified with the multifunction

$$\widetilde{f} \ :\ \boldsymbol{a} \mapsto \begin{cases} \{f(\boldsymbol{a})\} & \text{if } \boldsymbol{a} \in \operatorname{dom} f \\ \emptyset & \text{else} \end{cases}.$$

The set $\mathbf{P}_D$ corresponds to the set $\{f \in \mathbf{M}_D \mid (\forall \boldsymbol{a} \in D^{\mathsf{ar}(f)}) |f(\boldsymbol{a})| \leq 1\}$, and $\mathbf{O}_D$ is the set of all partial functions with $\operatorname{dom} f = D^n$.

For these and other sets $\mathcal{M}$ the operators $\mathcal{M}$-mPol have special abbreviations: $\mathsf{pPol} := \mathbf{P}_D$-$\mathsf{mPol}$, $\mathsf{wAut} := \mathbf{S}_D$-$\mathsf{mPol}$, $\mathsf{tmPol} := \mathbf{tM}_D$-$\mathsf{mPol}$, $\mathsf{Pol} = \mathbf{O}_D$-$\mathsf{mPol}$, $\mathsf{End} := \mathbf{O}_D^{(1)}$-$\mathsf{mPol}$.

Please note that every subset $C \subseteq \mathbf{O}_D$ is $\mathbf{O}_D$-strong. Therefore an $\mathbf{O}_D$-strong clone $C \subseteq \mathbf{O}_D$ is just a clone. Consequently, for $\mathcal{M} = \mathbf{O}_D$ we reobtain Theorem 2.9 from Corollary 5.1.

Approximating Closure Operators on $\mathbf{R}_D$ with Galois connections

We want to use these restricted Galois connections to approximate closure operators $\mathcal{H}$ on $\mathbf{R}_D$ with operators of the form $\mathsf{Inv}\ \mathcal{M}$-$\mathsf{mPol}$ and therefore we adapt the constructions in 1.10 and 1.11. We assume that $\mathcal{H}$ satisfies $\langle \Gamma_0 \rangle_{\wedge,\mathsf{t},\mathsf{f}} \subseteq \mathcal{H}\Gamma_0$ for all $\Gamma_0 \subseteq \mathbf{R}_D$, i.e. $\langle\ \rangle_{\wedge,\mathsf{t},\mathsf{f}} \leq \mathcal{H}$. According to 1.10 we have to consider the *approximation set* for $\mathcal{H}$:

$$\mathbf{App}(\mathcal{H}) = \{f \in \mathbf{M}_D \mid \mathcal{H}\,\mathsf{Inv}\,\{f\} = \mathsf{Inv}\,\{f\}\}.$$

We collect some properties.

Lemma 5.2. *1.* $\mathcal{H}\,\mathsf{Inv}\ F = \mathsf{Inv}\ F$ *for all* $F \subseteq \mathbf{App}(\mathcal{H})$.

2. $\Downarrow \widetilde{\mathbf{J}_D} \subseteq \mathbf{App}(\mathcal{H})$

3. Let $\mathcal{H} := \mathcal{H}_1 \vee \mathcal{H}_2$ *denote the least closure operator with* $\mathcal{H}_1 \leq \mathcal{H}$ *and* $\mathcal{H}_2 \leq \mathcal{H}$. *(This means, the* $\mathcal{H}$*–closed subsets of* $\mathbf{R}_D$ *are exactly the sets that are closed under* $\mathcal{H}_1$ *and under* $\mathcal{H}_2$*.) Then* $\mathbf{App}(\mathcal{H}) = \mathbf{App}(\mathcal{H}_1) \cap \mathbf{App}(\mathcal{H}_2)$.

4. $\mathcal{H} \leq \mathsf{Inv}\ \mathcal{M}\text{-}\mathsf{mPol}$ *iff* $\mathcal{M} \subseteq \mathbf{App}(\mathcal{H})$

Proof. (1.)–(3.) are immediate from the Definition. (4.) follows from Theorem 1.11. □

Unless (2.) holds, the approximation sets are usually not clones. So sometimes instead of $\mathcal{M} = \mathbf{App}(\mathcal{H})$ one chooses a more convenient proper subset $\mathcal{M} \subset \mathbf{App}(\mathcal{H})$.

It remains to answer the question, under which conditions we can guarantee the equality $\mathcal{H} = \mathsf{Inv}\ \mathcal{M}\text{-}\mathsf{mPol}$.

Theorem 5.3. *Let* $\mathcal{H} : \mathcal{P}\mathbf{R}_D \to \mathcal{P}\mathbf{R}_D$ *be a closure operator and* $\mathcal{M} \subseteq \mathbf{M}_D$. *Then* $\mathcal{H} = \mathsf{Inv}\ \mathcal{M}\text{-}\mathsf{mPol}$ *if and only if the following three conditions hold.*

1. $\mathcal{H}\Gamma_0 = \langle \mathcal{H}\Gamma_0 \rangle_{\wedge,\mathsf{t},\mathsf{f}}$ *for all* $\Gamma_0 \subseteq \mathbf{R}_D$. *(I.e. the* $\mathcal{H}$*–closed relation sets are closed under intersections and contain the empty and full relations,* $\langle\ \rangle_{\wedge,\mathsf{t},\mathsf{f}} \leq \mathcal{H}$*.)*

2. $\mathcal{M} \subseteq \mathbf{App}(\mathcal{H})$

3. For all $\Gamma_0 \subseteq \mathbf{R}_D$ *and all* $\boldsymbol{a}_1, \ldots, \boldsymbol{a}_n \in D^m$ *holds*

$$\mu_{\mathcal{H}\Gamma_0}(\boldsymbol{a}_1, \ldots, \boldsymbol{a}_n) \subseteq \bigcup \{f(\boldsymbol{a}_1, \ldots, \boldsymbol{a}_n) \mid f \in \mathsf{clone}^{(n)}\ \mathcal{M}\text{-}\mathsf{mPol}\ \Gamma_0\}$$

Proof. "$\Rightarrow$" Let $\mathcal{H} = \mathsf{Inv}\ \mathcal{M}\text{-}\mathsf{mPol}$.

1. Because of 4.5 and by 1.4(3) we have

$$\begin{aligned}\mathcal{H}\Gamma_0 \subseteq \langle \mathcal{H}\Gamma_0 \rangle_{\wedge,\mathsf{t},\mathsf{f}} &\subseteq \mathsf{Inv\ mPol}\ \mathcal{H}\Gamma_0 = \mathsf{Inv\ mPol\ Inv}\ \mathcal{M}\text{-}\mathsf{mPol}\ \Gamma_0 \\ &= (\mathsf{Inv\ mPol\ Inv}\,)\,\mathcal{M}\text{-}\mathsf{mPol}\ \Gamma_0 = \mathsf{Inv}\ \mathcal{M}\text{-}\mathsf{mPol}\ \Gamma_0 = \mathcal{H}\Gamma_0.\end{aligned}$$

2. If $f \in \mathcal{M}$, then $\mathsf{Inv}\,\{f\}$ is closed under $\mathsf{Inv}\ \mathcal{M}\text{-}\mathsf{mPol}$ and therefore under $\mathcal{H}$.
3. Because of $\mathcal{H} = \mathsf{Inv}\ \mathcal{M}\text{-}\mathsf{mPol}$ and of 4.6(1) we have

$$\begin{aligned}\mu_{\mathcal{H}\Gamma_0}(\boldsymbol{a}_1, \ldots, \boldsymbol{a}_n) &= \mu_{\mathsf{Inv}\ \mathcal{M}\text{-}\mathsf{mPol}\ \Gamma_0}(\boldsymbol{a}_1, \ldots, \boldsymbol{a}_n) \\ &= \bigcup \{f(\boldsymbol{a}_1, \ldots, \boldsymbol{a}_n) \mid f \in \mathsf{clone}^{(n)}\ \mathcal{M}\text{-}\mathsf{mPol}\ \Gamma_0\}\end{aligned}$$

"$\Leftarrow$" Because of 5.2(4) we have $\mathcal{H}\Gamma_0 \subseteq \mathsf{Inv}\ \mathcal{M}\text{-}\mathsf{mPol}\ \Gamma_0$ for all $\Gamma_0 \subseteq \mathbf{R}_D$. It remains to show $\mathsf{Inv}\ \mathcal{M}\text{-}\mathsf{mPol}\ \Gamma_0 \subseteq \mathcal{H}\Gamma_0$. Let $\varrho \in \mathsf{Inv}\ \mathcal{M}\text{-}\mathsf{mPol}\ \Gamma_0$. If $\varrho = \emptyset^{(m)}$ then (1.) implies $\varrho \in \mathcal{H}\Gamma_0$.

Otherwise put $n := |\varrho| \geq 1$ and $\varrho = \{\boldsymbol{a}_1, \ldots, \boldsymbol{a}_n\}$. Let $\boldsymbol{b} \in \mu_{\mathcal{H}\Gamma_0}(\varrho) = \mu_{\mathcal{H}\Gamma_0}(\boldsymbol{a}_1, \ldots, \boldsymbol{a}_n)$. Then because of (3) we have $\boldsymbol{b} \in f(\boldsymbol{a}_1, \ldots, \boldsymbol{a}_n)$ for some $f \in \mathsf{clone}^{(n)}\ \mathcal{M}\text{-}\mathsf{mPol}\ \Gamma_0$. The relation ϱ is invariant under all multifunctions in $\mathcal{M}\text{-}\mathsf{mPol}\ \Gamma_0$, therefore it is also invariant for all $f \in \mathsf{clone}\,\mathcal{M}\text{-}\mathsf{mPol}\ \Gamma_0$. Consequently, $\boldsymbol{b} \in \varrho = \{\boldsymbol{a}_1, \ldots, \boldsymbol{a}_n\}$. Therefore $\mu_{\mathcal{H}\Gamma_0}(\varrho) = \varrho$. Because of (1) $\mathcal{H}\Gamma_0$ is closed under intersections, and so $\mu_{\mathcal{H}\Gamma_0}(\varrho) = \varrho$ implies $\varrho \in \mathcal{H}\Gamma_0$. This shows $\mathsf{Inv}\ \mathcal{M}\text{-}\mathsf{mPol}\ \Gamma_0 \subseteq \mathcal{H}\Gamma_0$. □

Now we want to discuss some approximation sets for extensions of the closure operator $\langle\ \rangle_{\wedge,\mathsf{t},\mathsf{f}}$. Let $S_1, S_2, \ldots$ be a list of logical symbols in Symb (Eq. 1). Then instead of $\mathbf{App}(\langle\ \rangle_{\wedge,\mathsf{f},\mathsf{t},S_1,S_2,\ldots})$ we write $\mathbf{App}(\wedge, \mathsf{f}, \mathsf{t}, S_1, S_2, \ldots)$.

- Because of $\Delta_D \in \mathsf{Inv}\,\{f\} \iff f \in \mathsf{mPol}\,\{\Delta_D\}$ (see 1.4(2)) we have $\mathbf{App}(\wedge, \mathsf{f}, \mathsf{t}, =) = \mathsf{mPol}\,(\{\Delta_D\})$. In particular, $\mathsf{mPol}\,\{\Delta_D\} = \widetilde{\mathbf{P}_D}$, and
$$\begin{aligned}\langle \Gamma_0 \rangle_{\wedge,\mathsf{f},\mathsf{t},=} &= \langle \Gamma_0 \cup \{\Delta_D\} \rangle_{\wedge,\mathsf{t},\mathsf{f}} = \mathsf{Inv}\ \mathsf{mPol}\,(\Gamma_0 \cup \{\Delta_D\}) \\ &= \mathsf{Inv}\,(\mathsf{mPol}\,\{\Delta_D\} \cap \mathsf{mPol}\ \Gamma_0) = \mathsf{Inv}\ \mathsf{pPol}\,\Gamma_0\end{aligned}$$
holds for all $\Gamma_0 \subseteq \mathbf{R}_D$.
- Similarly we find $\mathbf{App}(\wedge, \mathsf{f}, \mathsf{t}, \neq) = \mathsf{mPol}\,(\mathsf{Co}\,\Delta_D)$ and with $\mathcal{M}_0 := \mathsf{mPol}\,(\mathsf{Co}\,\Delta_D)$ we obtain $\langle \Gamma_0 \rangle_{\wedge,\mathsf{f},\mathsf{t},\neq} = \mathsf{Inv}\ \mathcal{M}_0\text{-}\ \mathsf{mPol}\ \Gamma_0$.
- It is easy to see that $\mathbf{M}_D^{(1)} \subseteq \mathbf{App}(\wedge, \mathsf{f}, \mathsf{t}, \vee)$ and that the conditions (1.) and (2.) of 5.3 are satisfied for $\mathcal{H} = \langle\ \rangle_{\wedge,\mathsf{f},\mathsf{t},\vee}$ and $\mathcal{M} = \mathbf{M}_D^{(1)}$. In order to see 5.3 (3.), note that $\langle \Gamma_0 \rangle_{\wedge,\mathsf{f},\mathsf{t},\vee}$ is closed under union and therefore
$$\mu_{\mathcal{H}\Gamma_0}(\boldsymbol{a}_1, \ldots, \boldsymbol{a}_n) = \bigcup_{i=1}^{n} \mu_{\mathcal{H}\Gamma_0}(\boldsymbol{a}_i).$$
Then $\langle \Gamma_0 \rangle_{\wedge,\mathsf{t},\mathsf{f}} \subseteq \mathcal{H}\Gamma_0$ implies $\mu_{\mathcal{H}\Gamma_0}(\boldsymbol{a}) \subseteq \mu_{\langle \Gamma_0 \rangle_{\wedge,\mathsf{t},\mathsf{f}}}(\boldsymbol{a})$ and by 4.6 (2.) we have $\mu_{\langle \Gamma_0 \rangle_{\wedge,\mathsf{t},\mathsf{f}}}(\boldsymbol{a}) = \bigcup\{f(\boldsymbol{a}) \mid f \in \mathsf{mPol}^{(1)}\,\Gamma_0\}$. Now 5.3 implies $\mathcal{H} = \mathsf{Inv}\ \mathsf{mPol}^{(1)}$.
- "$\neg$": If $f \in \mathbf{App}(\wedge, \mathsf{f}, \mathsf{t}, \neg)$, then $\mathsf{Inv}\,\{f\}$ has to be closed under complementation, i.e. for all $\varrho \in \mathbf{R}_D$, $f\,\mathbf{mpres}\,\varrho \iff f\,\mathbf{mpres}\,\mathsf{Co}\,\varrho$. Clearly $\mathbf{S}_D \subseteq \mathbf{App}(\wedge, \mathsf{f}, \mathsf{t}, \neg)$. As we will see below, in this case f even *strongly preserves* ϱ and it seems to be more natural to use a stronger Galois connection for the description – see Def. 5.4.
- "$\exists$": If we add the existential quantifier then $\mathbf{tM}_D \subseteq \mathbf{App}(\wedge, \mathsf{f}, \mathsf{t}, \exists)$. The restriction set $\mathcal{M} = \mathbf{tM}_D$ suffices to describe the closure operator $\langle\ \rangle_{\wedge,\mathsf{t},\mathsf{f},\exists}$.
- "$\forall$": A multifunction $f \in \mathbf{M}_D^{(n)}$ is called *surjective*, if $\bigcup_{\boldsymbol{a} \in D^n} f(\boldsymbol{a}) = D$. We put
$$\mathbf{sur}\text{-}\mathbf{M}_D := \{f \in \mathbf{M}_D \mid f \text{ is surjective }\}.$$

It is not hard to see that $\mathbf{sur\text{-}M}_D \subseteq \mathbf{App}(\wedge, \mathsf{f}, \mathsf{t}, \forall)$, and with some effort it can be shown that this set suffices to describe $\langle\ \rangle_{\wedge,\mathsf{t},\mathsf{f},\forall}$.

– If we add more than one symbol from Symb to our list $\mathcal{S}$, then 5.2 (3.) gives a hint how to construct a possible approximation set for $\langle\ \rangle_{\mathcal{S}}$. For example, if we add $=$, $\exists$ and $\forall$ to $\wedge, \mathsf{t}, \mathsf{f}$, then the intersection

$$\mathbf{sur\text{-}O}_D := \widetilde{\mathbf{P}_D} \cap \mathbf{tM}_D \cap \mathbf{sur\text{-}M}_D$$

of the restriction sets for $=$, $\exists$, $\forall$ is a good candidate. ($\mathbf{sur\text{-}O}_D$ corresponds to the set of all surjective total functions.) It turns out, that this set really suffices to model the corresponding closure operator (see 6.2 and 6.3).

Table 1. Some Galois Connections, derived from Inv – mPol

Restriction set $\mathcal{M}$	Abbreviation for $\mathcal{M} \cap$ mPol	closed sets of multifunctions	closure operator	closed sets of relations
$\mathbf{M}_D$	mPol	strong clones of multifunctions	$\langle \Gamma_0 \rangle_{\wedge,\mathsf{f},\mathsf{t}}$	weak systems with zero ([4],[28])
$\mathbf{tM}_D$	tmPol	strong clones of total multif.	$\langle \Gamma_0 \rangle_{\exists,\wedge,\mathsf{f},\mathsf{t}}$	weak systems with projections and zero ([32],[27],[6])
$\mathbf{P}_D$	pPol	strong clones of partial functions	$\langle \Gamma_0 \rangle_{\wedge,=,\mathsf{f}}$	weak systems with zero and identity ([26],[18])
$\mathbf{O}_D$	Pol	clones of functions	$\langle \Gamma_0 \rangle_{\exists,\wedge,=,\mathsf{f}}$	relational clones ([2],[3],[20],[13])

The Tables 1 and 2 on pages 60 and 61 give an overview of certain Galois connections of the form Inv–$\mathcal{M}$-mPol . In Table 2 some Galois connections with unary restriction sets are listet. The basic set D is always assumed to be finite. (For a more general survey, including the infinite case, we refer to [22].) Note that $\mathsf{t} \iff (x = x)$; so we can omit the t whenever the equality sign $=$ is involved. Moreover, the word "strong" always means $\mathcal{M}$-strong w.r.t. the appropriate restriction set $\mathcal{M}$.

A Remark on the Empty Relations

In the definitions of clones of functions or multifunctions we considered only (multi)functions with arities greater or equal than 1. This is the reason why empty relations are always invariant for all (multi)functions under consideration, and therefore we always have to carry the f-symbol in the description of our logical closure operators. But it is not difficult to extend the notion of a clone to a "clone with constants". In the case of functions, we have to add a set $\mathbf{O}_D^{(0)}$ of nullary functions to $\mathbf{O}_D$ and we obtain $\mathsf{c}\mathbf{O}_D := \mathbf{O}_D \cup \mathbf{O}_D^{(0)}$. The nullary functions are just the constants, i.e. the elements of D. In the definition of a *clone with constants* $C \subseteq \mathsf{c}\mathbf{O}_D$ we must extend the superposition to the case $f[c_1, \ldots, c_n]$

Table 2. Some Galois Connections, derived from $\mathsf{Inv} - \mathsf{mPol}^{(1)}$

Restriction set $\mathcal{M}$	Abbreviation for $\mathcal{M} \cap \mathsf{mPol}^{(1)}$	closed sets of multifunctions	closure operator	closed sets of relations
$\mathbf{M}_D^{(1)}$	$\mathsf{mPol}^{(1)}$	strong monoids of unary multif.	$\langle \Gamma_0 \rangle_{\wedge,\vee,\mathsf{f},\mathsf{t}}$	distributive systems with zero ([6])
$\mathbf{tM}_D^{(1)}$	$\mathsf{tmPol}^{(1)}$	strong monoids of unar. total multif.	$\langle \Gamma_0 \rangle_{\exists,\wedge,\vee,\mathsf{f},\mathsf{t}}$	distributive systems with project. and zero ([6])
$\mathbf{P}_D^{(1)}$	$\mathsf{pPol}^{(1)}$	strong monoids of unary part. funct.	$\langle \Gamma_0 \rangle_{\wedge,\vee,=,\mathsf{f}}$	distributive systems with zero and identity ([6])
$\mathbf{O}_D^{(1)}$	End	transformation monoids	$\langle \Gamma_0 \rangle_{\exists,\wedge,\vee,=,\mathsf{f}}$	weak Krasner algebras ([17],[21])
$\mathbf{S}_D$	wAut (= Aut if D finite)	permutation groups	$\langle \Gamma_0 \rangle_{\exists,\forall,\wedge,\vee,=,\neq}$ (= $\langle \Gamma_0 \rangle_{\exists,\wedge,\neg,=}$)	Krasner algebras ([16],[20],[6])

$(= f(c_1, \ldots, c_n))$, where $c_1, \ldots, c_n \in \mathbf{O}_D^{(0)}$. And we have to require for a clone with constants C, that for each $c \in C^{(0)}$ there exists a unary function $\mathrm{cx}_c \in C^{(1)}$ with constant value $\mathrm{cx}_c(x) = c$ for all $x \in D$. (The usual clones also are clones with constants – with an empty set of constants.)

The notion of an invariant relation can easily be extended: ϱ is invariant for a constant c if $\varrho(c, c, \ldots, c)$. Then all the definitions and proofs of section 2 can be extended to this case. We obtain a Galois connection $\mathsf{Inv} - \text{c-}\mathsf{Pol}$ with $\mathsf{Inv}\ \text{c-}\mathsf{Pol} = \langle\ \rangle_{\exists,\wedge,=}$.

Similarly, for multifunctions a constant in $\mathbf{M}_D^{(0)}$ is a unary relation $B \subseteq D$. ($B = \emptyset$ is possible.) A relation $\varrho \in \mathbf{R}_D^{(m)}$ is invariant for such a constant B if $B^m \subseteq \varrho$. Then, following the same line as before, we obtain a G.C. $\mathsf{Inv} - \text{c-}\mathsf{mPol}$ with $\mathsf{Inv}\ \text{c-}\mathsf{mPol} = \langle\ \rangle_{\wedge,\mathsf{t}}$.

Therefore we also can describe the closure operators $\langle\ \rangle_{\wedge,\mathsf{t}}$, $\langle\ \rangle_{\exists,\wedge,\mathsf{t}}$, $\langle\ \rangle_{\wedge,=}$, etc. with appropriate Galois connections.

For the Galois connection $\mathsf{Inv} - \text{c-}\mathsf{Pol}$ it follows that $\Gamma \subseteq \mathbf{R}_D$ is closed under $\mathsf{Inv}\ \text{c-}\mathsf{Pol}$ iff $\Gamma = \langle \Gamma \rangle_{\exists,\wedge,=}$. And $C \subseteq \mathsf{cO}_D$ is closed under $\text{c-}\mathsf{Pol}\ \mathsf{Inv}$ iff C is a clone with constants.

The lattice of all clones with constants on D is uniquely determined by the lattice $\mathcal{L}_{\mathbf{O}_D}$ of all (usual) clones. If we have a clone with constants C and $C^{(0)} \neq \emptyset$, then $C' := C \setminus C^{(0)}$ is an ordinary clone, and C covers C' in the lattice of all clones of constants. Vice versa, to an ordinary clone C' we can add exactly the set of constants c with $\mathrm{cx}_c \in C'$ to obtain another clone with constants. Figure 3 on page 62 shows the lattice of all clones with constants for the Boolean case. The white circles denote the additional clones with constants.

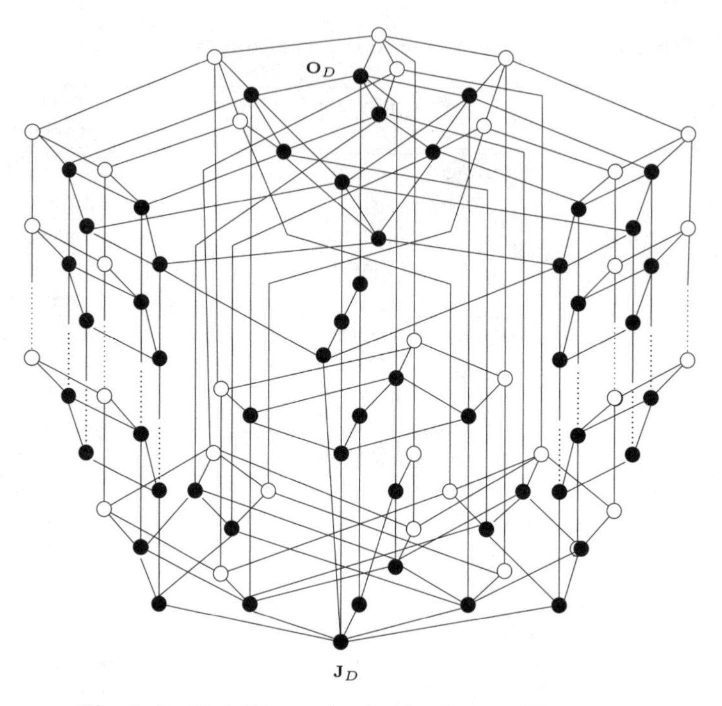

Fig. 3. Post's lattice, extended to clones with constants

Strongly Invariant Relations

The negation symbol did not occur in the closure operators of Tables 1 and 2. (But note that for finite D the Galois closed sets $\langle \Gamma_0 \rangle_{\exists,\forall,\wedge,\vee,=,\neq}$ for the G.C. $\mathsf{Inv} - \mathsf{wAut}$ are also closed under complementation.) We already have noted that $f \in \mathbf{App}(\wedge, \mathrm{t}, \neg)$ implies

$$f\, \mathbf{mpres}\, \varrho \iff f\, \mathbf{mpres}\, \mathsf{Co}\, \varrho.$$

This is a hard restriction to f and therefore, if we want to investigate relational systems closed under complementation, it is more natural to use another kind of Galois connections.

We restrict the functional side to the set $\mathbf{M}_D^{(1)}$ of all unary multifunctions. Superposition then reduces to composition $f \circ g$ with

$$(f \circ g)(a) = \bigcup \{f(b) \mid b \in g(a)\}.$$

The only elementary multifunction in $\mathbf{M}_D^{(1)}$ is $\widetilde{e}_1^1$. The set $\mathbf{M}_D^{(1)}$ with the operation $\circ$ and unit $\widetilde{e}_1^1$ is a monoid.

Definition 5.4. *([7]) Let $f \in \mathbf{M}_D^{(1)}$ and $\varrho \in \mathbf{R}_D$. We say that f* strongly preserves the relation *ϱ, or that ϱ is a* strongly invariant relation for *f, if f preserves both, ϱ and its complement $\mathsf{Co}\,\varrho$. In this case we write $f\, \mathbf{spres}\, \varrho$.*

$$f\, \mathbf{spres}\, \varrho :\iff f\, \mathbf{mpres}\, \varrho \text{ and } f\, \mathbf{mpres}\, \mathsf{Co}\, \varrho$$

Based on the defining relation **spres** *we define a Galois connection* sInv – smEnd*:*

$$\mathsf{sInv} : \mathcal{P}\mathbf{M}_D^{(1)} \to \mathcal{P}\mathbf{R}_D, \quad F \mapsto \{\varrho \in \mathbf{R}_D \mid (\forall f \in F) f \,\mathbf{spres}\, \varrho\}$$

$$\mathsf{smEnd} : \mathcal{P}\mathbf{R}_D \to \mathcal{P}\mathbf{M}_D^{(1)}, \quad \Gamma_0 \mapsto \{f \in \mathbf{M}_D^{(1)} \mid (\forall \varrho \in \Gamma_0) f \,\mathbf{spres}\, \varrho\}. \qquad \square$$

For unary multifunctions $f \in \mathbf{M}_D^{(1)}$ we can define a new operation, the *involution* $f \mapsto f^{-1} \in \mathbf{M}_D^{(1)}$ by $f^{-1}(a) := \{b \in D \mid a \in f(b)\}$.

Lemma 5.5. *([7]) A unary multifunction f strongly preserves a relation ϱ iff both, f and f^{-1} preserve ϱ.* $\square$

Consequently, $\mathsf{smEnd}\, \Gamma_0$ is always closed under involution, and $\mathsf{sInv}\, F$ is always closed under complementation.

Definition 5.6. *A set $\Gamma \subseteq \mathbf{R}_D$ is called a* Boolean system *if $\Gamma = \langle \Gamma \rangle_{\wedge,\neg}$.*

A set $C \subseteq \mathbf{M}_D^{(1)}$ is a strong involuted monoid *if $C = \langle C; \circ, ^{-1}, \widetilde{e}_1^1 \rangle$ is an involuted monoid and C is $\Downarrow$-closed, $\Downarrow C = C$.* $\square$

Now we give the characterization Theorems for the Galois connection sInv–smEnd. For the proofs we refer to [7].

Theorem 5.7. *Let D be a finite basic set.*

1. *A subset $C \subseteq \mathbf{M}_D^{(1)}$ is Galois closed for the G.C.* sInv–smEnd*, (i.e. $C = \mathsf{smEnd}\,\mathsf{sInv}\, C$) if and only if C is a strong involuted monoid.*
2. *For all $F \subseteq \mathbf{M}_D^{(1)}$, $\mathsf{smEnd}\,\mathsf{sInv}\, F$ is the least strong involuted monoid that contains F.*
3. *A subset $\Gamma \subseteq \mathbf{R}_D$ is Galois closed for the G.C.* sInv–smEnd*, (i.e. $\Gamma = \mathsf{sInv}\,\mathsf{smEnd}\, \Gamma$) if and only if Γ is a Boolean system.*
4. *For all $\Gamma_0 \subseteq \mathbf{R}_D$, $\langle \Gamma_0 \rangle_{\wedge,\neg} = \mathsf{sInv}\,\mathsf{smEnd}\, \Gamma_0$.* $\square$

Now we can proceed in the same way as before. If we choose a suitable restriction set $\mathcal{M} \subseteq \mathbf{M}_D^{(1)}$ then by 1.8 the operator pair

$$\mathsf{sInv} \;:\; \mathcal{P}\mathcal{M} \to \mathcal{P}\mathbf{R}_D, \quad \mathcal{M} \supseteq F \mapsto \mathsf{sInv}\, F$$

$$\mathcal{M}\text{-}\mathsf{smEnd} \;:\; \mathcal{P}\mathbf{R}_D \to \mathcal{P}\mathcal{M}, \quad \mathbf{R}_D \supseteq \Gamma_0 \mapsto \mathcal{M} \cap \mathsf{smEnd}\, \Gamma_0$$

forms a Galois connection. The Galois closed subsets of $\mathcal{M}$ are the restrictions of strong involuted monoids in $\mathbf{M}_D^{(1)}$ to $\mathcal{M}$.

Again we obtain several Galois connections in this way. Here all the Galois closed relation sets are closed under complementation. Table 3 on page 64 shows some examples. As restriction sets in Table 3 we choose the following subsets of $\mathbf{M}_D^{(1)}$:

- The set $\mathbf{pP}_D := \{f \in \mathbf{P}_D^{(1)} \mid f$ injective on $\operatorname{dom} f\}$ of all *partial permutations.*
- The set $\mathbf{B}_D := \{f \in \mathbf{M}_D^{(1)} \mid \operatorname{dom} f = D$ and $\operatorname{dom} f^{-1} = D\}$ of all *bitotal multifunctions.*
- The set $\mathbf{O}_D^{(1)}$ of all unary functions.
- The set $\mathbf{S}_D$ of all permutations on D.

As before, the word "strong" always means "$\mathcal{M}$-strong". Again, for the proofs we refer to [7].

Table 3. Some Galois Connections, derived from sInv — smEnd

Restriction set $\mathcal{M}$	Abbreviation for $\mathcal{M} \cap$ smEnd	closed sets of multifunctions	closure operator sInv $\mathcal{M}$- smEnd Γ_0	closed sets of relations
$\mathbf{M}_D^{(1)}$	smEnd	str. invol. monoids of unary multifunctions	$\langle \Gamma_0 \rangle_{\wedge,\neg}$	Boolean systems
$\mathbf{pP}_D$	spmEnd	str. invol. monoids of unar. part. perm.	$\langle \Gamma_0 \rangle_{\wedge,\neg,=}$	Boolean systems with identity
$\mathbf{B}_D$	sbmEnd	str. invol. monoids of bitotal unar. multif.	$\langle \Gamma_0 \rangle_{\exists,\wedge,\neg}$	Boolean systems with projections
$\mathbf{O}_D^{(1)}$	sEnd	special monoids of transformations	$\langle \Gamma_0 \rangle_{\exists,\wedge,\neg}$	Boolean systems with projections
$\mathbf{S}_D$	Aut	permutation groups	$\langle \Gamma_0 \rangle_{\exists,\wedge,\neg,=}$	Krasner algebras

6 The QCSP and Surjectively Generated Clones

The Quantified Constraint Satisfaction Problem is a generalization of the CSP (Def. 3.1). Besides existential quantification we also allow universal quantification in our input sentences.

Definition 6.1. *Let D be a finite basic set and let $\Gamma_0 \subseteq \mathbf{R}_A$ be a finite set of relations. Then the* Quantified Constraint Satisfaction Problem on Γ_0, denoted by QCSP(Γ_0), *is the following decision problem:*

Input: *A sentence of the form*

$$(Q_1 x_1) \ldots (Q_n x_n) \left(\varrho_1(x_{i_{1,1}}, \ldots) \wedge \ldots \wedge \varrho_m(x_{i_{m,1}}, \ldots) \right)$$

where the ϱ_j are (symbols for the) relations $\varrho_j \in \Gamma_0$, the indices $i_{s,t}$ are in $\{1, \ldots, n\}$ and the Q_l are quantifiers from the set $\{\exists, \forall\}$.
Problem: *Is this sentence true?* □

We want to treat this problem in a similar way as in the case of the CSP in Section 3. So let $\Gamma_1 = \{\sigma_1, \ldots \sigma_k\} \subseteq \mathbf{R}_D$ and $\Gamma_2 = \{\varrho_1, \ldots, \varrho_n\} \subseteq \mathbf{R}_D$ be two finite sets of relations with $\Gamma_1 \subseteq \langle \Gamma_2 \rangle_{\exists,\forall,\wedge,=,\mathrm{f}}$. Then for each nonempty relation $\sigma_j \in \Gamma_1$ there exists a formula $\varphi_j \in \Phi(\exists, \forall, \wedge, =)$ such that $\sigma_j = L_\varphi(\varrho_1, \ldots, \varrho_n)$. If an instance of QCSP(Γ_1) is given, then the relation symbols σ_j in this input sentence can be replaced by the formulas $\varphi_j(\varrho_1, \ldots, \varrho_n)$. Then the equality signs can be omitted and the new sentence can be converted into prenex normal form, and we obtain an instance of QCSP(Γ_2). Again it was shown (in [8]) that this reduction can be done in polynomial time. Consequently, QCSP(Γ_1) is polynomially time reducible to QCSP(Γ_2).

Now we want to proceed as in Section 3. But then we need a Galois connection that describes the closure operator $\langle\ \rangle_{\exists,\forall,\wedge,=,\mathrm{f}}$. If possible, this should be a G.C., derived from Inv – Pol. So we are looking for a subset $\mathcal{M} \subseteq \mathbf{O}_D$ with

$\langle\ \rangle_{\exists,\forall,\wedge,=,\mathsf{f}} = \mathsf{Inv}\ \mathcal{M}\text{-}\mathsf{Pol}$. Because of 1.11(2.), the subset $\mathcal{M}$ must satisfy $\mathcal{M} \subseteq$ $\mathbf{App}(\exists,\forall,\wedge,=,\mathsf{f})$. The discussion of the approximation sets in Section 5 shows that the set $\mathcal{M} = \mathbf{sur\text{-}O}_D$ of all surjective functions on D is a candidate for the restriction set. Because of 1.11 this choice guarantees $\langle\ \rangle_{\exists,\forall,\wedge,=,\mathsf{f}} \leq \mathsf{Inv}\ \mathcal{M}\text{-}\mathsf{Pol}$. But to ensure the equality, we have to investigate whether the conditions (1)–(3) of Theorem 5.3 are satisfied. Obviously, the conditions (1) and (2) of 5.3 are true for $\mathcal{H} = \langle\ \rangle_{\exists,\forall,\wedge,=,\mathsf{f}}$ and $\mathcal{M} = \mathbf{sur\text{-}O}_D$. The difficult part is condition (3).

Lemma 6.2. *([8]) Let $\Gamma_0 \subseteq \mathbf{R}_D$ and $\boldsymbol{a}_1,\ldots,\boldsymbol{a}_n \in D^m$. Then*

$$\mu_{\langle\Gamma_0\rangle_{\exists,\forall,\wedge,=,\mathsf{f}}}(\boldsymbol{a}_1,\ldots,\boldsymbol{a}_n) \subseteq \{f(\boldsymbol{a}_1,\ldots,\boldsymbol{a}_n) \mid f \in \mathbf{sur\text{-}O}_D \cap \mathsf{Pol}\ \Gamma_0\}. \qquad \square$$

Consequently, condition (3) of Theorem 5.3 is also satisfied and as a consequence we obtain the following result.

Theorem 6.3. *Let* sPol *denote the operator* $\mathbf{sur\text{-}O}_D\text{-}\mathsf{Pol}$, *i.e.*

$$\mathsf{sPol}\ :\ \mathcal{P}\mathbf{R}_D \to \mathcal{P}\,\mathbf{sur\text{-}O}_D,\ \Gamma_0 \mapsto \mathbf{sur\text{-}O}_D \cap \mathsf{Pol}\ \Gamma_0.$$

Then the pair $\mathsf{Inv}\ :\ \mathcal{P}\,\mathbf{sur\text{-}O}_D \to \mathcal{P}\mathbf{R}_D$ *and* sPol *forms a Galois connection. For all $\Gamma_0 \subseteq \mathbf{R}_D$ holds* $\mathsf{Inv}\ \mathsf{sPol}\ \Gamma_0 = \langle\Gamma_0\rangle_{\exists,\forall,\wedge,=,\mathsf{f}}$. $\square$

Now we obtain an analogue to Theorem 3.2.

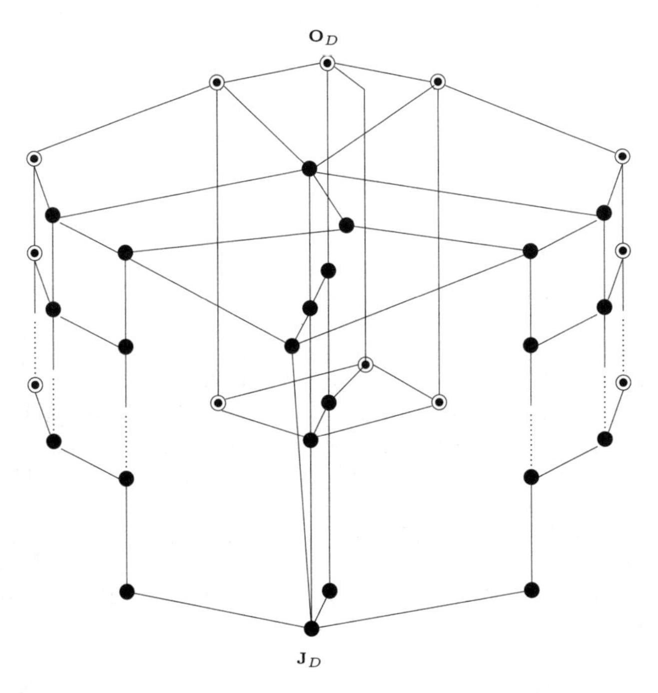

Fig. 4. The lattice of all surjectively generated clones for $|D| = 2$

Theorem 6.4. *([15]) Let $\Gamma_1, \Gamma_2 \subseteq \mathbf{R}_D$ be finite sets of relations. If* $\mathsf{sPol}\,\Gamma_2 \subseteq \mathsf{sPol}\,\Gamma_1$*, then* QCSP($\Gamma_1$) *is polynomially time reducible to* QCSP(Γ_2)*. If* $\mathsf{sPol}\,\Gamma_2 = \mathsf{sPol}\,\Gamma_1$*, then* QCSP($\Gamma_1$) *and* QCSP($\Gamma_2$)*, are polynomially time equivalent.* □

So we have a similar starting point as in the case of the CSP. One disadvantage is the fact that $\mathbf{sur\text{-}O}_D$ is not a clone. Because of 1.8 the Galois closed subsets are just the intersections of clones on D with $\mathbf{sur\text{-}O}_D$, and these subsets are in general not closed under superposition. But here helps the construction of Theorem 1.9. To each Galois closed subset $F \subseteq \mathbf{sur\text{-}O}_D$ we assign $\mathsf{Pol}\;\mathsf{Inv}\;F = \mathsf{clone}\,F$. By 1.9 this assignment is injective, we reobtain F as $\mathbf{sur\text{-}O}_D \cap \mathsf{clone}\,F$. Moreover, we have $\mathsf{Inv}\;\mathsf{clone}\,\mathsf{sPol}\,\Gamma_0 = \mathsf{Inv}\;\mathsf{sPol}\,\Gamma_0$ for all Γ_0.

The clones C with $C = \mathsf{clone}\,F$ for some $F = \mathsf{sPol}\,\mathsf{Inv}\;F$ are exactly the clones with a surjective generating system. We call such clones *surjectively generated* and we denote the set of all these clones with sur-$\mathcal{L}_{\mathbf{O}_D}$. This set is a lattice, isomorphic to $\mathcal{C}^{\mathsf{sPol}\,\mathsf{Inv}}_{\mathbf{sur\text{-}O}_D}$, but for $|D| \geq 3$ it is not a sublattice of the lattice $\mathcal{L}_{\mathbf{O}_D}$ of all clones on D. Now we can use our knowledge about $\mathcal{L}_{\mathbf{O}_D}$ to obtain information on sur-$\mathcal{L}_{\mathbf{O}_D}$. (It turns out that the structure of sur-$\mathcal{L}_{\mathbf{O}_D}$ is not essentially easier than the structure of $\mathcal{L}_{\mathbf{O}_D}$. E.g., sur-$\mathcal{L}_{\mathbf{O}_D}$ is also uncountable for $|D| \geq 3$.)

So we have related the complexity of QCSP–problems with a special lattice of clones. For the complexity theoretic results obtained with this "algebraic approach" we refer to [8, 9].

A case by case investigation of the clones in Post's lattice (Figure 1 on page 47) yields the picture of the lattice of all surjectively generated clones in the Boolean case. Figure 4 on page 65 shows this lattice, black points denote clones, consisting only of surjective functions, half-black points denote the other surjectively generated clones .

References

[1] Blyth, T.S., Janowitz, M.F.: Residuation Theory. Pergamon Press, Oxford (1972)

[2] Bodnaržuk, V.G., Kalužnin, L.A., Kotov, N.N., Romov, B.A.: Galois theory for Post algebras I. Kibernetika 3, 1–10 (1969) (Russian)

[3] Bodnaržuk, V.G., Kalužnin, L.A., Kotov, N.N., Romov, B.A.: Galois theory for Post algebras II. Kibernetika 5, 1–9 (1969) (Russian)

[4] Börner, F.: Operationen auf Relationen. Diss., Universität Leipzig (1989)

[5] Börner, F.: Krasneralgebren. Logos-Verlag (2000)

[6] Börner, F.: Total multifunctions and relations. Contributions to General Algebra 13, 23–36 (2001)

[7] Börner, F., Pöschel, R., Sushchansky, V.: Boolean systems of relations and Galois connections. Acta Sci. Math (Szeged) 68, 535–560 (2002)

[8] Börner, F., Krokhin, A., Bulatov, A., Jeavons, P.: Quantified constraints and surjective polymorphisms. Technical Report PRG-RR-02-11, Comp. Lab., Univ. of Oxford, UK (2002)

[9] Börner, F., Bulatov, A., Jeavons, P., Krokhin, A.: Quantified Constraints: Algorithms and Complexity. In: Baaz, M., Makowsky, J.A. (eds.) Proc. 17th Int. Workshop on Computer Science Logic CSL, pp. 58–70. Springer, Heidelberg (2003)

[10] Bulatov, A., Krokhin, A., Jeavons, P.: The complexity of maximal constraint languages. In: Proc. 33rd ACM Symp. on Theory of Computing, STOC 2001, pp. 667–674 (2001)
[11] Creignou, N., Khanna, S., Sudan, M.: Complexity Classifications of Boolean Constraint Satisfaction Problems. SIAM Monographs on Discrete Math. and Appl. 7 (2001)
[12] Freivald, R.V.: Functional completeness for not everywhere defined functions of the algebra of logic. Diskr. Analiz. 8, 55–68 (1966)
[13] Geiger, D.: Closed systems of functions and predicates. Pacific J. Math. 27, 95–100 (1968)
[14] Jeavons, P., Cohen, D., Gyssens, M.: Closure Properties of Constraints. J. of the ACM 44, 527–548 (1997)
[15] Jeavons, P.: On the algebraic structure of combinatorial problems. Theoretical Computer Science 200, 185–204 (1998)
[16] Krasner, M.: Une généralisation de la notion de corps. J. Math. pure et appl. 17, 367–385 (1938)
[17] Krasner, M.: Généralisation abstraite de la théorie de Galois. Colloq. internat. Centre nat. Rech. Sci., Algèbre et Théorie des Nombres 24, 163–168 (1950)
[18] Lau, D.: Function Algebras on Finite Sets. Springer, Heidelberg (2006)
[19] McKenzie, R.N., McNulty, G.F., Taylor, W.F.: Algebras, Lattices, Varieties, vol. 1. Wadsworth & Brooks/Cole, Monterey (1987)
[20] Pöschel, R., Kaluznin, L.A.: Funktionen- und Relationenalgebren. Verlag der Wissenschaften, Berlin (1979)
[21] Pöschel, R.: Closure properties for relational systems with given endomorphism structure. Beitr. Alg. Geom. 18, 153–166 (1984)
[22] Pöschel, R.: Galois connections for operations and relations. Technical Report MATH-AL-8-2001, TU Dresden (2001)
[23] Post, E.L.: Determination of all closed systems of truth tables. Bull. Amer. Math. Soc. 26, 437 (1920)
[24] Post, E.L.: Introduction to a general theory of elementary propositions. Amer. J. Math. 43, 163–185 (1921)
[25] Post, E.L.: The two-valued iterative system of mathematical logic. Annals Math. Stud. 5 (1941)
[26] Romov, B.A.: The algebras of partial functions and their invariants. Kibernetika 2, 1–11 (1981); Engl. transl.: Cybernetics 17, 157–167 (1981)
[27] Romov, B.A.: Hyperclones on a finite set. In: Mult. Val. Logic 1998, vol. 3, pp. 285–300 (1998)
[28] Romov, B.A.: Partial hyperclones on a finite set. In: Proc. 32nd IEEE Int. Symp. Multiple-Valued Logic, pp. 17–27 (2002)
[29] Rosenberg, I.G.: Galois theory for partial algebras. In: Freese, R., Garcia, O. (eds.) Universal Algebra and Lattice Theory. Lect. Notes in Math., pp. 257–272. Springer, Heidelberg (1883)
[30] Rosenberg, I.G.: Minimal clones I: the five types. Lectures in Universal Algebra. Colloq. Math. Soc. J. Bolyai 43, 405–427 (1983)
[31] Rosenberg, I.G.: Composition of functions on finite sets, completeness and relations, a short survey. In: Rine, D. (ed.) Multiple-valued Logic and Computer Science, 2nd edn., pp. 150–192. North-Holland, Amsterdam (1984)
[32] Rosenberg, I.G.: An algebraic approach to hyperalgebras. In: Proc. 26th ISMVL, pp. 203–207. IEEE, Los Alamitos (1996)
[33] Schaefer, T.: The complexity of satisfyability problems. In: Proc. 10th ACM Symp. on Theory of Computing, STOC 1978, pp. 216–226 (1978)

Recent Results on the Algebraic Approach to the CSP

Andrei A. Bulatov[1] and Matthew A. Valeriote[2]

[1] School of Computing Science
Simon Fraser University
Burnaby, BC, Canada, V5A 1S6
abulatov@cs.sfu.ca
[2] Department of Mathematics & Statistics
McMaster University
Hamilton, Ontario, Canada L8S 4K1
matt@math.mcmaster.ca

Abstract. We describe an algebraic approach to the constraint satisfaction problem (CSP) and present recent results on the CSP that make use of, in an essential way, this algebraic framework.

1 Introduction

This paper presents material from the talks that the authors gave at the Dagstuhl seminar on the Complexity of Constraints held in 2006. The primary goals of the talks were to describe an algebraic approach to the constraint satisfaction problem and to present, within the algebraic context, recent results relating to two of the main motivating conjectures in the field.

During our talks, by necessity, a fair amount of time was occupied in describing basic and advanced universal algebra. In particular, overviews of two approaches to analyzing the local structure of finite algebras were given. The first, known as tame congruence theory, was developed in the 1980s by David Hobby and Ralph McKenzie and has played an important role in the development of universal algebra ever since. The second is a much more recent approach developed by Bulatov specifically to address questions relating to the CSP. For readers who wish to learn more about basic universal algebra we recommend [17] and [36]. For more information on tame congruence, the works [26] or [19] can be consulted. The paper [10] contains details of the theory developed by Bulatov.

2 Constraint Satisfaction and Algebra

2.1 Constraint Satisfaction

We use the homomorphism definition of the CSP. A *vocabulary* τ is a finite set of *relational symbols*; each symbol has an associated *arity*. A (finite) *relational structure* $\mathcal{H}$ with vocabulary τ consists of a finite set H, its *universe*, and, for

N. Creignou et al. (Eds.): Complexity of Constraints, LNCS 5250, pp. 68–92, 2008.

every relational symbol $R \in \tau$ of arity n, an n-ary relation $R^{\mathcal{H}}$ on H, the *interpretation* of R by $\mathcal{H}$. A *homomorphism* of a structure $\mathcal{G}$ to a structure $\mathcal{H}$ with the same vocabulary τ is a mapping $\varphi\colon G \to H$ from the universe of $\mathcal{G}$ to the universe of $\mathcal{H}$ such that for each (n-ary) relational symbol $R \in \tau$ and any tuple $(a_1, \ldots, a_n) \in R^{\mathcal{G}}$ the tuple $(\varphi(a_1), \ldots, \varphi(a_n))$ belongs to $R^{\mathcal{H}}$.

For a finite structure $\mathcal{H}$ the *non-uniform constraint satisfaction problem*, denoted CSP($\mathcal{H}$), is the following combinatorial problem: Given a structure $\mathcal{G}$ of the same vocabulary as $\mathcal{H}$, decide whether or not there is a homomorphism from $\mathcal{G}$ to $\mathcal{H}$. The structure $\mathcal{H}$ is called the *template*, and $\mathcal{G}$ is called the *instance*. For a class $\mathfrak{H}$ of relational structures, in the *uniform constraint satisfaction problem*, denoted CSP($\mathfrak{H}$), the question is: given a structure $\mathcal{H} \in \mathfrak{H}$ and a structure $\mathcal{G}$ over the same vocabulary as $\mathcal{H}$, decide whether there exists a homomorphism from $\mathcal{G}$ to $\mathcal{H}$. Sometimes it is convenient to think of a uniform problem as of the union or collection of non-uniform problems CSP($\mathcal{H}$) for $\mathcal{H} \in \mathfrak{H}$.

*Example 1 (*NAE*,* Linear Equations*, and* H-Colouring*).*

1. Let $\mathcal{H}_{NAE}$ be a relational structure with universe $\{0,1\}$ and one ternary relation $R^{\mathcal{H}_{NAE}} = \{0,1\}^3 \setminus \{(0,0,0),(1,1,1)\}$. It is easy to see that the problem CSP($\mathcal{H}_{NAE}$) is the same as the Not-All-Equal Satisfiability problem, in which, given a set of propositional variables and a set of triples of these variables, the question is whether or not it is possible to assign values to the variables such that the variables from each of the specified triples take both possible values, 0 and 1.
2. Let F be a finite field and Γ the set of all relations over F that can be represented as the set of solutions of a linear equation over F. Let $\mathfrak{H}_{LQ}(F)$ denote the set of all structures with universe F, whose relations are from Γ. Then the uniform problem CSP($\mathfrak{H}_{LQ}(F)$) is equivalent in a certain sense to the problem of solving systems of linear equations over F.[1]
3. Let H be a (directed) graph. In the H-Colouring problem we are asked whether there is a homomorphism from a given graph G to H. So, the H-Colouring problem is just the problem CSP(H).

Two major issues have arisen in the study of the study of the constraint satisfaction problem. The first one is the computational complexity of solving such problems. Although constraint satisfaction problems may belong to and be complete in many complexity classes, see, e.g. [1,32,33], in this paper we concentrate

[1] The size of a CSP instance is defined to be the length of a reasonable encoding of the structures involved, that is the instance in the case of a non-uniform problem, and the source structure and the template in the case of a uniform problem. Usually such an encoding includes a list of elements of the structures and a list of tuples in all relations. In some cases such a general representation is not the most natural. For example, the natural representation of a CSP($\mathfrak{H}_{LQ}(F)$) instance is a list of equations defining relations of the template. Although no example is known, different representation may affect the complexity of uniform problems. However, for the sake of generality throughout the paper we use the explicit representation of relational structures. The choice of representation does not affect the complexity of non-uniform problems.

on problems solvable in polynomial time (such problems are often said to be *tractable*). The remaining problems are called *intractable*. (Throughout the paper we assume P≠NP.) All the intractable problems known so far turn out to be NP-complete. This prompted Feder and Vardi [24] to suggest the *Dichotomy Conjecture*: Every non-uniform CSP is either tractable or NP-complete.

The second issue is the descriptive complexity of non-uniform problems. Let $\mathcal{H}$ be a relational structure. The class of structures homomorphic to $\mathcal{H}$ is often denoted by CSP($\mathcal{H}$) (this does not cause any confusion, because CSP($\mathcal{H}$) is the class of yes-instances of the corresponding constraint satisfaction problem, and therefore the *language* associated to this problem). In many cases the class CSP($\mathcal{H}$) can be characterized as the class of all structures satisfying some formula in a certain logic. The goal is to describe structures $\mathcal{H}$ such that CSP($\mathcal{H}$) is expressible in this logic. We concentrate on the logic corresponding to Datalog. For definitions of Datalog, Datalog expressibility, related properties of structures and problems, as well as results on other important logical languages the reader is referred to [16] from the same volume.

Example 2 (continued).

1. NAE is NP-complete, [38].
2. LINEAR EQUATIONS is not expressible in Datalog, [24].
3. H-COLORING is tractable if and only if H is a bipartite graph. In this case it is expressible in Datalog. Otherwise it is NP-complete, [25].

2.2 Polymorphisms and Algebras

In this section we provide a brief overview of the algebraic approach to the constraint satisfaction problem.

At the core of this approach is the concept of a polymorphism. Let R be a relation on a set A. An (n-ary) operation f on the same set is said to be a *polymorphism* of R if for any tuples $\mathbf{a}_1, \ldots, \mathbf{a}_n \in R$ the tuple $f(\mathbf{a}_1, \ldots, \mathbf{a}_n)$ obtained by applying f component-wise also belongs to R. The relation R is called an invariant of f. An operation f is a polymorphism of a relational structure $\mathcal{H}$ if it is a polymorphism of each relation of the structure. The set of all polymorphisms of $\mathcal{H}$ is denoted by $\mathsf{Pol}(\mathcal{H})$. For a collection C of operations $\mathsf{Inv}(C)$ denotes the set of invariants of all operations from C.

Example 3 ([39]). Let R be the solution space of a system of linear equations over a finite field F. Then the operation $m(x, y, z) = x - y + z$ is a polymorphism of R. Indeed, let $A \cdot \mathbf{x} = \mathbf{b}$ be the system defining R, and $\mathbf{x}, \mathbf{z}, \mathbf{y} \in R$. Then

$$A \cdot m(\mathbf{x}, \mathbf{z}, \mathbf{y}) = A \cdot (\mathbf{x} - \mathbf{z} + \mathbf{y}) = A \cdot \mathbf{x} - A \cdot \mathbf{z} + A \cdot \mathbf{y} = \mathbf{b} - \mathbf{b} + \mathbf{b} = \mathbf{b}.$$

In fact, the converse can also be shown: if R is invariant under m then it is the solution space of a certain system of linear equations.

The following theorem relates polymorphisms, complexity, and expressibility in Datalog

Theorem 1 ([28,30,34]). *Let $\mathcal{H}_1$ and $\mathcal{H}_2$ be two structures with a common universe.*

1. *If* $\mathsf{Pol}(\mathcal{H}_1) \subseteq \mathsf{Pol}(\mathcal{H}_2)$ *then* $\mathrm{CSP}(\mathcal{H}_2)$ *is log-space reducible to* $\mathrm{CSP}(\mathcal{H}_1)$.
2. *If* $\mathsf{Pol}(\mathcal{H}_1) \subseteq \mathsf{Pol}(\mathcal{H}_2)$ *and* $\mathrm{CSP}(\mathcal{H}_1)$ *is expressible in Datalog, then* $\mathrm{CSP}(\mathcal{H}_2)$ *is expressible in Datalog.*

An *algebra* is a pair $\mathbb{A} = (A; F)$ consisting of a set A, the *universe* of $\mathbb{A}$, and a set F of operations on A, the *basic operations* of $\mathbb{A}$. Operations that can be obtained from the basic operations of $\mathbb{A}$ and the *projection* operations on A, that is operations of the form $f(x_1, \dots, x_n) = x_i$, by means of compositions are called *term* operations of $\mathbb{A}$. $\mathsf{Term}(\mathbb{A})$ denotes the set of all term operations of $\mathbb{A}$. Operations that can be obtained from term operations by substituting constants are called *polynomial operations* (or just *polynomials*) of $\mathbb{A}$.

Any relational structure $\mathcal{H}$ and therefore any non-uniform constraint satisfaction problem can be associated with an algebra $\mathsf{Alg}(\mathcal{H}) = (H; \mathsf{Pol}(\mathcal{H}))$ where H is the universe of $\mathcal{H}$. Conversely, any algebra $\mathbb{A} = (A; F)$ corresponds to a class of structures $\mathsf{Str}(\mathbb{A})$ that includes all the structures $\mathcal{H}$ with universe A, having a finite vocabulary, and such that $\mathsf{Term}(\mathbb{A}) \subseteq \mathsf{Pol}(\mathcal{H})$. Therefore every algebra gives rise to a uniform constraint satisfaction problem $\mathrm{CSP}(\mathsf{Str}(\mathbb{A}))$, which we will denote by $\mathrm{CSP}(\mathbb{A})$.

An algebra $\mathbb{A}$ is called *tractable* if $\mathrm{CSP}(\mathcal{H})$ is tractable for each $\mathcal{H} \in \mathsf{Str}(\mathbb{A})$ and is called *NP-complete* if $\mathrm{CSP}(\mathcal{H})$ for some $\mathcal{H} \in \mathsf{Str}(\mathbb{A})$ is. Theorem 1 implies that if $\mathrm{CSP}(\mathcal{H})$ is tractable then the algebra $\mathsf{Alg}(\mathcal{H})$ is tractable. We make two observations. First, if an algebra $\mathbb{A}$ is not tractable, it does not mean that $\mathrm{CSP}(\mathcal{H})$ is intractable for all $\mathcal{H} \in \mathsf{Str}(\mathbb{A})$; this class always contains poor structures whose associated class of constraint satisfaction problems is very easy. Second, if $\mathbb{A}$ is tractable it does not necessarily mean that the uniform problem $\mathrm{CSP}(\mathbb{A})$ is tractable. Although no example is known, it may be the case that the time complexity of problems $\mathrm{CSP}(\mathcal{H})$, $\mathcal{H} \in \mathsf{Str}(\mathbb{A})$, does not have a uniform polynomial bound, even though the complexity of each problem is polynomially bounded. To distinguish these two potential situations we sometimes call tractable algebras *locally tractable* and algebras for which $\mathrm{CSP}(\mathbb{A})$ is tractable, *globally tractable.* In other words, $\mathbb{A}$ is locally tractable if every non-uniform problem from $\mathrm{CSP}(\mathbb{A})$ is solvable in polynomial time.

The relational width of an algebra $\mathbb{A}$ is a parameter related to certain properties of Datalog programs or propagation algorithms that solve the problems $\mathrm{CSP}(\mathcal{H})$ for $\mathcal{H} \in \mathsf{Str}(\mathbb{A})$. The algebra $\mathbb{A}$ is said to be of *bounded width* if $\mathrm{CSP}(\mathcal{H})$ is expressible in Datalog for any structure $\mathcal{H} \in \mathsf{Str}(\mathbb{A})$. For complete definitions and discussion of this concept see [16] in the same volume.

The tractability and relational width of an algebra usually follows from the presence of a certain polymorphism of a structure (or a term operation of an algebra).

Example 4 ([5,13,20,22,28,30]). If one of the following operations is a term operation of an algebra $\mathbb{A}$ [a polymorphism of a relational structure $\mathcal{H}$] then $\mathrm{CSP}(\mathbb{A})$ [$\mathrm{CSP}(\mathcal{H})$] is tractable:

- a *semilattice* operation, that is a binary operation f satisfying the equations: (a) $f(x,x) \approx x$ (idempotency); (b) $f(x,y) \approx f(y,x)$ (commutativity); (c) $f(f(x,y),z) \approx f(x,f(y,z))$ (associativity);
- a *2-semilattice* operation, that is a binary operation f satisfying the equations $f(x,x) \approx x$, $f(x,y) \approx f(y,x)$, and $f(x,f(x,y)) \approx f(x,y)$;
- a *near-unanimity* (*NU*) operation, that is an operation f satisfying the equations $f(x,\dots,x,y) \approx f(x,\dots,x,y,x) \approx \dots \approx f(y,x,\dots,x) \approx x$.
- a *majority* operation, that is a ternary operation g satisfying the equations $g(x,x,y) \approx g(x,y,x) \approx g(y,x,x) \approx x$ (thus a majority operation is a ternary near-unanimity operation).
- a *Mal'tsev* operation, that is a ternary operation h satisfying the equations $h(x,x,y) \approx h(y,x,x) \approx y$.
- a *generalized majority-minority* (*GMM*) operation, that is an operation f such that for any $a,b \in A$ one of the following two conditions holds:
$f(x,\dots,x,y) = f(x,\dots,x,y,x) = \dots = f(y,x,\dots,x) = x$, for $x,y \in \{a,b\}$;
or
$f(x,\dots,x,y) = f(y,x,\dots,x) = x$, for $x,y \in \{a,b\}$.

Example 5. If one of the following operations is a polymorphism of a relational structure $\mathcal{H}$, then CSP($\mathcal{H}$) is expressible in Datalog:

- a semilattice operation;
- a 2-semilattice operation;
- a near-unanimity operation;
- a majority operation.

On the other hand, the intractability of a relational structure (or an algebra) seems to imply that it has rather uninteresting polymorphisms (term operations, respectively). An operation f on a set A is said to be an *essentially unary surjective operation* if $f(x_1,\dots,x_n) = g(x_i)$ for some i and some surjective map $g(x)$ of A.

Example 6 (continued).

1. An operation f is a polymorphism of $\mathcal{H}_{NAE}$ if and only if f is an essentially unary surjective operation, [30,31].
2. An operation f is a polymorphism of all relations representable by linear equations over a field F if and only if $f = \alpha_1 x_1 + \dots + \alpha_n x_n$ where $\alpha_1,\dots,\alpha_n \in F$ are such that $\alpha_1 + \dots + \alpha_n = 1$, [39].
3. If $\mathcal{H} = K_\ell$, a complete graph on $\ell > 2$ vertices, then an operation f is a polymorphism of $\mathcal{H}$ if and only if f is an essentially unary surjective operation. If $\mathcal{H} = K_2$ then $\mathcal{H}$ has a majority polymorphism.

The examples above and Theorem 1 provide necessary conditions for tractability and expressibility in Datalog.

Corollary 1

1. *If every polymorphism of a structure $\mathcal{H}$ [every term operation of an algebra $\mathbb{A}$] is an essentially unary surjective operation then* CSP($\mathcal{H}$) *[*CSP($\mathbb{A}$)*, respectively] is NP-complete.*
2. *If there is a field F such that every polymorphism of a structure $\mathcal{H}$ [every term operation of an algebra $\mathbb{A}$] is of the form $f = \alpha_1 x_1 + \ldots + \alpha_n x_n$, where $\alpha_1, \ldots, \alpha_n \in F$ are such that $\alpha_1 + \ldots + \alpha_n = 1$, then* CSP($\mathcal{H}$) *[*CSP($\mathbb{A}$)*, respectively] is not expressible in Datalog [is not of bounded relational width].*

If every term operation of a finite algebra $\mathbb{A}$ is essentially unary surjective then $\mathbb{A}$ is said to be a *G-set.* If there is a module M over a ring R such that every term operation of $\mathbb{A}$ can be represented as $\alpha_1 x_1 + \ldots + \alpha_n x_n$ for $\alpha_1, \ldots, \alpha_n \in R$ and $\alpha_1 + \ldots + \alpha_n = 1$, then $\mathbb{A}$ is called an *idempotent reduct of a module.*

Example 4 allows one to classify 2-element algebras in terms of complexity.

Proposition 1 (Schaefer's Dichotomy Theorem, [38]). *For any 2-element algebra $\mathbb{A}$, the problem* CSP($\mathbb{A}$) *is (globally) tractable if and only if* $\mathsf{Term}(\mathbb{A})$ *contains one of the following:*

- *the constant 0 or constant 1 operation;*
- *the conjunction or disjunction operations (which are semilattice);*
- *the majority operation $(x \vee y) \wedge (y \vee z) \wedge (z \vee x)$;*
- *the Mal'tsev operation $x - y + z \pmod 2$.*

In all other cases CSP($\mathbb{A}$) *is NP-complete.*

2.3 Varieties

For the purposes of settling the Dichotomy Conjecture and related questions, the class of algebras to be considered can be significantly reduced. An algebra $\mathbb{A}$ is called *surjective* if every one of its unary term operations is surjective. One way to transform an algebra into a surjective algebra is as follows: Let g be a unary term operation of $\mathbb{A} = (A; F)$ with a minimal range. Then $g(\mathbb{A})$ denotes the algebra $(g(A); F_g)$ where $F_g = \{gf|_{g(A)} \mid f \in \mathsf{Term}(\mathbb{A})\}$. It is not difficult to see that this algebra is surjective. The algebra $\mathbb{A}$ is called *idempotent* if every one of its term operations f satisfies the equation $f(x, \ldots, x) \approx x$. The *full idempotent reduct* of $\mathbb{A}$ is the algebra $\mathsf{Id}(\mathbb{A}) = (A; F_{id})$ where F_{id} is the set of all idempotent operations from $\mathsf{Term}(\mathbb{A})$.

Theorem 2 ([14]). *Let $\mathbb{A}$ be an algebra.*

1. *If g is a unary term operation of $\mathbb{A}$ with minimal range then $\mathbb{A}$ is tractable if and only if $g(\mathbb{A})$ is tractable. The algebra $\mathbb{A}$ is NP-complete if and only if $g(\mathbb{A})$ is NP-complete.*
2. *If $\mathbb{A}$ is surjective then $\mathbb{A}$ is tractable if and only if $\mathsf{Id}(\mathbb{A})$ is tractable. The algebra $\mathbb{A}$ is NP-complete if and only if $\mathsf{Id}(\mathbb{A})$ is NP-complete.*

The main idea of the algebraic approach is to use some properties of an algebra in order to determine the complexity of the associated constraint satisfaction problem. To identify these properties, some connections between the complexity of an algebra and the complexity of those algebras that can be obtained from it by some standard algebraic constructions will be very helpful.

- Let $\mathbb{A} = (A; F)$ be an algebra. The *k-th direct power* of $\mathbb{A}$ is the algebra $\mathbb{A}^k = (A^k; F)$ where we treat each (n-ary) operation $f \in F$ as acting on A^k component-wise.
- Let $\mathbb{A} = (A; F)$ be an algebra, and let B be a subset of A such that, for any (n-ary) $f \in F$, and for any $b_1, \ldots, b_n \in B$, we have $f(b_1, \ldots, b_n) \in B$. Such a subset is called a *subuniverse* of $\mathbb{A}$. When B is non-empty, the algebra $\mathbb{B} = (B; F|_B)$, where $F|_B$ consists of restrictions of operations $f \in F$ to B, is called a *subalgebra* of $\mathbb{A}$.
- Let $\mathbb{A}_1 = (A_1; F_1)$ and $\mathbb{A}_2 = (A_2; F_2)$ such that $F_1 = \{f_i^1 \mid i \in I\}$, $F_2 = \{f_i^2 \mid i \in I\}$, and f_i^1, f_i^2 are of the same arity, for some set I and each $i \in I$. A mapping $\varphi : A_1 \to A_2$ is called a *homomorphism* from $\mathbb{A}_1$ to $\mathbb{A}_2$ if $\varphi f_i^1(a_1, \ldots, a_{n_i}) = f_i^2(\varphi(a_1), \ldots, \varphi(a_{n_i}))$ holds for all $i \in I$ and all $a_1, \ldots, a_{n_i} \in A_1$. If the mapping φ is onto then $\mathbb{A}_2$ is said to be a *homomorphic image* of $\mathbb{A}_1$.

By a classic result of Birkhoff (see Theorem 11.9 from [17]), properties of algebras that are preserved under the taking of subalgebras, homomorphic images, and direct products (a natural generalization of the direct power construction) are precisely those that can be defined by equations. We note that except for the last one, all of the properties of the operations listed in Example 4 are defined by equations. Equationally defined classes of algebras, also known as varieties of algebras, are fundamental objects of study in universal algebra [26,36]. The following theorems thus provide an important link between the constraint satisfaction problem and universal algebra.

Theorem 3 ([14,8]). *Let $\mathbb{A}$ be a finite algebra. Then*

1. *if $\mathbb{A}$ is tractable then so is every subalgebra, homomorphic image, and direct power of $\mathbb{A}$.*
2. *if $\mathbb{A}$ has an NP-complete subalgebra, homomorphic image, or direct power, then $\mathbb{A}$ is NP-complete itself.*

Theorem 4 ([34]). *Let $\mathbb{A}$ be a finite algebra. If $\mathbb{A}$ has bounded width then every subalgebra, homomorphic image, and direct power of $\mathbb{A}$ has bounded width.*

Using Birkhoff's Theorem, the variety that an algebra $\mathbb{A}$ determines, denoted by $\mathsf{var}(\mathbb{A})$, can be defined either as the class of all algebras that satisfy the same equations that $\mathbb{A}$ does, or as the class of all algebras that arise as homomorphic images of subalgebras of direct powers of $\mathbb{A}$.

Corollary 2

1. *If* $\mathbb{A}$ *is tractable then so is every finite algebra from* $\mathsf{var}(\mathbb{A})$*. If* $\mathsf{var}(\mathbb{A})$ *contains an NP-complete algebra then* $\mathbb{A}$ *is NP-complete.*
2. *If* $\mathbb{A}$ *has bounded width then every finite algebra from* $\mathsf{var}(\mathbb{A})$ *has bounded width. If* $\mathsf{var}(\mathbb{A})$ *contains an algebra of unbounded width then* $\mathbb{A}$ *does not have bounded width.*
3. *Tractability, NP-completeness, and bounded width are properties of an algebra that depend only on the equations satisfied by the algebra.*

Using Corollary 2 we can strengthen Corollary 1 as follows.

Theorem 5 ([14,34]). *Let* $\mathbb{A}$ *be an algebra*

1. *If* $\mathsf{var}(\mathbb{A})$ *contains a G-set then* $\mathbb{A}$ *is NP-complete.*
2. *If* $\mathsf{var}(\mathbb{A})$ *contains a reduct of a module then* $\mathbb{A}$ *does not have bounded width.*

To date no NP-complete or unbounded width algebra is known that does not satisfy the corresponding condition of Theorem 5. It is widely believed that these necessary conditions are also sufficient, at least for idempotent algebras.

Conjecture 1 (complexity dichotomy conjecture). *An idempotent algebra* $\mathbb{A}$ *is tractable if and only if* $\mathsf{var}(\mathbb{A})$ *does not contain a G-set. Otherwise it is NP-complete.*

Conjecture 2 (bounded width conjecture). *An idempotent algebra* $\mathbb{A}$ *has bounded width if and only if* $\mathsf{var}(\mathbb{A})$ *does not contain a reduct of a module.*

Conjectures 1 and 2 have been proved in a number of particular cases: 2-element algebras ([38]), 3-element algebras ([12]), semigroups ([9,23]). The following example shows that the undirected graphs dichotomy theorem by Hell and Nešetřil [25] also fits Conjecture 1.

Example 7 ([11]). Let H be an undirected graph, $\mathbb{A} = \mathsf{Alg}(H)$, and g a unary term operation of $\mathbb{A}$ with a minimal range. Then H is non-bipartite if and only if $\mathsf{var}(g(\mathbb{A}))$ contains a G-set. Otherwise $g(H)$ is K_2 and $g(\mathbb{A})$ has a majority term operation.

3 Alternate Versions of the Conjectures

The goal of this section is to present new formulations of Conjectures 1 and 2 that have emerged over the past several years. Central to our first formulation is the notion of a *congruence* of an algebra $\mathbb{A}$. A congruence θ of $\mathbb{A}$ is an equivalence relation on A that is invariant under all basic (and therefore term) operations of $\mathbb{A}$. Every algebra $\mathbb{A}$ has two distinguished congruences 0_A and 1_A corresponding to the smallest and largest equivalence relations on the set A. For θ a congruence of $\mathbb{A} = (A; F)$ and $a \in A$ by $a/_\theta$ we denote the θ-class containing a; and denote $\{a/_\theta \mid a \in A\}$, the set of all θ-classes, by $A/_\theta$. The *quotient algebra* $\mathbb{A}/_\theta$ is the algebra with universe $A/_\theta$ and whose basic operations are $\{f/_\theta : f \in F\}$, where for $f \in F$,

$$f/_\theta(a_1/_\theta, \dots, a_n/_\theta) = (f(a_1, \dots, a_n))/_\theta.$$

It is elementary that the mapping $\varphi\colon \mathbb{A} \to \mathbb{A}/_\theta$ that maps an element $a \in A$ to $a/_\theta$ is a surjective homomorphism and so it follows that $\mathbb{A}/_\theta$ is a homomorphic image of $\mathbb{A}$.

3.1 Tame Congruence Theory

In the early 1980's Hobby and McKenzie developed a theory of the local structure of finite algebras called tame congruence theory [26]. At the heart of the theory is a notion of a neighbourhood of a finite algebra, relativized to certain congruences of the algebra. The local structure of a finite algebra that emerges from their theory is surprisingly well-behaved and has been used to prove many striking theorems in universal algebra.

Definition 1. *Let $\mathbb{A}$ be a finite algebra and α a minimal congruence of $\mathbb{A}$ (i.e., $0_A < \alpha$ and if β is a congruence of $\mathbb{A}$ with $0_A < \beta \leq \alpha$ then $\beta = \alpha$.)*

- *An α-minimal set of $\mathbb{A}$ is a subset U of A such that*
  - *$U = p(A)$ for some unary polynomial $p(x)$ of $\mathbb{A}$ that is not constant on at least one α-class, and*
  - *with respect to containment, U is minimal with this property.*
- *An α-neighbourhood (or α-trace) of A is a subset N of A such that*
  - *$N = U \cap (a/_\alpha)$ for some α-minimal set U and α-class $a/_\alpha$, and*
  - *$|N| > 1$.*

It follows from the definition that a given α-minimal set U contains within it at least one (and possibly several) α-neighbourhoods. The union of all of the α-neighbourhoods in U is called the *body* of U, while the remaining elements of U form the *tail* of U. What is surprising is that the structure that $\mathbb{A}$ induces on any one of its α-neighbourhoods is quite uniform and is restricted to one of five possible types. What is meant by induced structure is given in the next definition.

Definition 2. *Let $\mathbb{A}$ be an algebra and $U \subseteq A$. The algebra induced by $\mathbb{A}$ on U is the algebra with universe U whose basic operations consist of the restriction to U of all polynomials of $\mathbb{A}$ under which U is closed. We denote this induced algebra by $\mathbb{A}|_U$.*

Note the difference between this notion and the more familiar one of subuniverse (recall that a subuniverse of an algebra $\mathbb{A}$ is a subset of A that is closed under all term operations of $\mathbb{A}$). In the theory developed by Hobby and McKenzie, the polynomials of an algebra play a central role and in fact, two finite polynomially equivalent algebras (i.e., two algebras over the same universe whose sets of polynomials coincide) are, for the most part, indistinguishable using tame congruence theory.

Theorem 6. *Let $\mathbb{A}$ be a finite algebra and α a minimal congruence of $\mathbb{A}$.*

- *If U and V are α-minimal sets then $\mathbb{A}|_U$ and $\mathbb{A}|_V$ are isomorphic and in fact there is a polynomial $p(x)$ of $\mathbb{A}$ that maps U bijectively on to V.*
- *If N and M are α-neighbourhoods then $\mathbb{A}|_N$ and $\mathbb{A}|_M$ are isomorphic via the restriction of some polynomial of $\mathbb{A}$.*
- *If N is an α-neighbourhood then $\mathbb{A}|_N$ is polynomially equivalent to one of:*
  1. *A unary algebra whose basic operations are all permutations (unary type);*
  2. *A one-dimensional vector space over some finite field (affine type);*
  3. *A 2-element boolean algebra (boolean type);*
  4. *A 2-element lattice (lattice type);*
  5. *A 2-element semilattice (semilattice type).*

Much more can be said about the α-neighbourhoods and minimal sets of an algebra but for now we point out that the previous theorem allows us to assign a type to each minimal congruence α of an algebra according to the behaviour of the α-neighbourhoods. For example, a minimal congruence whose α-neighbourhoods are all polynomially equivalent to a vector-space is said to have affine type (or to have type 2).

In Figure 1 two α-minimal sets of an algebra $\mathbb{A}$, U and V, of a minimal congruence α are pictured, along with two α-neighbourhoods, N and M, contained in them. The dashed lines delineate the α-blocks of the algebra.

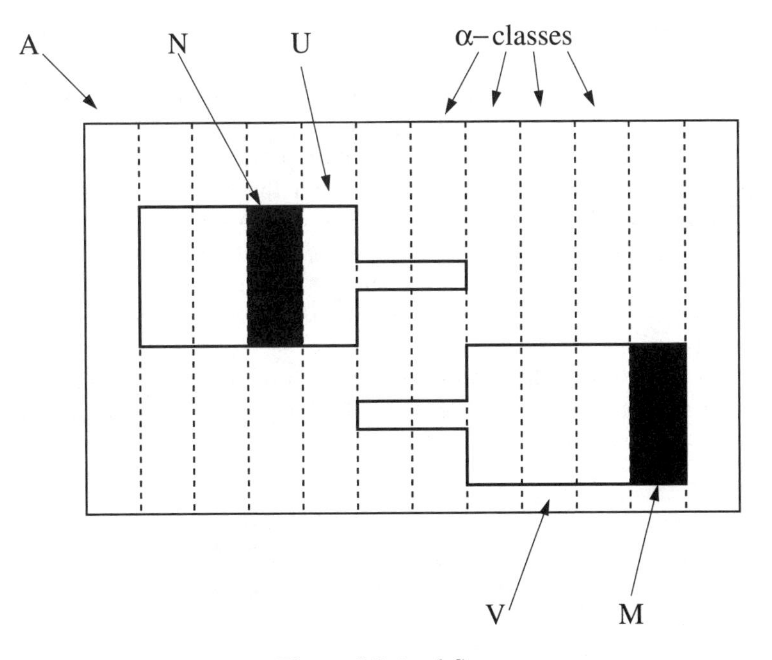

Fig. 1. Minimal Sets

Taking this idea one step further, given a pair of congruences (α, β) of $\mathbb{A}$ with β covering α (i.e., $\alpha < \beta$ and there are no congruences of $\mathbb{A}$ strictly between the two), one can form the quotient algebra $\mathbb{A}/_{\alpha}$ and then consider the congruence $\beta/_{\alpha} = \{(a/_{\alpha}, b/_{\alpha}) : (a,b) \in \beta\}$. Since β covers α in the congruence lattice of $\mathbb{A}$ then $\beta/_{\alpha}$ is a minimal congruence of $\mathbb{A}/_{\alpha}$ and so can be assigned one of the five types. In this way we can assign to each covering pair of congruences of $\mathbb{A}$ a type and so end up with a labelled congruence lattice for $\mathbb{A}$.

For modestly sized algebras, one can, without too much effort, compute their labelled congruence lattices. Since in general, the size of the congruence lattice of a finite algebra can be much larger than the algebra, the task of computing the labelled congruence lattice of an algebra is by no means tractable. If one is just interested in determining the type of a given covering pair of congruences or in the set of labels that appear in the labelled congruence lattice of an algebra, polynomial time algorithms exist (see [3]).

Example 8. Consider the algebra $\mathbb{A}$ with universe $\{0,1,2,3\}$ having a single binary basic operation $x \cdot y$ defined by:

$\cdot$	0	1	2	3
0	0	0	0	3
1	0	1	0	1
2	0	0	2	3
3	3	1	3	3

Besides the two congruence 0_A and 1_A, $\mathbb{A}$ only has two other (minimal) congruences, α and β, pictured in Figure 2 as partitions (using the dotted lines) of the universe of $\mathbb{A}$.

We claim that the type of α is boolean and the type of β is semilattice. To see this, consider the polynomials $p(x) = x \cdot 1$ and $q(x) = x \cdot 2$. The range of p is $\{0,1\}$ and so $N = \{0,1\}$ is both an α-minimal set and an α-neighbourhood (since p is non constant on some α-class and has minimal range subject to this property). On the other hand, the range of q is $\{0,2,3\}$ and so is either a β-minimal set or properly contains one since q is not constant on the only non-trivial β-class. By analyzing the set of unary polynomials of $\mathbb{A}$ it can be seen that in fact $V = \{0,2,3\}$ is indeed a β-minimal set and hence that $M = \{0,2\}$ is a β-neighbourhood.

Now that α and β-neighbourhoods have been identified, we need only determine the types of the algebras that $\mathbb{A}$ induces on each of them to determine the

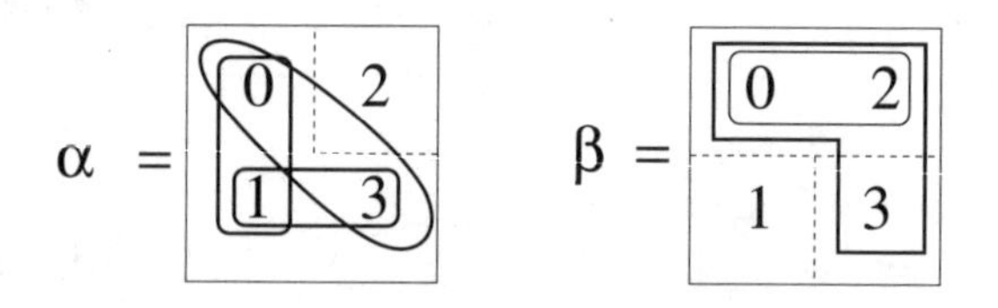

Fig. 2. The Congruences α and β, with their minimal sets

types of α and β. We see that the restriction of $x \cdot y$ to N provides a semilattice operation on N and so the type of α cannot be unary or affine since algebras of these types do not support a semilattice polynomial. Since all boolean operations can be obtained by composition from a boolean semilattice operation and complementation, it suffices to produce a unary polynomial of $\mathbb{A}$ that maps 0 to 1 and 1 to 0 in order to establish that the type of α is boolean. It can be checked that the polynomial $(((x \cdot 3) \cdot 2) \cdot 1)$ fits the bill. We leave the details of the calculation of the type of β to the reader and conclude the presentation of this example by claiming that the types of the covering pairs $(\beta, 1_A)$ and $(\alpha, 1_A)$ are boolean and semilattice, respectively.

While the type-labelled congruence lattice of a finite algebra carries much information about the algebra, it turns out that just knowing the set of labels that appear in the labelled congruence lattice of a finite algebra or the variety that it generates is useful.

Definition 3

1. *The typeset of a finite algebra* $\mathbb{A}$, *denoted* $\text{typ}\{\mathbb{A}\}$, *is the set of labels that appear in its labelled congruence lattice, and so is a subset of* {*unary, affine, boolean, lattice, semi-lattice*}.
2. *The typeset of a class of algebras* $\mathcal{K}$ *is the union of the typesets of all of its finite members and is denoted by* $\text{typ}\{\mathcal{K}\}$.
3. *We say that a finite algebra or a class of algebras omits a particular type if that type does not appear in its typeset.*

The following result, found in [8] provides a connection with Conjecture 1, the Complexity Dichotomy Conjecture.

Theorem 7. *Let* $\mathbb{A}$ *be a finite idempotent algebra and* $\mathcal{V}$ *the variety generated by* $\mathbb{A}$. *Then* $\mathcal{V}$ *omits the unary type if and only if* $\mathsf{var}(\mathbb{A})$ *does not contain a* G*-set. In fact, this condition holds if and only if there is no algebra in* $\mathsf{HS}(\mathbb{A})$ *that is term equivalent to a set (i.e., whose basic operations are just projections).*

This theorem allows us to restate the Complexity Dichotomy Conjecture in terms of types:

Conjecture 1 (the complexity dichotomy conjecture, version 2). *A finite idempotent algebra* $\mathbb{A}$ *is tractable if and only if the variety generated by* $\mathbb{A}$ *omits the unary type (or equivalently, that every subalgebra of* $\mathbb{A}$ *omits the unary type).*

Something similar occurs when considering Conjecture 2, the Bounded Width Conjecture, namely we can express it in terms of omitting tame congruence theoretic types.

Theorem 8 ([40]). *Let* $\mathbb{A}$ *be a finite idempotent algebra and* $\mathcal{V}$ *the variety generated by* $\mathbb{A}$. *Then* $\mathcal{V}$ *omits the unary and affine types if and only if* $\mathsf{var}(\mathbb{A})$ *does* *not contain an algebra that is term equivalent to a reduct of a module over some finite ring. In fact, this condition holds if and only if there is no algebra in* $\mathsf{HS}(\mathbb{A})$ *that is term equivalent to a set or to the full idempotent reduct of a module over some finite ring.*

In the language of tame congruence theory, the Bounded Width Conjecture becomes:

Conjecture 2 (the bounded width conjecture, version 2). *A finite idempotent algebra* $\mathbb{A}$ *has bounded width if and only if the variety generated by* $\mathbb{A}$ *omits the unary and affine types (or equivalently, that every subalgebra of* $\mathbb{A}$ *omits these types).*

3.2 Weak Near-Unanimity Operations

Recall that a near-unanimity operation on a set A is a function $t(x_1, \dots, x_n)$, for $n > 1$, that satisfies the equations

$$t(y, x, x, \dots, x) \approx t(x, y, x, \dots, x) \approx \cdots \approx t(x, x, \dots, x, y) \approx x$$

From [29] we know that if a relational structure $\mathcal{H}$ has a near-unanimity polymorphism then $\mathrm{CSP}(\mathcal{H})$ is tractable. The following variation of this notion was developed by E. Kiss and Valeriote while investigating the Bounded Width Conjecture.

Definition 4. *An operation* $t(x_1, \dots, x_n)$*, for* $n > 1$*, on a set* A *is a weak near-unanimity operation if it is idempotent and satisfies the equations*

$$t(y, x, x, \dots, x) \approx t(x, y, x, \dots, x) \approx \cdots \approx t(x, x, \dots, x, y)$$

Clearly any near-unanimity operation is also a weak near-unanimity operation but there are algebras that have term operations of the latter sort but not of the former. For example, for any positive integer n, the term operation $x_1 + x_2 + \cdots + x_{n+1}$ of the group of integers modulo n is a weak near-unanimity operation. It is not difficult to show that this group fails to have a near-unanimity term operation in any number of variables. We leave it as an exercise to show that the operation $x \cdot (y \cdot z)$ on our 4-element example is a weak near-unanimity operation (and that this algebra does not have a near unanimity term operation).

While it is not too difficult to show that if a finite algebra has a weak near-unanimity term operation then the variety that it generates must omit the unary type, the converse is much more difficult to show. A recent result of Maroti and McKenzie establishes this, along with a characterization of finite algebras that generate varieties that omit both the unary and affine types.

Theorem 9 ([37]). *Let* $\mathbb{A}$ *be a finite algebra and* $\mathcal{V}$ *the variety that it generates.*

1. $\mathcal{V}$ *omits the unary type if and only if* $\mathbb{A}$ *has a weak near-unanimity term operation.*

2. *$\mathcal{V}$ omits the unary and affine types if and only if there is some $N > 0$ such that for all $k \geq N$, $\mathbb{A}$ has a weak near unanimity term of arity k.*

This surprising result allows us to provide restatements of the conjectures.

Conjecture 1 (the complexity dichotomy conjecture, version 3). *A finite idempotent algebra $\mathbb{A}$ is tractable if and only if $\mathbb{A}$ has a weak near-unanimity term.*

Conjecture 2 (the bounded width conjecture, version 3). *A finite idempotent algebra $\mathbb{A}$ has bounded width if and only if for all but finitely many $k > 0$, $\mathbb{A}$ has a k-ary weak near unanimity term.*

4 Tractability Via Few Subpowers

In this section we discuss a thread of tractability results that culminates in a theorem that unifies them in terms of a notion of a finite algebra having few subpowers.

Definition 5 ([2]). *A finite algebra $\mathbb{A}$ is said to have few subpowers if there is some polynomial $p(n)$ such that for each $n > 0$,*

$$s_{\mathbb{A}}(n) = \log_2 |\{B : B \text{ is a subuniverse of } \mathbb{A}^n\}| \leq p(n).$$

It is not difficult to see that for any finite algebra $\mathbb{A}$ of size m, the function $s_{\mathbb{A}}(n)$ is bounded above by m^n. In general $s_{\mathbb{A}}(n)$ will grow exponentially and so the few subpowers condition imposes certain restrictions on the algebra $\mathbb{A}$. One consequence of a finite algebra $\mathbb{A}$ having few subpowers is the existence of a polynomial $g(n)$ such that for any $n > 0$, every subalgebra of $\mathbb{A}^n$ has a generating set of size bounded above by $g(n)$. In fact this "small generating set" property is equivalent to having few subpowers. Before characterizing such algebras, we present some examples.

Using a theorem of Baker and Pixley from [4] it follows that if $\mathbb{A}$ is a finite algebra that has a k-ary near unanimity term operation (see Example 4) then the function $s_{\mathbb{A}}(n)$ is bounded above by a polynomial of degree $k-1$ and so such algebras have few subpowers. An early tractability result of Jeavons, Cohen and Cooper [29] establishes that algebras having near unanimity terms are tractable, and it is no coincidence that this tractability result can be proved using the Baker-Pixley theorem.

In [24], Feder and Vardi prove that if a relational structure $\mathcal{H}$ has a polymorphism of the form $x \cdot y^{-1} \cdot z$ for some group operation $x \cdot y$ on H then CSP($\mathcal{H}$) is tractable. Generalizing this, Bulatov [5] proves that if a finite algebra $\mathbb{A}$ has a term $p(x, y, z)$ that satisfies the equations $p(x, x, y) \approx p(y, x, x) \approx y$ for all x, $y \in A$ then $\mathbb{A}$ is also tractable (any operation that satisfies these equations is known as a Mal'tsev operation, see Example 4). The proof of this theorem found in [13] exploits the fact that any finite algebra with a Mal'tsev term has the small generating sets property (and hence, few subpowers).

While Mal'tsev and near unanimity operations are of quite different character, Dalmau in [22] managed to find a common generalization of them via the generalized majority-minority operation (see Example 4 for the definition). In a modification of the algorithm presented in [13], Dalmau shows in [22] that any finite algebra that has a GMM term is tractable. As in the case of algebras with Mal'tsev terms, these algebras have few subpowers and the small generating sets property and it is this latter property that plays a crucial role in the proof.

In [2] a characterization of finite algebras with few subpowers is given in terms of the presence of a special type of operation.

Definition 6. *A k-edge operation on a set A is a $k+1$-variable operation t that satisfies the equations:*

$$
\begin{aligned}
t(x,x,y,y,y,\ldots,y,y) &\approx y\\
t(x,y,x,y,y,\ldots,y,y) &\approx y\\
t(y,y,y,x,y,\ldots,y,y) &\approx y\\
t(y,y,y,y,x,\ldots,y,y) &\approx y\\
&\vdots\\
t(y,y,y,y,y,\ldots,y,x) &\approx y.
\end{aligned}
$$

Theorem 10 ([2]). *A finite algebra $\mathbb{A}$ has few subpowers if and only if it has a k-edge term for some $k > 0$. If this condition fails to hold then the function $s_{\mathbb{A}}(n)$ grows exponentially.*

Using this characterization the tractability of algebras with few subpowers can be deduced.

Corollary 3 ([27]). *If the finite algebra $\mathbb{A}$ has few subpowers then it is globally tractable.*

We note that the proof of this corollary closely follows the GMM tractability proof of Dalmau. We also note that the theorem and corollary settle conjectures posed by Chen [18] and Dalmau [21] on the nature of algebras with few subpowers.

We conclude this section with a result of Marković and McKenzie [2,35] that highlights the singular position that algebras with near unanimity term operations occupy. We have already noted that if a finite idempotent algebra has a near unanimity operation, then it has bounded width and few subpowers and so can be shown to be tractable in two distinct ways. The following theorem provides a converse to this.

Theorem 11. *Let $\mathbb{A}$ be a finite idempotent algebra. If $\mathbb{A}$ is of bounded width and has few subpowers then it has a near unanimity term operation.*

5 Coloured Graphs and Finite Algebras

The conditions of tractability and bounded width that appear in Conjectures 1 and 2 are known to be necessary. In order to prove that they are also sufficient for the complexity dichotomy conjecture one needs to design an algorithm (or algorithms) that solves CSPs satisfying the tractability condition, and for the bounded width conjecture, that the constraint propagation algorithm solves CSPs satisfying the bounded width condition. In all known cases algorithms (of proofs of the soundness of algorithms) use some local structure of algebras. Usually this structure can be explained in terms of the action of term operations of algebras on small subsets. In this section we propose an approach to the local structure of a finite idempotent algebra that is based on term operations of the algebra.

5.1 Coloured Graphs of Algebras

5.2 The Graph

The results of this section were first presented in [10]. We relate to every idempotent finite algebra $\mathbb{A}$ an edge-coloured graph $\mathsf{Gr}(\mathbb{A})$. If $\mathbb{A} = (A; F)$ and $B \subseteq A$, then by $\langle B \rangle$ we denote the *subalgebra generated by* B, that is the smallest subalgebra of $\mathbb{A}$ containing B.

Definition 7. *Let $\mathbb{A} = (A; F)$ be a finite idempotent algebra. The vertex set of the graph $\mathsf{Gr}(\mathbb{A})$ is the universe A of $\mathbb{A}$. A pair ab of vertices is an edge if and only if there exists a congruence θ of $\langle a, b \rangle$ and a term operation f of $\mathbb{A}$ such that either $f/_\theta$ is an affine operation on $\langle a, b \rangle/_\theta$, or $f/_\theta$ is a semilattice operation on $\{a/_\theta, b/_\theta\}$, or $f/_\theta$ is a majority operation on $\{a/_\theta, b/_\theta\}$ (see Figure 3).*

The color of an edge is defined as follows.

- *If there exists a congruence θ and a term operation $f \in \mathsf{Term}(\mathbb{A})$ such that $f/_\theta$ is a semilattice operation on $\{a/_\theta, b/_\theta\}$ then ab is said to have the* semilattice type.
- *An edge ab is of the* majority type *if it is not of the semilattice type and there are a congruence θ and a term operation f of $\mathbb{A}$ such that $f/_\theta$ is a majority operation on $\{a/_\theta, b/_\theta\}$.*
- *An edge ab is of the* affine type *if it is not of the semilattice or majority type and there are a congruence θ and a term operation f of $\mathbb{A}$ such that $f/_\theta$ is an affine operation on $\langle a, b \rangle/_\theta$.*

We sometimes call the set $a/_\theta \cup b/_\theta$ a *thick edge*.

Example 9. Let $\mathbb{A} = (\{0, 1, 2\}; f)$ be an algebra, where the operation f is defined by its Cayley table

$f(x,y)$	0	1	2
0	0	1	2
1	1	1	0
2	2	0	2

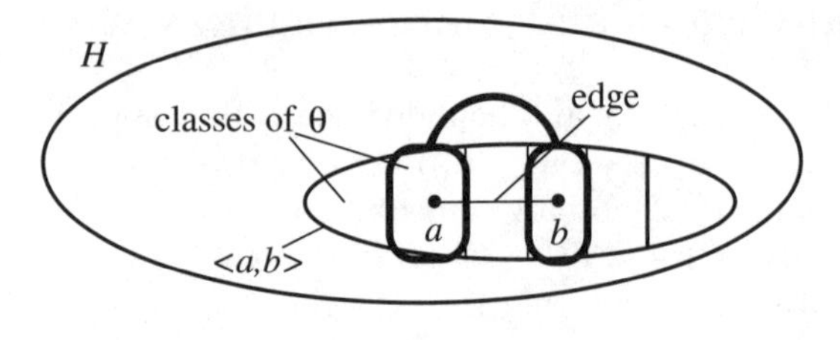

Fig. 3. Edges

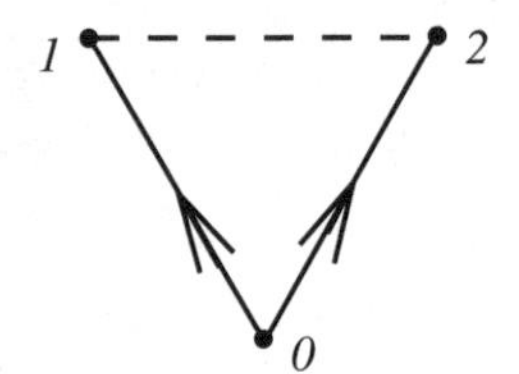

Fig. 4. $\mathsf{Gr}(\mathbb{A})$; edges of the semilattice type are drawn solid, edges of the majority type are dotted

(In fact, f occurs in [12]; in that paper it is called operation (6).) We have: $\langle 0,1\rangle = \{0,1\}$, $\langle 0,2\rangle = \{0,2\}$, $\langle 1,2\rangle = \{1,2,3\}$; the only congruence of $\langle 1,2\rangle$ such that $1,2$ belong to distinct classes is the equality relation; f witnesses that 01 and 02 are edges of semilattice type; 12 cannot be an edge of the semilattice type because no term operation of $\mathbb{A}$ is semilattice on $\{1,2\}$; however, the operation $g(x,y,z) = f(f(x,f(y,z)),f(f(x,y),z))$ is a majority operation on $\{1,2\}$. Thus, $\mathsf{Gr}(\mathbb{A})$ is the graph shown in Figure 4. Note also that this graph was implicitly used in [12] to prove the tractability of $\mathbb{A}$.

Observe that it is possible that for some pair a,b different congruences of $\langle a,b\rangle$ witness different types of the edge ab. Following the definition we always choose the 'strongest' type of the edge. Thus, the semilattice type is stronger than the majority type, which, in turn, is stronger than the affine type.

Example 10. Let $\mathbb{A}$, $\mathbb{B}$ be algebras with universes $\{0,1\}$ and $\{a,b\}$, respectively, and operations f,g. These operations are defined as follows:

– f is a semilattice operation on $\mathbb{A}$, i.e. $f(0,0) = f(0,1) = f(1,0) = 0$, $f(1,1)=1$;
– f is the first projection on $\mathbb{B}$, i.e. $f(x,y) = x$ for all $x,y \in \{a,b\}$;
– g is the ternary first projection on $\mathbb{A}$, i.e. $g(x,y,z) = x$ for all $x,y,z \in \{0,1\}$;
– g is a majority operation on $\mathbb{B}$; note that there is only one majority operation on a 2-element set.

Then let $\mathbb{C}$ denote the direct product of $\mathbb{A}$ and $\mathbb{B}$, that is the algebra with universe $C = \{(x,y) \mid x \in \{0,1\}, y \in \{a,b\}\}$, and operations f,g on C acting as follows:

$$f((x_1,y_1),(x_2,y_2)) = (f(x_1,x_2),f(y_1,y_2))$$

and

$$g((x_1,y_1),(x_2,y_2),(x_3,y_3)) = (g(x_1,x_2,x_3),g(y_1,y_2,y_3)).$$

As is easily seen, $\langle(0,a),(1,b)\rangle = \mathbb{C}$ and the equivalence relations η_1, η_2 defined by $((x_1,y_1),(x_2,y_2)) \in \eta_1$ if and only if $x_1 = x_2$, and $((x_1,y_1),(x_2,y_2)) \in \eta_2$ if and only if $y_1 = y_2$, are congruences of $\mathbb{C}$. Observe that $f/_{\eta_1}$ is a semilattice operation on $\mathbb{C}/_{\eta_1} = \{(0,a)/_{\eta_1}, (1,b)/_{\eta_1}\}$; and that $\mathbb{C}/_{\eta_2} = \{(0,a)/_{\eta_2}, (1,b)/_{\eta_2}\}$ is isomorphic to $\mathbb{B}$. Thus, congruence η_1 witnesses that $(0,a)(1,b)$ is an edge of semilattice type, while η_2 witnesses that the same edge has majority type. Since the semilattice type is stronger, this edge has semilattice type.

5.3 Connectedness and Omitting Types

We show that connectedness of the graph $\mathsf{Gr}(\mathbb{A})$ and the colours of edges that appear in it are closely related to omitting types in the sense of tame congruence theory, and to Conjectures 1 and 2.

Theorem 12 ([10]). *For an idempotent algebra $\mathbb{A}$ the following conditions are equivalent:*

(1) $\mathsf{var}(\mathbb{A})$ omits the unary type;

(2) $\mathsf{var}(\mathbb{A})$ does not contain a G-set;

(3) for any subalgebra $\mathbb{B}$ of $\mathbb{A}$ the graph $\mathsf{Gr}(\mathbb{B})$ is connected.

We shall refer to condition (3) from Theorem 12 as to the *connectedness condition.*

Theorem 13. *Let $\mathbb{A}$ be an idempotent algebra. The following conditions are equivalent:*

(1) $\mathsf{var}(\mathbb{A})$ omits the unary and affine types;

(2) $\mathsf{var}(\mathbb{A})$ does not contain an algebra that is term equivalent to a reduct of a module over some finite ring;

(3) $\mathbb{A}$ satisfies the connectedness condition, and $\mathsf{Gr}(\mathbb{A})$ does not contain edges of the affine type.

Since this result appears here for the first time we give a proof of it. We shall use an improved version of Lemma 1 from [10].

Lemma 1. *Let $\mathbb{A}$ be a finite idempotent algebra, and let ab be an edge of the affine type in $\mathsf{Gr}(\mathbb{A})$. Then there are a maximal congruence θ of $\langle a,b\rangle$ (that is there is no congruence strictly between θ and the total congruence) and a module M with the universe $\langle a,b\rangle/_\theta$ over a ring R such that every term operation of $\langle a,b\rangle/_\theta$ can be represented as an operation $\alpha_1x_1 + \ldots + \alpha_nx_n$ of M with $\alpha_1,\ldots,\alpha_n \in R$, $\alpha_1 + \ldots + \alpha_n = 1$.*

Proof (of Theorem 13). The equivalence of (1) and (2) is follows from Theorem 8. We show that (3) is equivalent to (1).

If for some subalgebra $\mathbb{B}$ of $\mathbb{A}$ the graph $\mathsf{Gr}(\mathbb{B})$ is not connected then by Theorem 12 $\mathsf{var}(\mathbb{B}) \subseteq \mathsf{var}(\mathbb{A})$ contains a G-set that is term equivalent to a reduct of any module, because in an idempotent variety any G-set is term equivalent to an algebra whose basic operations are projections. If $\mathsf{Gr}(\mathbb{A})$ contains an edge of the affine type ab then by Lemma 1 the algebra $\langle a, b\rangle/_\theta$ for a certain θ is a reduct of a module.

By Theorem 8 if $\mathsf{var}(\mathbb{A})$ contains an algebra term equivalent to a reduct of a module, then there is a subalgebra $\mathbb{B}$ of $\mathbb{A}$ and a congruence θ of $\mathbb{B}$ such that $\mathbb{B}/_\theta$ is term equivalent to a reduct of a module. If this algebra is a G-set then $\mathsf{Gr}(\mathbb{B})$ is not connected by Theorem 12. Otherwise we assume that $\mathbb{B}$ is a minimal (with respect to containment) subalgebra with this property and θ is a maximal congruence of $\mathbb{B}$. Then θ is the only maximal congruence of $\mathbb{B}$. Indeed, if η is another maximal congruence of $\mathbb{B}$, then any class C of η that is not contained in a class of θ induces a proper subalgebra $\mathbb{C}$ of $\mathbb{B}$, and $\mathbb{C}/_\theta$ is still term equivalent to a reduct of a module; a contradiction with minimality of $\mathbb{B}$. It is not hard to see, that, for any $a, b \in \mathbb{B}$ such that $(a, b) \notin \theta$, the pair ab is an edge of the affine type. $\square$

Using Theorems 12 and 13 we can give yet another formulation of the complexity and bounded width conjectures.

Conjecture 1 (the complexity dichotomy conjecture, version 4). *A finite idempotent algebra is tractable if and only if it satisfies the connectedness condition.*

Conjecture 2 (the bounded width conjecture, version 4). *A finite idempotent algebra $\mathbb{A}$ has bounded width if and only if it satisfies the connectedness condition and the graph $\mathsf{Gr}(\mathbb{A})$ does not contain edges of the affine type.*

5.4 Improving an Algebra

The study of finite algebras in the context of the complexity of the CSP does not necessarily suppose investigation of the exact structure of finite algebras. Therefore we can transform algebras under consideration as long as such a transformation preserves properties supposedly responsible for tractability, e.g. omitting the unary type. In this subsection we show two such transformations.

We say that the graph $\mathsf{Gr}(\mathbb{A})$ is *semilattice-connected*, if for any two vertices $a, b \in \mathbb{A}$ there is a path in $\mathsf{Gr}(\mathbb{A})$ consisting of edges of the semilattice type. The *semilattice/majority connectedness* of $\mathsf{Gr}(\mathbb{A})$ is defined similarly.

Proposition 2. *Let $\mathbb{A}$ be an idempotent algebra satisfying the connectedness condition, let ab be an edge of $\mathsf{Gr}(\mathbb{A})$ of the semilattice or majority type, and let $R_{ab} = (a/_\theta \cup b/_\theta)$ be the corresponding thick edge, where θ is a congruence certifying the type of ab.*

(1) $\mathbb{A}_{ab} = (A; F')$, where F' is the set of all term operations of $\mathbb{A}$ preserving R_{ab}, satisfies the connectedness condition.

(2) If ab is has the semilattice type and $\mathsf{Gr}(\mathbb{A})$ is semilattice-connected, then $\mathsf{Gr}(\mathbb{A}_{ab})$ is semilattice-connected.

(3) If ab has the majority type and $\mathsf{Gr}(\mathbb{A})$ is semilattice/majority-connected, then $\mathsf{Gr}(\mathbb{A}_{ab})$ is semilattice/majority-connected.

As the following example shows, constructing a reduct by adding an edge of the affine type can destroy the connectedness condition and even make a tractable algebra NP-complete.

Example 11. Let $\mathbb{A} = (\{0,1,2\}; h)$, where $h(x,y,z) = x - y + z$ and $+, -$ denote the operation of addition and subtraction modulo 3. It is well known (see e.g. [39]) that the term operations of $\mathbb{A}$ are the operations of the form $\alpha_1 x_1 + \ldots + \alpha_n x_n$, where $\alpha_1, \ldots, \alpha_n$ are integers and $\alpha_1 + \ldots + \alpha_n = 1 \pmod 3$. Therefore, for any $a, b \in A$, $\langle a, b\rangle = A$, the only maximal congruence of $\langle a, b\rangle$ is the equality relation, and ab is an edge of the affine type.

Since the affine operation $x - y + z$ is an operation of $\mathbb{A}$, the problem CSP($\mathbb{A}$) can be solved by Gaussian elimination [30]. Take an edge of $\mathsf{Gr}(\mathbb{A})$, say 01 and a term operation $f(x_1, \ldots, x_n) = \alpha_1 x_1 + \ldots + \alpha_n x_n$ of $\mathbb{A}$. If f preserves $\{0,1\}$, then, for any $i \in \{1, \ldots, n\}$, we have $f(0, \ldots, 0, 1, 0, \ldots, 0) = \alpha_i \in \{0,1\}$ (1 is on the ith place). Furthermore, if $\alpha_i, \alpha_j = 1$, then $f(0, \ldots, 0, 1, 0, \ldots, 0, 1, 0, \ldots, 0) = \alpha_i + \alpha_j = 2 \notin \{0,1\}$ (1s are on the ith and jth places). Thus, only one of the αs is non-zero, which means that f is a projection. Hence, every term operation of $\mathbb{A}_{01}$ is a projection and CSP($\mathbb{A}_{01}$) is NP-complete.

Proposition 2 amounts to saying that we may restrict our attention to algebras $\mathbb{A}$ such that every thick edge of the semilattice or majority type of $\mathsf{Gr}(\mathbb{A})$ is a subalgebra.

The second transformation preserving the connectedness condition is based on the following statement that shows that the term operations certifying the type of edges can be significantly unified.

Proposition 3. *Let $\mathbb{A}$ be an idempotent algebra. For an edge, θ always denotes a congruence certifying its type. There are term operations f, g, h of $\mathbb{A}$ such that*

$f|_{\{a/\theta, b/\theta\}}$ *is a semilattice operation if ab is an edge of the semilattice type, it is the first projection if ab is an edge of the majority or affine type;*

$g|_{\{a/\theta, b/\theta\}}$ *is a majority operation if ab is an edge of the majority type, it is the first projection if ab is an edge of the affine type, and $g|_{\{a/\theta, b/\theta\}}(x,y,z) = f|_{\{a/\theta, b/\theta\}}(x, f|_{\{a/\theta, b/\theta\}}(y,z))$ if ab has the semilattice type;*

$h|_{\langle ab\rangle/\theta}$ *is an affine operation operation if ab is an edge of the affine type, it is the first projection if ab is an edge of the majority type, and $h|_{\{a/\theta, b/\theta_{ab}\}}(x,y,z) = f|_{\{a/\theta, b/\theta\}}(x, f|_{\{a/\theta, b/\theta\}}(y,z))$ if ab has the semilattice type.*

Example 9 (continued). Let us reconsider the algebra $\mathbb{A}$ from Example 9. By Proposition 2, since 12 is an edge of the majority type, the algebra $\mathbb{A}_{12}$ satisfies the connectedness condition. The operations f, g, h satisfying the conditions of Proposition 3 can be chosen as follows: g is the operation obtained in Example 9, $f(x,y) = g(x,x,y)$ (the binary operation defined in Example 9 does not fit, because it does not preserve $\{1,2\}$) and $h(x,y,z) = f(x, f(y,z))$.

Propositions 2 and 3 together allow us to restrict ourselves to the study of idempotent algebras that have at most three basic operations, one binary and two ternary, and such that, for any edge of the semilattice or majority ab and a congruence θ certifying this, the thick edge $a/_\theta \cup b/_\theta$ is a subalgebra. In the next section we shall see that the class of algebras to be studied can be further narrowed down.

Edges of the semilattice type. In this section we focus on edges of the semilattice type of the graph $\mathsf{Gr}(\mathbb{A})$. Note first that if one fixes a congruence θ_{ab} for each edge of $\mathsf{Gr}(\mathbb{A})$ that certifies its type, and a term operation f such that f is a semilattice operation on $\{a/_{\theta_{ab}}, b/_{\theta_{ab}}\}$ for every edge ab of the semilattice type of $\mathsf{Gr}(\mathbb{A})$, then one can define an orientation of every such edge. An edge ab of the semilattice type is oriented from a to b if $f(a/_{\theta_{ab}}, b/_{\theta_{ab}}) = f(b/_{\theta_{ab}}, a/_{\theta_{ab}}) = b/_{\theta_{ab}}$. For instance, the edges 01, 02 of the graph from Example 9 are oriented from 0 to 1 and 2 respectively. Clearly, orientation strongly depends on the choice of the operation f. The graph $\mathsf{Gr}(\mathbb{A})$ oriented accordingly to a term operation f will be denoted by $\mathsf{Gr}_f(\mathbb{A})$. We then can define *semilattice-connected* and *strongly semilattice-connected* components of $\mathsf{Gr}_f(\mathbb{A})$. We will also use the natural order on the set of strongly semilattice-connected components of $\mathsf{Gr}_f(\mathbb{A})$: for components A, B, $A \leq B$ if there is a directed path in $\mathsf{Gr}_f(\mathbb{A})$ consisting of edges of the semilattice type and connecting a vertex from A with a vertex from B. Later we show that certain restrictions on the set of strongly semilattice-connected components of $\mathsf{Gr}_f(\mathbb{A})$ yield the tractability of CSP($\mathbb{A}$).

First we show that if for an edge ab of the semilattice type there is no semilattice term operation on the set $\{a, b\}$ then ab can be thrown out of the graph $\mathsf{Gr}(\mathbb{A})$ such that the connectedness condition is preserved in the remaining graph. Therefore, we can assume that for any edge of the semilattice type ab there is a semilattice term operation on $\{a, b\}$.

Proposition 4. *Let $\mathbb{A}$ be an algebra and $\mathsf{Gr}'(\mathbb{A})$ the subgraph of $\mathsf{Gr}(\mathbb{A})$ obtained by omitting edges ab of the semilattice type such that there is no semilattice operation on $\{a, b\}$. Then $\mathsf{Gr}'(\mathbb{A})$ is connected. If $\mathsf{Gr}(\mathbb{A})$ is semilattice-connected then $\mathsf{Gr}'(\mathbb{A})$ is semilattice-connected. If $\mathsf{Gr}(\mathbb{A})$ is semilattice/majority-connected then $\mathsf{Gr}'(\mathbb{A})$ is semilattice/majority-connected.*

The graph $\mathsf{Gr}'(\mathbb{A})$ oriented according to a binary term operation f will be denoted by $\mathsf{Gr}'_f(\mathbb{A})$.

We conclude this subsection with a result that shows how properties of the graph $\mathsf{Gr}(\mathbb{A})$ can help in establishing the tractability and bounded width of the algebra $\mathbb{A}$. Let us consider algebras $\mathbb{A}$ with a binary term operation f such that, for every subalgebra $\mathbb{B}$ of $\mathbb{A}$, the subgraph of $\mathsf{Gr}'_f(\mathbb{A})$ induced by $\mathbb{B}$ has a unique maximal strongly semilattice-connected component. This condition we shall call the *maximal semilattice component condition.*

Theorem 14. *If an algebra $\mathbb{A}$ satisfies the maximal semilattice component condition, then* CSP($\mathbb{A}$) *is of relational width 3.*

Observe that a 2-semilattice, that is a groupoid with a 2-semilattice basic operation, satisfies the maximal semilattice component condition. Indeed, if $\mathbb{A}$ has a 2-semilattice term operation f, then f is a semilattice operation on $\{a, f(a,b)\}$ and $\{b, f(a,b) = f(b,a)\}$. This means that $\mathsf{Gr}'_f(\mathbb{A})$ is semilattice-connected. Moreover, if a, b belong to different maximal strongly semilattice-connected components B and C, then $f(a,b)$ belongs to a strongly semilattice-connected component D such that $B \leq D$ and $C \leq D$, a contradiction with the maximality of B, C. The same argument is valid for any subalgebra of $\mathbb{A}$, thus, $\mathbb{A}$ satisfies the maximal semilattice component condition. Since every semilattice operation is also a 2-semilattice operation, the same holds for algebras with a semilattice term operation. Thus, by Theorem 14, we obtain the main result of [6], and also the results of [7], since a *binary commutative conservative* operation is a 2-semilattice operation, and also the results of [30,28] concerning semilattice operations.

5.5 Conservative Algebras and Their Graphs

Let $\mathcal{H}$ be a relational structure. In the *conservative* (*list*) *constraint satisfaction problem*, denoted CCSP($\mathcal{H}$), the question is, given a structure $\mathcal{G}$ and, for each element $g \in \mathcal{G}$, a *list* $L(g)$ of elements of $\mathcal{H}$, whether there exists a homomorphism $\varphi \colon \mathcal{G} \to \mathcal{H}$ such that $\varphi(g) \in L(g)$ for all $g \in \mathcal{G}$.

Example 12 (List-H-Colouring). Let H be a (directed) graph. In the List H-Colouring problem we are given a graph G and, for each vertex v of G, a set $L(v)$ of vertices of H. The question is whether there is a homomorphism φ from G to H such that $\varphi(v) \in L(v)$ for every vertex v of G. Clearly, List H-Colouring can be represented in the form of the conservative CSP.

Notice that, for any structure $\mathcal{H}$, the problem CCSP($\mathcal{H}$) is equivalent to CSP($\mathcal{H}^*$), where $\mathcal{H}^*$ is an expansion of $\mathcal{H}$ obtained by adding all unary relations. A structure $\mathcal{H}$ such that for each subset $S \subseteq H$ there is a relational symbol R in the vocabulary with $R^{\mathcal{H}} = S$ is said to be *conservative*. Thus, instead of conservative CSPs we may study ordinary constraint satisfaction problems corresponding to conservative structures.

On the algebraic side, every term operation f of an algebra $\mathbb{A}$ that gives rise to a conservative CSP must be *conservative*, that is $f(x_1, \dots, x_n) \in \{x_1, \dots, x_n\}$ for all $x_1, \dots, x_n$. Algebras satisfying this condition are also called conservative.

If $\mathbb{A}$ is a conservative algebra, then in particular every 2-element subset of A induces a subalgebra of $\mathbb{A}$. Therefore, $\mathbb{A}$ satisfies the connectedness condition if and only if every pair of its elements constitutes an edge of $\mathsf{Gr}(\mathbb{A})$. Moreover, every edge of this graph is 2-element, implying that the operations f, g, h constructed in Proposition 3 are a semilattice (that is conjunction or disjunction) operation, the majority operation $(x \vee y) \wedge (y \vee z) \wedge (z \vee x)$, and the Mal'tsev operation $x - y + z (\text{mod } 2)$ on each 2-element subset from $\mathbb{A}$, respectively (we denote the elements of this subset by 0 and 1).

Theorem 15. *A conservative algebra $\mathbb{A}$ is tractable if and only if it satisfies the connectedness condition, that is, for any 2-element subalgebra $\mathbb{B}$ of $\mathbb{A}$ (we assume*

$B = \{0,1\}$), *there exists a term operation* t *such that* $t|_B$ *is either a semilattice operation* $x \vee y$ *or* $x \wedge y$, *or the majority operation* $(x \vee y) \wedge (y \vee z) \wedge (z \vee x)$, *or the Mal'tsev operation* $x - y + z(\text{mod } 2)$. *In this case* $\mathbb{A}$ *is also globally tractable. Otherwise* $\mathbb{A}$ *is NP-complete.*

Observe that by Proposition 3 the tractability of a conservative algebra is witnessed by term operations of arity at most 3. This observation implies a stronger version of Theorem 15. An algebra such that each of its k-element subsets induces a subalgebra is called *k-conservative.*

Corollary 4. *If* $\mathbb{A}$ *is a 3-conservative algebra then* $\mathbb{A}$ *is globally tractable if and only if it satisfies the connectedness condition. Otherwise it is NP-complete.*

References

1. Allender, E., Bauland, M., Immerman, N., Schnoor, H., Vollmer, H.: The complexity of satisfiability problems: Refining schaefer's theorem. In: Jedrzejowicz, J., Szepietowski, A. (eds.) MFCS 2005. LNCS, vol. 3618, pp. 71–82. Springer, Heidelberg (2005)
2. Berman, J., Idziak, P., Marković, P., McKenzie, R., Valeriote, M., Willard, R.: Varieties with few subalgebras of powers. Journal of the AMS (to appear)
3. Berman, J.D., Kiss, E.W., Pröhle, P., Szendrei, Á.: The set of types of a finitely generated variety. Discrete Math. 112(1-3), 1–20 (1993)
4. Baker, K.A., Pixley, A.F.: Polynomial interpolation and the chinese remainder theorem. Mathematische Zeitschrift 143, 165–174 (1975)
5. Bulatov, A.A.: Mal'tsev constraints are tractable. Technical Report PRG-RR-02-05, Computing Laboratory, University of Oxford, Oxford, UK (2002)
6. Bulatov, A.A.: Combinatorial problems raised from 2-semilattices. Journal of Algebra 298(2), 321–339 (2006)
7. Bulatov, A.A., Jeavons, P.G.: Tractable constraints closed under a binary operation. Technical Report PRG-TR-12-00, Computing Laboratory, University of Oxford, Oxford, UK (2000)
8. Bulatov, A.A., Jeavons, P.G.: Algebraic structures in combinatorial problems. Technical Report MATH-AL-4-2001, Technische universität Dresden, Dresden, Germany (2001), `http://web.comlab.ox.ac.uk/oucl/research/areas/constraints/publications/index.html`
9. Bulatov, A.A., Jeavons, P.G., Volkov, M.V.: Finite semigroups imposing tractable constraints. In: Gomes, G.M.S., Pin, J.-E., Silva, P.V. (eds.) Semigroups, Algorithms, Automata and Languages, pp. 313–329. World Scientific, Singapore (2002)
10. Bulatov, A.A.: A graph of a relational structure and constraint satisfaction problems. In: LICS, pp. 448–457 (2004)
11. Bulatov, A.A.: H-coloring dichotomy revisited. Theor. Comput. Sci. 349(1), 31–39 (2005)
12. Bulatov, A.A.: A dichotomy theorem for constraint satisfaction problems on a 3-element set. J. ACM 53(1), 66–120 (2006)
13. Bulatov, A.A., Dalmau, V.: A simple algorithm for Mal'tsev constraints. SIAM J. Comput. 36(1), 16–27 (2006)
14. Bulatov, A.A., Jeavons, P., Krokhin, A.A.: Classifying the complexity of constraints using finite algebras. SIAM J. Comput. 34(3), 720–742 (2005)

15. Atserias, A., Bulatov, A., Dawar, A.: Affine systems of equations and counting infinitary logic. In: Arge, L., Cachin, C., Jurdziński, T., Tarlecki, A. (eds.) ICALP 2007. LNCS, vol. 4596. Springer, Heidelberg (2007)
16. Bulatov, A.A., Krokhin, A.A., Larose, B.: Dualities for Constraint Satisfaction Problems. In: Creignou, N., Kolaitis, P., Vollmer, H. (eds.) Complexity of Constraints. LNCS, vol. 5250, pp. 93–124. Springer, Heidelberg (2008)
17. Burris, S., Sankappanavar, H.P.: A course in universal algebra. Graduate Texts in Mathematics, vol. 78. Springer, New York (1981)
18. Chen, H.: The expressive rate of constraints. Ann. Math. Artif. Intell. 44(4), 341–352 (2005)
19. Clasen, M., Valeriote, M.: Tame congruence theory. In: Lectures on algebraic model theory. Fields Inst. Monogr., vol. 15, pp. 67–111. Amer. Math. Soc., Providence (2002)
20. Dalmau, V.: A new tractable class of constraint satisfaction problems. Annals of Mathematics and Artificial Intelligence 44(1-2), 61–85 (2005)
21. Dalmau, V.: Computational Complexity of Problems over Generalised Formulas. Ph.D thesis, Department LSI of the Universitat Politecnica de Catalunya (UPC), Barcelona (March 2000)
22. Dalmau, V.: Generalized majority-minority operations are tractable. In: LICS, pp. 438–447 (2005)
23. Dalmau, V., Gavaldà, R., Tesson, P., Thérien, D.: Tractable clones of polynomials over semigroups. In: van Beek, P. (ed.) CP 2005. LNCS, vol. 3709, pp. 196–210. Springer, Heidelberg (2005)
24. Feder, T., Vardi, M.Y.: The computational structure of monotone monadic SNP and constraint satisfaction: A study through datalog and group theory. SIAM Journal on Computing 28, 57–104 (1998)
25. Hell, P., Nešetřil, J.: On the complexity of H coloring. Journal of Combinatorial Theory, Ser. B 48, 92–110 (1990)
26. Hobby, D., McKenzie, R.N.: The Structure of Finite Algebras. Contemporary Mathematics, vol. 76. American Mathematical Society, Providence (1988)
27. Idziak, P., Marković, P., McKenzie, R., Valeriote, M., Willard, R.: Tractability and learnability arising from algebras with few subpowers. In: LICS 2007: Proceedings of the 22nd Annual IEEE Symposium on Logic in Computer Science, Washington, DC, USA, pp. 213–224. IEEE Computer Society, Los Alamitos (2007)
28. Jeavons, P.G.: On the algebraic structure of combinatorial problems. Theoretical Computer Science 200, 185–204 (1998)
29. Jeavons, P.G., Cohen, D.A., Cooper, M.C.: Constraints, consistency and closure. Artificial Intelligence 101(1-2), 251–265 (1998)
30. Jeavons, P.G., Cohen, D.A., Gyssens, M.: Closure properties of constraints. Journal of the ACM 44, 527–548 (1997)
31. Jeavons, P.G., Cohen, D.A., Pearson, J.K.: Constraints and universal algebra. Annals of Mathematics and Artificial Intelligence 24, 51–67 (1998)
32. Larose, B., Loten, C., Tardif, C.: A characterisation of first-order constraint satisfaction problems. In: LICS, pp. 201–210 (2006)
33. Larose, B., Tesson, P.: Universal algebra and hardness results for constraint satisfaction problems. In: Arge, L., Cachin, C., Jurdziński, T., Tarlecki, A. (eds.) ICALP 2007. LNCS, vol. 4596, pp. 267–278. Springer, Heidelberg (2007)
34. Larose, B., Zadori, L.: Bounded width problems and algebras. Algebra Universalis 56(3-4), 439–466 (2007)
35. Marković, P., McKenzie, R.: Few subpowers, congruence distributivity and near-unanimity. Algebra Universalis 58(2), 119–128 (2008)

36. McKenzie, R.N., McNulty, G.F., Taylor, W.F.: Algebras, Lattices and Varieties, vol. I. Wadsworth and Brooks, California (1987)
37. McKenzie, R., Maróti, M.: Existence theorems for weakly symmetric operations. In: Algebra Universalis (to appear, 2006)
38. Schaefer, T.J.: The complexity of satisfiability problems. In: Proceedings of the 10th ACM Symposium on Theory of Computing (STOC 1978), pp. 216–226 (1978)
39. Szendrei, A.: Clones in Universal Algebra. Seminaires de Mathematiques Superieures, vol. 99. Université de Móntreal (1986)
40. Valeriote, M.: A subalgebra intersection property for congruence distributive varieties. Canadian Journal of Mathematics (accepted, 2006)

Dualities for Constraint Satisfaction Problems

Andrei A. Bulatov[1], Andrei Krokhin[2], and Benoit Larose[3]

[1] School of Computing Science
Simon Fraser University
Burnaby, BC, Canada, V5A 1S6
abulatov@cs.sfu.ca
[2] Department of Computer Science
Durham University
Durham, DH1 3LE, UK
andrei.krokhin@durham.ac.uk
[3] Department of Mathematics and Statistics
Concordia University
Montréal, Qc, Canada, H3G 1M8
larose@mathstat.concordia.ca

Abstract. In a nutshell, a duality for a constraint satisfaction problem equates the existence of one homomorphism to the non-existence of other homomorphisms. In this survey paper, we give an overview of logical, combinatorial, and algebraic aspects of the following forms of duality for constraint satisfaction problems: finite duality, bounded pathwidth duality, and bounded treewidth duality.

1 Introduction

The constraint satisfaction problem (CSP) provides a framework in which it is possible to express, in a natural way, many combinatorial problems encountered in artificial intelligence, computer science, discrete mathematics, and elsewhere [19,34,61]. An instance of the constraint satisfaction problem is represented by a finite set V of variables, a (finite) domain D of values for each variable, and a set of constraints $\{(\overline{s}_1, R_1), \ldots, (\overline{s}_q, R_q)\}$. Each constraint consists of a constraint scope $\overline{s}_i$, which is an m_i-tuple of variables, and a constraint relation $R_i \subseteq D^{m_i}$. The aim is then to decide whether there is an assignment $h : V \to D$ that satisfies the constraints, i.e., such that $h(\overline{s}_i) \in R_i$ for all i.

It has been observed [28] (see also [42]) that the constraint satisfaction problem can be recast as the following fundamental problem: given two finite relational structures $\mathbf{A}$ and $\mathbf{B}$, is there a homomorphism from $\mathbf{A}$ to $\mathbf{B}$? One of the most studied restrictions on the CSP is the non-uniform CSP – when the structure $\mathbf{B}$ is fixed, and only $\mathbf{A}$ is part of the input. The obtained problem is denoted by CSP($\mathbf{B}$). Examples of such problems include various versions of k-Sat, Graph Colouring, and Systems of Equations (see [17,34,42,50]). Strong motivation for studying this framework was given in [28] where it was shown that such problems can be used in attempts to identify a largest natural subclass of **NP** that avoids problems of intermediate complexity.

N. Creignou et al. (Eds.): Complexity of Constraints, LNCS 5250, pp. 93–124, 2008.

The two main general classification problems about the class of problems of the form CSP($\mathbf{B}$) are:

1. classify the problems CSP($\mathbf{B}$) with respect to computational complexity, that is, for a given complexity class $\mathcal{K}$, characterise (under suitable complexity-theoretic assumptions) structures $\mathbf{B}$ such that CSP($\mathbf{B}$) is in $\mathcal{K}$;
2. classify the problems CSP($\mathbf{B}$) with respect to descriptive complexity, that is, for a given logic L, characterise structures $\mathbf{B}$ such that CSP($\mathbf{B}$), as the class of all structures admitting a homomorphism to $\mathbf{B}$, is definable in L.

   In addition, there is a so-called meta-problem:

3. Determine the (computational) complexity of deciding whether, for a given structure $\mathbf{B}$, CSP($\mathbf{B}$) has a certain (computational or descriptive) complexity.

A variety of mathematical approaches to study problems CSP($\mathbf{B}$) has been recently suggested. The most advanced approaches use logic (e.g., [48]), combinatorics (e.g., [32,34,51]), universal algebra (e.g., [7,10,12,17,41,50]), or combinations of those (e.g., [2,8,20,28,52]). In this survey, we will discuss a combinatorial idea that has a bearing on all the above problems, and has strong links with the three approaches — the idea of *homomorphism duality*.

The concept of duality has been much used to study homomorphism problems. In essence, a duality equates the existence of one homomorphism to the non-existence of some other homomorphism(s). The idea is to provide a set $\mathcal{O}_\mathbf{B}$ of *obstructions* for $\mathbf{B}$ such that, for any relational structure $\mathbf{A}$, $\mathbf{A}$ homomorphically maps to $\mathbf{B}$ if and only if $\mathbf{A}$ does not admit a homomorphism from any structure from $\mathcal{O}_\mathbf{B}$. Of course, the set $\mathcal{O}_\mathbf{B}$ can always be chosen to consist of all structures that do not homomorphically map to $\mathbf{B}$, but this choice does not give any information about CSP($\mathbf{B}$). If, however, $\mathcal{O}_\mathbf{B}$ can be chosen so that it has certain nice properties then this can tell us much about the computational or descriptive complexity of CSP($\mathbf{B}$).

Most of the early studies of dualities were restricted to the case of (di)graphs (see survey [36], also [34,35,38,39,49,58]). For general relational structures, the main forms of duality that have been considered in the literature are finite duality, bounded pathwidth duality, and bounded treewidth duality. We give the necessary combinatorial, logical, and algebraic preliminaries in Section 2, and then consider the three dualities in Sections 3, 4, and 5, respectively. Sections 6 and 7 contain some remarks and a list of open problems concerning dualities.

2 Preliminaries

2.1 Basic Definitions

Most of the terminology introduced in this section is fairly standard. A *vocabulary* is a finite set of relation symbols or predicates. In what follows, τ always denotes a vocabulary. Every relation symbol R in τ has an *arity* $r = \rho(R) \geq 0$ associated to it. We also say that R is an r-ary relation symbol.

A τ-structure $\mathbf{A}$ consists of a set A, called the *universe* of $\mathbf{A}$, and a relation $R^{\mathbf{A}} \subseteq A^r$ for every relation symbol $R \in \tau$ where r is the arity of R. Let maxar($\mathbf{A}$) denote the maximum arity of a relation in $\mathbf{A}$. Unless specified otherwise, all structures in this paper are assumed to be *finite*, i.e., structures with a finite universe. Throughout the paper we use the same boldface and slanted capital letters to denote a structure and its universe, respectively.

Let $\mathbf{A}$ and $\mathbf{A}'$ be τ-structures. We say that $\mathbf{A}'$ is a *substructure* of $\mathbf{A}$, denoted by $\mathbf{A}' \subseteq \mathbf{A}$, if $A' \subseteq A$ and for every $R \in \tau$, $R^{\mathbf{A}'} \subseteq R^{\mathbf{A}}$. If $\mathbf{A}$ is a τ-structure and $I \subseteq A$, then $\mathbf{A}_{|I}$ denotes the substructure *induced* by $\mathbf{A}$ on I, i.e., the τ-structure $\mathbf{I}$ with universe I and $R^{\mathbf{I}} = R^{\mathbf{A}} \cap I^r$ for every r-ary $R \in \tau$.

A *homomorphism* from a τ-structure $\mathbf{A}$ to a τ-structure $\mathbf{B}$ is a mapping $h : A \to B$ such that for every r-ary $R \in \tau$ and every $(a_1, \ldots, a_r) \in R^{\mathbf{A}}$, we have $(h(a_1), \ldots, h(a_r)) \in R^{\mathbf{B}}$. We denote this by $h : \mathbf{A} \to \mathbf{B}$, and the set of all homomorphisms from $\mathbf{A}$ to $\mathbf{B}$ is denoted by hom($\mathbf{A}, \mathbf{B}$). We also say that $\mathbf{A}$ homomorphically maps to $\mathbf{B}$, and write $\mathbf{A} \to \mathbf{B}$ if there is a homomorphism from $\mathbf{A}$ to $\mathbf{B}$ and $\mathbf{A} \not\to \mathbf{B}$ if there is no homomorphism. Now CSP($\mathbf{B}$) can be defined to be the class of all structures $\mathbf{A}$ such that $\mathbf{A} \to \mathbf{B}$. The class of all structures $\mathbf{A}$ such that $\mathbf{A} \not\to \mathbf{B}$ will be denoted by co-CSP($\mathbf{B}$).

Example 1. If $\mathbf{B}_{hc}$ is a digraph $\mathbf{H}$ then CSP($\mathbf{B}_{hc}$) is the much-studied problem, $\mathbf{H}$-COLOURING, of deciding whether there is a homomorphism from a given digraph to $\mathbf{H}$ [34]. If $\mathbf{H}$ is the complete graph $\mathbf{K}_k$ on k vertices then it is well known (and easy to see) that CSP($\mathbf{B}_{hc}$) is precisely the k-COLOURING problem.

Example 2. If $\mathbf{B}_{lhc}$ is a structure obtained from a digraph $\mathbf{H}$ by adding, for each non-empty subset U of H, a unary relation U then CSP($\mathbf{B}_{lhc}$) is exactly the LIST $\mathbf{H}$-COLOURING problem, in which every vertex v of the input digraph $\mathbf{G}$ gets a list L_v of vertices of $\mathbf{H}$, and the question is whether there is a homomorphism $h : \mathbf{G} \to \mathbf{H}$ such that $h(v) \in L_v$ for all $v \in G$ (see [34]).

Example 3. If $\mathbf{B}_{unr}$ is the Boolean (i.e., with universe $\{0, 1\}$) structure with one binary relation Eq, which is the equality relation, and two unary relations $\{0\}$ and $\{1\}$ then CSP($\mathbf{B}_{unr}$) is the (undirected) UNREACHABILITY problem where one is given a graph and two sets of vertices in it, S and T, and the question is whether there is no path in the graph from any vertex in S to a vertex in T.

Example 4. In the PATH SYSTEM ACCESSIBILITY problem [31], one is given a relational structure $\mathbf{A}$ with one ternary relation $P^{\mathbf{A}}$, and two unary relations $S^{\mathbf{A}}$ and $T^{\mathbf{A}}$. The unary relations represent "source" and "terminal" nodes, respectively. The question is whether there is an "accessible" terminal node, where a node x is accessible if $x \in S^{\mathbf{A}}$ or $(a, b, x) \in P^{\mathbf{A}}$ for some accessible $a, b \in A$.

Let $\mathbf{B}_{ps}$ be the Boolean structure with one ternary relation $P^{\mathbf{B}_{ps}} = \{(x, y, z) \mid x \wedge y \to z\}$ and two unary relations $S^{\mathbf{B}_{ps}} = \{1\}$ and $T^{\mathbf{B}_{ps}} = \{0\}$. Then it is easy to verify that the PATH SYSTEM ACCESSIBILITY problem is precisely co-CSP($\mathbf{B}_{ps}$).

Example 5. Let $\mathbf{B}_{3H}$ be the structure with universe $\{0, 1\}$, one unary relation $U^{\mathbf{B}_{3H}} = \{1\}$ and two ternary relations $P^{\mathbf{B}_{3H}} = \{0, 1\}^3 \setminus \{(1, 1, 0)\}$ and

$N^{\mathbf{B}_{3H}} = \{0,1\}^3 \setminus \{(1,1,1)\}$. It is easy to see that every Horn 3-CNF formula φ with variables $x_1, \ldots, x_n$ can be represented as a structure $\mathbf{A}_\varphi$ with universe $\{x_1, \ldots, x_n\}$ and relations $U^{\mathbf{A}_\varphi}$, $P^{\mathbf{A}_\varphi}$, $N^{\mathbf{A}_\varphi}$ where $U^{\mathbf{A}_\varphi}$ is the set of all unit clauses (in φ), $P^{\mathbf{A}_\varphi}$ is the set of all clauses of the form $(\neg x \vee \neg y \vee z)$, and $N^{\mathbf{A}_\varphi}$ is the set of all clauses of the form $(\neg x \vee \neg y \vee \neg z)$. Clearly, we have $\mathbf{A}_\varphi \to \mathbf{B}_{3H}$ if and only if φ is satisfiable. Hence HORN 3-SAT is precisely CSP($\mathbf{B}_{3H}$).

Example 6. Let $\mathbf{B}_{le}$ be a structure with universe $\{0,1\}$, one ternary relation $\{(x,y,z) \mid x+y+z = 1 (\text{mod } 2)\}$, and one unary relation $\{0\}$. It is well known, and easy to verify, that CSP($\mathbf{B}_{le}$) is the problem of solving systems of linear equations (with at most 3 variables per equation) over the two-element field.

For any subset I of A, any homomorphism from $\mathbf{A}_{|I}$ to $\mathbf{B}$ is called a *partial* homomorphism from $\mathbf{A}$ to $\mathbf{B}$. A *projective* homomorphism from $\mathbf{A}$ to $\mathbf{B}$ is a partial mapping h from A to B such that, for any $R \in \tau$ (say, of arity n) and any tuple $(a_1, \ldots, a_n) \in R^{\mathbf{A}}$, there exists a tuple $(b_1, \ldots, b_n) \in R^{\mathbf{B}}$ such that $h(a_i) = b_i$ for every a_i in dom(h), the domain of h. Clearly, every projective homomorphism is also a partial homomorphism.

A *retract* of a structure $\mathbf{B}$ is an induced substructure $\mathbf{B}'$ of $\mathbf{B}$ such that there is a homomorphism $g : \mathbf{B} \to \mathbf{B}'$ with $g(b) = b$ for every $b \in B'$. In this case we (trivially) have that CSP($\mathbf{B}$) and CSP($\mathbf{B}'$) coincide. A structure is called a *core* if it has no homomorphism to any of its proper substructures. A retract of $\mathbf{B}$ that has minimal size among all retracts of $\mathbf{B}$ is called a core of $\mathbf{B}$. It is well known that all cores of a structure are isomorphic, and so one speaks of *the core*, core($\mathbf{B}$), of a structure $\mathbf{B}$.

2.2 Obstructions and Dualities

In order to define some of our dualities, we will need the notions of pathwidth and treewidth of relational structures.

Definition 1. *For $0 \leq j \leq k$, a τ-structure $\mathbf{A}$ is said to have* treewidth *at most (j,k) if there is a tree T, called a* tree-decomposition *of $\mathbf{A}$, such that*

1. *the nodes of T are subsets of A of size at most k,*
2. *adjacent nodes can share at most j elements,*
3. *nodes containing any given element of A form a subtree,*
4. *for any tuple in any relation in $\mathbf{A}$, there is a node in T containing all elements from that tuple.*

If T is a path then it is called a path-decomposition *of $\mathbf{A}$, and $\mathbf{A}$ is said to have* pathwidth *at most (j,k).*

Example 7. 1. Consider the graph $\mathbf{G}$ from Fig. 1. The top-left decomposition shows that $\mathbf{G}$ has treewidth at most (1,3), the top-right and the bottom decompositions imply that $\mathbf{G}$ has pathwidth at most (1,5) and at most (2,4), respectively.

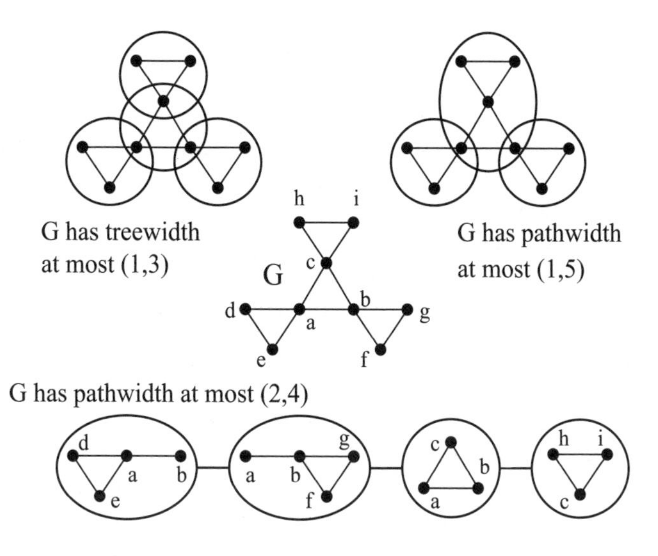

Fig. 1. Examples of pathwidth and treewidth

2. Any cycle has pathwidth at most (2,3). Indeed, assume that the nodes of the cycle are $0, 1, \ldots, n-1$ and the edges are $(i, i+1)$, where addition is modulo n. Consider a path with nodes $S_1, S_2, \ldots, S_{n-2}$ where $S_i = \{0, i, i+1\}$. It is easy to check that this is a path-decomposition of the cycle.
3. Any tree has treewidth at most (1,2). Indeed, take T to have the edges of the original tree as nodes and the adjacency relation given by the incidence relation of edges in the original tree.

Note that we use two numbers to parameterise treewidth and pathwidth, as is customary in the study of CSPs [20,28,54] (rather than one as is customary in graph theory [34]), for the following reason. The first parameter j gives a more convenient parameterisation of CSPs, since the second parameter k is bounded from below by the maximum arity of a relation in a structure, and hence it is less convenient to use for uniform treatment of structures of different vocabularies that behave essentially in the same way with respect to homomorphisms. Nevertheless, the notions of pathwidth and treewidth of relational structures are closely related to the corresponding notions from graph theory, as follows. The *Gaifman graph* $\mathbf{G}(\mathbf{A})$ of a structure $\mathbf{A}$ is defined to have the same universe (set of vertices) as $\mathbf{A}$ and the edges of $\mathbf{G}(\mathbf{A})$ are the pairs (a, a') of distinct elements such that a and a' appear in the same tuple in some relation in $\mathbf{A}$. Then it is not hard to check that the following numbers are equal:

- the minimum k such that $\mathbf{A}$ has pathwidth (treewidth) at most $(k, k+1)$,
- pathwidth (treewidth, respectively) of $\mathbf{G}(\mathbf{A})$ in the sense of graph theory.

Definition 2. *A set $\mathcal{O}$ of τ-structures is called an* obstruction set *for* $\mathbf{B}$ *if, for any τ-structure* $\mathbf{A}$, $\mathbf{A} \to \mathbf{B}$ *if and only if* $\mathbf{A}' \not\to \mathbf{A}$ *for all* $\mathbf{A}' \in \mathcal{O}$.

Note that sometimes such sets are called "complete obstruction sets".

Definition 3. *A structure* **B** *is said to have* finite duality *if it has a finite obstruction set.*

Example 8. Let $\mathbf{T}_n$ be the *transitive tournament* on n vertices, that is, the universe of $\mathbf{T}_n$ is $\{0, 1, \ldots, n-1\}$, and the only relation is the binary relation $\{(i, j) \mid 0 \leq i < j \leq n-1\}$. Also, let $\mathbf{P}_n$ be the *directed path* on $n+1$ vertices, that is the structure with universe $\{0, 1, \ldots, n\}$ and the relation $\{(i, i+1) \mid 0 \leq i \leq n-1\}$. It is well known (see, e.g., Proposition 1.20 of [34]) and easy to show that, for any digraph $\mathbf{G}$, $\mathbf{G} \to \mathbf{T}_n$ if and only if $\mathbf{P}_n \not\to \mathbf{G}$. Hence, $\{\mathbf{P}_n\}$ is an obstruction set for $\mathbf{T}_n$, and $\mathbf{T}_n$ has finite duality.

Definition 4. *A τ-structure* **B** *is said to have* (j, k)-pathwidth duality[1] *if it has an obstruction set consisting of structures of pathwidth at most (j, k). In other words,* **B** *has (j, k)-pathwidth duality if, for any τ-structure* **A***, we have* $\mathbf{A} \to \mathbf{B}$ *if and only if* $\mathbf{C} \to \mathbf{A}$ *implies* $\mathbf{C} \to \mathbf{B}$ *for every τ-structure* **C** *of pathwidth at most (j, k).*

We say that **B** *has* j-pathwidth duality *if it has (j, k)-pathwidth duality for some $k \geq j$, and* **B** *has* bounded pathwidth duality *if it has j-pathwidth duality for some $j \geq 0$.*

Example 9. It is well known that a graph **G** is 2-colourable if and only if it contains no odd cycles, which is the same as to say that **G** does not admit a homomorphism from any odd cycle. Since the 2-COLOURABILITY problem is the same as CSP($\mathbf{K}_2$), we obtain that the family of all odd cycles forms an obstruction set for $\mathbf{K}_2$. By Example 7, any cycle has pathwidth at most (2,3), so the structure $\mathbf{K}_2$ has (2,3)-pathwidth duality. It is easy to see that $\mathbf{K}_2$ does not have finite duality.

Definition 5. *By replacing "pathwidth" with "treewidth" throughout Definition 4, one obtains the corresponding definitions of* treewidth dualities.

Example 10. The structure $\mathbf{B}_{ps}$ from Example 4 has (1,3)-treewidth duality. To prove this, we need to show that, for any structure $\mathbf{A} \in$ co-CSP($\mathbf{B}_{ps}$), there exists a structure $\mathbf{C} \in$ co-CSP($\mathbf{B}_{ps}$) such that $\mathbf{C} \to \mathbf{A}$ and **C** has treewidth at most (1,3). If $\mathbf{A} \in$ co-CSP($\mathbf{B}_{ps}$) then we can choose some terminal node in **A** that can be "accessed" (or "derived") from the source nodes. It is clear that this derivation procedure can be represented as a "tree", as shown in Fig. 2. The substructure $\mathbf{A}'$ of **A** (corresponding to the derivation) is shown on the right; d and e are source nodes, t is a terminal node, and every oval depicts a unit derivation via a triple from the relation $P^{\mathbf{A}}$. Now modify the structure $\mathbf{A}'$ as follows: for every element $x \in A'$, give new names to the occurrences of x in A' so that each element in the obtained structure appears either in a single oval or else in two ovals such that this element is the intersection of the two ovals, and then modify the set of source nodes accordingly. Let **C** be the obtained structure (see Fig. 2, left). It is clear that **C** has treewidth at most (1,3). Furthermore, we have $\mathbf{C} \in$ co-CSP($\mathbf{B}_{ps}$) because a terminal node is still accessible from the

[1] Called (j, k)-path duality in [20].

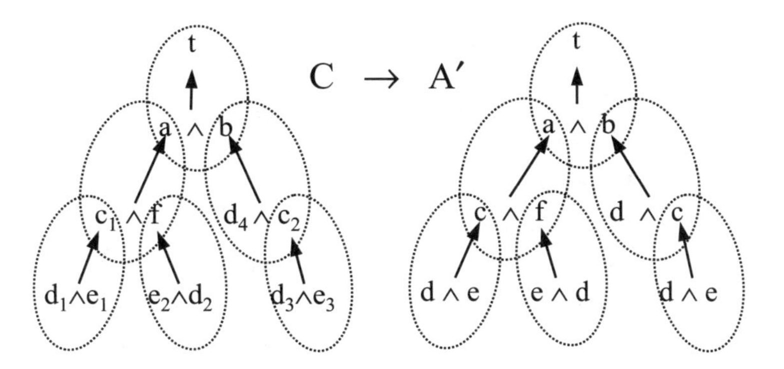

Fig. 2. (1,3)-treewidth duality for the structure $\mathbf{B}_{ps}$

source nodes, and we also have that $\mathbf{C} \to \mathbf{A}$ because the reverse renaming of elements is a homomorphism from $\mathbf{C}$ to $\mathbf{A}'$, and hence to $\mathbf{A}$.

2.3 Datalog and Infinitary Logics

For logical descriptions of the three dualities, we use first-order logic, the logic programming language Datalog, and its restriction, linear Datalog, and also some infinitary finite-variable logics. We assume that the reader is familiar with first-order logic, and we now briefly describe the basics of Datalog (for more details, see, e.g., [45]).

Fix a vocabulary τ. A *Datalog program* is a finite set of rules of the form $t_0 : - \ t_1, \ldots, t_n$ where each t_i is an atomic formula $R(x_{i_1}, \ldots, x_{i_k})$. Then t_0 is called the *head* of the rule, and the sequence $t_1, \ldots, t_n$ the *body* of the rule. The predicates occurring in the heads of the rules are not from τ and are called *IDBs* (from "intensional database predicates"), while all other predicates come from τ and are called *EDBs* (from "extensional database predicates"). One of the IDBs, which is usually 0-ary in our case, is designated as the *goal predicate* of the program. Since the IDBs may occur in the bodies of rules, each Datalog program is a recursive specification of the IDBs, with semantics obtained via least fixed-points of monotone operators. The goal predicate is assumed to be initially set to `false`, and we say that a Datalog program *accepts* a τ-structure $\mathbf{A}$ if its goal predicate evaluates to `true` on $\mathbf{A}$.

For $0 \leq j \leq k$, a (j,k)*-Datalog program* is a Datalog program with at most j variables in the head and at most k variables per rule. A Datalog program is called *linear* if every rule in it has at most one occurrence of an IDB in its body. A class $\mathcal{C}$ of structures is said to be definable in (linear) (j,k)-Datalog if there is a (linear) (j,k)-Datalog program which accepts precisely the structures from $\mathcal{C}$.

Note that, for any Datalog program, the class $\mathcal{C}$ of all structures accepted by the program is closed under extension (that is, if a structure $\mathbf{A}$ has a substructure $\mathbf{A}'$ which is in $\mathcal{C}$ then $\mathbf{A}$ is also in $\mathcal{C}$). Every class of the form co-CSP($\mathbf{B}$) has this monotonicity property, but it is not the case for CSP($\mathbf{B}$). Hence, when using

Datalog to study CSPs, one usually speaks of definability of co-CSP($\mathbf{B}$) in (some version of) Datalog.

Example 11. Consider the structure $\mathbf{B}_{3H}$ from Example 5. It is well known that Horn 3-Sat can be solved by the unit propagation algorithm which can be represented as the following Datalog program.

$$\begin{aligned} T(X) &:- U(X) \\ T(Z) &:- P(X,Y,Z), T(X), T(Y) \\ unsat &:- N(X,Y,Z), T(X), T(Y), T(Z) \end{aligned}$$

Hence, co-CSP($\mathbf{B}_{3H}$) is definable in (1,3)-Datalog.

Example 12. The following linear (2,4)-Datalog program accepts a graph (as a structure with one binary relation E) if and only if the graph is not 2-colourable

$$\begin{aligned} Odd(X,Y) &:- E(X,Y) \\ Odd(X,Y) &:- Odd(X,Z), E(Z,T), E(T,Y) \\ non2col &:- Odd(X,X) \end{aligned}$$

It is easy to see how to modify this program so that it accepts a digraph if and only if the digraph is not 2-colourable (just add all rules obtained from the second rule by permuting Z, T, and Y in the part $E(Z,T), E(T,Y)$). Hence, co-CSP($\mathbf{K}_2$) is definable in linear (2,4)-Datalog.

The notion of a *canonical* (j,k)-Datalog program for a τ-structure $\mathbf{B}$ has proved to be useful in the study of dualities [28]. Let $\tau = \{R_1, \ldots, R_n\}$, and let S_0, $S_1, \ldots, S_p$ be an enumeration of relations of arity j on B that can be expressed by a first-order $\exists\wedge$-formula over $\mathbf{B}$. Assume that S_0 is the empty relation. For each S_i, introduce a j-ary IDB I_i. Then the canonical (j,k)-Datalog program for $\mathbf{B}$ involves the IDBs $I_0, \ldots, I_p$ and EDBs $R_1, \ldots, R_n$, and contains all the rules with at most k variables with the following property: if every I_i in the rule is replaced by S_i and every R_s by $R_s^{\mathbf{B}}$, then every assignment of elements of B to the variables that satisfies the conjunction of atomic formulas in the body must also satisfy the atomic formula in the head. Finally, introduce one 0-ary IDB G together with the rule $G :- I_0(x_1, \ldots, x_j)$, and make G the goal predicate of the program. The *canonical linear* (j,k)-Datalog program for $\mathbf{B}$ consists of all linear rules from the canonical program described above.

Our definitions of *infinitary logics* are inspired by [20,45,47]. Let $L_{\infty\omega}$ be the first-order logic extended with infinitary conjunctions $\bigwedge$ and infinitary disjunctions $\bigvee$. For every $k \geq 0$, let $\exists L^k_{\infty\omega}$ be the existential positive (i.e., without negation and universal quantifiers) fragment of $L_{\infty\omega}$ with at most k different variables. A (possibly infinitary) conjunction $\bigwedge \Phi$ of $L_{\infty\omega}$-formulas is said to be *j-restricted* if every formula from Φ that contains more than j free variables is quantifier-free, and it is said to be *strongly j-restricted* if, in addition, at most one formula in Φ having quantifiers is not a sentence. Then $\exists L^{j,k}_{\infty\omega}$ is the fragment of $\exists L^k_{\infty\omega}$ obtained by using atomic formulas, existential quantification, arbitrary disjunctions, and j-restricted conjunctions. It is known that every class of structures definable in (j,k)-Datalog is also definable in $\exists L^{j,k}_{\infty\omega}$. The logic $\exists M^{j,k}_{\infty\omega}$ is defined similarly to $\exists L^{j,k}_{\infty\omega}$, but with strongly j-restricted conjunctions, and definability in linear (j,k)-Datalog implies definability in this logic.

We will also need the infinitary *counting* logics. Let $C_{\infty\omega}$ (see [3,55]) be the logic whose formulas are obtained from atomic formulas by using negation, infinitary conjunction and disjunction, and counting quantifiers ($\exists^i x$ for any $i \geq 0$). The fragment $C^k_{\infty\omega}$ consists of those formulas of $C_{\infty\omega}$ in which at most k distinct variables appear, and $C^{\omega}_{\infty\omega} = \bigcup_{k\in\omega} C^k_{\infty\omega}$.

2.4 Pebble Games

We will now define two pebble games, the pebble-relation game and the existential pebble game, which have proved to be very useful in the analysis of pathwidth and treewidth dualities. These games have been introduced in [20] and [46], respectively.

Let $0 \leq j \leq k$, and let $\mathbf{A}$ and $\mathbf{B}$ be τ-structures. The *(j,k)-pebble-relation* (or *(j,k)-PR) game* on $(\mathbf{A},\mathbf{B})$ is played between two players, the *Spoiler* and the *Duplicator.* A configuration of the game consists of a subset $I \subseteq A$ with $|I| \leq k$ and a collection of partial homomorphisms $T \subseteq \hom(\mathbf{A}_{|I},\mathbf{B})$. If $T \subseteq \hom(\mathbf{A}_{|I},\mathbf{B})$ then we say that I is the *domain* of T. For a subset $J \subseteq I$, let $T_{|J}$ denote the set $\{f_{|J} \mid f \in T\}$.

Initially, $I = \emptyset$ and T contains the (unique) homomorphism from $\mathbf{A}_{|\emptyset}$ to $\mathbf{B}$. Each round of the game consists of a move of the Spoiler and a move of the Duplicator. Intuitively, the Spoiler has control on the domain I of T, which can be regarded as placing some pebbles on the elements of A that constitute I, whereas the Duplicator decides the content of T after the domain I has been set by the Spoiler. There are two types of rounds: *shrinking* and *blowing* rounds.

Let T^n be the configuration after the n-th round. The Spoiler decides whether the following round is a blowing or shrinking round.

- If the $(n+1)$-th round is a shrinking round, the Spoiler sets I^{n+1} to be a non-empty subset of the domain I^n of T^n. The Duplicator responds by restricting every function in T^n onto I^{n+1}, that is, $T^{n+1} = T^n_{|I^{n+1}}$.
- A blowing round only can be performed if $|I^n| \leq j$. In this case the Spoiler sets I^{n+1} to be a superset of I^n with $|I^{n+1}| \leq k$. The Duplicator responds by providing a family $T^{n+1} \subseteq \hom(\mathbf{A}_{|I^{n+1}},\mathbf{B})$ such that $T^{n+1}_{|I^n} \subseteq T^n$.

The Spoiler wins the game if the response of the Duplicator sets T^{n+1} to $\emptyset$, i.e., the Duplicator cannot extend successfully any of the partial homomorphisms from T^n. Otherwise, the game resumes. The Duplicator wins the game if he has a strategy that allows him to play "forever", i.e., if the Spoiler can never win a round of the game. The notion of winning strategy for the Duplicator can be conveniently formalised as follows.

Definition 6. *Let $0 \leq j < k$, and let $\mathbf{A}$ and $\mathbf{B}$ be τ-structures. We say that* the Duplicator has a winning strategy for the (j,k)-pebble-relation game *on $(\mathbf{A}$,$\mathbf{B})$ if there is a non-empty family $\mathcal{H}$ of sets of partial homomorphisms such that:*

1. *for every* $T \in \mathcal{H}$, $T \subseteq \hom(\mathbf{A}_{|I}, \mathbf{B})$ *for some* $I \subseteq A$, $|I| \leq k$, *and* $\emptyset \notin T$,
2. $\mathcal{H}$ *is closed under restrictions: for every* $T \in \mathcal{H}$ *with domain* I *and every* $I' \subseteq I$, *we have that* $T_{|I'} \in \mathcal{H}$,
3. $\mathcal{H}$ *has the* (j,k)*-forth property: for every* $T \in \mathcal{H}$ *with domain* I, $|I| \leq j$, *and every superset* I' *of* I *with* $|I'| \leq k$, *there exists* $T' \in \mathcal{H}$ *with domain* I' *such that* $T'_{|I} \subseteq T$.

The intuition behind the above definition is that every set T in a winning strategy corresponds to a winning configuration for the Duplicator in the game.

If we impose the restriction that every configuration in the (j,k)-PR game consists of a single function (i.e., in every round, the Duplicator commits to a particular partial homomorphism) then the obtained game is known as the *existential* (j,k)*-pebble game*. The notion of a winning strategy for the Duplicator in this game is obtained in a natural way from the one in Definition 6, by restricting each set T to consist of a single partial homomorphism.

Note that if we have a homomorphism $h : \mathbf{A} \to \mathbf{B}$ then the Duplicator always has a winning strategy in any PR or existential pebble game on $(\mathbf{A}, \mathbf{B})$: to win, the Duplicator only has to always include the suitable restriction of the homomorphism h in his response. However, the converse does not always hold. That is, the existence of a winning strategy for the Duplicator on $(\mathbf{A}, \mathbf{B})$ does not, in general, imply that $\mathbf{A} \to \mathbf{B}$ (see Example 13 below). Thus, the structures $\mathbf{B}$, for which the converse also holds (for a particular type of game), must have some special properties. These properties are closely related with dualities, as we will discuss in Sections 4 and 5.

Example 13. Let $\mathbf{A}$ be the undirected cycle with 5 nodes and $\mathbf{B}$ the undirected cycle with 6 nodes. Obviously, we have $\mathbf{A} \not\to \mathbf{B}$, but the Duplicator still wins the existential (1,2)-pebble game. Indeed, fix any two adjacent elements, b_1 and b_2 in $\mathbf{B}$, and let the winning strategy simply contain all partial homomorphisms that have at most two-element domains and range $\{b_1, b_2\}$. It is straightforward to check that this is indeed a winning strategy. However, it is not hard to verify that the Spoiler wins the existential (2,3)-pebble game on $(\mathbf{A}, \mathbf{B})$.

2.5 Algebraic Background

The algebraic approach to constraint satisfaction (see, e.g., [11,12,13,17,50]) has proved to be extremely successful. It provides a convenient dual language to analyse CSPs, and, more importantly, allows one to use powerful machinery from universal algebra.

First, let us formally define polymorphisms of relations and structures.

Definition 7. *Let* f *be an* n*-ary operation on* B, *and* R *a relation on* B. *Then* f *is said to be a* polymorphism *of* R *(or* R *is* invariant *under* f*) if, for any tuples* $\bar{a}_1, \ldots, \bar{a}_n \in R$, *the tuple obtained by applying* f *componentwise also belongs to the relation* R.

An operation is called a polymorphism of a relational structure if it is a polymorphism of every relation in the structure. Let $\mathrm{Pol}(\mathbf{B})$ *denote the set of all polymorphisms (of all arities) of a structure* $\mathbf{B}$.

For τ-structures $\mathbf{B}_1, \ldots, \mathbf{B}_n$, define the *direct product* structure $\mathbf{C} = \prod_{i=1}^n \mathbf{B}_i$ to be a τ-structure with base set $C = B_1 \times \ldots \times B_n$, and, for any m-ary $R \in \tau$, let $(\mathbf{a}_1, \ldots, \mathbf{a}_m) \in R^{\mathbf{C}}$ if and only if $(\mathbf{a}_1[i], \ldots, \mathbf{a}_m[i]) \in R^{\mathbf{B}_i}$ for each $1 \leq i \leq n$. As usual, the direct product of n copies of a structure $\mathbf{B}$ is called the *n-th power* of $\mathbf{B}$, and is denoted $\mathbf{B}^n$. It is easy to check that the n-ary polymorphisms of $\mathbf{B}$ are precisely the homomorphisms from $\mathbf{B}^n$ to $\mathbf{B}$.

Example 14. It is straightforward to verify that the Boolean relation $OR = \{0,1\}^2 \setminus \{(0,0)\}$ is invariant under the binary operation max on $\{0,1\}$, but is not invariant under the operation min.

One nice feature of the polymorphisms is that they allow one to simultaneously deal with structures over different vocabularies. For example, it is known (see [12] or [42]) that if τ_1-structure $\mathbf{B}_1$ and τ_2-structure $\mathbf{B}_2$ have the same universe and $\text{Pol}(\mathbf{B}_1) \subseteq \text{Pol}(\mathbf{B}_2)$ then every relation in $\mathbf{B}_2$ can be defined by a primitive positive first-order formula (i.e., $\exists\wedge$-formula with equality) in $\mathbf{B}_1$, and hence the problem $\text{CSP}(\mathbf{B}_2)$ is polynomial-time (even logarithmic-space) reducible to $\text{CSP}(\mathbf{B}_1)$. In particular, if $\text{Pol}(\mathbf{B}_1) = \text{Pol}(\mathbf{B}_2)$ then $\text{CSP}(\mathbf{B}_1)$ and $\text{CSP}(\mathbf{B}_2)$ are equivalent. Hence, it is very convenient to group relational structures according to their polymorphisms. Note that sets of operations of the form $\text{Pol}(\mathbf{B})$ are *clones* of operations, they are well-studied objects in universal algebra (see, e.g., [63]).

We will now define some types of operations which will be useful in the subsequent sections.

Definition 8. *An n-ary operation f on B is called* idempotent *if it satisfies the identity $f(x, \ldots, x) = x$.*

- *A binary commutative idempotent operation f is called a* 2-semilattice *operation if it satisfies the identity $f(x, f(x,y)) = f(x,y)$.*
- *An n-ary ($n \geq 2$) operation f is called* totally symmetric *if $f(x_1, \ldots, x_n) = f(y_1, \ldots, y_n)$ whenever $\{x_1, \ldots, x_n\} = \{y_1, \ldots, y_n\}$. If, in addition, f is idempotent then we say that it is a TSI operation.*
- *An n-ary ($n \geq 3$) operation is called an* NU (near-unanimity) *operation if it satisfies the identities*

$$f(y, x, \ldots, x, x) = f(x, y, \ldots, x, x) = \ldots = f(x, x, \ldots, x, y) = x.$$

- *A ternary NU operation is called a* majority *operation.*
- *An n-ary ($n \geq 2$) idempotent operation is called a* WNU (weak NU) *operation if it satisfies the identities*

$$f(y, x, \ldots, x, x) = f(x, y, \ldots, x, x) = \ldots = f(x, x, \ldots, x, y).$$

Example 15. 1. For any binary idempotent operation f, the following conditions are equivalent: (a) f is a TSI operation, (b) f is a WNU operation, and (c) f is commutative.

2. A binary operation g is called conservative if $g(a,b) \in \{a,b\}$ for all a,b. Any binary commutative conservative operation is a 2-semilattice operation.
3. Let f be a binary idempotent commutative associative operation. Then f is called a semilattice operation. It is easy to see that f is also a 2-semilattice operation, and, for any $n \geq 2$, the operation $f(x_1, f(x_2, f(\ldots, f(x_{n-1}, x_n)\ldots)$ is a TSI operation.
4. It is easy to check that the (ternary) median operation on a totally ordered set is a majority operation.
5. Any TSI operation and any NU operation is a WNU operation. Also, the Boolean affine operation $f(x,y,z) = x+y+z(mod\ 2)$ is a WNU operation.

Example 16. Schaefer's celebrated dichotomy theorem for Boolean CSP can be restated (see, e.g., [12,17,50]) as follows. For a Boolean core structure $\mathbf{B}$, if $\mathbf{B}$ has a semilattice polymorphism, or a majority polymorphism, or the affine polymorphism, then CSP($\mathbf{B}$) is in **PTIME**. In all other cases, CSP($\mathbf{B}$) is **NP**-complete. A refinement of this theorem, including a classification for definability in Datalog and its restrictions, can be found in [52].

The subsequent definitions in this subsection are sketchy, for more details see the surveys [13,16] or the monograph [40].

Definition 9. *A* finite algebra *is a pair* $\mathbb{A} = (A, F)$ *where* A *is a finite set and* $F = (f_i)_{i \in I}$ *is a family of finitary operations on* A. *For a relational structure* $\mathbf{B}$, *the algebra* $\mathbb{A}_{\mathbf{B}} = (B, \mathrm{Pol}(\mathbf{B}))$ *is called the* algebra associated with $\mathbf{B}$.

Definition 10. *A* variety *is a class of algebras closed under taking homomorphic images, subalgebras, and (possibly infinite) direct products. The variety* generated *by a finite algebra* $\mathbb{A}$, *denoted* var($\mathbb{A}$), *consists of all homomorphic images of subalgebras of direct powers of* $\mathbb{A}$.

Every finite algebra $\mathbb{A}$ can be assigned a set of *types*. The types are numbers from **1** to **5**, and they correspond to different possible basic "local behaviours" of the algebra. The correspondence is as follows:

type **1** – unary algebra,
type **2** – vector space over a finite field,
type **3** – 2-element Boolean algebra,
type **4** – 2-element lattice,
type **5** – 2-element semilattice.

A variety is said to *admit* a type **i** if this type occurs in some finite algebra in the variety, and it *omits* type **i** otherwise.

It is known (see [12,13,50]) that if, for a core structure $\mathbf{B}$, the variety var($\mathbb{A}_{\mathbf{B}}$) admits type **1** (or, equivalently, $\mathbf{B}$ has no WNU polymorphism of any arity [57]) then CSP($\mathbf{B}$) is **NP**-complete. Moreover, all core structures $\mathbf{B}$ that are known to give rise to **NP**-complete problems CSP($\mathbf{B}$) do satisfy this condition. It has been conjectured that all other core structures give rise to problems in **PTIME**, and this conjecture has been confirmed in many important cases (see, e.g., [7,10,12,13]). For other results about the correspondence between the type set of var($\mathbb{A}_{\mathbf{B}}$) on one side and the computational and descriptive complexity of CSP($\mathbf{B}$) on the other side, see [3,13,52].

3 Finite Duality

Arguably, the simplest case of duality is that of finite duality. In this section, we outline several characterisations of constraint satisfaction problems with this property. We shall address, in particular, questions about the relationship of finite duality to definability in first-order logic (FO), the nature of the obstruction set of a structure with finite duality, and the (meta-)problem of recognising such structures.

Recall from Example 8 that the transitive tournament $\mathbf{T}_n$ has an obstruction set consisting of a single structure $\mathbf{P}_n$. In general, a structure with finite duality might not have a set of obstructions that consists of a single structure:

Example 17. Let $\mathbf{B}=\langle\{0,1\};R,\{0\},\{1\}\rangle$ where $R=\{(0,0),(0,1),(1,0)\}$. Viewing structures of this type as coloured digraphs (with colours given by the unary relations), it is easy to see that $\mathbf{A}\not\to\mathbf{B}$ if and only if there exists a vertex v of $\mathbf{A}$ which is coloured with both colours 0 and 1, or an edge (a,b) with both endpoints coloured 1. Consequently $\mathbf{B}$ has a two-element obstruction set, one structure $\mathbf{A}_1$ consisting of a single vertex with two colours, the other structure $\mathbf{A}_2$ consisting of one directed edge with both ends coloured 1. It is easy to see that $\mathbf{B}$ does not have a one-element obstruction set.

Example 18. Recall the problem UNREACHABILITY, or CSP($\mathbf{B}_{unr}$) from Example 3. It is not difficult to see that $\mathbf{B}_{unr}$ does not have finite duality. As in Example 17, we can view structures as coloured digraphs. Note that any path with ends coloured 0 and 1 does not have a homomorphism to $\mathbf{B}_{unr}$, but any proper substructure of the path does. If $\mathcal{O}$ is a finite obstruction set for $\mathbf{B}_{unr}$, then one can find a long enough path $\mathbf{P}$ (with coloured ends) such that every structure in $\mathcal{O}$ can have only non-surjective homomorphisms (if any) to $\mathbf{P}$. Hence, either none of the structures in $\mathcal{O}$ has a homomorphism to $\mathbf{P}$ or some structure in $\mathcal{O}$ has a homomorphism to $\mathbf{B}_{unr}$. In either case, $\mathcal{O}$ cannot be an obstruction set for $\mathbf{B}_{unr}$.

It is easy to see that if a structure $\mathbf{B}$ has finite duality then CSP($\mathbf{B}$) is FO-definable; in fact, co-CSP($\mathbf{B}$) is definable in existential positive FO, or said differently, it is definable in Datalog without IDBs other than the goal predicate. Indeed, let $\mathbf{C}$ be a τ-structure with $C=\{c_1,\ldots,c_l\}$, and consider the following sentence $T_{\mathbf{C}}=\exists x_1\ldots\exists x_l\bigwedge_{R\in\tau}\bigwedge_{(c_{i_1},\ldots,c_{i_r})\in R^{\mathbf{C}}}R(x_{i_1},\ldots,x_{i_r})$. It is well known and easy to check that, for any τ-structure $\mathbf{A}$, we have $\mathbf{C}\to\mathbf{A}$ if and only if $\mathbf{A}$ satisfies $T_{\mathbf{C}}$. Hence, if $\mathcal{O}$ is a finite obstruction set for a τ-structure $\mathbf{B}$, then a τ-structure $\mathbf{A}$ belongs to co-CSP($\mathbf{B}$) if and only if the sentence $\bigvee_{\mathbf{C}\in\mathcal{O}}T_{\mathbf{C}}$ holds true in $\mathbf{A}$.

Atserias ([2], see also [62]) has shown that the converse also holds: if CSP($\mathbf{B}$) is FO-definable then $\mathbf{B}$ has finite duality. We now show how this result follows from other, more recent, results.

Theorem 1 ([52]). *If a structure* $\mathbf{B}$ *does not have finite duality then* CSP($\mathbf{B}$) *is* **LOGSPACE**-*hard under first-order reductions.*

Recall that the complexity class **non-uniform AC^0** consists of all languages accepted by polynomial-size constant-depth families of Boolean circuits (see, e.g., [55]). It is known that any FO-definable class of structures belongs to this complexity class (see Theorem 6.4 of [55]). Moreover, any problem which is **LOGSPACE**-hard under first-order reductions cannot lie in **non-uniform AC^0** because there are problems in **LOGSPACE** which are not in **non-uniform AC^0** (see [30]) and **non-uniform AC^0** is closed under first-order reductions. These facts and Theorem 1 imply the following result.

Theorem 2. *For any structure* $\mathbf{B}$*, the following conditions are equivalent:*

1. $\mathbf{B}$ *has finite duality.*
2. CSP($\mathbf{B}$) *is* FO*-definable.*
3. CSP($\mathbf{B}$) *is in* **non-uniform AC^0**.

Let us now consider the question about the nature of finite obstruction sets.

Definition 11. *Let* $\mathbf{A}$ *be a* τ*-structure. The* incidence multigraph *of* $\mathbf{A}$*, denoted* $Inc(\mathbf{A})$*, is defined as the bipartite multigraph with parts* A *and* $Block(\mathbf{A})$*, where* $Block(\mathbf{A})$ *consists of all pairs* $(R, \overline{a})$ *such that* $R \in \tau$ *and* $\overline{a} \in R^{\mathbf{A}}$*, and with edges* $e_{a,i,Z}$ *joining* $a \in A$ *to* $Z = (R, (a_1, \ldots, a_r)) \in Block(\mathbf{A})$ *when* $a_i = a$*. We say that the structure* $\mathbf{A}$ *is a* τ-tree *(or simply a* tree*) if its incidence multigraph is a tree (in particular, it has no multiple edges).*

Theorem 3 ([59,60]). *If a finite structure has finite duality, then it admits an obstruction set consisting of finitely many trees. Conversely, for any finite set* $\mathcal{O}$ *of trees, there is a structure* $\mathbf{B}$ *that can be explicitly constructed from* $\mathcal{O}$ *such that* $\mathcal{O} = \mathcal{O}_{\mathbf{B}}$*.*

Note that the structure $\mathbf{B}$ obtained in the above theorem may not be a core; in fact, it may be much larger than its core.

We now give an algebraic characterisation of structures with finite duality.

Definition 12. *Let* R *be a relation on the set* A*. An* n*-ary operation* f *on* A *is a* 1-tolerant *polymorphism of* R *if, for any tuples* $\bar{a}_1, \ldots, \bar{a}_n$ *at least* $n-1$ *of which belong to* R*, the tuple obtained by applying* f *componentwise also belongs to* R*.*

Theorem 4 ([51]). *A structure* $\mathbf{B}$ *has finite duality if and only if its core has a 1-tolerant NU polymorphism.*

In fact, the arity of such a 1-tolerant NU polymorphism is determined by the total number of tuples in the relations of minimal obstructions. A structure $\mathbf{A}$ is a *critical obstruction* of $\mathbf{B}$ if $\mathbf{A} \not\to \mathbf{B}$ and $\mathbf{A}' \to \mathbf{B}$ for any proper substructure $\mathbf{A}'$ of $\mathbf{A}$. Call any tuple of any relation of a structure $\mathbf{A}$ a *hyperedge* of this structure. Then we have the following:

Theorem 5 ([51]). *The core of* $\mathbf{B}$ *admits a 1-tolerant NU polymorphism of arity* $n+1$ *if and only if each critical obstruction of* $\mathbf{B}$ *has at most* n *hyperedges.*

Example 19. 1. The transitive tournament $\mathbf{T}_n$ of Example 8 admits a 1-tolerant NU polymorphism of arity $n+1$, but none of smaller arity (even though it has a majority polymorphism).
2. The structure $\mathbf{B}$ of Example 17 admits a 1-tolerant NU polymorphism of arity 4, but not 3. Indeed, if m was a ternary 1-tolerant NU polymorphism of the relation R of $\mathbf{B}$, we would have $(1,1) = (m(1,1,0), m(1,0,1)) \in R$, which is false. On the other hand, it is straightforward to check that the 4-ary operation f such that $f(x_1,\ldots,x_4) = 1$ if and only if at most one x_i is equal to 0 is a 1-tolerant NU polymorphism of $\mathbf{B}$. This structure also has a majority polymorphism.

Core structures with finite duality that admit a majority polymorphism were described in [56]. For a τ-tree $\mathbf{A}$, we say that an element of A is a *leaf* if it is incident to exactly one block in $Inc(\mathbf{A})$. A block of $\mathbf{A}$ (i.e., a member of $Block(\mathbf{A})$) is said to be *pendant* if it is incident to at most one non-leaf element, and it is said to be *non-pendant* otherwise. We say that a τ-tree is a *τ-caterpillar* (or simply a caterpillar) if each of its blocks is incident to at most two non-leaf elements, and each element is incident to at most two non-pendant blocks.

Theorem 6 ([56]). *Let* $\mathbf{B}$ *be a core with finite duality. Then* $\mathbf{B}$ *has a majority polymorphism if and only if it has an obstruction set consisting of finitely many caterpillars.*

Call a relation R on B *biredundant* if the projection of R onto some two coordinates is the equality relation on some subset $C \subseteq B$ with $|C| \geq 2$.

Theorem 7 ([25,51]). *Let* $\mathbf{B}_1$ *and* $\mathbf{B}_2$ *be structures such that* $\mathbf{B}_1$ *is a core with finite duality and* $\mathrm{Pol}(\mathbf{B}_1) \subseteq \mathrm{Pol}(\mathbf{B}_2)$*. Then the following holds.*

1. *If* $\mathbf{B}_2$ *does not have finite duality then* $\mathrm{CSP}(\mathbf{B}_2)$ *is* **LOGSPACE***-complete.*
2. *If none of the relations in* $\mathbf{B}_2$ *is biredundant then* $\mathbf{B}_2$ *also has finite duality. If* $\mathbf{B}_2$ *is a core then the converse holds as well.*

Example 20. We will now describe Boolean structures that are cores with finite duality. (Boolean non-core structures trivially have this property). It can be derived from [51,52] that these are precisely the (Boolean) core structures $\mathbf{B}$ without biredundant relations and such that (at least) one of the ternary operations $x \vee (y \wedge \bar{z})$ and $x \wedge (y \vee \bar{z})$ is a polymorphism of $\mathbf{B}$.

We shall now describe a simple algorithm to determine if a structure $\mathbf{B}$ has finite duality. A slight modification of this algorithm also provides a way of producing solutions of a CSP with finite duality. First, we require a few straightforward definitions.

Definition 13. *Let* $\mathbf{A}$ *be a structure and let* $a, b \in A$*. We say that the element* a dominates *the element* b *if, in any tuple* $\bar{t}$ *in any relation* R *in* $\mathbf{A}$*, replacement in* $\bar{t}$ *of any number of occurrences of* b *by* a *yields a tuple also in* R*.*

For example, if a dominates b and $(b, c, b) \in R$ then (a, c, b), (b, c, a), and (a, c, a) are all in R. Note that this notion is a direct generalisation of the notion of domination in graph theory.

Recall from Section 2.5 the definition of the n-th power of a structure. Obviously, the second power $\mathbf{B}^2$ of a structure $\mathbf{B}$ is called the *square* of $\mathbf{B}$. The *diagonal* $\Delta(\mathbf{B}^2)$ of the square $\mathbf{B}^2$ is the substructure of $\mathbf{B}^2$ induced by the set $\{(b, b) \mid b \in B\}$. Note that $\Delta(\mathbf{B}^2)$ is isomorphic to $\mathbf{B}$.

Definition 14. *A structure* $\mathbf{A}$ *is said to* dismantle *to its substructure* $\mathbf{C}$ *if there exists a sequence of induced substructures* $\mathbf{A}_0, \ldots, \mathbf{A}_k$ *of* $\mathbf{A}$ *such that (i)* $\mathbf{A}_0 = \mathbf{A}$, *(ii)* $\mathbf{A}_k = \mathbf{C}$ *and (iii) for each* $0 \leq j < k$ *the structure* $\mathbf{A}_{j+1}$ *is obtained from* $\mathbf{A}_j$ *by removal of a dominated element of* $\mathbf{A}_j$*.*

It is known [51] that the procedure of dismantling can always be done greedily, by successively removing arbitrary dominated elements in substructures of $\mathbf{A}$ to eventually obtain $\mathbf{C}$.

Theorem 8 ([51]). *A structure* $\mathbf{B}$ *has finite duality if and only if it has a retract* $\mathbf{A}$ *whose square* $\mathbf{A}^2$ *dismantles to its diagonal* $\Delta(\mathbf{A}^2)$*.*

Example 21. Consider the tournament $\mathbf{T}_3$ (see Example 8). We know that $\mathbf{T}_3$ is a core with finite duality, so its square $\mathbf{T}_3^2$ should dismantle to its diagonal. We will now show that this is indeed the case. The process of dismantling is shown on Fig. 3. The digraph $\mathbf{T}_3^2$ is shown in Fig. 3, top-left. The vertices $(2, 0)$ and $(0, 2)$ are dominated by all vertices, so they are removed, and the resulting digraph is shown in Fig. 3, top-right. Next, the vertices $(1, 0)$ and $(0, 1)$ are now dominated by $(0, 0)$, so they are are removed (see Fig. 3, bottom-left). Finally, the vertices $(1, 2)$ and $(2, 1)$ are now dominated by $(2, 2)$, so they are removed as well, which leaves only the diagonal $\Delta(\mathbf{T}_3^2)$, shown in Fig. 3, bottom-right.

From Theorem 8, the problem of recognising structures with finite duality is in **NP**. Indeed, one only needs to guess a mapping ϕ from B onto its subset A, and then to check that the induced (by A) substructure $\mathbf{A}$ of $\mathbf{B}$ is a retract of $\mathbf{B}$ (via ϕ), then to form the square $\mathbf{A}^2$ and, finally, to check (greedily) that $\mathbf{A}^2$ dismantles to its diagonal, which clearly can all be done in polynomial time.

Theorem 9 ([51])

1. *The problem of deciding whether a given structure* $\mathbf{B}$ *has finite duality is* **NP***-complete.*
2. *The problem of deciding whether a given structure* $\mathbf{B}$ *is a core with finite duality is in* **PTIME***.*

We now present a slight modification of this algorithm which will yield a solution to the CSP when one exists. In a product $\mathbf{A} \times \mathbf{B}$, an element (a, b) is said to be *dominated in the second coordinate* if it is dominated by an element of the form (a, b'). We say that $\mathbf{A} \times \mathbf{B}$ *dismantles in the second coordinate* to its substructure $\mathbf{C}$ if $\mathbf{C}$ can be obtained from $\mathbf{A} \times \mathbf{B}$ by successively removing elements that are dominated in the second coordinate.

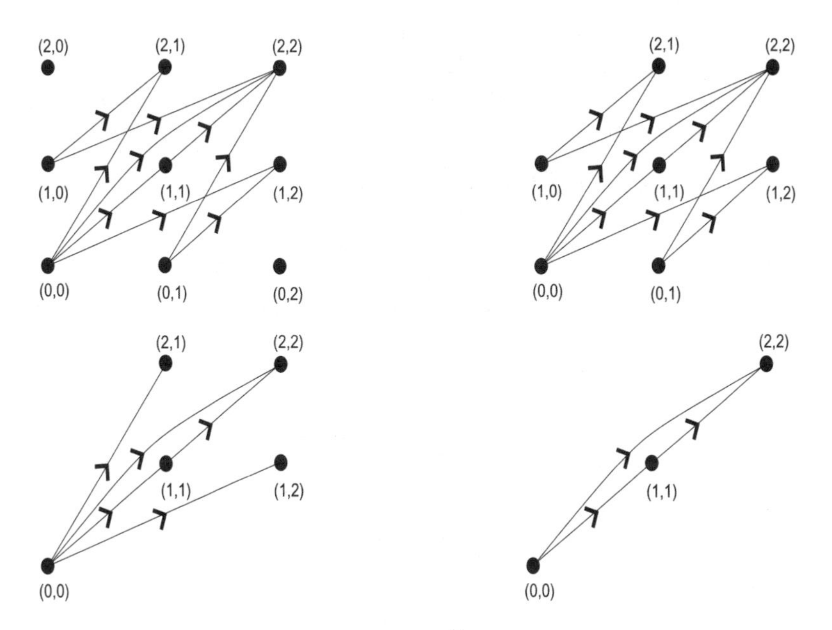

Fig. 3. Dismantling $\mathbf{T}_3^2$ to its diagonal

Theorem 10 ([51]). *Let* **B** *be a core with finite duality and let* **A** *be a structure similar to* **B**. *Let* **C** *be a structure with no dominations which is obtained from* $\mathbf{A} \times \mathbf{B}$ *by dismantling in the second coordinate. Then* $\mathbf{A} \to \mathbf{B}$ *if and only if* **C** *is the graph of a homomorphism from* **A** *to* **B**.

In other words, the procedure is as follows: (i) dismantle greedily the product $\mathbf{A} \times \mathbf{B}$ in the second coordinate until no dominations are left; (ii) check if the resulting set is of the form $C = \{(a, \phi(a)) : a \in A\}$ for some map $\phi : A \to B$; if it is, verify that ϕ is a homomorphism. Then either it is and ϕ is the desired solution, or else there is no homomorphism from **A** to **B**. Note that the result remains valid not only for cores, but for any structure **B** whose square dismantles to the diagonal.

4 Bounded Pathwidth Duality

In this section we consider bounded pathwidth duality, which is a property shared by all structures **B** such that CSP(**B**) is currently known to belong to **NL**. The following result ties together pathwidth dualities, linear Datalog, and PR games.

Theorem 11 ([20]). *For any structure* **B**, *the following conditions are equivalent:*

1. **B** *has* (j,k)*-pathwidth duality.*
2. co-CSP(**B**) *is definable in linear* (j,k)*-Datalog.*

3. co-CSP(**B**) *is definable by the canonical linear* (j,k)*-Datalog program for* **B**.
4. co-CSP(**B**) *is definable in* $\exists M^{j,k}_{\infty\omega}$.
5. CSP(**B**) *is the class of all structures* **A** *such that the Duplicator wins the* (j,k)*-PR game on* $(\mathbf{A},\mathbf{B})$.

If these conditions hold then CSP(**B**) *is in* **NL**.

Dalmau [20] also provides other equivalent conditions, including definability in other infinitary finite-variable logics and in fragments of second-order logic.

We will now give examples of structures with bounded pathwidth duality.

Example 22. An *oriented path* is a digraph obtained from a path by orienting its edges in some way. A digraph is called a *local tournament* if the set of out-neighbours of any vertex induces a tournament. For example, all transitive tournaments and all directed paths (see Example 8) are local tournaments. It was shown in [37,38] that any digraph **H** that is an oriented path or an acyclic local tournament has an obstruction set consisting of oriented paths. Since any oriented path has pathwidth at most (1,2), it follows that **H** has (1,2)-pathwidth duality.

Example 23. An *oriented cycle* is a digraph obtained from a cycle by orienting its edges in some way. An oriented cycle is called *balanced* if it has the same number of edges in one direction and in the other, and it is *unbalanced* otherwise. It was shown in [39] that any unbalanced oriented cycle **H** has an obstruction set consisting of oriented paths and oriented cycles. Since oriented cycles have pathwidth at most (2,3) (see Example 7), such a digraph **H** has (2,3)-pathwidth duality. Moreover, if the difference between the number of edges in **H** going in one direction and the number of edges in the other direction is exactly one then **H** has an obstruction set consisting only of oriented paths [39], and so it has (1,2)-pathwidth duality.

Example 24. A binary relation on B is called *implicational* (or $0/1/all$) if it has one of the following three forms: (1) $C \times D$ for some $C, D \subseteq B$, (2) $\{(c, f(c)) \mid c \in C\}$ for some $C \subseteq B$ and some permutation f on B, (3) $(\{c\} \times D) \cup (C \times \{d\})$ for some $C, D \subseteq B$, $c \in C$, and $d \in D$. A structure is called implicational if all of its relations are such. For example, it is easy to show (or see [20]) that the 2-SAT problem can be represented as CSP(**B**) for an implicational structure **B** (with universe $\{0,1\}$). It was shown in [20] that every implicational structure has (2,3)-pathwidth duality.

Example 25. The class of *implicative hitting-set bounded* (IHS-B) relations was introduced in [19]. For $k \geq 2$, a Boolean relation is in k-IHS-B+ if it can be expressed as a CNF where each clause is of the form $\neg x$, $\neg x \vee y$, or $x_1 \vee \ldots \vee x_k$. Dually, a Boolean relation is in k-IHS-B$-$ if it can be expressed as a CNF where each clause is of the form x, $\neg x \vee y$, or $\neg x_1 \vee \ldots \vee \neg x_k$. It was shown in [20] that any structure $\mathbf{B}_{ihs}$ (with universe $\{0,1\}$) all whose relations are in k-IHS-B+ (or in k-IHS-B$-$) has $(k, k-1+\mathrm{maxar}(\mathbf{B}_{ihs}))$-pathwidth duality.

We mentioned in Section 2.5 that the polymorphisms of a structure $\mathbf{B}$ determine the complexity of CSP($\mathbf{B}$). Similarly, the polymorphisms determine whether a structure has bounded pathwidth duality.

Theorem 12 ([25,52]). *Let* $\mathbf{B}_1$ *and* $\mathbf{B}_2$ *be relational structures with the same universe and such that* $\mathrm{Pol}(\mathbf{B}_1) \subseteq \mathrm{Pol}(\mathbf{B}_2)$. *If* co-CSP($\mathbf{B}_1$) *is definable in linear Datalog, then so is* co-CSP($\mathbf{B}_2$).

For a structure $\mathbf{B}$, let $\mathbf{B}_c$ denote the structure obtained from $\mathbf{B}$ by adding all elements of B as singleton unary relations.

Theorem 13 ([52]). *For a core structure* $\mathbf{B}$, co-CSP($\mathbf{B}$) *is definable in linear Datalog if and only if* co-CSP($\mathbf{B}_c$) *is.*

Note that the polymorphisms of the structure $\mathbf{B}_c$ in the above theorem are the idempotent polymorphisms of $\mathbf{B}$. Hence, for core structures, the idempotent polymorphisms determine whether a structure has bounded pathwidth duality.

The only currently known sufficient algebraic condition for general structures to have bounded pathwidth duality is given by the following result:

Theorem 14 ([22]). *If* $|B| = k$ *and* $\mathbf{B}$ *has a majority polymorphism then* $\mathbf{B}$ *has* $(3k+2, 3k+\mathrm{maxar}(\mathbf{B}))$*-pathwidth duality.*

Note that Theorem 14 can be used to obtain bounded pathwidth duality for all structures from Examples 22-24 (though, with worse bounds). For example, it was shown in [26] that any oriented path and any unbalanced oriented cycle has a majority polymorphism, and the same can be shown for any acyclic local tournament. If $\mathbf{B}$ is an implicational structure then, as shown in [18], $\mathbf{B}$ has a majority polymorphism of a very specific form, the so-called dual discriminator.

For certain types of structures $\mathbf{B}$, the presence of a majority polymorphism is the dividing line, for CSP($\mathbf{B}$), between membership in **PTIME** (which, by Theorems 11 and 14, becomes membership in **NL**) and **NP**-completeness, that is, either $\mathbf{B}$ has a majority polymorphism or else CSP($\mathbf{B}$) is **NP**-complete. For example, this is the case when $\mathbf{B}$ is a structure $\mathbf{B}_{lhc}$ (from Example 2) whose underlying digraph $\mathbf{H}$ is undirected [27]. A combinatorial description of the boundary for the classification can also be found in [27].

The structures $\mathbf{B}_{ihs}$ from Example 25, with $k \geq 3$ have bounded pathwidth duality, but do not have a majority polymorphism. However these structures are known to have an NU polymorphism of arity $k+1$. Furthermore, it follows from known algebraic results (see, e.g., [63]) that a Boolean core structure $\mathbf{B}$ has an NU polymorphism (of some arity) if and only if $\mathbf{B}$ has a majority polymorphism or it is a structure of the form $\mathbf{B}_{ihs}$. Moreover, it can be derived from [52] that in all other cases $\mathbf{B}$ does not have bounded pathwidth duality. That is, we obtain the following result:

Theorem 15. *Let* $|B| = 2$. *Then* $\mathbf{B}$ *has bounded pathwidth duality if and only* $\mathbf{B}$ *has an NU polymorphism of some arity.*

It is not known whether the presence of an NU polymorphism is a sufficient condition for general structures to have bounded pathwidth duality. However, it is known that, in general, this condition is not necessary, i.e., there exist structures **B** such that **B** has bounded pathwidth duality, but no NU polymorphism of any arity. The simplest (known) structure **B** with these properties is obtained as follows. Take the poset **Q** whose Hasse diagram is shown in Fig. 4. Then **B** is obtained from this poset by adding all elements of the universe as singleton unary relations. An explicit description of the minimal obstructions for **B** in which the binary relation is a partial order (i.e., the so-called **Q**-zigzags) can be found in [65]. It can be easily derived from this description that **B** has the required properties.

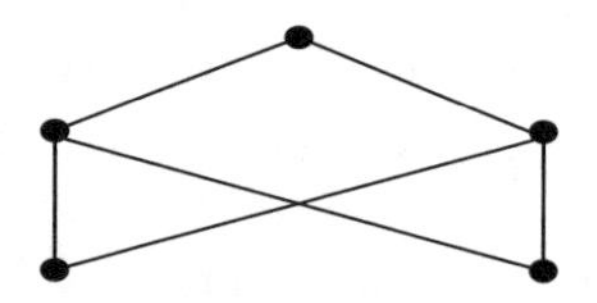

Fig. 4. A poset without NU polymorphisms

Let us now consider the question of which structures do not have bounded pathwidth duality. Trivially, any structure without bounded treewidth duality cannot have bounded pathwidth duality.

Example 26. Reconsider the structure $\mathbf{B}_{ps}$ from Example 4. This structure has 1-treewidth duality, as shown in Example 10. By using the game technique (see below), it can be shown that $\mathbf{B}_{ps}$ does not have bounded pathwidth duality (an alternative proof of this can be found in [1]).

At present, the most general algebraic necessary condition for the presence of bounded pathwidth duality is given by the following result.

Theorem 16 ([52]). *If a core structure* **B** *has bounded pathwidth duality then the variety* $\mathrm{var}(\mathbb{A}_{\mathbf{B}})$ *omits types* **1**, **2**, *and* **5**.

By using Theorem 9.11 of [40], the previous theorem can be re-stated as follows.

Theorem 17. *If a core structure* **B** *has bounded pathwidth duality then* **B** *has ternary polymorphisms* $d_0, \ldots, d_n$, $n \geq 2$, *satisfying the following identities:*

$$\begin{aligned}
d_0(x, y, z) &= x, \\
d_n(x, y, z) &= z, \\
d_i(x, y, x) &= d_{i+1}(x, y, x) \quad \textit{for all even } i < n, \\
d_i(x, y, y) &= d_{i+1}(x, y, y) \quad \textit{for all even } i < n, \\
d_i(x, x, y) &= d_{i+1}(x, x, y) \quad \textit{for all odd } i < n.
\end{aligned}$$

Moreover, if a core structure $\mathbf{B}$ does not have the polymorphisms described above (or, equivalently, the variety var($\mathbb{A}_{\mathbf{B}}$) admits at least one of the types **1**,**2**, and **5**) then CSP($\mathbf{B}$) is hard for **PTIME** or for $\mathbf{Mod_pL}$ (for some prime p) under first-order reductions [52], and thus is unlikely to belong to **NL**.

One very natural question about pathwidth dualities is whether they form a proper hierarchy or the hierarchy collapses to some level. That is, the question is whether there exists a number j such that, for any $j' \geq j$, every structure with j'-pathwidth duality also has j-pathwidth duality.

It follows from Theorem 11 that, in order to prove that a certain structure $\mathbf{B}$ does not have (j,k)-pathwidth duality, one only needs to provide a structure $\mathbf{A}$ such that the Duplicator has a winning strategy in the (j,k)-PR game on $(\mathbf{A},\mathbf{B})$, but it holds that $\mathbf{A} \not\to \mathbf{B}$. This game technique was used in [22] to give a negative answer to the above question. Let us now describe the structures that were used in [22].

Let $n \geq 1$ and let $\mathbf{B}_n$ be the structure with universe B_n and relations R^l_n, $1 \leq l \leq n$, defined as follows. The universe B_n is the set $\{1,\dots,n\} \times \{1,2\}$. For every $1 \leq l \leq n$, R^l_n is a binary symmetric relation on B_n that consists of all pairs $((i,m),(i',m'))$ satisfying at least one of the following conditions:

- $i > l$, $i = i'$, $m = m'$
- $i = i' = l$, $m \neq m'$,
- $i < l$ and $i' \leq l$,
- $i \leq l$ and $i' < l$.

Theorem 18 ([22]). *For every $n \geq 2$, the structure $\mathbf{B}_n$ does not have n-pathwidth duality, but $\mathbf{B}_n$ has a majority polymorphism (and hence $(6n+2)$-pathwidth duality).*

The only known fact concerning the meta-problem for bounded pathwidth duality is that, for any fixed $k \geq 1$, the problem of recognising structures with $(1,k)$-pathwidth duality is decidable [20].

5 Bounded Treewidth Duality

In this section we consider bounded treewidth duality. Arguably, it is the most important duality because it is one of the two most general basic properties of relational structures $\mathbf{B}$ that are known to guarantee that the problem CSP($\mathbf{B}$) is solvable in polynomial time (the other property can be found, e.g., in [41]), and the vast majority of such structures have bounded treewidth duality. The notion of bounded treewidth duality has strong links with methods of solving constraint satisfaction problems based on establishing local consistency (see, e.g., [17,24,34,43]).

The following result links together treewidth dualities, Datalog, infinitary logics, and existential pebble games.

Theorem 19 ([28,47]). *For any structure* $\mathbf{B}$, *the following conditions are equivalent:*

1. $\mathbf{B}$ *has* (j,k)*-treewidth duality.*
2. co-CSP($\mathbf{B}$) *is definable in* (j,k)*-Datalog.*
3. co-CSP($\mathbf{B}$) *is definable by the canonical* (j,k)*-Datalog program for* $\mathbf{B}$.
4. co-CSP($\mathbf{B}$) *is expressible in* $\exists L^{j,k}_{\infty\omega}$.
5. CSP($\mathbf{B}$) *consists of all structures* $\mathbf{A}$ *such that Duplicator has a winning strategy in the existential* (j,k)*-pebble game on (A,B).*

If these conditions hold then CSP($\mathbf{B}$) *is in* **PTIME**.

To prevent possible confusion, we note that the paper [47] speaks about definability of co-CSP($\mathbf{B}$) in k-Datalog meaning (k,k)-Datalog (in our notation). Hence, this does not exactly correspond to k-treewidth duality in our sense.

In [28] and some subsequent papers (e.g., in [54]), problems CSP($\mathbf{B}$) that have j-treewidth duality (or bounded treewidth duality) are called *width-j* (or *bounded width*, respectively) problems.

We will now give some examples of structures with and without bounded treewidth duality. Note that, trivially, every structure with bounded pathwidth duality also has bounded treewidth duality.

Example 27. Recall the structure $\mathbf{B}_{3H}$ from Example 5; the problem CSP($\mathbf{B}_{3H}$) is precisely HORN-3-SAT. It follows from Example 11 that $\mathbf{B}_{3H}$ has (1,3)-treewidth duality. By replacing relations $P^{\mathbf{B}_{3H}}$ and $N^{\mathbf{B}_{3H}}$ in $\mathbf{B}_{3H}$ with k-ary relations $P^{\mathbf{B}_{kH}} = \{0,1\}^k \backslash \{(1,\ldots,1,1)\}$ and $N^{\mathbf{B}_{kH}} = \{0,1\}^k \backslash \{(1,\ldots,1,0)\}$, respectively, one obtains a structure $\mathbf{B}_{kH}$ such that CSP($\mathbf{B}_{kH}$) is exactly HORN-k-SAT. An obvious modification of Example 11 shows that $\mathbf{B}_{kH}$ has $(1,k)$-treewidth duality.

Example 28. Recall from Example 23 that each unbalanced oriented cycle has (2,3)-pathwidth duality. It was shown in [26] that if $\mathbf{H}$ is a balanced oriented cycle then either it has bounded treewidth duality or CSP($\mathbf{H}$) is **NP**-complete. The description of the boundary between the cases is rather involved.

Example 29. Let τ be the vocabulary (P,S,T,E) where P is ternary, E is binary, and S and T are unary relation symbols. Consider the τ-structure $\mathbf{B}$ with 4-element universe $\{0,1,a,b\}$ and relations defined as follows. Reconsider structures $\mathbf{B}_{ps}$ from Example 4 and $\mathbf{K}_2$ from Example 9, and assume that the universe of $\mathbf{K}_2$ is $\{a,b\}$. If $R \in \{P,S,T\}$ then define $R^{\mathbf{B}} = R^{\mathbf{B}_{ps}}$, and let $E^{\mathbf{B}} = E^{\mathbf{K}_2}$. We claim that the structure $\mathbf{B}$ has 2-treewidth duality, but neither 1-treewidth duality nor bounded pathwidth duality. It is easy to see that if a connected τ-structure $\mathbf{A}$ homomorphically maps to $\mathbf{B}$ then either $E^{\mathbf{A}}$ is empty or else the other three relations in $\mathbf{A}$ are empty. Hence, $\mathbf{B}$ has an obstruction set consisting of structures from $\mathcal{O}_{\mathbf{B}_{ps}}$ and $\mathcal{O}_{\mathbf{K}_2}$ (suitably expanded with empty relations) and of finitely many structures in which some element appears both in the binary relation and in one of the other three relations. Since $\mathbf{B}_{ps}$ has 1-treewidth duality and $\mathbf{K}_2$ has (2,3)-pathwidth duality, we conclude that $\mathbf{B}$ has 2-treewidth duality. On the other hand, $\mathbf{B}_{ps}$ does not have bounded pathwidth duality (see Example 26) and it is straightforward to show that $\mathbf{K}_2$ does not have 1-treewidth duality. Hence, $\mathbf{B}$ cannot have either of these two properties.

Example 30. As we know from Example 1, the **H**-COLORING problem coincides with the problem CSP($\mathbf{B}_{hc}$) where $\mathbf{B}_{hc}$ is the (di)graph **H**. If **H** is a bipartite graph, then core(**H**) = $\mathbf{K}_2$ and CSP($\mathbf{B}_{hc}$) coincides with the 2-COLOURABILTY problem, and $\mathbf{B}_{hc}$ hence has (2,3)-pathwidth duality (see Example 9). If **H** is a non-bipartite graph then CSP($\mathbf{B}_{hc}$) is **NP**-complete [33]. It is known (see, e.g., [34]) that in this case $\mathbf{B}_{hc}$ does not have bounded treewidth duality (without any complexity-theoretic assumptions).

Example 31. A *triad* is a digraph obtained from three oriented paths by choosing one end of each path and identifying these three vertices. It is shown in [2] that there exists a triad **H** such that CSP($\mathbf{B}_{hc}$) is in **PTIME**, but **H** does not have bounded treewidth duality.

Within the algebraic approach to the CSP, a different concept, *relational width*, is often very useful, see, e.g., [10]. This concept is applicable even for infinite sets of relations, but in the case of relational structures (with finite vocabulary) relational width is strongly related to treewidth duality, as we shall now see. Relational width is usually (e.g., in [7,9,10,13]) defined using the "variable-value" form (as given in Section 1) of the constraint satisfaction problem. A straightforward translation into the homomorphism form goes as follows.

Definition 15. *For $k \geq 1$, a family $\mathcal{M} = \{H_I \mid I \subseteq A, |I| \leq k\}$, where each H_I is a non-empty set of mappings from I to B, is called a* k-minimal *family for* ($\mathbf{A}, \mathbf{B}$) *if*

1. *for any $I' \subseteq I \subseteq A$, $|I| \leq k$, we have $H_{I'} = \{h_{|I'} \mid h \in H_I\}$, and*
2. *for any $I \subseteq A$ with $|I| \leq k$, any $h \in H_I$, any (n-ary) $R \in \tau$, and any $(a_1, \ldots, a_n) \in R^{\mathbf{A}}$, there exists a tuple $(b_1, \ldots, b_n) \in R^{\mathbf{B}}$ such that*
   (a) *$h(a_i) = b_i$ for all $a_i \in I$,*
   (b) *for any $J \subseteq A$ with $|J| \leq k$, there exists $h' \in H_J$ such that $h'(a_i) = b_i$ for all $a_i \in J$.*

A structure $\mathbf{B}$ *is said to have* relational width k *if, for any structure* $\mathbf{A}$ *such that there is a k-minimal family for* ($\mathbf{A}, \mathbf{B}$)*, we have* $\mathbf{A} \to \mathbf{B}$*. A structure* $\mathbf{B}$ *has* bounded relational width *if it has relational width k for some k.*

Note that property 2(a) shows that every member of every $H_I \in \mathcal{M}$ is a projective homomorphism, while properties 1 and 2(b) show that there is strong compatibility between different sets in $\mathcal{M}$.

A duality characterisation of structures of relational width k was obtained in [21]. Call a structure **A** a *k-reltree* if it has a tree-decomposition in which (i) each node with more than k elements consists of all elements in some tuple in a relation in **A**, and (ii) two adjacent nodes can share at most k elements.

Theorem 20 ([21]). *A structure has relational width k if and only if it has an obstruction set consisting of k-reltrees.*

The following result can be easily derived from the above theorem.

Corollary 1. *For any structure* $\mathbf{B}$, *the following holds.*

1. *If* $\mathbf{B}$ *has* (j,k)*-treewidth duality, then it has relational width* k.
2. *If* $\mathbf{B}$ *has relational width* k *and* $\mathrm{maxar}(\mathbf{B}) = r$, *then it has* (k,k')*-treewidth duality where* $k' = \max(k,r)$.

In particular, $\mathbf{B}$ *has bounded treewidth duality if and only if it has bounded relational width.*

Note that Corollary 1 shows that there is a correspondence between the parameters of relational width and treewidth duality, but it does not show how optimal parameters for these widths are related in general.

We will state results about bounded treewidth duality and bounded relational width in the way they were stated originally. By the above theorem, one can translate such results between the widths.

Similarly to Theorems 12 and 13, the (idempotent) polymorphisms determine whether a (core) structure has bounded treewidth duality.

Theorem 21 ([54]). *Let* $\mathbf{B}_1$ *and* $\mathbf{B}_2$ *be relational structures with the same universe and such that* $\mathrm{Pol}(\mathbf{B}_1) \subseteq \mathrm{Pol}(\mathbf{B}_2)$. *If* co-CSP$(\mathbf{B}_2)$ *is definable in Datalog, then so is* co-CSP$(\mathbf{B}_1)$.

Theorem 22 ([52]). *For a core structure* $\mathbf{B}$, co-CSP$(\mathbf{B})$ *is definable in Datalog if and only if* co-CSP$(\mathbf{B}_c)$ *is.*

We will now give examples of polymorphisms that guarantee that a structure has bounded treewidth duality.

Theorem 23 ([28], see also [43]). *If a structure* $\mathbf{B}$ *has a* $(l{+}1)$*-ary NU polymorphism then* $\mathbf{B}$ *has* l*-treewidth duality.*

Tree duality is just a shorter name for 1-treewidth duality. It is known [28,51] that every structure with tree duality has an obstruction set consisting of trees in the sense of Definition 11. (Note that this fact does not follow trivially from the definition of 1-treewidth duality.) In particular, if a structure $\mathbf{B}$ has tree duality then it has $(1, \mathrm{maxar}(\mathbf{B}))$-treewidth duality. Structures with tree duality have been completely characterised in [28]. To state this result, we need to give a certain construction. For a τ-structure $\mathbf{B}$, its *power structure* is a τ-structure $\mathcal{P}_1(\mathbf{B})$ with universe consisting of all non-empty subsets of B, and, for each r-ary $R \in \tau$, we have $(A_1,\ldots,A_r) \in R^{\mathcal{P}_1(\mathbf{B})}$ if and only if, for each $1 \le i \le r$ and each $a \in A_i$, there is $(a_1,\ldots,a_r) \in R^{\mathbf{B}}$ such that $a_i = a$.

Theorem 24 ([28], see also [24]). *For any structure* $\mathbf{B}$, *the following conditions are equivalent:*

1. $\mathbf{B}$ *has tree duality.*
2. *The structure* $\mathcal{P}_1(\mathbf{B})$ *admits a homomorphism to* $\mathbf{B}$.
3. *For every* $n \ge 2$, $\mathbf{B}$ *has an* n*-ary totally symmetric polymorphism.*

Example 32. It is not difficult to see that, for $n \geq 2$, the operation $f_n = \bigwedge_{i=1}^{n} x_i$ is a TSI polymorphism of the structure $\mathbf{B}_{kH}$ (see Example 27).

Structures with caterpillar duality (see Theorem 6 for finite caterpillar duality) can be characterised in the spirit of Theorem 24, see [14] for details.

Theorem 20 implies that every structure has relational width 1 if and only if it has 1-treewidth duality. Note that, in general, the optimal parameters for relational width and treewidth duality need not be equal. For example, the structure $\mathbf{B} = \mathbf{K}_2$ of Example 9 has (2,3)-treewidth duality, but not tree duality (since it has no binary TSI polymorphism). On the other hand, $\mathbf{K}_2$ has relational width 3 by Corollary 1, but not 2 (which can be seen by taking $\mathbf{K}_3$ as $\mathbf{A}$). In fact, it is shown in [21] that if a structure $\mathbf{B}$ has relational width 2 then it has relational width 1.

Theorem 25 ([9]). *If a structure* $\mathbf{B}$ *has a 2-semilattice polymorphism then* $\mathbf{B}$ *has relational width 3.*

Some of the most studied varieties in universal algebra are the so-called *congruence distributive* varieties (see, e.g., [40,44,64]). For a core structure $\mathbf{B}$, the algebra $\mathbb{A}_{\mathbf{B}}$ belongs to a congruence distributive variety if, for some $n \geq 2$, $\mathbf{B}$ has ternary polymorphisms $d_0, \ldots, d_n$ (called *Jónsson* operations) satisfying the identities from Theorem 17, and, in addition, such that $d_i(x, y, x) = x$ for all $0 \leq i \leq n$. In this case we say that $\mathbb{A}_{\mathbf{B}}$ is in the class $CD(n)$. Note that $\mathbb{A}_{\mathbf{B}}$ is in $CD(2)$ if and only if $\mathbf{B}$ has a majority polymorphism (which is d_1 in this case).

Theorem 26 ([44]). *For any structure* $\mathbf{B}$*, if the algebra* $\mathbb{A}_{\mathbf{B}}$ *is in* $CD(3)$ *then* $\mathbf{B}$ *has relational width* $\min(|B|^2, \max(3, \mathrm{maxar}(\mathbf{B})))$.

Theorem 27 ([15]). *For any structure* $\mathbf{B}$*, if the algebra* $\mathbb{A}_{\mathbf{B}}$ *is in* $CD(4)$ *then* $\mathbf{B}$ *has* $(k-1, k)$*-treewidth duality where* $k = \max(3, \mathrm{maxar}(\mathbf{B}))$.

Theorem 21 makes it possible to introduce *algebras* having bounded treewidth duality: An algebra $\mathbb{A} = (B; F)$ has bounded treewidth duality if every structure $\mathbf{B}$ with universe B such that $F \subseteq \mathrm{Pol}(\mathbf{B})$ has bounded treewidth duality. The following result shows that bounded treewidth duality can be lifted further to varieties of algebras.

Theorem 28 ([54]). *If* $\mathbb{A}$ *is an algebra with bounded treewidth duality then every finite algebra from the variety* $\mathrm{var}(\mathbb{A})$ *also has bounded treewidth duality.*

Clearly, if CSP($\mathbf{B}$) is **NP**-complete, then it does not have bounded treewidth duality unless **PTIME** = **NP**. Systems of linear equations (see Example 6 in this paper or the proof of Theorem 1 of [8]), as well as problems that can "simulate" them, provide benchmark examples of structures $\mathbf{B}$ such that CSP($\mathbf{B}$) is in **PTIME**, but $\mathbf{B}$ does not have bounded treewidth duality [28]. Combining these two reasons for not having bounded treewidth duality, one obtains the following equivalent necessary conditions for bounded treewidth duality.

Theorem 29. *If a core structure* **B** *has bounded treewidth duality then the following equivalent conditions hold:*

1. *The variety* $\mathrm{var}(\mathbb{A}_{\mathbf{B}})$ *omits types* **1** *and* **2**.
2. *There is* $k \geq 2$ *such that* **B** *has n-ary weak NU polymorphisms for all* $n \geq k$.

In the above theorem, the necessity of condition (1) was proved in [54], and the equivalence of conditions (1) and (2) in [57]. It is shown in [53] that (the complement of) condition (1) is very closely related with the so-called property of "ability to count" which was introduced in [28] and conjectured there to be the main obstacle for a structure to have bounded treewidth duality.

Conjecture 1 ([54]). A core structure **B** has bounded treewidth duality if and only if the equivalent conditions from Theorem 29 hold.

A somewhat different way of applying algebras to analyse a relational structure **B**, via an *edge-coloured graph* $Gr(\mathbf{B})$ of the structure, was introduced in [8] (see also [13]). The conditions in Theorem 29 can be equivalently expressed in terms of properties of this graph, and a conjecture equivalent to Conjecture 1 was made in [8,13].

Conjecture 1 was confirmed in the following important cases, and, interestingly, the best possible bound for some width turns out to be quite small.

Theorem 30 ([28]). *If* **B** *is a 2-element core structure then* **B** *has bounded treewidth if and only if* **B** *has a semilattice polymorphism or a majority polymorphism. Moreover, in this case* **B** *has 2-treewidth duality.*

A *factor* of an algebra $\mathbb{A}$ is a homomorphic image of a subalgebra of $\mathbb{A}$.

Theorem 31 ([10]). *If* **B** *is a core structure with* $|B| \leq 3$ *then* **B** *has bounded relational width if and only if the algebra* $\mathbb{A}_{\mathbf{B}}$ *itself or each of its factors have an operation (depending on a factor) which is a majority operation or a 2-semilattice operation. Moreover, in this case* **B** *has relational width 3.*

Theorem 32 ([7]). *Let* **B** *be a structure containing all unary relations. Then* **B** *has bounded relational width if and only if, for each two-element subset* $C \subseteq B$, *there is a polymorphism* $f \in \mathrm{Pol}(\mathbf{B})$ *(depending on C) such that* $f|_C$ *is either a semilattice operation or a majority operation. Moreover, in this case* **B** *has relational width 3.*

Conjecture 1 can be strengthened in the following sense. As we saw above, bounded treewidth duality is equivalent to expressibility in a certain infinitary logic. The expressive power of this logic is relatively weak, and it is natural to ask if it possible to express constraint satisfaction problems in terms of a more powerful logic. One such logic is $C^{\omega}_{\infty\omega}$ (see Section 2.3). This logic can express a number of undecidable problems (e.g., the HALTING problem). However, if Conjecture 1 is true than its expressive power for constraint satisfaction problems is no greater than that of Datalog.

Theorem 33 ([3]). *Let* **B** *be a structure. If the variety* var($\mathbb{A}_{\mathbf{B}}$) *admits type* **1** *or* **2** *then* CSP(**B**) *is not expressible in* $C^{\omega}_{\infty\omega}$.

In the direction of solving the meta-problem for bounded treewidth, the following is known.

Theorem 34 ([64]). *There is a polynomial time algorithm which, given a finite idempotent algebra* $\mathbb{A}$, *checks whether the variety* var($\mathbb{A}$) *omits types* **1** *and* **2**.

Following a strategy from [11] where the case of omitting type **1** is treated, Theorem 34 can be used to show that, for a core structure **B** with at most n elements (where n is any fixed number), it can be checked in polynomial time whether the variety var($\mathbb{A}_{\mathbf{B}}$) omits types **1** and **2**. Thus, we have the following corollary.

Corollary 2. *Assuming Conjecture 1 holds, the meta-problem for bounded treewidth duality is tractable for structures of bounded size.*

It is a natural question to determine the complexity of recognising structures with j-treewidth duality for a fixed j. For $j = 1$, it follows from Theorem 24 that this problem is decidable, while the proof of Theorem 6.1 of [51] implies the following lower bound.

Theorem 35 ([51]). *It is* **NP***-hard to decide whether a given structure* **B** *has tree duality.*

In Section 4, we have considered the hierarchy problem for j-pathwidth dualities and found (see Theorem 18) that the hierarchy does not collapse. We now consider a similar problem for j-treewidth dualities. Let TW_j be the class of all structures with j-treewidth duality. Clearly, we have a hierarchy $TW_1 \subseteq TW_2 \subseteq TW_3 \subseteq TW_4 \subseteq \ldots$.

It is easy to show that $TW_1 \subsetneq TW_2$. Consider the problem 2-COLOURABILITY, or CSP($\mathbf{K}_2$), from Example 9. The structure $\mathbf{K}_2$ has (2,3)-pathwidth duality, and hence (2,3)-treewidth duality. On the other hand, it is easy to see that $\mathbf{K}_2$ does not have a binary commutative polymorphism, and hence, by Theorem 24 (see also Example 15(1)), it cannot have 1-treewidth duality. Surprisingly, the question whether any other inclusion in the treewidth duality hierarchy is strict remains open. It may seem that Theorems 23, 26 and 27 contradict this claim. However, they give only an upper bound for the treewidth duality. For instance, every 2-element structure with an NU polymorphism has 2-treewidth duality.

6 Additional Remarks

6.1 Symmetric Datalog

A restriction of linear Datalog, *symmetric Datalog*, has been recently introduced in [25]. A linear Datalog program is called symmetric, if, for every rule of the form $t_0 :- t_1, t_2, \ldots, t_n$ ($n \geq 1$), where t_0 and t_1 are IDBs, that appears in the program, the program also contains its "symmetric" rule $t_1 :- t_0, t_2, \ldots, t_n$, obtained by formally swapping the IDBs in the rule. We say that co-CSP($\mathbf{B}$) is definable in symmetric Datalog if it is accepted by a symmetric Datalog program. In broad terms, symmetric Datalog for CSP is to **LOGSPACE** what linear Datalog for CSP is to **NL**: if co-CSP($\mathbf{B}$) is definable in symmetric Datalog then CSP($\mathbf{B}$) is in **LOGSPACE**, and, for all problems CSP($\mathbf{B}$) that are known to be in **LOGSPACE**, co-CSP($\mathbf{B}$) is definable in symmetric Datalog [25]. In particular, this holds for all Boolean problems CSP($\mathbf{B}$) in **LOGSPACE**. If $\mathbf{B}_1$ and $\mathbf{B}_2$ are structures such that Pol($\mathbf{B}_1$) $\subseteq$ Pol($\mathbf{B}_2$) and co-CSP($\mathbf{B}_1$) is definable in symmetric Datalog, then co-CSP($\mathbf{B}_2$) is also definable in symmetric Datalog (compare with Theorems 12 and 21). The following analog of Theorems 16 and 29 holds for symmetric Datalog: for a core structure $\mathbf{B}$, if co-CSP($\mathbf{B}$) is definable in symmetric Datalog then the variety var($\mathbb{A}_{\mathbf{B}}$) omits types **1**, **2**, **4**, and **5** (i.e., it admits only type **3**) [52]. It is proved in [23] that if co-CSP($\mathbf{B}$) is definable in Datalog and $\mathbf{B}$ has a Mal'tsev polymorphism (i.e., a ternary polymorphism m with $m(x, y, y) = m(y, y, x) = x$ for all x, y) then co-CSP($\mathbf{B}$) is definable in symmetric Datalog. It is shown in [25] that definability of co-CSP($\mathbf{B}$) in symmetric Datalog is equivalent to definability in a certain fragment of second order logic (this parallels a result in [20]). It would be interesting to find a convenient pebble game and an appropriate notion of duality for symmetric Datalog, in the spirit of Theorems 11 and 19.

6.2 Extending Datalog with Inequality and Negation

One can extend (j, k)-Datalog, and the logic $\exists L^{j,k}_{\infty\omega}$ by allowing the use of inequalities ($\neq$) and negated atomic formulas (which must be EDBs in the case of Datalog). The obtained logics are denoted (j, k)-Datalog($\neq, \neg$) and $\exists L^{j,k}_{\infty\omega}(\neq, \neg)$, respectively. It was shown in [29] that these extensions do not add any expressive power for homomorphism-closed classes (e.g., such as co-CSP($\mathbf{B}$)). In other words, if a class co-CSP($\mathbf{B}$) is definable in (j, k)-Datalog($\neq, \neg$) then it is also definable in (j, k)-Datalog, and the same holds for $\exists L^{j,k}_{\infty\omega}(\neq, \neg)$. Moreover, a closer inspection of the proof reveals that this result remains true for linear (j, k)-Datalog and the logic $\exists M^{j,k}_{\infty\omega}$.

6.3 Infinite CSP

Up until now we have considered only finite structures. However, one can also consider the problem of deciding whether a given finite τ-structure admits a homomorphism to a fixed infinite τ-structure $\mathbf{B}$ (see survey [4]). Some natural problems such as BETWEENNESS (see [31]) and the ACYCLICITY problem for digraphs can be represented as CSP($\mathbf{B}$) for suitable infinitely countable structures $\mathbf{B}$ (but not for any finite structure $\mathbf{B}$). Bounded treewidth duality for infinitely countable structures has been considered in [5,6]. It was shown in these papers that, for general countable structures, Theorem 19 fails. However, there is a large class of structures (ω-categorical structures), for which Theorem 19 holds. Recall that a structure **B** is called *ω-categorical* if, for each $n \geq 1$, there are only finitely many inequivalent first-order formulas with n free variables over **B**. Moreover, analogs of Theorems 23 and 24 hold for such structures.

7 A List of Open Questions

1. If **B** is a core structure with finite duality, how large can the minimal arity of its 1-tolerant NU polymorphism be?
2. Is the property of having j-pathwidth and j-treewidth duality for fixed j determined by the polymorphisms of a structure?
3. Is it true that a structure **B** has bounded pathwidth duality whenever CSP(**B**) is in **NL**?
4. Prove that every structure with an NU polymorphism has bounded pathwidth duality.
5. Are the conditions in Theorems 16 and 17 necessary and sufficient for a core structure to have bounded pathwidth duality?
6. Prove Conjecture 1 (that the conditions in Theorem 29 are necessary and sufficient for a core structure to have bounded treewidth duality).[2]
7. For $j \geq 2$, is there a structure $\mathcal{P}_j(\mathbf{B})$ such that **B** has j-treewidth duality if and only if $\mathcal{P}_j(\mathbf{B}) \to \mathbf{B}$ (particularly, for $j = 2$)?
8. Does the treewidth duality hierarchy collapse (in particular, to its second level) or not?
9. Are there structures that have bounded relational width, but not relational width 3?
10. Is it true that the number k from Theorem 29 can always be chosen to be equal to 3?
11. Prove that a structure **B** has bounded treewidth (or even bounded pathwidth) duality whenever the algebra $\mathbb{A}_\mathbf{B}$ is in $CD(n)$ for some n?
12. Find a pebble-game and a duality characterisation for structures **B** such that co-CSP(**B**) is definable in symmetric Datalog.
13. Is it true that co-CSP(**B**) is definable in symmetric Datalog whenever CSP(**B**) is in **LOGSPACE**?
14. Is it true that, for a core structure **B**, co-CSP(**B**) is definable in symmetric Datalog whenever the variety var($\mathbb{A}_\mathbf{B}$) admits only type **3**?
15. Are there other naturally arising dualities for the CSP?

Acknowledgements

The authors would like to thank Catarina Carvalho, Víctor Dalmau, Pavol Hell, Jarik Nešetřil, Claude Tardif and Pascal Tesson for useful discussions and helpful comments. Remarks from an anonymous reviewer are also appreciated.

[2] While this article was in press, a solution to problem 6 was announced by L. Barto and M. Kozik.

References

1. Afrati, F., Cosmodakis, S.: Expressiveness of restricted recursive queries. In: STOC 1989, pp. 113–126 (1989)
2. Atserias, A.: On digraph coloring problems and treewidth duality. European Journal of Combinatorics 29(4), 796–820 (2008)
3. Atserias, A., Bulatov, A., Dawar, A.: Affine systems of equations and counting infinitary logic. In: Arge, L., Cachin, C., Jurdziński, T., Tarlecki, A. (eds.) ICALP 2007. LNCS, vol. 4596, pp. 558–570. Springer, Heidelberg (2007)
4. Bodirsky, M.: Constraint satisfaction problems with infinite templates. In: Creignou, N., Kolaitis, P.G., Vollmer, H. (eds.) Complexity of Constraints. LNCS, vol. 5250. Springer, Heidelberg (2008)
5. Bodirsky, M., Dalmau, V.: Datalog and constraint satisfaction with infinite templates. In: Durand, B., Thomas, W. (eds.) STACS 2006. LNCS, vol. 3884, pp. 646–659. Springer, Heidelberg (2006)
6. Bodirsky, M., Dalmau, V.: Datalog and constraint satisfaction with infinite templates (2008) arXiv: 0809.2386v1
7. Bulatov, A.: Tractable conservative constraint satisfaction problems. In: LICS 2003, pp. 321–330 (2003)
8. Bulatov, A.: A graph of a relational structure and constraint satisfaction problems. In: LICS 2004, pp. 448–457 (2004)
9. Bulatov, A.: Combinatorial problems raised from 2-semilattices. Journal of Algebra 298(2), 321–339 (2006)
10. Bulatov, A.: A dichotomy theorem for constraint satisfaction problems on a 3-element set. Journal of the ACM 53(1), 66–120 (2006)
11. Bulatov, A., Jeavons, P.: Algebraic structures in combinatorial problems. Technical Report MATH-AL-4-2001, Technische Universität Dresden, Germany (2001)
12. Bulatov, A., Jeavons, P., Krokhin, A.: Classifying complexity of constraints using finite algebras. SIAM Journal on Computing 34(3), 720–742 (2005)
13. Bulatov, A., Valeriote, M.: Recent results on the algebraic approach to the CSP. In: Creignou, N., Kolaitis, P., Vollmer, H. (eds.) Complexity of Constraints. LNCS, vol. 5250, pp. 68–92. Springer, Heidelberg (2008)
14. Carvalho, C., Dalmau, V., Krokhin, A.: Caterpillar duality for constraint satisfaction problems. In: LICS 2008, pp. 307–316 (2008)
15. Carvalho, C., Dalmau, V., Marković, P., Maróti, M.: CD(4) has bounded width. Algebra Universalis (accepted)
16. Clasen, M., Valeriote, M.: Tame congruence theory. In: Lectures on Algebraic Model Theory. Fields Institute Monographs, vol. 15, pp. 67–111 (2002)
17. Cohen, D., Jeavons, P.: The complexity of constraint languages. In: Rossi, F., van Beek, P., Walsh, T. (eds.) Handbook of Constraint Programming, ch. 8. Elsevier, Amsterdam (2006)
18. Cooper, M.C., Cohen, D.A., Jeavons, P.G.: Characterising tractable constraints. Artificial Intelligence 65, 347–361 (1994)
19. Creignou, N., Khanna, S., Sudan, M.: Complexity Classifications of Boolean Constraint Satisfaction Problems. SIAM Monographs on Discrete Mathematics and Applications, vol. 7 (2001)
20. Dalmau, V.: Linear Datalog and bounded path duality for relational structures. Logical Methods in Computer Science 1(1) (2005) (electronic)
21. Dalmau, V.: There are no pure relational width 2 constraint satisfaction problems (submitted, 2008)

22. Dalmau, V., Krokhin, A.: Majority constraints have bounded pathwidth duality. European Journal of Combinatorics 29(4), 821–837 (2008)
23. Dalmau, V., Larose, B.: Maltsev + Datalog ⇒ Symmetric Datalog. In: LICS 2008, pp. 297–306 (2008)
24. Dalmau, V., Pearson, J.: Set functions and width 1 problems. In: Jaffar, J. (ed.) CP 1999. LNCS, vol. 1713, pp. 159–173. Springer, Heidelberg (1999)
25. Egri, L., Larose, B., Tesson, P.: Symmetric Datalog and constraint satisfaction problems in Logspace. In: LICS 2007, pp. 193–202 (2007)
26. Feder, T.: Classification of homomorphisms to oriented cycles and of k-partite satisfiability. SIAM Journal on Discrete Mathematics 14(4), 471–480 (2001)
27. Feder, T., Hell, P., Huang, J.: Bi-arc graphs and the complexity of list homomorphisms. Journal of Graph Theory 42, 61–80 (2003)
28. Feder, T., Vardi, M.Y.: The computational structure of monotone monadic SNP and constraint satisfaction: A study through Datalog and group theory. SIAM Journal on Computing 28, 57–104 (1998)
29. Feder, T., Vardi, M.Y.: Homomorphism closed vs. existential positive. In: Proc. 18th IEEE Symp. on Logic in Computer Science, LICS 2003, pp. 311–320 (2003)
30. Furst, M., Saxe, J., Sipser, M.: Parity, circuits, and the polynomial-time hierarchy. Mathematical Systems Theory 17(1), 13–27 (1984)
31. Garey, M., Johnson, D.S.: Computers and Intractability: A Guide to the Theory of NP-Completeness. Freeman, San Francisco (1979)
32. Hell, P.: From graph colouring to constraint satisfaction: there and back again. In: Topics in Discrete Mathematics. Algorithms and Combinatorics, vol. 26, pp. 407–432. Springer, Heidelberg (2006)
33. Hell, P., Nešetřil, J.: On the complexity of H-coloring. Journal of Combinatorial Theory, Ser. B 48, 92–110 (1990)
34. Hell, P., Nešetřil, J.: Graphs and Homomorphisms. Oxford University Press, Oxford (2004)
35. Hell, P., Nešetřil, J., Zhu, X.: Duality and polynomial testing of tree homomorphisms. Trans. Amer. Math. Soc. 348, 147–156 (1996)
36. Hell, P., Nešetřil, J., Zhu, X.: Duality of graph homomorphisms. In: Combinatorics, Paul Erdös is Eighty. Bolyai Soc. Math. Stud., vol. 2, pp. 271–282. János Bolyai Math. Soc. (1996)
37. Hell, P., Zhou, H., Zhu, X.: On homomorphisms to acyclic local tournaments. Journal of Graph Theory 20(4), 467–471 (1995)
38. Hell, P., Zhu, X.: Homomorphisms to oriented paths. Discrete Mathematics 132, 107–114 (1994)
39. Hell, P., Zhu, X.: The existence of homomorphisms to oriented cycles. SIAM Journal on Discrete Mathematics 8, 208–222 (1995)
40. Hobby, D., McKenzie, R.N.: The Structure of Finite Algebras. Contemporary Mathematics, vol. 76. American Mathematical Society, Providence (1988)
41. Idziak, P., Markovic, P., McKenzie, R., Valeriote, M., Willard, R.: Tractability and learnability arising from algebras with few subpowers. In: LICS 2007, pp. 213–222 (2007)
42. Jeavons, P.: On the algebraic structure of combinatorial problems. Theoretical Computer Science 200, 185–204 (1998)
43. Jeavons, P.G., Cohen, D.A., Cooper, M.C.: Constraints, consistency and closure. Artificial Intelligence 101(1-2), 251–265 (1998)
44. Kiss, E.W., Valeriote, M.: On tractability and congruence distributivity. Logical Methods in Computer Science 3(2) (2007) (electronic)

45. Kolaitis, P.G.: On the expressive power of logics on finite models. In: Finite Model Theory and its Applications. EATCS Series: Texts in Theoretical Computer Science, pp. 27–124. Springer, Heidelberg (2007)
46. Kolaitis, P.G., Vardi, M.Y.: On the expressive power of Datalog: tools and a case study. Journal of Computer and System Sciences 51, 110–134 (1995)
47. Kolaitis, P.G., Vardi, M.Y.: Conjunctive-query containment and constraint satisfaction. Journal of Computer and System Sciences 61, 302–332 (2000)
48. Kolaitis, P.G., Vardi, M.Y.: A logical approach to constraint satisfaction. In: Finite Model Theory and its Applications. EATCS Series: Texts in Theoretical Computer Science, pp. 339–370. Springer, Heidelberg (2007)
49. Komárek, P.: Some new good characterisations of directed graphs. Časopis Pěst. Mat. 51, 348–354 (1984)
50. Krokhin, A., Bulatov, A., Jeavons, P.: The complexity of constraint satisfaction: an algebraic approach. In: Structural Theory of Automata, Semigroups, and Universal Algebra. NATO Science Series II: Math., Phys., Chem., vol. 207, pp. 181–213. Springer, Heidelberg (2005)
51. Larose, B., Loten, C., Tardif, C.: A characterisation of first-order constraint satisfaction problems. Logical Methods in Computer Science 3(4) (2007) (electronic)
52. Larose, B., Tesson, P.: Universal algebra and hardness results for constraint satisfaction problems. In: Arge, L., Cachin, C., Jurdziński, T., Tarlecki, A. (eds.) ICALP 2007. LNCS, vol. 4596, pp. 267–278. Springer, Heidelberg (2007)
53. Larose, B., Valeriote, M., Zádori, L.: Omitting types, bounded width and the ability to count (submitted, 2008)
54. Larose, B., Zádori, L.: Bounded width problems and algebras. Algebra Universalis 56(3-4), 439–466 (2007)
55. Libkin, L.: Elements of Finite Model Theory. EATCS Series: Texts in Theoretical Computer Science. Springer, Heidelberg (2004)
56. Loten, C., Tardif, C.: Majority functions on structures with finite duality. European Journal of Combinatorics 29(4), 979–986 (2008)
57. Maróti, M., McKenzie, R.: Existence theorems for weakly symmetric operations. In: Algebra Universalis (to appear, 2007)
58. Nešetřil, J., Pultr, A.: On classes of relations and graphs determined by subobjects and factorobjects. Discrete Mathematics 22, 287–300 (1978)
59. Nešetřil, J., Tardif, C.: Duality theorems for finite structures (characterising gaps and good characterisations). Journal of Combinatorial Theory, Ser. B 80, 80–97 (2000)
60. Nešetřil, J., Tardif, C.: Short answers to exponentially long questions: extremal aspects of homomorphism duality. SIAM Journal on Discrete Mathematics 19(4), 914–920 (2005)
61. Rossi, F., van Beek, P., Walsh, T. (eds.): Handbook of Constraint Programming. Elsevier, Amsterdam (2006)
62. Rossman, B.: Existential positive types and preservation under homomorphisms. In: LICS 2005, pp. 467–476 (2005)
63. Szendrei, A.: Clones in Universal Algebra. Seminaires de Mathematiques Superieures, vol. 99. University of Montreal (1986)
64. Valeriote, M.: A subalgebra intersection property for congruence-distributive varieties. Canadian Journal of Mathematics (to appear, 2007)
65. Zádori, L.: Posets, near-unanimity functions and zigzags. Bulletin of the Australian Mathematical Society 47, 79–93 (1993)

A Logical Approach to Constraint Satisfaction

Phokion G. Kolaitis[1] and Moshe Y. Vardi[2]

[1] IBM Almaden Research Center, Computer Science Principles and Methodologies and Computer Science Department, University of California, Santa Cruz
kolaitis@almaden.ibm.com, kolaitis@cs.ucsc.edu
[2] Department of Computer Science, Rice University, 6100 S. Main Street, Houston, TX 77005-1892
vardi@rice.edu

1 Introduction

Since the early 1970s, researchers in artificial intelligence (AI) have investigated a class of combinatorial problems that became known as *constraint-satisfaction problems* (CSP). The input to such a problem consists of a set of variables, a set of possible values for the variables, and a set of constraints between the variables; the question is to determine whether there is an assignment of values to the variables that satisfies the given constraints. The study of constraint satisfaction occupies a prominent place in artificial intelligence, because many problems that arise in different areas can be modelled as constraint-satisfaction problems in a natural way; these areas include Boolean satisfiability, temporal reasoning, belief maintenance, machine vision, and scheduling (cf. [Dec92a, Kum92, Mes89, Tsa93]). In its full generality, constraint satisfaction is an NP-complete problem. For this reason, researchers in artificial intelligence have pursued both heuristics for constraint-satisfaction problems and tractable cases obtained by imposing various restrictions on the input (cf. [MF93, Dec92a, DM94, Fro97, PJ97]).

Over the past decade, it has become clear that there is an intimate connection between constraint satisfaction and various problems in database theory and finite-model theory. The goal of this chapter is to describe several such connections. We start in Section 2 by defining the constraint-satisfaction problem and showing how it can be phrased also as a homomorphism problem, conjunctive-query evaluation problem, or join-evaluation problem. In Section 3, we discuss the computational complexity of constraint satisfaction and show that it can be studied from two perspectives, a uniform perspective and a non-uniform perspective. We relate both perspectives to the study of the computational complexity of query evaluation. In Section 4, we focus on the non-uniform case and describe a Dichotomy Conjecture, asserting that every non-uniform constraint-satisfaction problem is either in PTIME or NP-complete. In Section 5, we examine the complexity of non-uniform constraint satisfaction from a logical perspective and show that it is related to the data complexity of a fragment of existential second-order logic. We then go on in Section 6 and offer a logical approach, via definability in Datalog, to establishing the tractability of non-uniform constraint-satisfaction problems. In Section 7, we leverage the connection between Datalog and certain

N. Creignou et al. (Eds.): Complexity of Constraints, LNCS 5250, pp. 125–155, 2008.

pebble games, and show how these pebble games offer an algorithmic approach to solving uniform constraint-satisfaction problems. In Section 8, we relate these pebble games to consistency properties of constraint-satisfaction instances, a well-known approach in constraint solving. Finally, in Section 9, we show how the same pebble games can be used to identify large "islands of tractability" in the constraint-satisfaction terrain that are based on the concept of bounded treewidth.

Much of the logical machinery used in this chapter is described in detail in [GKL+05, Chapter 2]. For a book-length treatment of constraint satisfaction from the perspective of graph homomorphism, see [HN04]. Two books on constraint programming and constraint processing are [Apt03, Dec03].

2 Preliminaries

The standard terminology in AI formalizes an instance $\mathcal{P}$ of constraint satisfaction as a triple $(V, D, \mathcal{C})$, where

1. V is a set of variables;
2. D is a set of values, referred to as the *domain*;
3. $\mathcal{C}$ is a collection of *constraints* $C_1, \ldots, C_q$, where each constraint C_i is a pair $(\mathbf{t}, R)$ with $\mathbf{t}$ a k-tuple over V, $k \geq 1$, referred to as the *scope* of the constraint, and R a k-relation on D.

A *solution* of such an instance is a mapping $h : V \to D$ such that, for each constraint $(\mathbf{t}, R)$ in $\mathcal{C}$, we have that $h(\mathbf{t}) \in R$, where h is defined on tuples component-wise, that is, if $\mathbf{t} = (a_1, \ldots, a_k)$, then $h(\mathbf{t}) = (h(a_1), \ldots, h(a_k))$. The CONSTRAINT-SATISFACTION PROBLEM asks whether a given instance is *solvable*, i.e., whether it has a solution. Note that, without loss of generality, we may assume that all constraints $(\mathbf{t}, R_i)$ involving the same scope $\mathbf{t}$ have been consolidated to a single constraint $(\mathbf{t}, R)$, where R is the intersection of all relations R_i constraining $\mathbf{t}$. Thus, we can assume that each tuple $\mathbf{t}$ of variables occurs at most once in the collection $\mathcal{C}$.

Consider the Boolean satisfiability problem 3-SAT: given a 3CNF-formula φ with variables $x_1, \ldots, x_n$ and clauses $c_1, \ldots, c_m$, is φ satisfiable? Such an instance of 3-SAT can be thought of as the constraint-satisfaction instance in which the set of variables is $V = \{x_1, \ldots, x_n\}$, the domain is $D = \{0, 1\}$, and the constraints are determined by the clauses of φ. For example, a clause of the form $(\neg x \vee \neg y \vee z)$ gives rise to the constraint $((x, y, z), \{0, 1\}^3 - \{(1, 1, 0)\})$. In an analogous manner, 3-COLORABILITY can be modelled as a constraint-satisfaction problem. Indeed, an instance $\mathbf{G} = (V, E)$ of 3-COLORABILITY can be thought of as the constraint-satisfaction instance in which the set of variables is the set V of the nodes of the graph $\mathbf{G}$, the domain is the set $D = \{r, b, g\}$ of three colors, and the constraints are the pairs $((u, v), Q)$, where $(u, v) \in E$ and $Q = \{(r, b)(b, r), (r, g)(g, r), (b, g)(g, b)\}$ is the disequality relation on D.

Let $\mathbf{A}$ and $\mathbf{B}$ be two relational structures[1] over the same vocabulary. A *homomorphism* h from $\mathbf{A}$ to $\mathbf{B}$ is a mapping $h : A \to B$ from the universe A of $\mathbf{A}$

[1] We consider only finite structures in this chapter.

to the universe B of $\mathbf{B}$ such that, for every relation $R^{\mathbf{A}}$ of $\mathbf{A}$ and every tuple $(a_1, \ldots, a_k) \in R^{\mathbf{A}}$, we have that $(h(a_1), \ldots, h(a_k)) \in R^{\mathbf{B}}$. The existence of a homomorphism from $\mathbf{A}$ to $\mathbf{B}$ is denoted by $\mathbf{A} \to \mathbf{B}$, or by $\mathbf{A} \to^h \mathbf{B}$, when we want to name the homomorphism h explicitly. An important observation made in [FV98][2] is that every such constraint-satisfaction instance $\mathcal{P} = (V, D, \mathcal{C})$ can be viewed as an instance of the HOMOMORPHISM PROBLEM, asking whether there is a homomorphism between two structures $\mathbf{A}_{\mathcal{P}}$ and $\mathbf{B}_{\mathcal{P}}$ that are obtained from $\mathcal{P}$ in the following way:

1. the universe of $\mathbf{A}_{\mathcal{P}}$ is V and the universe of $\mathbf{B}_{\mathcal{P}}$ is D;
2. the relations of $\mathbf{B}_{\mathcal{P}}$ are the distinct relations R occurring in $\mathcal{C}$;
3. the relations of $\mathbf{A}_{\mathcal{P}}$ are defined as follows: for each distinct relation R on D occurring in $\mathcal{C}$, we have the relation $R^{\mathbf{A}} = \{\mathbf{t} : (\mathbf{t}, R) \in \mathcal{C}\}$. Thus, $R^{\mathbf{A}}$ consists of all scopes associated with R.

We call $(\mathbf{A}_{\mathcal{P}}, \mathbf{B}_{\mathcal{P}})$ the *homomorphism instance* of $\mathcal{P}$. Conversely, it is also clear that every instance of the homomorphism problem between two structures $\mathbf{A}$ and $\mathbf{B}$ can be viewed as a constraint-satisfaction instance CSP$(\mathbf{A}, \mathbf{B})$ by simply "breaking up" each relation $R^{\mathbf{A}}$ on $\mathbf{A}$ as follows: we generate a constraint $(\mathbf{t}, R^{\mathbf{B}})$ for each $\mathbf{t} \in R^{\mathbf{A}}$. We call CSP$(\mathbf{A}, \mathbf{B})$ the *constraint-satisfaction instance* of $(\mathbf{A}, \mathbf{B})$. Thus, as pointed out in [FV98], the constraint-satisfaction problem can be identified with the homomorphism problem.

To illustrate the passage from the constraint-satisfaction problem to the homomorphism problem, let us consider 3-SAT. A 3CNF-formula φ with variables $x_1, \ldots, x_n$ and clauses $c_1, \ldots, c_m$ gives rise to a homomorphism instance $(\mathbf{A}_\varphi, \mathbf{B}_\varphi)$ defined as follows:

- $\mathbf{A}_\varphi = (\{x_1, \ldots, x_n\}, R_0^\varphi, R_1^\varphi, R_2^\varphi, R_3^\varphi)$, where R_i^φ is the ternary relation consisting of all triples (x, y, z) of variables that occur in a clause of φ with i negated literals, $0 \leq i \leq 3$; for instance, R_2^φ consists of all triples (x, y, z) of variables such that $(\neg x \vee \neg y \vee z)$ is a clause of φ (here, we assume without loss of generality that the negated literals precede the positive literals).
- $\mathbf{B}_\varphi = (\{0, 1\}, R_0, R_1, R_2, R_3)$, where R_i consists of all triples that satisfy a 3-clause in which the first i literals are negated; for instance, $R_2 = \{0, 1\}^3 - \{1, 1, 0\}$.

Note that $\mathbf{B}_\varphi$ does not depend on φ. It is clear that φ is satisfiable if and only if there is a homomorphism from $\mathbf{A}_\varphi$ to $\mathbf{B}_\varphi$ (in symbols, $\mathbf{A}_\varphi \to \mathbf{B}_\varphi$).

As another example, 3-COLORABILITY is equivalent to the problem of deciding whether there is a homomorphism h from a given graph $\mathbf{G}$ to the complete graph $\mathbf{K}_3 = (\{r, b, g\}, \{(r, b)(b, r), (r, g)(g, r), (b, g)(g, b)\}$ with 3 nodes. More generally, k-COLORABILITY, $k \geq 2$, amounts to the existence of a homomorphism from a given graph $\mathbf{G}$ to the complete graph $\mathbf{K}_k$ with k nodes (also known as the k-clique).

Numerous other important NP-complete problems can be viewed as special cases of the HOMOMORPHISM PROBLEM (and, hence, also of the CONSTRAINT-SATISFACTION PROBLEM). For example, consider the CLIQUE problem: given a

[2] An early version appeared in [FV93].

graph $\mathbf{G}$ and an integer k, does $\mathbf{G}$ contain a clique of size k? As a homomorphism instance this is equivalent to asking if there is a homomorphism from the complete graph $\mathbf{K}_k$ to $\mathbf{G}$. As a constraint-satisfaction instance, the set of variables is $\{1, 2, \ldots, k\}$, the domain is the set V of nodes of $\mathbf{G}$, and the constraints are the pairs $((i, j), E)$ such that $i \neq j$, $1 \leq i, j \leq k$, and E is the edge relation of $\mathbf{G}$. For another example, consider the HAMILTONICITY PROBLEM: given a graph $\mathbf{G} = (\mathbf{V}, \mathbf{E})$ does it have a Hamiltonian cycle? This is equivalent to asking if there is a homomorphism from the structure $(V, C_V, \neq)$ to the structure $(V, E, \neq)$, where C_V is some cycle on the set V of nodes of $\mathbf{G}$ and $\neq$ is the disequality relation on V. NP-completeness of the Homomorphism problem was pointed out explicitly in [Lev73]. In this chapter, we use both the traditional AI formulation of constraint satisfaction and the formulation as the homomorphism problem, as each has its own advantages.

It turns out that in both formulations constraint satisfaction can be expressed as a database-theoretic problem. We start with the homomorphism formulation, which is intimately related to *conjunctive-query evaluation* [KV00a]. A *conjunctive query* Q of arity n is a query definable by a positive existential first-order formula $\varphi(X_1, \ldots, X_n)$ having conjunction as its only propositional connective, that is, by a formula of the form

$$\exists Z_1 \ldots \exists Z_m \psi(X_1, \ldots, X_n, Z_1, \ldots, Z_m),$$

where $\psi(X_1, \ldots, X_n, Z_1, \ldots, Z_m)$ is a conjunction of (positive) atomic formulas. The free variables $X_1, \ldots, X_n$ of the defining formula are called the *distinguished variables* of Q. Such a conjunctive query is usually written as a rule, whose head is $Q(X_1, \ldots, X_n)$ and whose body is $\psi(X_1, \ldots, X_n, Z_1, \ldots, Z_m)$. For example, the formula

$$\exists Z_1 \exists Z_2 (P(X_1, Z_1, Z_2) \wedge R(Z_2, Z_3) \wedge R(Z_3, X_2))$$

defines a binary conjunctive query Q, which as a rule becomes

$$Q(X_1, X_2) \text{ :- } P(X_1, Z_1, Z_2), R(Z_2, Z_3), R(Z_3, X_2).$$

If the formula defining a conjunctive query Q has no free variables (i.e., if it is a sentence), then Q is a *Boolean* conjunctive query. For example, the sentence

$$\exists Z_1 \exists Z_2 \exists Z_3 (E(Z_1, Z_2) \wedge E(Z_2, Z_3) \wedge E(Z_3, Z_1))$$

defines the Boolean conjunctive query "is there a cycle of length 3?".

If D is a databaseand Q is a n-ary query, then $Q(D)$ is the n-ary relation on D obtained by evaluating the query Q on D, that is, the collection of all n-tuples from D that satisfy the query (cf. [GKL$^+$05, Chapter 2]). The CONJUNCTIVE-QUERY EVALUATION PROBLEM asks: given a n-ary query Q, a database D, and a n-tuple $\mathbf{a}$ from D, does $\mathbf{a} \in Q(D)$? Let Q_1 and Q_2 be two n-ary queries having the same tuple of distinguished variables. We say that Q_1 is *contained in* Q_2, and write $Q_1 \subseteq Q_2$, if $Q_1(D) \subseteq Q_2(D)$ for every database D. The CONJUNCTIVE-QUERY CONTAINMENT PROBLEM asks: given two conjunctive queries Q_1 and Q_2, is $Q_1 \subseteq Q_2$? These concepts can be defined for Boolean conjunctive queries in an analogous manner. In particular, if Q is a Boolean query and D is a database, then $Q(D) = 1$ if D satisfies Q; otherwise, $Q(D) = 0$. Moreover, the containment problem for Boolean queries Q_1 and Q_2 is equivalent to asking whether Q_1 logically implies Q_2.

It is well known that conjunctive-query containment can be reformulated both as a *conjunctive-query evaluation* problem and as a *homomorphism* problem. What links these problems together is the *canonical* database D^Q associated with Q. This database is defined as follows. Each variable occurring in Q is considered a distinct element in the universe of D^Q. Every predicate in the body of Q is a predicate of D^Q as well; moreover, for every distinguished variable X_i of Q, there is a distinct monadic predicate P_i (not occurring in Q). Every subgoal in the body of Q gives rise to a tuple in the corresponding predicate of D^Q; moreover, if X_i is a distinguished variable of Q, then $P_i(X_i)$ is also a (monadic) tuple of D^Q. Thus, returning to the preceding example, the canonical database of the conjunctive query $\exists Z_1 \exists Z_2 (P(X_1, Z_1, Z_2) \wedge R(Z_2, Z_3) \wedge R(Z_3, X_2))$ consists of the facts $P(X_1, Z_1, Z_2)$, $R(Z_2, Z_3)$, $R(Z_3, X_2)$, $P_1(X_1)$, $P_2(X_2)$. The relationship between conjunctive-query containment, conjunctive-query evaluation, and homomorphisms is provided by the following classical result, due to Chandra and Merlin.

Theorem 2.1. [CM77] *Let Q_1 and Q_2 be two conjunctive queries having the same tuple $(X_1, \ldots, X_n)$ of distinguished variables. Then the following statements are equivalent.*

- $Q_1 \subseteq Q_2$.
- $(X_1, \ldots, X_n) \in Q_2(D^{Q_1})$.
- *There is a homomorphism $h : D^{Q_2} \rightarrow D^{Q_1}$.*

It follows that the homomorphism problem can be viewed as a conjunctive-query evaluation problem or as a conjunctive-query containment problem. For this, with every structure $\mathbf{A}$, we view the universe $A = \{X_1, \ldots, X_n\}$ of $\mathbf{A}$ as a set of individual variables and associate with $\mathbf{A}$ the Boolean conjunctive query $\exists X_1 \ldots \exists X_n \bigwedge_{\mathbf{t} \in R^{\mathbf{A}}} R(\mathbf{t})$; we call this query the *canonical conjunctive query* of $\mathbf{A}$ and denote it by $Q_{\mathbf{A}}$. It is clear that $\mathbf{A}$ is isomorphic to the canonical database associated with $Q_{\mathbf{A}}$.

Corollary 2.2. *Let $\mathbf{A}$ and $\mathbf{B}$ be two structures over the same vocabulary. Then the following statements are equivalent.*

- $\mathbf{A} \rightarrow \mathbf{B}$.
- $\mathbf{B} \models Q_{\mathbf{A}}$.
- $Q_{\mathbf{B}} \subseteq Q_{\mathbf{A}}$.

As an illustration, we have that a graph $\mathbf{G}$ is 3-colorable iff $\mathbf{K}_3 \models Q_{\mathbf{G}}$ iff $Q_{\mathbf{K}_3} \subseteq Q_{\mathbf{G}}$.

A *relational join*, denoted by the symbol $\bowtie$, is a conjunctive query with no existentially quantified variables. Thus, relational-join evaluation is a special case of conjunctive-query evaluation. For example, $E(Z_1, Z_2) \wedge E(Z_2, Z_3) \wedge E(Z_3, Z_1)$ is a relational join that, when evaluated on a graph $\mathbf{G} = (\mathbf{V}, \mathbf{E})$, returns all triples of nodes forming a 3-cycle. There is a well known connection between the traditional AI formulation of constraint satisfaction and relational-join evaluation that we describe next. Suppose we are given a constraint-satisfaction instance $(V, D, \mathcal{C})$. We can assume without loss of generality that in every constraint $(\mathbf{t}, R) \in \mathcal{C}$ the elements in $\mathbf{t}$ are distinct. (Suppose to the contrary that $t_i = t_j$. Then we can delete from R every tuple in which the ith and jth entries disagree, and then project out that j-th column from t and R.) We can thus view every element of V as a relational *attribute*, every tuple of distinct elements of V as a *relational schema*, and every constraint $(\mathbf{t}, R)$ as a relation R over the schema t (cf. [AHV95]). It now follows from the definition of constraint satisfaction that CSP can be viewed as a relational-join evaluation problem.

Proposition 2.3. [Bib88, GJC94] *A constraint-satisfaction instance* $(V, D, \mathcal{C})$ *is solvable if and only if* $\bowtie_{(\mathbf{t},R)\in\mathcal{C}} R$ *is nonempty.*

Note that Proposition 2.3 is essentially the same as Corollary 2.2. Indeed, the condition $\mathbf{B} \models Q_{\mathbf{A}}$ amounts to the non-emptiness of the relational join obtained from $Q_{\mathbf{A}}$ by dropping all existential quantifiers and using the relations from $\mathbf{B}$ as interpretations of the relational symbols in $Q_{\mathbf{A}}$. Moreover, the homomorphisms from $\mathbf{A}$ to $\mathbf{B}$ are precisely the tuples in the relational join associated with the constraint-satisfaction instance CSP$(\mathbf{A}, \mathbf{B})$.

3 Computational Complexity of Constraint Satisfaction

The Constraint-Satisfaction Problem is NP-complete, because it is clearly in NP and also contains NP-hard problems as special cases, including 3-Sat, 3-Colorability, and Clique. As explained in Garey and Johnson's classic monograph [GJ79], one of the main ways to cope with NP-completeness is to identify polynomial-time solvable cases of the problem at hand that are obtained by imposing restrictions on the possible inputs. For instance, Horn 3-Sat, the restriction of 3-Sat to Horn 3CNF-formulas, is solvable in polynomial-time using a unit-propagation algorithm. Similarly, it is known that 3-Colorability restricted to graphs of bounded treewidth is solvable in polynomial time (see [DF99]). In the case of constraint satisfaction, the pursuit of tractable cases has evolved over the years from the discovery of isolated cases to the discovery of large "islands of tractability" of constraint satisfaction. In what follows, we will give an account of some of the progress made in this area. Using the fact that the Constraint-Satisfaction Problem can be identified with the Homomorphism Problem, we begin by introducing some terminology and notation that will enable us to formalize the concept of an "island of tractability" of constraint satisfaction.

In general, an instance of the Homomorphism Problem consists of two relational structures $\mathbf{A}$ and $\mathbf{B}$. Thus, all restricted cases of this problem can be obtained by imposing restrictions on the input structures $\mathbf{A}$ and $\mathbf{B}$.

Definition 3.1. *Let $\mathcal{A}$, $\mathcal{B}$ be two classes of relational structures. We write* CSP($\mathcal{A}, \mathcal{B}$) *to denote the restriction of the* HOMOMORPHISM PROBLEM *to input structures from $\mathcal{A}$ and $\mathcal{B}$. In other words,*

$$\mathrm{CSP}(\mathcal{A}, \mathcal{B}) = \{(\mathbf{A}, \mathbf{B}) : \mathbf{A} \in \mathcal{A},\ \mathbf{B} \in \mathcal{B} \text{ and } \mathbf{A} \to \mathbf{B}\}.$$

An island of tractability *of constraint satisfaction is a pair* ($\mathcal{A}, \mathcal{B}$) *of classes of relational structures such that* CSP($\mathcal{A}, \mathcal{B}$) *is in the complexity class PTIME of all decision problems solvable in polynomial time.*

(A more general definition of islands of tractability of constraint satisfaction would consider classes of pairs ($\mathbf{A}, \mathbf{B}$) of structures, cf. [FF05]; we do not pursue this more general definition here.)

The ultimate goal in the pursuit of islands of tractability of constraint satisfaction is to identify or characterize classes $\mathcal{A}$ and $\mathcal{B}$ of relational structures such that CSP($\mathcal{A}, \mathcal{B}$) is in PTIME. The basic starting point in this investigation is to consider the cases in which one of the two classes $\mathcal{A}$, $\mathcal{B}$ is as small as possible, while the other is as large as possible. This amounts to considering the cases in which one of $\mathcal{A}$, $\mathcal{B}$ is the class *All* of all relational structures over some arbitrary, but fixed, relational vocabulary, while the other is a singleton, consisting of some fixed structure over that vocabulary. Thus, the starting points of the investigation is to determine, for fixed relational structures $\mathbf{A}, \mathbf{B}$, the computational complexity of the decision problems CSP($\{\mathbf{A}\}$, *All*) and CSP(*All*, $\{\mathbf{B}\}$).

Clearly, for each fixed $\mathbf{A}$, the decision problem CSP($\{\mathbf{A}\}$, *All*) can be solved in polynomial time, because, given a structure $\mathbf{B}$, the existence of a homomorphism from $\mathbf{A}$ to $\mathbf{B}$ can be checked by testing all functions h from the universe A of $\mathbf{A}$ to the universe B of $\mathbf{B}$ (the total number of such functions is $|B|^{|A|}$, which is a polynomial number in the size of the structure $\mathbf{B}$ when $\mathbf{A}$ is fixed). Thus, having a singleton structure "on the left' is of little interest.

At the other extreme, however, the situation is quite different, since the computational complexity of CSP(*All*, $\{\mathbf{B}\}$) may very well depend on the particular structure $\mathbf{B}$. Indeed, CSP(*All*, $\{\mathbf{K}_3\}$) is NP-complete, because it is the 3-COLORABILITY problem; in contrast, CSP(*All*, $\{\mathbf{K}_2\}$) is in P, because it is the 2-COLORABILITY problem. For simplicity, in what follows, for every fixed structure $\mathbf{B}$, we define CSP($\mathbf{B}$) = CSP(*All*, $\{\mathbf{B}\}$) and call this the *non-uniform* constraint-satisfaction problem associated with $\mathbf{B}$. For such problems, we refer to $\mathbf{B}$ as the *template*. Thus, the first major goal in the study of the computational complexity of constraint satisfaction is to identify those templates $\mathbf{B}$ for which CSP($\mathbf{B}$) is in PTIME. This goals gives rise to an important open decision problem.

THE TRACTABILITY CLASSIFICATION PROBLEM: Given a relational structure $\mathbf{B}$, decide if CSP($\mathbf{B}$) is in PTIME.

In addition to the family of non-uniform constraint-satisfaction problems CSP($\mathbf{B}$), where $\mathbf{B}$ is a relational structure, we also study decision problems of the form CSP($\mathcal{A}$, *All*), where $\mathcal{A}$ is a class of structures. We refer to such problems as *uniform* constraint-satisfaction problems.

It is illuminating to consider the complexity of uniform and non-uniform constraint satisfaction from the perspective of query evaluation. As argued in [Var82] (see [GKL+05, Chapter 2]), there are three ways to measure the complexity of evaluating queries (we focus here on Boolean queries) expressible in a query language L:

- The *combined complexity* of L is the complexity of the following decision problem: given an L-query Q and a structure $\mathbf{A}$, does $\mathbf{A} \models Q$? In symbols,
$$\{\langle Q, \mathbf{A}\rangle : Q \in L \text{ and } \mathbf{A} \models Q\}.$$
- The *expression complexity* of L is the complexity of the following decision problems, one for each fixed structure $\mathbf{A}$:
$$\{Q : Q \in L \text{ and } \mathbf{A} \models Q\}.$$
- The *data complexity* of L is the complexity of the following decision problems, one for each fixed query $Q \in L$:
$$\{\mathbf{A} : \mathbf{A} \models Q\}.$$

As discussed in [GKL+05, Chapter 2], the data complexity of first-order logic is in LOGSPACE, which means that, for each first-order query Q, the problem $\{\mathbf{A} : \mathbf{A} \models Q\}$ is in LOGSPACE. In contrast, the combined complexity for first-order logic is PSPACE-complete. Furthermore, the expression complexity for first-order logic is also PSPACE-complete. In fact, for all but trivial structures $\mathbf{A}$, the problem $\{Q : Q \in FO \text{ and } \mathbf{A} \models Q\}$ is PSPACE-complete. This exponential gap between data complexity, on one hand, and combined and expression complexity, on the other hand, is typical [Var82]. For conjunctive queries, on the other hand, both combined and expression complexity are NP-complete.

Consider now the uniform constraint-satisfaction problem $\mathrm{CSP}(\mathcal{A}, All) = \{(\mathbf{A}, \mathbf{B}) : \mathbf{A} \in \mathcal{A}, \text{ and } \mathbf{A} \rightarrow \mathbf{B}\}$, where $\mathcal{A}$ is a class of structures. By Corollary 2.2, we have that
$$\mathrm{CSP}(\mathcal{A}, All) = \{(\mathbf{A}, \mathbf{B}) : \mathbf{A} \in \mathcal{A},\ \mathbf{B} \text{ is a structure and } \mathbf{B} \models Q_{\mathbf{A}}\}.$$

Thus, studying the complexity of uniform constraint satisfaction amounts to studying the combined complexity for a class of conjunctive queries, as, for example, in [CR97, GLS99b, Sar91]. In contrast, consider the non-uniform constraint-satisfaction problem $\mathrm{CSP}(\mathbf{B}) = \{\mathbf{A} : \mathbf{A} \rightarrow \mathbf{B}\}$. By Corollary 2.2 we have that $\mathrm{CSP}(\mathbf{B}) = \{\mathbf{A} : \mathbf{B} \models Q_{\mathbf{A}}\}$. Thus, studying the complexity of non-uniform constraint satisfaction amounts to studying the expression complexity of conjunctive queries with respect to different structures. This is a problem that has not been studied in the context of database theory.

4 Non-uniform Constraint Satisfaction

The first major result in the study of non-uniform constraint-satisfaction problems was obtained by Schaefer [Sch78], who, in effect, classified the computational complexity of all Boolean non-uniform constraint-satisfaction problems.

A *Boolean* structure is simply a relational structure with a 2-element universe, that is, a structure of the form $\mathbf{B} = (\{0,1\}, R_1^{\mathbf{B}}, \ldots, R_m^{\mathbf{B}})$. A *Boolean non-uniform constraint-satisfaction problem* is a problem of the form CSP($\mathbf{B}$) with a Boolean template $\mathbf{B}$. These problems are also known as GENERALIZED-SATISFIABILITY PROBLEMS, because they can be viewed as variants of Boolean-satisfiability problems in which the formulas are conjunctions of generalized connectives [GJ79]. In particular, they contain the well known problems k-SAT, $k \geq 2$, 1-IN-3-SAT, POSITIVE 1-IN-3-SAT, NOT-ALL-EQUAL 3-SAT, and MONOTONE 3-SAT as special cases. For example, as seen earlier, 3-SAT is CSP($\mathbf{B}$), where $\mathbf{B} = (\{0,1\}, R_0, R_1, R_2, R_3)$ and R_i is the set of all triples that satisfy a 3-clause in which the first i-literals are negated, $i = 0,1,2,3$ (thus, $R_0 = \{0,1\}^3 - \{(0,0,0)\}$). Similarly, MONOTONE 3-SAT is CSP($\mathbf{B}$), where $\mathbf{B} = (\{0,1\}, R_0, R_3)$.

Ladner [Lad75] showed that if PTIME $\neq$ NP, then there are decision problems in NP that are neither NP-complete, nor belong to PTIME. Such problems are called *intermediate* problems. Consequently, it is conceivable that a given family of NP-problems contains intermediate problems. Schaefer [Sch78], however, showed that the family of all Boolean non-uniform constraint-satisfaction problems contains no intermediate problems.

Theorem 4.1. (Schaefer's Dichotomy Theorem [Sch78])

- *If* $\mathbf{B} = (\{0,1\}, R_1^{\mathbf{B}}, \ldots, R_m^{\mathbf{B}})$ *is Boolean structure, then either* CSP($\mathbf{B}$) *is in* PTIME *or* CSP($\mathbf{B}$) *is* NP*-complete.*
- *The* TRACTABILITY CLASSIFICATION PROBLEM *for Boolean structures is decidable; in fact, there is a polynomial-time algorithm to decide, given a Boolean structure* $\mathbf{B}$, *whether* CSP($\mathbf{B}$) *is in* PTIME *or is* NP*-complete.*

Schaefer's Dichotomy Theorem can be described pictorially as follows:

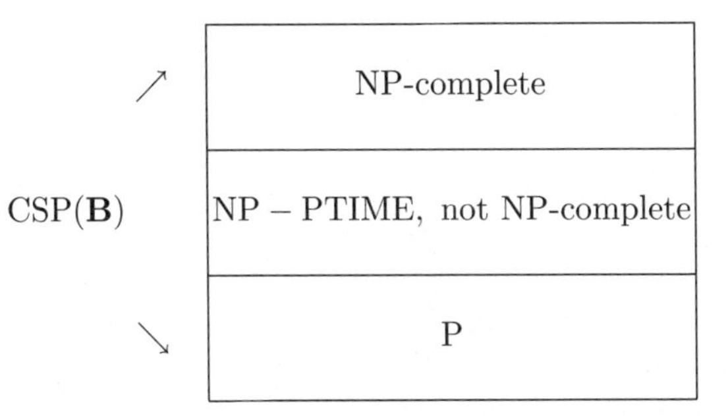

Schaefer [Sch78] actually showed that there are exactly six types of Boolean structures such that CSP($\mathbf{B}$) is in PTIME, and provided explicit descriptions of them. Specifically, he showed that CSP($\mathbf{B}$) is in PTIME precisely when at least one of the following six conditions is satisfied:

- Every relation $R_i^{\mathbf{B}}$, $1 \leq i \leq m$, of $\mathbf{B}$ is 0-*valid*, that is, $R_i^{\mathbf{B}}$ contains the all-zeroes tuple $(0, \ldots, 0)$.
- Every relation $R_i^{\mathbf{B}}$, $1 \leq i \leq m$, of $\mathbf{B}$ is 1-*valid*, that is, $R_i^{\mathbf{B}}$ contains the all-ones tuple $(1, \ldots, 1)$.

- Every relation $R_i^{\mathbf{B}}$, $1 \leq i \leq m$, of $\mathbf{B}$ is *bijunctive*, that is, $R_i^{\mathbf{B}}$ is the set of truth assignments satisfying some 2-CNF formula.
- Every relation $R_i^{\mathbf{B}}$, $1 \leq i \leq m$, of $\mathbf{B}$ is *Horn*, that is, $R_i^{\mathbf{B}}$ is the set of truth assignments satisfying some Horn formula.
- Every relation $R_i^{\mathbf{B}}$, $1 \leq i \leq m$, of $\mathbf{B}$ is *dual Horn*, that is, $R_i^{\mathbf{B}}$ is the set of truth assignments satisfying some dual Horn formula.
- Every relation $R_i^{\mathbf{B}}$, $1 \leq i \leq m$, of $\mathbf{B}$ is *affine*, that is, $R_i^{\mathbf{B}}$ is the set of solutions to a system of linear equations over the two-element field.

Schaefer's Dichotomy Theorem established a dichotomy and a decidable classification of the complexity of CSP($\mathbf{B}$) for Boolean templates $\mathbf{B}$. After this, Hell and Nešetřil [HN90] established a dichotomy theorem for CSP($\mathbf{B}$) problems in which the template $\mathbf{B}$ is an *undirected* graph: if $\mathbf{B}$ is bipartite, then CSP($\mathbf{B}$) is solvable in polynomial time; otherwise, CSP($\mathbf{B}$) is NP-complete. To illustrate this dichotomy theorem, let $\mathbf{C}_n$, $n \geq 3$, be a cycle with n elements. Then CSP($\mathbf{C}_n$) is in PTIME if n is even, and is NP-complete if n is odd.

The preceding two dichotomy results raise the challenge of classifying the computational complexity of CSP($\mathbf{B}$) for arbitrary relational templates $\mathbf{B}$. Addressing this question, Feder and Vardi [FV98] formulated the following conjecture.

Conjecture 4.2. (Dichotomy Conjecture) [FV98]
If $\mathbf{B} = (B, R_1^{\mathbf{B}}, \ldots, R_m^{\mathbf{B}})$ is an arbitrary relational structure, then either CSP($\mathbf{B}$) is in PTIME or CSP($\mathbf{B}$) is NP-complete.

In other words, the Dichotomy Conjecture says that the picture above describes the complexity of non-uniform constraint-satisfaction problems CSP($\mathbf{B}$) for arbitrary structures $\mathbf{B}$. The basis for the conjecture is not only the evidence from Boolean constraint satisfaction and undirected constraint satisfaction, but also from the seeming inability to carry out the diagonalization argument of [Lad75] using the constraint-satisfaction machinery [Fed06].

The Dichotomy Conjecture inspired intensive research efforts that significantly advanced our understanding of the complexity of non-uniform constraint satisfaction. In particular, Bulatov confirmed two important cases of this conjecture. We say that a structure $\mathbf{B} = (B, R_1^{\mathbf{B}}, \ldots, R_m^{\mathbf{B}})$ is a *3-element* structure if B contains at most three element. We say that $\mathbf{B}$ is *conservative* if all possible monadic relations on the universe included, that is, every non-empty subset of B is one of the relations $R_i^{\mathbf{B}}$ of $\mathbf{B}$.

Theorem 4.3. [Bul02, Bul03] *If* $\mathbf{B}$ *a 3-element structure or a conservative structure, then either* CSP($\mathbf{B}$) *is in* PTIME *or* CSP($\mathbf{B}$) *is* NP-*complete. Moreover, in both cases the* TRACTABILITY CLASSIFICATION PROBLEM *is decidable in polynomial time.*

In spite of the progress made, the Dichotomy Conjecture remains unresolved in general. The research efforts towards this conjecture, however, have also resulted into the discovery of broad sufficient conditions for tractability and intractability of non-uniform constraint satisfaction that have provided unifying explanations for numerous seemingly disparate tractability and intractability results and have also led to the discovery of new islands of tractability of CSP(**B**). These broad sufficient conditions are based on concepts and techniques from two different areas: universal algebra and logic.

The approach via universal algebra yields sufficient conditions for tractability of CSP(**B**) in terms of *closure* properties of the relations in **B** under certain functions on its universe B. Let R be a n-ary relation on a set B and $f : B^k \to B$ a k-ary function. We say that R *is closed under* f, if whenever $\mathbf{t}_1 = (t_1^1, t_1^2, \ldots, t_1^n), \ldots, \mathbf{t_k} = (t_k^1, t_k^2, \ldots, t_k^n)$ are k (not necessarily distinct) tuples in R, then the tuple

$$(f(t_1^1, \ldots, t_k^1), f(t_1^2, \ldots, t_k^2), \ldots, f(t_1^n, \ldots, t_k^n))$$

is also in R. We say that $f : B^k \to B$ is a *polymorphism* of a structure $\mathbf{B} = (B, R_1, \ldots, R_m)$ if each of the relations R_j, $1 \leq j \leq m$, is closed under f. It is easy to see that f is a polymorphism of **B** if and only if f is a homomorphism from $\mathbf{B}^k$ to **B**, where $\mathbf{B}^k$ is the k-th *power* of **B**. By definition, the k-th power $\mathbf{B}^k$ is the structure $(B^k, R'_1 \ldots, R'_m)$ over the same vocabulary as **B** with universe B^k and relations R'_j, $1 \leq j \leq m$, defined as follows: if R_j is of arity n, then $R'_j(\mathbf{s}_1, \ldots, \mathbf{s}_n)$ holds in $\mathbf{B}^k$ if and only if $R_j(s_1^i, \ldots, s_n^i)$ holds in **B** for $1 \leq i \leq n$.

We write Pol(**B**) for the set of all polymorphisms of **B**. As it turns out, the complexity of CSP(**B**) is intimately connected to the kinds of functions that Pol(**B**) contains. This connection was first unveiled in [FV98], and explored in depth by Jeavons and his collaborators; for a recent survey see [BJK05]. In particular, they showed that if Pol($\mathbf{B_1}$) = Pol($\mathbf{B_2}$) for two structures $\mathbf{B_1}$ and $\mathbf{B_2}$ (over *finite* vocabularies), then CSP($\mathbf{B_1}$) and CSP($\mathbf{B_2}$) are polynomially reducible to each other. Thus, the polymorphisms of a template **B** characterize the complexity of CSP(**B**). The above mentioned dichotomy results for 3-element and conservative constraint satisfaction are based on a rather deep analysis of the appropriate sets of polymorphisms.

5 Monotone Monadic SNP and Non-uniform Constraint Satisfaction

We discussed earlier how non-uniform constraint satisfaction is related to the study of the expression complexity of conjunctive queries. We now show that it can also be viewed as the study of the data complexity of second-order logic. This will suggest a way to identify islands of tractability via logic.

As described in [GKL+05, Chapter 3] and [GKL+05, Chapter 2], existential second-order logic ESO defines, by Fagin's Theorem, precisely the complexity class NP. The class SNP (for *strict* NP) [KV87, PY91] is a fragment of ESO, consisting of all existential second-order sentences with a universal first-order part, namely, sentences of the form $(\exists S')(\forall \mathbf{x})\Phi(\mathbf{x}, S, S')$, where Φ is a first-order quantifier-free formula. We refer to the relations over the input vocabulary S as *input relations*, while the relations over the quantified vocabulary S' are referred to as *existential relations.* 3-SAT is an example of an SNP problem. The input structure consists of four ternary relations C_0, C_1, C_2, C_3, on the universe $\{0,1\}$, where C_i corresponds to a clause on three variables with the first i of them negated. There is a single existential monadic relation T describing a truth assignment. The condition that must be satisfied states that for all x_1, x_2, x_3, if $C_0(x_1, x_2, x_3)$ then $T(x_1)$ or $T(x_2)$ or $T(x_3)$, and similarly for the remaining C_i by negating $T(x_j)$ if $j \leq i$. Formally, we can express 3-SAT with the SNP sentence:

$$\begin{aligned}(\exists T)(\forall x_1, x_2, x_3)\ &((C_0(x_1, x_2, x_3) \rightarrow T(x_1) \vee T(x_2) \vee T(x_3)) \wedge \\ &(C_1(x_1, x_2, x_3) \rightarrow \neg T(x_1) \vee T(x_2) \vee T(x_3)) \wedge \\ &(C_2(x_1, x_2, x_3) \rightarrow \neg T(x_1) \vee \neg T(x_2) \vee T(x_3)) \wedge \\ &(C_3(x_1, x_2, x_3) \rightarrow \neg T(x_1) \vee \neg T(x_2) \vee \neg T(x_3))).\end{aligned}$$

It is easy to see that CSP(**B**) is in SNP for each structure **B**. For each element a in the universe of **B**, we introduce an existentially quantified monadic relation T_a; intuitively, $T_a(x)$ indicates that a variable x has been assigned value a by the homomorphism. The sentence $\varphi_{\mathbf{B}}$ says that the sets T_a cover all elements in the universe[3], and that the tuples in the input relations satisfy the constraints imposed by the structure **B**. Thus, if $R(a_1, \ldots, a_n)$ does *not* hold in **B**, then $\varphi_{\mathbf{B}}$ contains the conjunct $\neg(R(x_1, \ldots, x_n) \wedge \bigwedge_{i=1}^{n} T_{a_i}(x_i))$. For example, 3-COLORABILITY over a binary input relation E can be expressed by the following sentence:

$$\begin{aligned}(\exists C_1, C_2, C_3)(\forall x, y)\ &((C_1(x) \vee C_2(x) \vee C_3(x)) \wedge \\ &\neg(E(x,y) \wedge C_1(x) \wedge C_1(y)) \wedge \\ &\neg(E(x,y) \wedge C_2(x) \wedge C_2(y)) \wedge \\ &\neg(E(x,y) \wedge C_3(x) \wedge C_3(y))).\end{aligned}$$

It follows that CSP(**B**) $= \{\mathbf{A} : \mathbf{A} \models \varphi_{\mathbf{B}}\}$. Thus, the study of the complexity of non-uniform constraint satisfaction can be viewed as the study of the *data complexity* of certain SNP sentences.

A close examination of $\varphi_{\mathbf{B}}$ above shows that it actually resides in a syntactic fragment of SNP. For *monotone* SNP, we require that all occurrences of an input relation C_i in Φ have the same polarity (the polarity of a relation is positive if it is contained in an even number of subformulas with a negation applied to it, and it is negative otherwise); by convention, we assume that this polarity is negative, so that the C_i can be interpreted as constraints, in the sense that imposing C_i on more elements of the input structure can only make the instance "less satisfiable". For *monadic* SNP we require that the existential structure S' consist of monadic relations only. Normally we assume that the language contains also the equality relation, so both equalities and inequalities are allowed in Φ, unless we say *without inequality*, which means that the $\neq$ relation cannot be used (note that equalities can always be eliminated here). We refer to the class

[3] It is not necessary to require disjointness.

when all restrictions hold, that is, monotone monadic SNP without inequality, as MMSNP. It is clear then that non-uniform constraint satisfaction can be expressed in MMSNP.

What is the precise relationship between non-uniform constraint satisfaction and MMSNP? It is easy to see that MMSNP is more expressive than non-uniform constraint satisfaction. The property asserting that the input graph is triangle-free is clearly in MMSNP (in fact, it can be expressed by a universal first-order sentence), but it can be easily shown that there is no graph $\mathbf{G}$ such that CSP($\mathbf{G}$) consists of all triangle-free graphs [FV98]. From a computational point of view, however, MMSNP and non-uniform constraint satisfaction turn out to be equivalent.

Theorem 5.1. [FV98] *Every problem in MMSNP is polynomially equivalent to* CSP($\mathbf{B}$) *for some template* $\mathbf{B}$. *The equivalence is by a randomized Turing reduction*[4] *from CSP to MMSNP and by a deterministic Karp reduction from MMSNP to CSP.*

An immediate corollary is that the Dichotomy Conjecture holds for CSP if and only if it holds for MMSNP. At the same time, MMSNP seems to be a maximal class with this property. Specifically, any attempt to relax the syntactical restrictions of MMSNP yields a class that is polynomially equivalent to NP, and, consequently, a class for which dichotomy fails.

Theorem 5.2. [FV98]

- *Every problem in NP has a polynomially equivalent problem in monotone monadic SNP with inequality.*
- *Every problem in NP has a polynomially equivalent problem in monadic SNP without inequality.*
- *Every problem in NP has a polynomially equivalent problem in monotone SNP without inequality.*

By Ladner's Theorem it follows that if PTIME $\neq$ NP, then there are intermediate problems, which are neither in PTIME nor NP-complete, in each of monotone monadic SNP with inequality, monadic SNP without inequality, and monotone SNP without inequality. This is the sense in which MMSNP is a maximal class for which we would expect a dichotomy theorem to hold.

The fact that each constraint-satisfaction problem CSP($\mathbf{B}$) can be expressed by the MMSNP sentence $\varphi_{\mathbf{B}}$ suggests a way to identify templates $\mathbf{B}$ for which CSP($\mathbf{B}$) is tractable: characterize those templates $\mathbf{B}$ for which $\varphi_{\mathbf{B}}$ is equivalent to a sentence in a logic whose data complexity is in PTIME. We discuss this approach in the next section.

6 Datalog and Non-uniform Constraint Satisfaction

Consider all tractable problems of the form CSP($\mathbf{B}$). In principle, it is conceivable that every such problem requires a completely different algorithm. In practice, however, there seem to be two basic algorithmic approaches for solving tractable

[4] G. Kun recently announced a derandomization of this reduction.

constraint-satisfaction problems: one based on a logical framework and one based on an algebraic framework.[5] Feder and Vardi [FV98] conjectured that these two algorithmic approaches cover all tractable constraint-satisfaction problems. Their group-theoretic approach, which extended the algorithm used to solve affine Boolean constraint-satisfaction problems [Sch78], has more recently been subsumed by a universal-algebraic approach [Bul02, Bul03]. We discuss here the logical approach.

As described in [GKL+05, Chapter 2], a Datalog program is a finite set of rules of the form t_0 :- $t_1, \ldots, t_m$, where each t_i is an atomic formula $R(x_1, \ldots, x_n)$. The relational predicates that occur in the heads of the rules are the *intensional database* predicates (IDBs), while all others are the *extensional database* predicates (EDBs). One of the IDBs is designated as the *goal* of the program. Note that IDBs may occur in the bodies of rules and, thus, a Datalog program is a recursive specification of the IDBs with semantics obtained via least fixed-points of monotone operators. Each Datalog program defines a query which, given a set of EDB predicates, returns the value of the goal predicate. Moreover, this query is computable in polynomial time, since the bottom-up evaluation of the least fixed-point of the program terminates within a polynomial number of steps (in the size of the given EDBs). It follows that Datalog has data complexity in PTIME. Thus, expressibility in Datalog is a sufficient condition for tractability of a query. This suggests trying to identify those templates $\mathbf{B}$ for which the MMSNP sentence $\varphi_{\mathbf{B}}$ is equivalent to a Boolean Datalog query.

It should be noted, however, that Datalog queries are *preserved under homomorphisms*. This means that if $\mathbf{A} \rightarrow^h \mathbf{A}'$ and $\mathbf{t} \in P(\mathbf{A})$ for a Datalog program M, with goal predicate P, then $h(\mathbf{t}) \in P(\mathbf{A}')$. In contrast, constraint-satisfaction problems are not preserved under homomorphism; however, their complements are. If $\mathbf{B}$ is a relational structure, then we write $\overline{\mathrm{CSP}(\mathbf{B})}$ for the *complement* of CSP($\mathbf{B}$), that is, the class of all structures $\mathbf{A}$ such that there is no homomorphism $h : \mathbf{A} \rightarrow \mathbf{B}$. If $\mathbf{A} \rightarrow^h \mathbf{A}'$ and $\mathbf{A} \in \mathrm{CSP}(\mathbf{B})$, then it does not follow that $\mathbf{A}' \in \mathrm{CSP}(\mathbf{B})$. On the other hand, if $\mathbf{A} \rightarrow^h \mathbf{A}'$ and $\mathbf{A} \in \overline{\mathrm{CSP}(\mathbf{B})}$, then $\mathbf{A}' \in \overline{\mathrm{CSP}(\mathbf{B})}$, since homomorphisms compose. Thus, rather then try to identify those templates $\mathbf{B}$ for which $\varphi_{\mathbf{B}}$ is equivalent to a Boolean Datalog query, we try to identify those templates $\mathbf{B}$ for which the negated sentence $\neg\varphi_{\mathbf{B}}$ is equivalent to a Boolean Datalog query.

Along this line of investigation, Feder and Vardi [FV98] provided a unifying explanation for the tractability of many non-uniform CSP($\mathbf{B}$) problems by showing that the complement of each of these problems is expressible in Datalog. It should be pointed out, however, that Datalog does not cover all tractable constraint-satisfaction problems. For example, it is shown in [FV98] that Datalog cannot express the complement of affine Boolean constraint-satisfaction problems; see also [Ats05]. Affine Boolean constraint-satisfaction problems and their generalizations require algebraic techniques to establish their tractability [Bul02, FV98]).

[5] The two approaches, however, are not alwasy cleanly separated; in fact, they can be fruitfully combined to yield new tractable classes, cf. [Dal05].

For every positive integer k, let k-Datalog be the collection of all Datalog programs in which the body of every rule has at most k distinct variables and also the head of every rule has at most k variables (the variables of the body may be different from the variables of the head). For example, the query NON-2-COLORABILITY is expressible in 3-Datalog, since it is definable by the goal predicate Q of the following Datalog program, which asserts that a cycle of odd length exists:

$$\begin{aligned} P_1(X,Y) &:- \ E(X,Y) \\ P_0(X,Y) &:- \ P_1(X,Z), E(Z,Y) \\ P_1(X,Y) &:- \ P_0(X,Z), E(Z,Y) \\ Q &:- \ P_1(X,X). \end{aligned}$$

The fact that expressibility in Datalog, and, more specifically, expressibility in k-Datalog provide sufficient conditions for tractability, gives rise to two classification problems:

- THE k-DATALOG CLASSIFICATION PROBLEM: Given a relational structure $\mathbf{B}$ and $k > 1$, decide if $\overline{\text{CSP}(\mathbf{B})}$ is expressible in k-Datalog?
- THE DATALOG CLASSIFICATION PROBLEM: Given a relational structure $\mathbf{B}$, decide if $\overline{\text{CSP}(\mathbf{B})}$ is expressible in k-Datalog for some $k > 1$.

The universal-algebraic approach does offer some sufficient conditions for $\overline{\text{CSP}(\mathbf{B})}$ to be expressible in Datalog. We mention here two examples. A k-ary function $f : B^k \to B$ with $k \geq 3$ is a *near-unanimity function* if $f(a_1, \ldots, a_k) = b$, for every k-tuple $(a_1, \ldots, a_k)$ such that at least $k-1$ of the a_i's are equal to b. Note that the ternary majority function from $\{0,1\}^3$ to $\{0,1\}$ is a near-unanimity function.

Theorem 6.1. [FV98] *Let* $\mathbf{B}$ *be relational structure, and* $k \geq 3$. *If* $\text{Pol}(\mathbf{B})$ *contains a* k*-ary near-unanimity function, then* $\overline{\text{CSP}(\mathbf{B})}$ *is expressible in* k*-Datalog.*

Since the number of k-ary functions over the universe B of $\mathbf{B}$ is finite, checking the condition of the preceding theorem for a given k is clearly decidable. It is not known, however, whether it is decidable to check, given $\mathbf{B}$, if $\text{Pol}(\mathbf{B})$ contains a k-ary near-unanimity function for *some* k.

A special class of Datalog consists of those programs whose IDB predicates are all monadic. We refer to such Datalog programs as *monadic* Datalog programs. It can easily be seen that the Horn case of Boolean constraint satisfaction can be dealt with by monadic programs. Consider, for example, a Boolean template with three relations: H_1 is a monadic relation corresponding to positive Horn clauses ("facts"), H_2 is a ternary relation corresponding to Horn clauses of the form $p \wedge q \rightarrow r$, and H_3 is a ternary relation corresponding to negative Horn clause of the form $\neg p \vee \neg q \vee \neg r$. Then, unsatisfiability of Horn formulas with at most three literals per clause is expressed by the following monadic Datalog program:

$$H(X) :- H_1(X)$$
$$H(X) :- H(X), H_2(Y, Z, X)$$
$$Q :- H(X), H(Y), H(Z), H_2(X, Y, Z)$$

It turns out that we can fully characterize expressibility in monadic Datalog. A k-ary function f is a *set function* if $f(a_1, \ldots, a_k) = f(b_1, \ldots, b_k)$ whenever $\{a_1, \ldots, a_k\} = \{b_1, \ldots, b_k\}$. In other words, a set function depends only the set of its arguments. As a concrete example, the binary Boolean functions $\wedge$ and $\vee$ are set functions.

Theorem 6.2. [FV98] *Let* **B** *be relational structure with universe* B. *Then the following two statements are equivalent.*

- $\overline{\mathrm{CSP}(\mathbf{B})}$ *is expressible in monadic Datalog.*
- $\mathrm{Pol}(\mathbf{B})$ *contains a* $|B|$*-ary set function.*

Since the number of $|B|$-ary functions over the universe B of **B** is finite, checking the condition of the theorem is clearly decidable; in fact, it is in NEXPTIME. Thus, the classification problem for monadic Datalog is decidable.

The main reason for the focus on Datalog as a language to solve constraint-satisfaction problems is that its data complexity is in PTIME. Datalog, however, is not the only logic with this property. We know, for example, that the data complexity of first-order logic is in LOGSPACE. Thus, it would be interesting to characterize the templates **B** such that CSP(**B**) is expressible in first-order logic. This turns out to have an intimate connection to expressibility in (non-recursive) Datalog.

Theorem 6.3. [Ats05, Ros95] *Let* **B** *be a relational structure. The following are equivalent:*

- $\overline{\mathrm{CSP}(\mathbf{B})}$ *is expressible in first-order logic.*
- $\overline{\mathrm{CSP}(\mathbf{B})}$ *is expressible by a finite union of conjunctive queries.*

It is known that a Datalog program is always equivalent to a (possibly infinite) union of conjunctive queries. A Datalog program is *bounded* if it is equivalent to a finite union of conjunctive queries [GMSV87]. It is known that a Datalog program is bounded if and only if it is equivalent to a first-order formula [AG94, Ros05]. Thus, expressibility of non-uniform CSP in first-order logic is a special case of expressibility in Datalog. Concerning the classification problem, Larose, Loten, and Tardif [LLT06] have shown that there is an algorithm to decide, given a structure **B**, whether CSP(**B**) is expressible in first-order logic; actually, this problem turns out to be NP-complete.

In another direction, we may ask if there are constraint-satisfaction problems that cannot be expressed by Datalog, but can be expressed in least fixed-point logic LFP, whose data complexity is also in PTIME. This is an open question. It is conjectured in [FV98] that if $\overline{\mathrm{CSP}(\mathbf{B})}$ is expressible in LFP, then it is also expressible in Datalog.

7 Datalog, Games, and Constraint Satisfaction

So far, we focused on using Datalog to obtain tractability for non-uniform constraint satisfaction. Kolaitis and Vardi [KV00a] showed how the logical framework also provides a unifying explanation for the uniform tractability of constraint-satisfaction problems. Note that, in general, non-uniform tractability results do not uniformize. Thus, tractability results for each problem in a collection of non-uniform CSP(**B**) problems do not necessarily yield a tractable case of the uniform constraint-satisfaction problem. The reason is that both structures **A** and **B** are part of the input to the constraint-satisfaction problem, and the running times of the polynomial-time algorithms for CSP(**B**) may very well be exponential in the size of **B**. We now leverage the intimate connection between Datalog and pebble games to shed new light on expressibility in Datalog, and show how tractability via k-Datalog does uniformize.

As discussed in [GKL+05, Chapter 2], Datalog can be viewed as a fragment of least fixed-point logic LFP; furthermore, on the class *All* of all finite structures, LFP is subsumed by the finite-variable infinitary logic $\mathcal{L}^{\omega}_{\infty\omega} = \bigcup_{k>0} \mathcal{L}^{k}_{\infty\omega}$ (see [GKL+05, Chapter 2]). Here we are interested in the existential positive fragments of $\exists\mathcal{L}^{k}_{\infty\omega}$, k a positive integer, which are tailored for the study of Datalog.

Theorem 7.1. [KV00a] *Let k be a positive integer. Every k-*Datalog *query over finite structures is expressible in $\exists\mathcal{L}^{k}_{\infty\omega}$. Thus, k-*Datalog $\subseteq \exists\mathcal{L}^{k}_{\infty\omega}$ *on finite structures.*

We make use here of the $(\exists, k)$-pebble games discussed in [GKL+05, Chapter 2]. We saw there that if k is a positive integer and Q a Boolean query on a class $\mathcal{C}$ of finite structures, then Q is expressible in $\exists\mathcal{L}^{k}_{\infty\omega}$ on $\mathcal{C}$ iff for all $\mathbf{A}, \mathbf{B} \in \mathcal{C}$ such that $\mathbf{A} \models Q$ and the Duplicator wins the $(\exists, k)$-pebble game on **A** and **B**, we have that $\mathbf{B} \models Q$. The next theorem establishes a connection between expressibility in k-Datalog and $(\exists, k)$-pebble games. (A closely related, but somewhat less precise, such connection was established in [FV98]). In what follows, if $\mathcal{A}$ is a class of structures and **B** is a structure, we write $\mathrm{CSP}(\mathcal{A}, \mathbf{B})$ to denote the class of structures **A** such that $\mathbf{A} \in \mathcal{A}$ and $\mathbf{A} \rightarrow \mathbf{B}$.

Theorem 7.2. [KV00a] *Let k be a positive integer, **B** a relational structure, and $\mathcal{A}$ a class of relational structures such that $\mathbf{B} \in \mathcal{A}$. Then the following statements are equivalent.*

1. $\overline{\mathrm{CSP}(\mathcal{A}, \mathbf{B})}$ *is expressible in k-*Datalog *on $\mathcal{A}$.*
2. $\overline{\mathrm{CSP}(\mathcal{A}, \mathbf{B})}$ *is expressible in $\exists\mathcal{L}^{k}_{\infty\omega}$ on $\mathcal{A}$.*
3. $\overline{\mathrm{CSP}(\mathcal{A}, \mathbf{B})} = \{\mathbf{A} \in \mathcal{A}$ *: The Spoiler wins the $(\exists, k)$-pebble game on **A** and **B**$\}$.*

Recall also from [GKL+05, Chapter 2] that the query "Given two structures **A** and **B**, does the Spoiler win the $(\exists, k)$-pebble on **A** and **B**?" is definable in LFP; as a result, there is a polynomial-time (in fact, $O(n^{2k})$) algorithm that, given two structures **A** and **B**, determines whether the Spoiler wins the $(\exists, k)$-pebble game on **A** and **B**.

By combining Theorem 7.2 with the results of [GKL+05, Chapter 2], we obtain the following uniform tractability result for classes of constraint-satisfaction problems expressible in Datalog.

Theorem 7.3. [KV00a] *Let k be a positive integer, $\mathcal{A}$ a class of relational structures, and $\mathcal{B} = \{\mathbf{B} \in \mathcal{A} : \neg\mathrm{CSP}(\mathcal{A}, \mathbf{B})$ is expressible in k-Datalog$\}$. Then the uniform constraint-satisfaction problem $\mathrm{CSP}(\mathcal{A}, \mathcal{B})$ is solvable in polynomial time. Moreover, the running time of the algorithm is $O(n^{2k})$, where n is the maximum of the sizes of the input structures $\mathbf{A}$ and $\mathbf{B}$.*

Intuitively, if we consider the class of all templates $\mathbf{B}$ for which k-Datalog solves $\mathrm{CSP}(\mathbf{B})$, then computing the winner in the existential k-pebble game offers a *uniform* polynomial-time algorithm. That is, the algorithm determining the winner in the existential k-pebble game is a uniform algorithm for all (non-uniform) constraint-satisfaction problems that can be expressed in k-Datalog.

The characterization in terms of pebble games turns also sheds light on non-uniform constraint satisfaction. As described in [GKL+05, Chapter 2], for every relational structure $\mathbf{B}$ and every positive integer k, there is a k-Datalog program $\rho^k_{\mathbf{B}}$ that expresses the query "Given a structure $\mathbf{A}$, does the Spoiler win the $(\exists, k)$ pebble game on $\mathbf{A}$ and $\mathbf{B}$?". As an immediate consequence of this fact, we get that $\overline{\mathrm{CSP}(\mathbf{B})}$ is expressible in k-Datalog if and only if it is expressible by a *specific* k-Datalog program.

Theorem 7.4. [FV98, KV00a] *$\overline{\mathrm{CSP}(\mathbf{B})}$ is expressible in k-Datalog if and only if it is expressible by $\rho^k_{\mathbf{B}}$.*

It follows that $\overline{\mathrm{CSP}(\mathbf{B})}$ is expressible in k-Datalog if and only if $\neg\varphi_{\mathbf{B}}$ is logically equivalent to $\rho^k_{\mathbf{B}}$, where $\varphi_{\mathbf{B}}$ is the MMSNP sentence expressing $\mathrm{CSP}(\mathbf{B})$. Unfortunately, it is not known if equivalence of complemented MMSNP to Datalog is decidable.

8 Games and Consistency

One of the most fruitful approaches to coping with the intractability of constraint satisfaction has been the introduction and use of various *consistency* concepts that make explicit additional constraints implied by the original constraints. The connection between consistency properties and tractability was first described in [Fre78, Fre82]. In a similar vein, the relationship between *local consistency* and *global consistency* is investigated in [Dec92b, vB94, vBD97]. Intuitively, local consistency means that any partial solution on a set of variables can be extended to a partial solution containing an additional variable, whereas global consistency means that any partial solution can be extended to a global solution. Note that if the inputs are such that local consistency implies global consistency, then there is a polynomial-time algorithm for constraint satisfaction; moreover, in this case a solution can be constructed via a backtrack-free search. We now describe this approach from the Datalog perspective. The crucial insight is that the key concept of *strong k-consistency* [Dec92b] is equivalent to a property of winning strategies for the Duplicator in the $(\exists, k)$-pebble game. Specifically, an instance of a constraint-satisfaction problem is strongly k-consistent if and only if the family of *all* k-partial homomorphisms f is a winning strategy for the Duplicator in the $(\exists, k)$-pebble game on the two relational structures that represent the given instance.

The connection between pebble games and consistency properties, however, is deeper than just a mere reformulation of the concept of strong k-consistency. Indeed, as mentioned earlier, consistency properties underly the process of making explicit new constraints that are implied by the original constraints. A key technical step in this approach is the procedure known as "establishing strong k-consistency", which propagates the original constraints, adds implied constraints, and transforms a given instance of a constraint-satisfaction problem to a strongly k-consistent instance with the same solution space [Coo89,Dec92b]. In fact, strong k-consistency can be established if and only if the Duplicator wins the $(\exists, k)$-pebble game. Moreover, whenever strong k-consistency can be established, one method for doing this is to first compute the largest winning strategy for the Duplicator in the $(\exists, k)$-pebble game and then modify the original problem by augmenting it with the constraints expressed by the largest winning strategy; this method gives rise to the least constrained instance that establishes strong k-consistency and, in addition, satisfies a natural *coherence* property. By combining this result with known results concerning the definability of the largest winning strategy, it follows that the algorithm for establishing strong k-consistency in this way (with k fixed) is actually expressible in least fixed-point logic; this strengthens the fact that strong k-consistency can be established in polynomial time, when k is fixed. If we consider non-uniform constraint satisfaction, it follows that for every relational structure $\mathbf{B}$, the complement of CSP($\mathbf{B}$) is expressible by a Datalog program with k variables if and only if CSP($\mathbf{B}$) coincides with the collection of all relational structures $\mathbf{A}$ such that establishing strong k-consistency on $\mathbf{A}$ and $\mathbf{B}$ implies that there is a homomorphism from $\mathbf{A}$ to $\mathbf{B}$.

We start the formal treatment by returning first to $(\exists, k)$-pebble games. Recall from [GKL$^+$05, Chapter 2] that a winning strategy for the Duplicator in the $(\exists, k)$-pebble game on $\mathbf{A}$ and $\mathbf{B}$ is a nonempty family of k-partial homomorphisms (that is, partial homomorphisms defined on at most k elements) from $\mathbf{A}$ to $\mathbf{B}$ that is closed under subfunctions and has the forth property up to k. A *configuration* for the $(\exists, k)$-pebble game on $\mathbf{A}$ and $\mathbf{B}$ is a $2k$-tuple $\mathbf{a}, \mathbf{b}$, where $\mathbf{a} = (a_1, \ldots, a_k)$ and $\mathbf{b} = (b_1, \ldots, b_k)$ are elements of A^k and B^k, respectively, such that if $a_i = a_j$, then $b_i = b_j$; this means that the correspondence $a_i \mapsto b_i$, $1 \leq i \leq k$, is a partial function from A to B, which we denote by $h_{\mathbf{a},\mathbf{b}}$. A *winning configuration* for the Duplicator in the existential k-pebble game on $\mathbf{A}$ and $\mathbf{B}$ is a configuration $\mathbf{a}, \mathbf{b}$ for this game such that $h_{\mathbf{a},\mathbf{b}}$ is a member of some winning strategy for the Duplicator in this game. We denote by $\mathcal{W}^k(\mathbf{A}, \mathbf{B})$ the set of all such configurations. The following results show that expressibility in $\exists\mathcal{L}^k_{\infty\omega}$ can be characterized in terms of the set $\mathcal{W}^k(\mathbf{A}, \mathbf{B})$.

Proposition 8.1. [KV00b] *If $\mathcal{F}$ and $\mathcal{F}'$ are two winning strategies for the Duplicator in the $(\exists, k)$-pebble game on two structures $\mathbf{A}$ and $\mathbf{B}$, then also the union $\mathcal{F} \cup \mathcal{F}'$ is a winning strategy for the Duplicator. Consequently, there is a largest winning strategy for the Duplicator in the $(\exists, k)$-pebble game, namely the union of all winning strategies, which is precisely the set $\mathcal{H}^k(\mathbf{A}, \mathbf{B}) = \{h_{\overline{a},\overline{b}} : (\overline{a}, \overline{b}) \in \mathcal{W}^k(\mathbf{A}, \mathbf{B})\}$.*

Corollary 8.2. [KV00a] *Let k be a positive integer and Q a k-ary query on a class $\mathcal{C}$ of finite structures. Then the following two statements are equivalent:*

1. *Q is expressible in $\exists\mathcal{L}^k_{\infty\omega}$ on $\mathcal{C}$.*
2. *If $\mathbf{A}$, $\mathbf{B}$ are two structures in $\mathcal{C}$, $(\mathbf{a}, \mathbf{b}) \in \mathcal{W}^k(\mathbf{A}, \mathbf{B})$, and $\mathbf{A} \models Q(\mathbf{a})$, then $\mathbf{B} \models Q(\mathbf{b})$.*

The following lemma is a crucial definability result.

Lemma 8.3. [KV00a] *There is a positive-in-S first-order formula $\varphi(\overline{x}, \overline{y}, S)$, where $\overline{x}$ and $\overline{y}$ are k-tuples of variables, such that the complement of its least fixed-point on a pair $(\mathbf{A}, \mathbf{B})$ of structures defines the set $\mathcal{W}^k(\mathbf{A}, \mathbf{B})$ of all winning configurations for the Duplicator in the $(\exists, k)$-pebble game on $\mathbf{A}, \mathbf{B}$.*

We now formally define the concepts of *i-consistency* and *strong k-consistency.*

Definition 8.4. *Let $\mathcal{P} = (V, D, \mathcal{C})$ be a constraint-satisfaction instance.*

- *A* partial solution *on a set $V' \subset V$ is an assignment $h : V' \to D$ that satisfies all the constraints whose scope is contained in V'.*
- *$\mathcal{P}$ is* i-consistent *if for every $i-1$ variables $v_1, \ldots, v_{i-1}$, for every partial solution on these variables, and for every variable $v_i \notin \{v_1, \ldots, v_{i-1}\}$, there is a partial solution on the variables $v_1, \ldots, v_{i-1}, v_i$ extending the given partial solution on the variables $v_1, \ldots, v_{i-1}$.*
- *$\mathcal{P}$ is* strongly k-consistent *if it is i-consistent for every $i \leq k$.* ∎

To illustrate these concepts, consider the Boolean formula

$$(\neg x_1 \vee x_3) \wedge (\neg x_2 \vee x_3) \wedge (x_2 \vee \neg x_3).$$

It is easy to verify that this formula, viewed as a constraint-satisfaction instance, is strongly 3-consistent. For instance, the partial solution $x_2 = 0$, $x_3 = 0$ can be extended to the solution $x_1 = 0$, $x_2 = 0$, $x_3 = 0$, while the partial solution $x_1 = 1$, $x_3 = 1$ can be extended to the solution $x_1 = 1$, $x_2 = 1$, $x_3 = 1$. In contrast, the Boolean formula

$$(x_1 \vee x_2) \wedge (\neg x_1 \vee x_3) \wedge (\neg x_2 \vee x_3) \wedge (x_2 \vee \neg x_3)$$

is satisfiable and strongly 2-consistent, but not 3-consistent (hence, it is not strongly 3-consistent either). The reason is that the partial solution $x_2 = 0$, $x_3 = 0$ cannot be extended to a solution, since the only solutions of this formula are $x_1 = 0$, $x_2 = 1$, $x_3 = 1$ and $x_1 = 1$, $x_2 = 1$, $x_3 = 1$. We note that the concepts of strong 2-consistency and strong 3-consistency were first studied in the literature under the names of *arc consistency* and *path consistency* (see [Dec03]).

A key insight is that the concepts of i-consistency and strong k-consistency can be naturally recast in terms of existential pebble games.

Proposition 8.5. [KV00b] *Let $\mathcal{P}$ be a CSP instance, and let $(\mathbf{A}_{\mathcal{P}}, \mathbf{B}_{\mathcal{P}})$ be the associated homomorphism instance.*

- *$\mathcal{P}$ is i-consistent if and only if the family of all partial homomorphisms from $\mathbf{A}_{\mathcal{P}}$ to $\mathbf{B}_{\mathcal{P}}$ with $i-1$ elements in their universe has the i-forth property.*
- *$\mathcal{P}$ is strongly k-consistent if and only if the family of all k-partial homomorphisms from $\mathbf{A}_{\mathcal{P}}$ to $\mathbf{B}_{\mathcal{P}}$ is a winning strategy for the Duplicator in the $(\exists, k)$-pebble game on $\mathbf{A}_{\mathcal{P}}$ and $\mathbf{B}_{\mathcal{P}}$.*

Let us now recall the concept of *establishing strong k-consistency*, as defined, for instance, in [Coo89,Dec92b]. This concept has been defined rather informally in the AI literature to mean that, given a constraint-satisfaction instance $\mathcal{P}$, we associate with it another instance $\mathcal{P}'$ that has the following properties: (1) $\mathcal{P}'$ has the same set of variables and the same set of values as $\mathcal{P}$ (2) $\mathcal{P}'$ is strongly k-consistent; (3) $\mathcal{P}'$ is at least as constrained as $\mathcal{P}$; and (4) $\mathcal{P}$ and $\mathcal{P}'$ have the same space of solutions. The next definition formalizes the above concept in the context of the homomorphism problem (cf. [DP99, KV00b]).

Definition 8.6. *Let $\mathbf{A}$ and $\mathbf{B}$ be two relational structures over a k-ary vocabulary σ (i.e., every relation symbol in σ has arity at most k).* Establishing strong k-consistency for $\mathbf{A}$ and $\mathbf{B}$ *means that we associate two relational structures $\mathbf{A}'$ and $\mathbf{B}'$ with the following properties:*

1. *$\mathbf{A}'$ and $\mathbf{B}'$ are structures over some k-ary vocabulary σ' (in general, different than σ); moreover, the universe of $\mathbf{A}'$ is the universe A of $\mathbf{A}$, and the universe of $\mathbf{B}'$ is the universe B of $\mathbf{B}$.*
2. CSP$(\mathbf{A}', \mathbf{B}')$ *is strongly k-consistent.*
3. *if h is a k-partial homomorphism from $\mathbf{A}'$ to $\mathbf{B}'$, then h is a k-partial homomorphism from $\mathbf{A}$ to $\mathbf{B}$.*
4. *If h is a function from A to B, then h is a homomorphism from $\mathbf{A}$ to $\mathbf{B}$ if and only if h is a homomorphism from $\mathbf{A}'$ to $\mathbf{B}'$.*

If the structures $\mathbf{A}'$ and $\mathbf{B}'$ have the above properties, then we say that $\mathbf{A}'$ and $\mathbf{B}'$ establish strong k-consistency for $\mathbf{A}$ and $\mathbf{B}$. ∎

A constraint-satisfaction instance $\mathcal{P}$ is *coherent* if every constraint $(\mathbf{t}, R)$ of $\mathcal{P}$ completely determines all constraints $(\mathbf{u}, Q)$ in which all variables occurring in $\mathbf{u}$ are among the variables of $\mathbf{t}$. We formalize this concept as follows.

Definition 8.7. *An instance $\mathbf{A}, \mathbf{B}$ of the homomorphism problem is* coherent *if its associated constraint-satisfaction instance* CSP$(\mathbf{A}, \mathbf{B})$ *has the following property: for every constraint $(\mathbf{a}, R)$ of* CSP$(\mathbf{A}, \mathbf{B})$ *and every tuple $\mathbf{b} \in R$, the mapping $h_{\mathbf{a},\mathbf{b}}$ is well defined and is a partial homomorphism from $\mathbf{A}$ to $\mathbf{B}$.* ∎

Note that a constraint-satisfaction instance can be made coherent in polynomial-time by constraint propagation.

The main result of this section is that strong k-consistency can be established precisely when the Duplicator wins the $(\exists, k)$-pebble game. Moreover, one method for establishing strong k-consistency is to first compute the largest winning strategy for the Duplicator in this game and then generate an instance of the constraint-satisfaction problem consisting of all the constraints embodied in the largest winning strategy. Furthermore, this method gives rise to the largest coherent instance that establishes strong k-consistency (and, hence, the least constrained such instance).

Theorem 8.8. [KV00b] *Let k be a positive integer, let σ be a k-ary vocabulary, and let $\mathbf{A}$ and $\mathbf{B}$ be two relational structures over σ with universes A and B, respectively. It is possible to establish strong k-consistency for $\mathbf{A}$ and $\mathbf{B}$ if and only if $\mathcal{W}^k(\mathbf{A}, \mathbf{B}) \neq \emptyset$. Furthermore, if $\mathcal{W}^k(\mathbf{A}, \mathbf{B}) \neq \emptyset$, then the following sequence of steps gives rise to two structures $\mathbf{A}'$ and $\mathbf{B}'$ that establish strong k-consistency for $\mathbf{A}$ and $\mathbf{B}$:*

1. *Compute the set $\mathcal{W}^k(\mathbf{A}, \mathbf{B})$.*
2. *For every $i \leq k$ and for every i-tuple $\mathbf{a} \in A^i$, form the set $R_{\mathbf{a}} = \{\mathbf{b} \in B^i : (\mathbf{a}, \mathbf{b}) \in \mathcal{W}^k(\mathbf{A}, \mathbf{B})\}$.*
3. *Form the constraint-satisfaction instance $\mathcal{P}$ with A as the set of variables, B as the set of values, and $\{(\mathbf{a}, R_{\mathbf{a}}) : \mathbf{a} \in \cup_{i=1}^{k} A^i\}$ as the collection of constraints.*
4. *Let $(\mathbf{A}', \mathbf{B}')$ be the homomorphism instance of $\mathcal{P}$.*

In addition, the structures $\mathbf{A}'$ and $\mathbf{B}'$ obtained above constitute the largest coherent instance establishing strong k-consistency for $\mathbf{A}$ and $\mathbf{B}$, that is, if $(\mathbf{A}'', \mathbf{B}'')$ is another such coherent instance, then for every constraint $(\mathbf{a}, R)$ of $\mathrm{CSP}(\mathbf{A}'', \mathbf{B}'')$, we have that $R \subseteq R_{\mathbf{a}}$.

The key step in the procedure described in Theorem 8.8 is the first step, in which the set $\mathcal{W}^k(\mathbf{A}, \mathbf{B})$ is computed. The other steps simply "re-format" $\mathcal{W}^k(\mathbf{A}, \mathbf{B})$. From Lemma 8.3 it follows that we can establish strong k-consistency by computing the fixed-point of a monotone first-order formula. We can now relate the concept of strong k-consistency to the results in [FV98] regarding Datalog and non-uniform CSP.

Theorem 8.9. [KV00b] *Let $\mathbf{B}$ be a relational structure over a vocabulary σ. Then the following two statements are equivalent.*

- *$\overline{\mathrm{CSP}(\mathbf{B})}$ is expressible in k-Datalog.*
- *For every structure $\mathbf{A}$ over σ, establishing strong k-consistency for $\mathbf{A}, \mathbf{B}$ implies that there is a homomorphism from $\mathbf{A}$ to $\mathbf{B}$.*

Given the fundamental role that the set $\mathcal{W}^k(\mathbf{A}, \mathbf{B})$ plays here, it is natural to ask about the complexity of computing it. To turn it into a decision problem, we just ask about the non-emptiness of this set.

Theorem 8.10. [KP03] *The problem $\{(\mathbf{A}, \mathbf{B}, k) : \mathcal{W}^k(\mathbf{A}, \mathbf{B}) \neq \emptyset\}$, with k encoded in unary, is* EXPTIME*-complete. In words, the following problem is* EXPTIME*-complete: given a positive integer k and two structures $\mathbf{A}$ and $\mathbf{B}$, does the Duplicator win the $(\exists, k)$-pebble game on $\mathbf{A}$ and $\mathbf{B}$?*

This result is rather surprising. After all, the complexity of constraint satisfaction is "only" NP-complete. In contrast, the complexity of establishing strong k-consistency is provably exponential and not in PTIME. This offers an a posteriori justification of the practice of establishing only a "low degree" of consistency, such as *arc consistency* or *path consistency* [Apt03, Dec03].

9 Uniform Constraint Satisfaction and Bounded Treewidth

So far, we focused on the pursuit of islands of tractability of non-uniform constraint satisfaction, that is, islands of the form $\mathrm{CSP}(\mathbf{B}) = \mathrm{CSP}(All, \{\mathbf{B}\})$, where $\mathbf{B}$ is a fixed template. Even when we discussed uniform constraint satisfaction, it was with respect to tractable templates. In this section we focus on uniform constraint satisfaction of the form $\mathrm{CSP}(\mathcal{A}, All)$, where $\mathcal{A}$ is a class of structures. The goal is to identify conditions on $\mathcal{A}$ that ensure uniform tractability.

As is well known, many algorithmic problems that are "hard" on arbitrary structures become "easy" on trees. This phenomenon motivated researchers to investigate whether the concept of a tree can be appropriately relaxed while maintaining good computational behavior. As part of their seminal work on graph minors, Robertson and Seymour introduced the concept of *treewidth*, which, intuitively, measures how "tree-like" a structure is; moreover, they showed that graphs of *bounded treewidth* exhibit such good behavior, cf. [NR90].

Definition 9.1. *A* tree decomposition *of a relational structure* $\mathbf{A}$ *is a labelled tree* T *such that the following conditions hold:*

1. *every node of* T *is labelled by a non-empty subset of the universe* A *of* $\mathbf{A}$*,*
2. *for every relation* R *of* $\mathbf{A}$ *and every tuple* $(a_1, \ldots, a_n)$ *in* R*, there is a node of* T *whose label contains* $\{a_1, \ldots, a_n\}$*,*
3. *for every* $a \in A$*, the set of nodes of* T *whose labels include* a *forms a subtree of* T*.*

The width *of a tree decomposition* T *is the maximum cardinality of a label of a node in* T *minus 1. The* treewidth *of* $\mathbf{A}$*, denoted* $\mathrm{tw}(\mathbf{A})$*, is the smallest positive integer* k *such that* A *has a tree decomposition of width* k*. We write* $\mathcal{T}(k)$ *to denote the class of all structures* $\mathbf{A}$ *such that* $\mathrm{tw}(\mathbf{A}) < k$*.*

Clearly, if $\mathbf{T}$ is a tree, then $\mathrm{tw}(\mathbf{T}) = 1$. Similarly, if $n \geq 3$ and $\mathbf{C}_n$ is the n-element (directed) cycle, then $\mathrm{tw}(\mathbf{C}) = 2$. At the other end of the scale, $\mathrm{tw}(\mathbf{K}_k) = k - 1$, for every $k \geq 2$. Computing the treewidth of a structure is an intractable problem. Specifically, the following problem is NP-complete [ACP87]: given a graph $\mathbf{H}$ and an integer $k \geq 1$, is $\mathrm{tw}(\mathbf{H}) \leq k$? Nonetheless, Bodlaender [Bod93] showed that for every fixed integer $k \geq 1$, there is a linear-time algorithm such that, given a structure $\mathbf{A}$, it determines whether or not $\mathrm{tw}(\mathbf{A}) < k$. In other words, each class $\mathcal{T}(k)$ is recognizable in polynomial time.

Dechter and Pearl [DP89] and Freuder [Fre90] showed that the classes of structures of bounded treewidth give rise to large islands of tractability of uniform constraint satisfaction.

Theorem 9.2. [DP89,Fre90] *If* $k \geq 2$ *is a positive integer, then* $\mathrm{CSP}(\mathcal{T}(k), All)$ *is in* PTIME.

The polynomial-time algorithm for $\mathrm{CSP}(\mathcal{T}(k), All)$ in the above theorem is often described as a *bucket-elimination algorithm* [Dec99]. It should be noted that it is not a constraint-propagation algorithm. Instead, this algorithm uses the bound on the treewidth to test if a solution to the constraint-satisfaction problem exists by solving a join-evaluation problem in which all intermediate relations are of bounded arity.

Kolaitis and Vardi [KV00a], and Dalmau, Kolaitis and Vardi [DKV02] investigated certain logical aspects of the treewidth of a relational structure and showed that this combinatorial concept is closely connected to the canonical conjunctive query of the structure being definable in a fragment of first-order logic with a fixed number of variables. This made it possible to show that the tractability of CSP($\mathcal{T}(k)$, *All*) can be explained in purely logical terms. Moreover, it led to the discovery of larger islands of tractability of uniform constraint satisfaction.

Definition 9.3. *Let $k \geq 2$ be a positive integer.*

- FO^k *is the collection of all first-order formulas with at most k distinct variables.*
- L^k *is the collection of all* FO^k*-formulas built using atomic formulas, conjunction, and existential first-order quantification only.*

Intuitively, queries expressible in FO^k and L^k are simply first-order queries and conjunctive queries, respectively, with a bound k on the number of distinct variables (each variable, however, may be reused any number of times).

As an example, it is easy to see that if $\mathbf{C}_n$ is the n-element cycle, $n \geq 3$, then the canonical conjunctive query $Q_{\mathbf{C}_n}$ is expressible in L^3. For instance, $Q_{\mathbf{C}_4}$ is logically equivalent to $(\exists x \exists y \exists z)(E(x,y) \wedge E(y,z) \wedge (\exists y)(E(z,y) \wedge E(y,x)))$. As mentioned earlier, for every $n \geq 3$, we have that $\mathrm{tw}(\mathbf{C}_n) = 2$.

The logics FO^k and L^k are referred to as *variable-confined logics* [KV96]. The complexity of query evaluation for such queries has been studied in [Var95]. Since in data complexity the queries are fixed, bounding the number of variables does not change data complexity. The change in expression and combined complexity, however, is quite dramatic, as the combined complexity of FO^k has been shown to be in PTIME [Var95]. (More generally, the exponential gap between data complexity and expression and combined complexity shrinks when the number of variables is bounded.)

The next result shows that the relationship we just saw in the example between treewidth and number of variables needed to express the canonical conjunctive query of a cycle is not an accident.

Theorem 9.4. [KV00a] *Let $k \geq 2$ be a positive integer. If $\mathbf{A} \in \mathcal{T}(k)$, then the canonical conjunctive query $Q_{\mathbf{A}}$ is expressible in L^k.*

Corollary 9.5. CSP($\mathcal{T}(k)$, *All*) *can be solved in polynomial time by determining, given a structure $\mathbf{A} \in \mathcal{T}(k)$ and an arbitrary structure $\mathbf{B}$, whether $\mathbf{B} \models Q_{\mathbf{A}}$.*

A precise complexity analysis of CSP($\mathcal{T}(k)$, *All*) is provided in [GLS98], where it is shown that the problem is LOGFCL-complete; by definition, LOGCFL is the class of decision problems that are logspace-reducible to a context-free language. Note that, in contrast, the combined complexity of evaluating FO^k-queries, for $k > 3$, is PTIME-complete [Var95].

Theorem 9.4 can be viewed as a logical recasting of the bucket-elimination algorithm. It derives the tractability of CSP($\mathcal{T}(k)$, *All*) from the fact that the canonical conjunctive query $Q_{\mathbf{A}}$ can be written using at most k variables. Consequently, evaluating this query amounts to solving a join-evaluation problem in which all intermediate relations are of bounded arity. For an investigation of how the ideas underlying Theorem 9.4 can be used to solve practical join-evaluation problems, see [MPPV04].

It turns out, however, that we can also approach solving CSP($\mathcal{T}(k)$, *All*) from the perspective of k-Datalog and $(\exists, k)$-pebble games. This is because L^k is a fragment of $\exists\mathcal{L}^k_{\infty\omega}$, whose expressive power, as seen earlier, can be characterized in terms of such games.

Theorem 9.6. [DKV02] *Let $k \geq 2$ be a positive integer.*

- *If* $\mathbf{B}$ *is an arbitrary, but fixed, structure, then* $\mathcal{T}(k) \cap \overline{\mathrm{CSP}(\mathcal{T}(k), \{\mathbf{B}\})}$ *is expressible in k-Datalog*[6].
- CSP($\mathcal{T}(k)$, *All*) *can be solved in polynomial time by determining whether, given a structure* $\mathbf{A} \in \mathcal{T}(k)$ *and an arbitrary structure* $\mathbf{B}$, *the Duplicator wins the* $(\exists, k)$*-pebble on* $\mathbf{A}$ *and* $\mathbf{B}$.

The situation for bounded treewidth structures, as described by Theorem 9.6, should be contrasted with the situation for bounded *cliquewidth* structures (cf. [CMR00]). Let $\mathcal{C}(k)$ be the class of structures of cliquewidth bounded by k. It is shown in [CMR00] that CSP($\mathcal{C}(k), \{\mathbf{B}\}$) is in PTIME for each structure $\mathbf{B}$. Since, however, complete graphs have bounded cliquewidth, it follows that the CLIQUE problem can be reduced to CSP($\mathcal{C}(k)$, *All*), implying NP-hardness of the latter.

As a consequence of Theorem 9.6, we see that CSP($\mathcal{T}(k)$, *All*) can be solved in polynomial time using a constraint-propagation algorithm that is quite different from the bucket-elimination algorithm in Theorem 9.2. It should be noted, however, that this requires knowing that we are given an instance $\mathbf{A}, \mathbf{B}$ where $\mathrm{tw}(\mathbf{A}) \leq k$. In contrast, the bucket-elimination algorithm can be used for arbitrary constraint-satisfaction instances (with no tractability guarantee, in general).

The classes CSP($\mathcal{T}(k)$, *All*) enjoy also nice tractability properties from the perspective of *Parameterized Complexity Theory* [DF99], as they are *fixed-parameter tractable*, and, in a precise technical sense, are maximal with this property under a certain complexity-theoretic assumption (see [GSS01]).

The development so far shows that $\mathcal{T}(k)$ provides an island of tractability for uniform constraint satisfaction. We now show that this island can be expanded.

Definition 9.7. *Let* $\mathbf{A}$ *and* $\mathbf{B}$ *be two relational structures.*

- *We say that* $\mathbf{A}$ *and* $\mathbf{B}$ *are* homomorphically equivalent, *denoted* $\mathbf{A} \sim_h \mathbf{B}$, *if both* $\mathbf{A} \to \mathbf{B}$ *and* $\mathbf{B} \to \mathbf{A}$ *hold.*

[6] The intersection with $\mathcal{T}(k)$ ensures that only structures with treewidth bounded by k are considered.

– *We say that* $\mathbf{B}$ *is the* core *of* $\mathbf{A}$*, and write* $\mathrm{core}(\mathbf{A}) = \mathbf{B}$*, if* $\mathbf{B}$ *is a substructure of* $\mathbf{A}$*,* $\mathbf{A} \rightarrow \mathbf{B}$ *holds, and* $\mathbf{A} \rightarrow \mathbf{B}'$ *fails for each proper substructure* $\mathbf{B}'$ *of* $\mathbf{B}$*.*

Clearly, $\mathrm{core}(\mathbf{K}_k) = \mathbf{K}_k$ and $\mathrm{core}(\mathbf{C}_n) = \mathbf{C}_n$. On the other hand, if $\mathbf{H}$ is a 2-colorable graph with at least one edge, then $\mathrm{core}(\mathbf{H}) = \mathbf{K}_2$. It should be noted that cores play an important role in database query processing and optimization (see [CM77]). The next result shows that they can also be used to characterize when the canonical conjunctive query is definable in L^k.

Theorem 9.8. [DKV02] *Let* $k \geq 2$ *be a positive integer and* $\mathbf{A}$ *a relational structure. Then the following are equivalent:*

– $Q_{\mathbf{A}}$ *is definable in* L^k*.*
– *There is a structure* $\mathbf{B} \in \mathcal{T}(k)$ *such that* $\mathbf{A} \sim_h \mathbf{B}$*.*
– $\mathrm{core}(\mathbf{A}) \in \mathcal{T}(k)$*.*

The tight connection between definability in L^k and the boundedness of the treewidth of the core suggests a way to expand the "island" $\mathcal{T}(k)$.

Definition 9.9. *If* $k \geq 2$ *is a positive integer, then* $\mathcal{H}(\mathcal{T}(k))$ *is the class of relational structures* $\mathbf{A}$ *such that* $\mathrm{core}(\mathbf{A})$ *has treewidth less than* k*.*

It should noted that $\mathcal{T}(k)$ is properly contained in $\mathcal{H}(\mathcal{T}(k))$, for every $k \geq 2$. Indeed, it is known that there are 2-colorable graphs of arbitrarily large treewidth. In particular, *grids* are known to have these properties (see [DF99]). Yet, these graphs are members of $\mathcal{H}(\mathcal{T}(2))$, since their core is $\mathbf{K}_2$.

Theorem 9.10. [DKV02] *Let* $k \geq 2$ *be a positive integer.*

– *If* $\mathbf{B}$ *is an arbitrary, but fixed, structure, then* $\mathcal{H}(\mathcal{T}(k)) \cap \overline{\mathrm{CSP}(\mathcal{H}(\mathcal{T}(k)), \{\mathbf{B}\})}$ *is expressible in k-Datalog.*
– $\mathrm{CSP}(\mathcal{H}(\mathcal{T}(k)), \mathit{All})$ *is in* PTIME*. Moreover,* $\mathrm{CSP}(\mathcal{H}(\mathcal{T}(k)), \mathit{All})$ *can be solved in polynomial time by determining whether, given a structure* $\mathbf{A} \in \mathcal{H}(\mathcal{T}(k))$ *and an arbitrary structure* $\mathbf{B}$*, the Spoiler or the Duplicator wins the* $(\exists, k)$*-pebble on* $\mathbf{A}$ *and* $\mathbf{B}$*.*

The preceding Theorem 9.10 yields new islands of tractability for uniform constraint satisfaction, which properly subsume the islands of tractability constituted by the classes of structures of bounded treewidth. This expansion of the tractability landscape comes, however, at a certain price. Specifically, as seen earlier, for every fixed $k \geq 2$, there is a polynomial-time algorithm for determining membership in $\mathcal{T}(k)$ [Bod93]. In contrast, it has been shown that, for every fixed $k \geq 2$, determining membership in $\mathcal{H}(\mathcal{T}(k))$ is an NP-complete problem [DKV02]. Thus, these new islands of tractability are, in some sense, "inaccessible".

Since $\mathcal{H}(\mathcal{T}(k))$ contains structures of arbitrarily large treewidth, the bucket-elimination algorithm cannot be used to solve $\mathrm{CSP}(\mathcal{H}(\mathcal{T}(k)), \mathit{All})$ in polynomial time. Thus, Theorem 9.10 also shows that determining the winner of the

$(\exists, k)$-pebble is a polynomial-time algorithm that applies to islands of tractability not covered by the bucket elimination algorithm.

It is now natural to ask whether there are classes $\mathcal{A}$ of relational structures that are larger than the classes $\mathcal{H}(\mathcal{T}(k))$ and CSP$(\mathcal{A}, All)$ is solvable in polynomial time. A remarkable result by Grohe [Gro03] essentially shows that, if we fix the vocabulary, *no* such classes exist, provided a certain complexity-theoretic hypothesis is true.

Theorem 9.11. [Gro03] *Assume that* FPT $\neq W[1]$. *If $\mathcal{A}$ is a recursively enumerable class of relational structures over some fixed vocabulary such that* CSP$(\mathcal{A}, All)$ *is in* PTIME, *then there is a positive integer k such that $\mathcal{A} \subseteq \mathcal{H}(\mathcal{T}(k))$.*

The hypothesis FPT $\neq W[1]$ is a statement in *parameterized complexity* that is analogous to the hypothesis PTIME $\neq$ NP, and it is widely accepted as being true (see [DF99]). In effect, Grohe's Theorem 9.11 is a converse to Theorem 9.10 for fixed vocabularies. Together, these two theorems yield a complete characterization of all islands of tractability of the form CSP$(\mathcal{A}, All)$, where $\mathcal{A}$ is a class of structures over some fixed vocabulary. Moreover, they reveal that all tractable cases of the form CSP$(\mathcal{A}, All)$ can be solved by the same polynomial-time algorithm, namely, the algorithm for determining the winner in the $(\exists, k)$-pebble game. In other words, all tractable cases of constraint satisfaction of the form CSP$(\mathcal{A}, All)$ can be solved in polynomial time using constraint propagation.

It is important to emphasize that the classes $\mathcal{H}(\mathcal{T}(k))$ are the largest islands of tractability for uniform constraint satisfaction only under the assumption in Theorem 9.11 of a fixed vocabulary. For variable vocabularies, there is a long line of research, studying the impact of the "topology" of conjunctive queries on the complexity of their evaluation; this line of research goes back to the study of *acyclic* joins in [Yan81], The connection between acyclic joins and acyclic constraints was pointed out in [GJC94]. This is still an active research area. Chekuri and Ramajaran [CR97] showed that the uniform constraint-satisfaction problem CSP$(\mathcal{Q}(k), All)$ is solvable in polynomial time, where $\mathcal{Q}(k)$ is the class of structures of *querywidth* k. Gottlob, Leone, and Scarcello [GLS99b] define another notion of width, called *hypertree width*. They showed that the querywidth of a structure $\mathbf{A}$ provides a strict upper bound for the hypertree width of $\mathbf{A}$, but that the class $\mathcal{H}(k)$ of structures of hypertree width at most k is polynomially recognizable (unlike the class $\mathcal{Q}(k)$), and that CSP$(\mathcal{H}(k), All)$ is tractable. For further discussion on the relative merit of various notions of "width", see [GLS99a]. This is an active area of research (see [CD05, CJG05]).

Acknowledgements. We are grateful to Benoit Larose and Scott Weinstein for helpful comments on a previous draft of this chapter. This work was supported in part by NSF grants CCR-9988322, CCR-0124077, CCR-0311326, ANI-0216467, and a Guggenheim Fellowship. Part of this work was done while the second author was visiting the Isaac Newton Institute for Mathematical Science, as part of a Special Programme on Logic and Algorithms.

References

[ACP87] Arnborg, S., Corneil, D.G., Proskurowski, A.: Complexity of finding embeddings in a k-tree. SIAM J. of Algebraic and Discrete Methods 8, 277–284 (1987)

[AG94] Ajtai, M., Gurevich, Y.: Datalog vs first-order logic. J. Comput. Syst. Sci. 49(3), 562–588 (1994)

[AHV95] Abiteboul, S., Hull, R., Vianu, V.: Foundations of Databases. Addison-Wesley, Reading (1995)

[Apt03] Apt, K.: Principles of Constraint Programming. Cambridge Univ. Press, Cambridge (2003)

[Ats05] Atserias, A.: On digraph coloring problems and treewidth duality. In: Proc. 20th IEEE Symp. on Logic in Computer Science, pp. 106–115. IEEE Computer Society Press, Los Alamitos (2005)

[Bib88] Bibel, W.: Constraint satisfaction from a deductive viewpoint. Artificial Intelligence 35, 401–413 (1988)

[BJK05] Bulatov, A.A., Jeavons, P., Krokhin, A.A.: Classifying the complexity of constraints using finite algebras. SIAM J. Comput. 34(3), 720–742 (2005)

[Bod93] Bodlaender, H.L.: A linear-time algorithm for finding tree-decompositions of small treewidth. In: Proc. 25th ACM Symp. on Theory of Computing, pp. 226–234 (1993)

[Bul02] Bulatov, A.A.: A dichotomy theorem for constraints on a three-element set. In: Proc. 43rd Symp. on Foundations of Computer Science, pp. 649–658. IEEE Computer Society, Los Alamitos (2002)

[Bul03] Bulatov, A.A.: Tractable conservative constraint satisfaction problems. In: Proc. 18th IEEE Symp. on Logic in Computer Science, pp. 321–330. IEEE Computer Society, Los Alamitos (2003)

[CD05] Chen, H., Dalmau, V.: Beyond hypertree width: Decomposition methods without decompositions. In: van Beek, P. (ed.) CP 2005. LNCS, vol. 3709, pp. 167–181. Springer, Heidelberg (2005)

[CJG05] Cohen, D.A., Jeavons, P., Gyssens, M.: A unified theory of structural tractability for constraint satisfaction and spread cut decomposition. In: Proc. 19th Int'l Joint Conf. on Artificial Intelligence, pp. 72–77 (2005)

[CM77] Chandra, A.K., Merlin, P.M.: Optimal implementation of conjunctive queries in relational databases. In: Proc. 9th ACM Symp. on Theory of Computing, pp. 77–90 (1977)

[CMR00] Courcelle, B., Makowsky, J.A., Rotics, U.: Linear time solvable optimization problems on graphs of bounded clique-width. Theory of Computing Systems 33, 125–150 (2000)

[Coo89] Cooper, M.C.: An optimal k-consistency algorithm. Artificial Intelligence 41(1), 89–95 (1989)

[CR97] Chekuri, C., Rajaraman, A.: Conjunctive query containment revisited. In: Afrati, F.N., Kolaitis, P.G. (eds.) ICDT 1997. LNCS, vol. 1186, pp. 56–70. Springer, Heidelberg (1996)

[Dal05] Dalmau, V.: Generalized majority-minority operations are tractable. In: Proc. 20th IEEE Symp. on Logic in Computer Science (LICS 2005), pp. 438–447 (2005)

[Dec92a] Dechter, R.: Constraint networks. In: Shapiro, S.C. (ed.) Encyclopedia of Artificial Intelligence, pp. 276–185. Wiley, Chichester (1992)

[Dec92b] Dechter, R.: From local to global consistency. Artificial Intelligence 55(1), 87–107 (1992)

[Dec99] Dechter, R.: Bucket elimination: a unifying framework for reasoning. Artificial Intelligence 113(1–2), 41–85 (1999)

[Dec03] Dechter, R.: Constraint Processing. Morgan Kaufmann, San Francisco (2003)

[DF99] Downey, R.G., Fellows, M.R.: Parametrized Complexity. Springer, Heidelberg (1999)

[DKV02] Dalmau, V., Kolaitis, P.G., Vardi, M.Y.: Constraint satisfaction, bounded treewidth, and finite-variable logics. In: Van Hentenryck, P. (ed.) CP 2002. LNCS, vol. 2470, pp. 310–326. Springer, Heidelberg (2002)

[DM94] Dechter, R., Meiri, I.: Experimental evaluation of preprocessing algorithms for constraint satisfaction problems. Artificial Intelligence 68, 211–241 (1994)

[DP89] Dechter, R., Pearl, J.: Tree clustering for constraint networks. Artificial Intelligence, 353–366 (1989)

[DP99] Dalmau, V., Pearson, J.: Closure functions and width 1 problems. In: Jaffar, J. (ed.) CP 1999. LNCS, vol. 1713, pp. 159–173. Springer, Heidelberg (1999)

[Fed06] Feder, T.: Constraint satisfaction: A personal perspective. Technical report, Electronic Colloquium on Computational Complexity, Report TR06-021 (2006)

[FF05] Feder, T., Ford, D.: Classification of bipartite boolean constraint satisfaction through delta-matroid intersection (2005)

[Fre78] Freuder, E.C.: Synthesizing constraint expressions. Communications of the ACM 21(11), 958–966 (1978)

[Fre82] Freuder, E.C.: A sufficient condition for backtrack-free search. Journal of the Association for Computing Machinery 29(1), 24–32 (1982)

[Fre90] Freuder, E.C.: Complexity of k-tree structured constraint satisfaction problems. In: Proc. AAAI 1990, pp. 4–9 (1990)

[Fro97] Frost, D.H.: Algorithms and Heuristics for Constraint Satisfaction Problems. Ph.D thesis, Department of Computer Science, University of California, Irvine (1997)

[FV93] Feder, T.A., Vardi, M.Y.: Monotone monadic SNP and constraint satisfaction. In: Proc. 25th ACM Symp. on Theory of Computing, pp. 612–622 (1993)

[FV98] Feder, T., Vardi, M.Y.: The computational structure of monotone monadic SNP and constraint satisfaction: a study through Datalog and group theory. SIAM J. on Computing 28, 57–104 (1998); Preliminary version in Proc. 25th ACM Symp. on Theory of Computing, pp. 612–622 (May 1993)

[GJ79] Garey, M.R., Johnson, D.S.: Computers and Intractability - A Guide to the Theory of NP-Completeness. W. H. Freeman and Co., New York (1979)

[GJC94] Gyssens, M., Jeavons, P.G., Cohen, D.A.: Decomposition constraint satisfaction problems using database techniques. Artificial Intelligence 66, 57–89 (1994)

[GKL+05] Grädel, E., Kolaitis, P.G., Libkin, L., Marx, M., Spencer, J., Vardi, M.Y., Venema, Y., Weinstein, S.: Finite Model Theory and Its Applications (Texts in Theoretical Computer Science. In: Finite Model Theory and Its Applications (Texts in Theoretical Computer Science. An EATCS Series). Springer, New York (2005)

[GLS98] Gottlob, G., Leone, N., Scarcello, F.: The complexity of acyclic conjunctive queries. In: Proc. 39th IEEE Symp. on Foundation of Computer Science, pp. 706–715 (1998)

[GLS99a] Gottlob, G., Leone, N., Scarcello, F.: A comparison of structural CSP decomposition methods. In: Proc. 16th Int'l. Joint Conf. on Artificial Intelligence (IJCAI 1999), pp. 394–399 (1999)

[GLS99b] Gottlob, G., Leone, N., Scarcello, F.: Hypertree decompositions and tractable queries. In: Proc. 18th ACM Symp. on Principles of Database Systems, pp. 21–32 (1999)

[GMSV87] Gaifman, H., Mairson, H., Sagiv, Y., Vardi, M.Y.: Undecidable optimization problems for database logic programs. In: Proc. 2nd IEEE Symp. on Logic in Computer Science, pp. 106–115 (1987)

[Gro03] Grohe, M.: The complexity of homomorphism and constraint satisfaction problems seen from the other side. In: Proc. 44th IEEE Symp. on Foundations of Computer Science, pp. 552–561. IEEE Computer Society, Los Alamitos (2003)

[GSS01] Grohe, M., Schwentick, T., Segoufin, L.: When is the evaluation of conjunctive queries tractable? In: Proc. 33rd ACM Symp. on Theory of Computing, pp. 657–666 (2001)

[HN90] Hell, P., Nešetřil, J.: On the complexity of H-coloring. Journal of Combinatorial Theory, Series B 48, 92–110 (1990)

[HN04] Hell, P., Nešetřil, J.: Graphs and Homomorphisms. Oxford Lecture Series in Mathematics and Its applications, vol. 28. Oxford Univ. Press, Oxford (2004)

[KP03] Kolaitis, P.G., Panttaja, J.: On the complexity of existential pebble games. In: Baaz, M., Makowsky, J.A. (eds.) CSL 2003. LNCS, vol. 2803, pp. 314–329. Springer, Heidelberg (2003)

[Kum92] Kumar, V.: Algorithms for constraint-satisfaction problems. AI Magazine 13, 32–44 (1992)

[KV87] Kolaitis, P.G., Vardi, M.Y.: The decision problem for the probabilities of higher-order properties. In: Proc. 19th ACM Symp. on Theory of Computing, pp. 425–435 (1987)

[KV96] Kolaitis, P.G., Vardi, M.Y.: On the expressive power of variable-confined logics. In: Proc. 11th IEEE Symp. on Logic in Computer Science, pp. 348–359 (1996)

[KV00a] Kolaitis, P.G., Vardi, M.Y.: Conjunctive-query containment and constraint satisfaction. Journal of Computer and System Sciences, 302–332 (2000); Earlier version in Proc. 17th ACM Symp. on Principles of Database Systems (PODS 1998)

[KV00b] Kolaitis, P.G., Vardi, M.Y.: A game-theoretic approach to constraint satisfaction. In: Proc. of the 17th National Conference on Artificial Intelligence (AAAI 2000), pp. 175–181 (2000)

[Lad75] Ladner, R.E.: On the structure of polynomial time reducibility. J. Assoc. Comput. Mach. 22, 155–171 (1975)

[Lev73] Levin, L.A.: Universal sorting problems. Problemy Peredaci Informacii 9, 115–116 (1973); English translation in Problems of Information Transmission 9, 265–266 (in Russian)

[LLT06] Larose, B., Loten, C., Tardiff, C.: A characterization of first-order constraint satisfaction problems. In: Proc. 21st IEEE Symp. on Logic in Computer Science (2006)

[Mes89] Meseguer, P.: Constraint satisfaction problem: an overview. AICOM 2, 3–16 (1989)

[MF93] Mackworth, A.K., Freuder, E.C.: The complexity of constraint satisfaction revisited. Artificial Intelligence 59(1-2), 57–62 (1993)

[MPPV04] McMahan, B.J., Pan, G., Porter, P., Vardi, M.Y.: Projection pushing revisited. In: Bertino, E., Christodoulakis, S., Plexousakis, D., Christophides, V., Koubarakis, M., Böhm, K., Ferrari, E. (eds.) EDBT 2004. LNCS, vol. 2992, pp. 441–458. Springer, Heidelberg (2004)

[NR90] Seymour, P.D., Robertson, N.: Graph minors iv: Tree-width and well-quasi-ordering. J. Combinatorial Theory, Ser. B 48(2), 227–254 (1990)

[PJ97] Pearson, J., Jeavons, P.: A survey of tractable constraint satisfaction problems. Technical Report CSD-TR-97-15, Royal Holloway University of London (1997)

[PY91] Papadimitriou, C., Yannakakis, M.: Optimization, approximation and complexity classes. J. Comput. System Sci. 43, 425–440 (1991)

[Ros95] Rosen, E.: Finite Model Theory and Finite Variable Logics. Ph.D thesis, University of Pennsylvania (1995)

[Ros05] Rossman, B.: Existential positive types and preservation under homomorphisisms. In: Proc. 20th IEEE Symp. on Logic in Computer Science, pp. 467–476. IEEE Computer Society, Los Alamitos (2005)

[Sar91] Saraiya, Y.: Subtree elimination algorithms in deductive databases. Ph.D thesis, Department of Computer Science, Stanford University (1991)

[Sch78] Schaefer, T.J.: The complexity of satisfiability problems. In: Proc. 10th ACM Symp. on Theory of Computing, pp. 216–226 (1978)

[Tsa93] Tsang, E.P.K.: Foundations of Constraint Satisfaction. Academic Press, London (1993)

[Var82] Vardi, M.Y.: The complexity of relational query languages. In: Proc. 14th ACM Symp. on Theory of Computing, pp. 137–146 (1982)

[Var95] Vardi, M.Y.: On the complexity of bounded-variable queries. In: Proc. 14th ACM Symp. on Principles of Database Systems, pp. 266–276 (1995)

[vB94] van Beek, P.: On the inherent tightness of local consistency in constraint networks. In: Proc. of National Conference on Artificial Intelligence (AAAI 1994), pp. 368–373 (1994)

[vBD97] van Beek, P., Dechter, R.: Constraint tightness and looseness versus local and global consistency. Journal of the ACM 44(4), 549–566 (1997)

[Yan81] Yannakakis, M.: Algorithms for acyclic database schemes. In: Proc. 7 Int'l. Conf. on Very Large Data Bases, pp. 82–94 (1981)

Uniform Constraint Satisfaction Problems and Database Theory

Francesco Scarcello[1], Georg Gottlob[2], and Gianluigi Greco[3]

[1] DEIS, Università della Calabria, Italy
scarcello@deis.unical.it
[2] Computing Laboratory, University of Oxford, UK
georg.gottlob@comlab.ox.ac.uk
[3] Dipartimento di Matematica, Università della Calabria, Italy
ggreco@mat.unical.it

Abstract. It is well-known that there is a close similarity between constraint satisfaction and conjunctive query evaluation. This paper explains this relationship and describes structural query decomposition methods that can equally be used to decompose CSP instances. In particular, we explain how "islands of tractability" can be achieved by decomposing the query on a database, or, equivalently, the scopes of a constraint satisfaction problem. We focus on advanced decomposition methods such as hypertree decompositions, which are hypergraph-based and subsume earlier graph-based decomposition methods. We also discuss generalizations thereof, such as weighted hypertree decompositions, and subedge-based decompositions. Finally, we report on an interesting new type of structural tractability results that, rather than explicitly computing problem decompositions, use algorithms that are guaranteed to find a correct solution in polynomial time if a decomposition exists.

1 Uniform CSPs and DBs

Constraint Satisfaction is a well-known framework to model and solve search problems, and has an impressive spectrum of applications [64].

An instance of a *constraint satisfaction problem (CSP)* (also *constraint network*) (e.g., [21]) is a triple $I = (\mathit{Var}, U, \mathcal{C})$, where Var is a finite set of variables, U is a finite domain of values, and $\mathcal{C} = \{C_1, C_2, \ldots, C_q\}$ is a finite set of constraints. Each constraint C_i is a pair (S_i, r_i), where S_i is a list of variables of length m_i called the *constraint scope*, and r_i is an m_i-ary relation over U, called the *constraint relation*. (The tuples of r_i indicate the allowed combinations of simultaneous values for the variables S_i). A *solution* to a CSP instance is a substitution $\theta : \mathit{Var} \longrightarrow U$, such that for each $1 \leq i \leq q$, $S_i\theta \in r_i$. The problem of deciding whether a CSP instance has any solution is called *constraint satisfiability (CS)*.

Observe that in this chapter we will focus on CSPs where the constraint relations are finite and explicitly given. If they are represented as functions to be computed, or in other succinct ways, then many results that we describe may

N. Creignou et al. (Eds.): Complexity of Constraints, LNCS 5250, pp. 156–195, 2008.

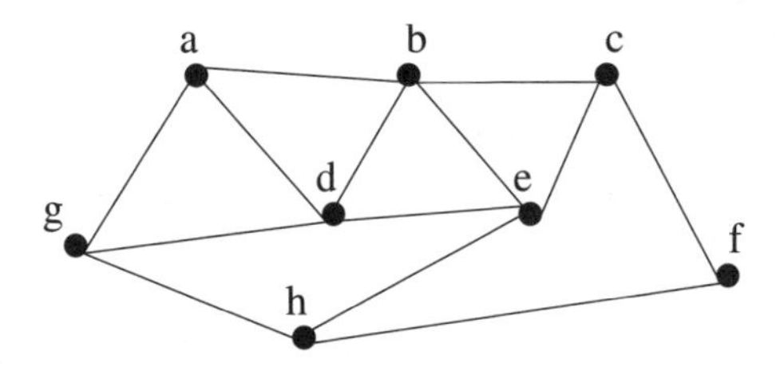

Fig. 1. The graph G_1

be not usable as they are, and requires suitable adaptations. In some cases, they cannot profitably be applied at all.

Many well-known problems in Computer Science and Mathematics can be formulated as CSPs.

Example 1. The famous *graph three-colorability (3COL)* problem, i.e., deciding whether the vertices of a graph $G = (\mathit{Vertices}, \mathit{Edges})$ can be colored by three colors (say: red, green, blue) such that no edge links two vertices having the same color, is formulated as follows as a CSP. The set *Var* contains a variable X_v for each vertex $v \in \mathit{Vertices}$. For each edge $e = \{v, w\} \in \mathit{Edges}$, where $v < w$ according to some ordering on *Vertices*, the set $\mathcal{C}$ contains a constraint $C_e = (S_e, r_e)$, where $S_e = (X_v, X_w)$ and r_e is the relation $r_{\neq}$ consisting of all pairs of different colors, i.e., $r_{\neq} = \{\langle \mathit{red}, \mathit{green}\rangle, \langle \mathit{red}, \mathit{blue}\rangle, \langle \mathit{green}, \mathit{red}\rangle, \langle \mathit{green}, \mathit{blue}\rangle, \langle \mathit{blue}, \mathit{red}\rangle, \langle \mathit{blue}, \mathit{green}\rangle\}$.

For instance, the set of constraints for the graph G_1 in Figure 1 is the following $\mathcal{C} = \{((A,B), r_{\neq}), ((A,D), r_{\neq}), ((A,G), r_{\neq}), ((B,C), r_{\neq}), \ldots, ((G,H), r_{\neq})\}$. ◁

Example 2. Figure 2 shows a combinatorial crossword puzzle, which is a typical CSP [20,64]. A set of legal words is associated to each horizontal or vertical array of white boxes delimited by black boxes. A solution to the puzzle is an assignment

Fig. 2. A crossword puzzle

of a letter to each white box such that to each white array is assigned a word from its set of legal words.

This problem is represented as follows. There is a variable X_i for each white box, and a constraint C for each array D of white boxes. (For simplicity, we just write the index i for variable X_i.) The scope of C is the list of variables corresponding to the white boxes of the sequence D; the relation of C contains the legal words for D. For the example in Figure 2, we have $C_{1H} = ((1,2,3,4,5), r_{1H})$, $C_{8H} = ((8,9,10), r_{8H})$, $C_{11H} = ((11,12,13), r_{11H})$, $C_{20H} = ((20,21,22,23,24,25,26), r_{20H})$, $C_{1V} = ((1,7,11,16,20), r_{1V})$, $C_{5V} = ((5,8,14,18,24), r_{5V})$, $C_{6V} = ((6,10,15,19,26), r_{6V})$, $C_{13V} = ((13,17,22), r_{13V})$. Subscripts H and V stand for "Horizontal" and "Vertical," respectively, resembling the usual naming of definitions in the crossword puzzles. A possible instance for the relation r_{1H} is $\{\langle h,o,u,s,e\rangle, \langle c,o,i,n,s\rangle, \langle b,l,o,c,k\rangle\}$. $\lhd$

It is well-known and easy to see that Constraint Satisfiability is an NP-complete problem. Membership in NP is obvious. NP-hardness follows, e.g., immediately from the NP hardness of 3COL [29].

Moreover, it has been observed [55] that constraint satisfiability can be recast as the fundamental *homomorphism problem (HOM)*. Recall that a relational structure A consists of a set U, called the universe of A, and a sequence of relations $R_1, R_2, \ldots, R_m$ over U. For example, a graph is a structure with a single binary relation that is symmetric and irreflexive. All structures in this chapter are assumed to be finite, i.e., having a finite universe.

Let $A = \langle U, R_1, R_2, \ldots, R_m\rangle$ and $B = \langle V, S_1, S_2, \ldots, S_m\rangle$ be two relational structures. A mapping between the two domains $h : U \mapsto V$ is a *homomorphism* if, for every tuple $\mathbf{X}$ and every $1 \leq i \leq m$, $\mathbf{X} \in R_i$ entails $h(\mathbf{X}) \in S_i$, where $h(\langle X_1, X_2, \ldots, X_a\rangle)$ denotes the tuple $\langle h(X_1), h(X_2), \ldots, h(X_a)\rangle$. A pair of relational structures (A, B) is a "yes" instance of the homomorphism problem if there exists such a mapping from A to B. If this is the case, we will simply write $A \mapsto B$, hereafter.

For instance, the CSP in Example 1 can be seen as the homomorphism instance (G, T), where $G = \langle \mathit{Vertices}, \mathit{Edges}\rangle$ is the graph relational structure, and T is the relational structure $\langle\{\mathit{red}, \mathit{green}, \mathit{blue}\}, r_{\neq}\rangle$ (that is in fact the triangle graph K_3, because any pair of different colors is a tuple of the binary relation $r_{\neq}$).

More in general, any CSP instance $(\mathit{Var}, U, \{C_1, C_2, \ldots, C_q\})$, with $C_i = (S_i, r_i)$, $1 \leq i \leq q$, corresponds to a homomorphism instance $(\langle \mathit{Var}, S_1, S_2, \ldots, S_q\rangle, \langle U, r_1, r_2, \ldots, r_q\rangle)$, and vice versa. Note that, in case of arbitrary (non-binary) relations, we cannot simply look at these structures as graphs, e.g., think of the crossword puzzle example in Figure 2.

Denote the former relational structure—with the constraint scopes—by $\mathcal{S}$, and the latter one—with the constraint relations—by $\mathcal{R}$. Then, we may formally define the Constraint Satisfiability language in terms of homomorphisms:

$$\mathrm{CSP} = \{(\mathcal{S}, \mathcal{R}) \mid \mathcal{S} \mapsto \mathcal{R}\},$$

where $(\mathcal{S}, \mathcal{R})$ is a suitable (string) encoding of the two relational structures. This general version of the problem, where the input is such a pair $(\mathcal{S}, \mathcal{R})$, and we have to decide whether $(\mathcal{S}, \mathcal{R}) \in$ CSP, is also known as the *Uniform Constraint Satisfaction Problem.* We will see some important restricted variants of this problem in the following sections.

It has been shown that Constraint Satisfiability and Homomorphism are in fact equivalent to various problems in Database theory [49,20,55], e.g., to the problem of conjunctive query containment [55], or to the problem of evaluating *Boolean conjunctive queries* over a relational database [58]. Recall that a *conjunctive query* Q on a database schema $\mathbf{DS} = \{R_1, \ldots, R_m\}$ consists of a formula of the form

$$Q: \quad ans(\mathbf{u_0}) \leftarrow r_1(\mathbf{u_1}) \wedge \cdots \wedge r_n(\mathbf{u_n}),$$

where $n \geq 0$; $r_1, \ldots r_n$ are relation names (not necessarily distinct) of $\mathbf{DS}$; *ans* is a relation name not in $\mathbf{DS}$; and $\mathbf{u_0}, \mathbf{u_1}, \ldots, \mathbf{u_n}$ are lists of terms (i.e., variables or constants) of appropriate length. The set of variables occurring in Q is denoted by $var(Q)$. The set of atoms contained in the body of Q is referred to as $atoms(Q)$.

The *answer* of Q on a database instance $\mathbf{db}$ (of the schema $\mathbf{DS}$), with associated universe U, consists of a relation *ans*, whose arity is equal to the length of $\mathbf{u_0}$, defined as follows. Relation *ans* contains all tuples $\mathbf{u_0}\theta$ such that $\theta : var(Q) \longrightarrow U$ is a substitution replacing each variable in $var(Q)$ by a value of U and such that for $1 \leq i \leq n$, $r_i(\mathbf{u_i})\theta \in \mathbf{db}$. (For an atom A, $A\theta$ denotes the atom obtained from A by uniformly substituting $\theta(X)$ for each variable X occurring in A.) If $\mathbf{u_0}$ is empty, Q is a Boolean conjunctive query.

Observe that answering a Boolean conjunctive query Q on $\mathbf{db}$ is exactly the same problem as CSP (Homomorphism). Indeed, consider the relational structures $\mathcal{Q} = \langle var(Q), atoms(Q) \rangle$ and $\mathcal{D} = \langle U, \mathbf{db} \rangle$ corresponding to query and database, respectively. Then, the answer of Q is "yes" if and only if $\mathcal{Q} \mapsto \mathcal{D}$, as the substitution θ in the answer definition is nothing else than a homomorphism between these relational structures. That is, answering Q on $\mathbf{db}$ means solving the CSP instance $(\mathcal{Q}, \mathcal{D})$. On the other hand, every CSP instance is associated with a query having the constraint scopes as its atoms. For instance, the CSP in Example 1 can be seen as the Boolean conjunctive query $ans \leftarrow \bigwedge_{e \in Edges}$, where the set of variables occurring in the query is the set *Nodes*, and the database has the colors as its universe and consists only of the relation $r_{\neq}$.

Therefore, cross fertilization among these different research fields was possible and led to major achievements both in the AI and the DB communities. However, observe that, even if in principle we are talking about equivalent problems, in practice the instances considered in the applications are very different, and thus one cannot simply take any technique from AI and apply it to DB, or vice-versa. Typical CSP instances are characterized by many constraints with relatively small constraint relations, while typical query-answering tasks involve relatively small queries on large (often huge) databases. In fact, e.g., the powerful backtracking-like procedures developed for CSPs does not work for typical queries, while the sophisticated indexing techniques developed for DBs are quite useless in typical CSPs.

It follows that, not surprisingly, the major interactions and most of the results applicable to both fields are about the CSP uniform problem and, correspondingly, about query answering tasks where both the query *and* the database are part of the problem input. Indeed, in the uniform case, we have to take care of both relational structures, as we are not allowed to assume either to be very small—and hence representable in the applications by a fixed parameter of the problem, as in the non-uniform case.

2 Islands of Tractability

As we have seen, in its general version Constraint Satisfiability is an intractable problem. Therefore, different restrictions of the basic fundamental problem have been studied, in order to circumvent its hardness and to model specific application domains, sometimes easier and tractable.

Let **A** and **B** be two classes (possibly infinite sets) of relational structures. Then, the *Uniform Constraint Satisfaction Problem* can be generalized as follows:

$$\mathrm{CSP}(\mathbf{A},\mathbf{B}) = \{(\mathcal{S},\mathcal{R}) \mid \mathcal{S} \in \mathbf{A}, \mathcal{S} \in \mathbf{B}, \text{ and } \mathcal{S} \mapsto \mathcal{R}\}.$$

That is, the input is a pair of relational structures $(\mathcal{S},\mathcal{R})$, and we have to decide whether $(\mathcal{S},\mathcal{R}) \in \mathrm{CSP}(\mathbf{A},\mathbf{B})$, with **A** and **B** parameters of the problem. Clearly, according to this notation, the standard Uniform Constraint Satisfaction Problem is $\mathrm{CSP}(\mathbf{All},\mathbf{All})$, where **All** denotes the class of all relational structures, and thus no restriction is given for the possible inputs of both sides of the homomorphism (constraints scopes and constraint relations). In database theory, the complexity of the general unrestricted problem is called the *combined complexity* of conjunctive queries, because the input of the problem consists of both the query and the database. However, in many applications, the query is very small or fixed, while the database is large and growing. Then, the query is modeled as a fixed parameter Q of the problem and the input consists of the database only. That is, the homomorphism problem to be solved is $\mathrm{CSP}(\{\mathcal{Q}\},\mathbf{All})$, where $\mathcal{Q}$ is the relational structure associated with the fixed query Q, and we actually have a single input because the pair of input structures reduces to the only structure $\mathcal{D} \in \mathbf{All}$ associated with the given database **db**. In these cases, we speak of the *data complexity* of conjunctive queries. Finally, the last combination where the database is fixed and the query is the input is called *expression complexity* [78].

Correspondingly, in constraint satisfaction research, an important role is played by the *Non-Uniform Constraint Satisfaction Problem*, defined for any fixed relational structure $\mathcal{R}$:

$$\mathrm{CSP}(\mathbf{All},\{\mathcal{R}\}).$$

In this case, the input is thus a single relational structure $\mathcal{S} \in \mathbf{All}$, and we have to decide whether there is a homomorphism from $\mathcal{S}$ to $\mathcal{R}$. Note that this is precisely the expression complexity of conjunctive queries, as the database (also, the set of constraint relations) is fixed. Clearly enough, this restriction of the general problem finds many more applications in CSPs than in databases, and thus received much more attention in the former research area, as witnessed by many chapters of this book, devoted to the study of the Non-Uniform CSP. Interestingly, this problem is also equivalent to the data complexity of existential MMSNP formulas (monotone monadic strict NP without inequality) [25,57].

The quest for tractable cases of constraint satisfiability and database problems leads the researchers to the identification of tractable classes of CSP instances. In fact, looking at the definition of the problem, we may naturally consider different ways of restricting the problem, focusing either on the class of structures **A** or on the class of structures **B**. The former approaches are usually classified as *left-hand restrictions*, while the latter ones are classified as *right-hand restrictions*, because of the side where their instances $\mathcal{S} \in \mathbf{A}$, $\mathcal{R} \in \mathbf{B}$ occur in the homomorphism relationship $\mathcal{S} \mapsto \mathcal{R}$.

The case of restrictions on both sides is clearly a very attracting option, but we do not have equally attracting results, up to now. Finding such *combined left-right* tractable classes—not trivially following from known results on either kind of restriction—is currently a challenging research issue.

In this chapter we deal with left-hand restrictions for the Uniform Constraint Satisfaction Problem, looking for the so-called *islands of tractability*, where tractability is based on structural properties of CSPs [54].

Definition 1. A class **A** of relational structures is an *island of tractability* if

1. deciding whether $\mathcal{S} \in \mathbf{A}$ belongs to P, i.e., checking membership to the class is feasible in polynomial time, and
2. CSP($\mathbf{A}, \mathbf{All}$) $\in$ P, that is, for each $\mathcal{S} \in \mathbf{A}$ and for each $\mathcal{R}$, deciding whether there is a homomorphism from $\mathcal{S}$ to $\mathcal{R}$ is feasible in polynomial time. □

Note that the two conditions above are only the basic requirements, and model the tractability of the decision problem. In fact, in real-world applications, we are mainly interested in the search problem of "computing" a homomorphism, i.e., a mapping for the variables of the CSP (or of the query). In many applications—almost always in databases, we actually need *all* solutions of the problem. However, in these cases a polynomial-time algorithm cannot exist, because in general there are exponentially many solutions, and thus the best we can do is to have an algorithm that compute all mappings in time polynomial in the combined size of the input and of the output, also called *input-output polynomial time*. Therefore, a highly desirable property for an island of tractability **A** is

2′. for each $\mathcal{S} \in \mathbf{A}$ and for each $\mathcal{R}$, the set $\mathcal{H}$ of all homomorphisms from $\mathcal{S}$ to $\mathcal{R}$ can be computed in $O((|\mathcal{S}| + |\mathcal{R}| + |\mathcal{H}|)^k)$, for some constant $k \geq 0$, and where $|\cdot|$ denotes the size of a structure. □

A slightly stronger property that entails both Condition 2 and Condition 2′ is called *polynomial-time output delay*. In this case, it is required to have an algorithm that is able to compute all solutions, with the delay to output any new solution being bounded by a polynomial of the input. This could be particularly useful if we do not actually need all solutions, but just look for some solution that meets specific requirements, or we are interested in a small number of solutions. Indeed, such an algorithm would be *any-time*, that is, you can stop it whenever you are satisfied with the solutions you obtained up to that moment (if any), with the guarantee that the time you have waited is (polynomially) proportional with the number of solutions in your hands. Instead, in the case of input-output polynomial time algorithms (without the polynomial-time output delay property), if there are exponentially many solutions, you can wait exponential time (in the input) before starting to get them. For instance, the algorithm may require some long preliminary step, proportional to the number of solutions, before starting its output phase.

2.1 Graphs and Hypergraphs

In order to identify islands of tractability, we have to find classes of relational structures **A**, whose elements have something in common that make the homomorphism problem CSP(**A**, **All**) easy. The idea is to look at their structure, looking for some feature to be exploited in the algorithms. In particular, we are interested in the way tuples belonging to different relations interact with each other. For a basic example, if there is no connection, that is, every value of the universe occurs in one relation only, the problem is trivial. We will see that there are much more interesting properties that characterize tractable classes.

The structure of a relational structure $\mathcal{S} = \langle V, R_1, R_2, \ldots, R_m \rangle$ is best represented by its associated hypergraph $\mathcal{H}(\mathcal{S}) = (V, H)$, whose set of vertices is the universe V of $\mathcal{S}$ and the set of hyperedges consists of the tuples of $\mathcal{S}$, i.e., $H = \{t \mid t \in R_i, 1 \leq i \leq m\}$.

Since we are dealing with left-hand structures, each relation corresponds to a constraint scope and contains just one tuple, with the list of variables of that scope. Thus, the vertices of the hypergraph are the variables of the CSP problem, and the hyperedges are its constraint scopes—actually, the sets of variables occurring therein. Figure 3 shows the hypergraph $\mathcal{H}_{cp}$ associated with the crossword puzzle in Example 2.

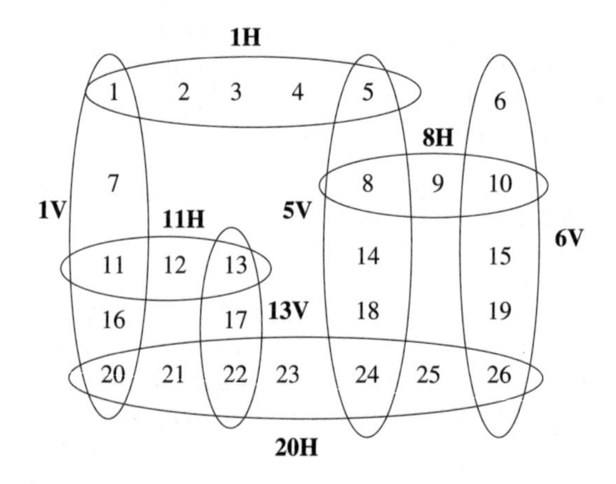

Fig. 3. Hypergraph $\mathcal{H}_{cp}$ of the crossword puzzle in Example 2

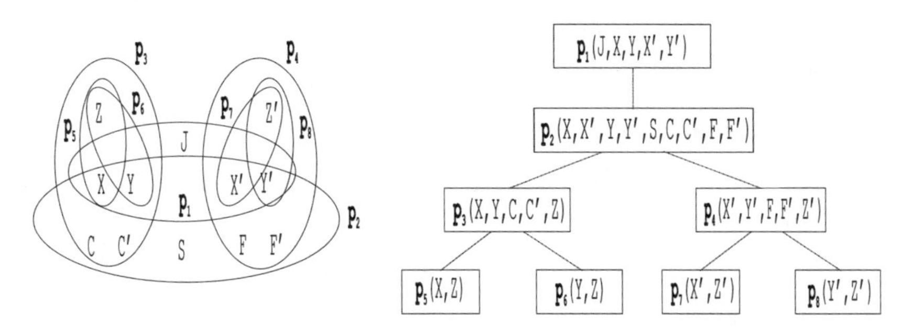

Fig. 4. Hypergraph $\mathcal{H}(Q_a)$ (right) and join tree JT_a (left) for query Q_a

Of course, for database queries we have the analogous situation: for a conjunctive query Q, its *query hypergraph* $\mathcal{H}(Q)$ has the set of variables occurring in Q as its vertices, and the query atoms as its hyperedges. More precisely, for each query atom A of Q, the set $var(A)$ of all variables occurring in A is a hyperedge of $\mathcal{H}(Q)$.

Example 3. Consider the following query Q_a: $ans \leftarrow p_1(J, X, Y, X', Y') \wedge p_2(X, X', Y, Y', S, C, C', F, F') \wedge p_3(X, Y, C, C', Z) \wedge p_4(X', Y', F, F', Z') \wedge p_5(X, Z) \wedge p_6(Y, Z) \wedge p_7(X', Z') \wedge p_8(Y', Z')$.

Figure 4 shows on the left its associated hypergraph $\mathcal{H}(Q_a)$. ◁

Since in this paper we always deal with hypergraphs corresponding to CSPs instances or queries, the vertices of any hypergraph $\mathcal{H} = (V, H)$ are always variables of some instance of these problems. Thus, we will often use the term *variable* as a synonym for vertex, when referring to elements of V. Moreover, $var(\mathcal{H})$ and $edges(\mathcal{H})$ will denote the sets V and H, respectively.

Before leaving this section we note that, historically, the first attempts to identify islands of tractability for constraint satisfaction problems have been done in the context of *binary* CSPs, i.e., when each relation R_i in the structure $\mathcal{S} = \langle V, R_1, R_2, \ldots, R_m \rangle$ is binary. In this case, it is easy to see that $\mathcal{H}(\mathcal{S}) = (V, H)$ is in fact a graph. Therefore, it comes with no surprise that several efforts have been spent to exploit the many existent methods to the case of general (non-binary) CSPs, by representing any CSP instance by some graph, rather than by its associated hypergraph. In particular, for a given hypergraph $\mathcal{H} = (V, H)$, the following graph representations have been proposed in the literature:

Primal-graph Representation. The *primal graph* of $\mathcal{H}$, denoted by $G(\mathcal{H}) = (N, E)$, is the graph whose set of vertices N is the set of variables V, and whose edges connect each pair of vertices (i.e., variables) occurring together in some hyperedge, that is $E = \{\{V_1, V_2\} \mid \exists h \in H \text{ s.t. } \{V_1, V_2\} \subseteq h\}$.

Dual-graph Representation [20]. The *dual graph* of $\mathcal{H}$, denoted by $dual(\mathcal{H}) = (N, E)$, is the graph whose set of vertices N is the set of hyperedges H, and whose edges connect each pair of vertices (i.e., hyperedges) having some variable in common, that is $E = \{\{h_1, h_2\} \mid h_1, h_2 \in H \text{ and } h_1 \cap h_2 \neq \emptyset\}$.

Hidden-variable Representation [72]. The *incidence graph* of $\mathcal{H}$, denoted by $inc(\mathcal{H}) = (N, E)$, is bipartite graph where $N = H \cup V$ and $E = \{ \{h, a\} \mid h \in H \text{ and } a \in h)\}$, i.e. it contains an edge from h to a if and only if the variable a occurs in the hyperedge h.

The relationships among the islands of tractability designed for binary problems, and hence based on graph properties of such *binarized* instances, as well as their relationships with hypergraph based techniques, have been studied in [34] and in [45]. In particular, the latter work focused on the relationships among islands of tractability—possibly based on the same graph property—when the above different graph representations are chosen.

It turned out that, using graphs instead of hypergraphs we have a substantial loss in information and, consequently, we are not able to identify some large tractable classes of instances. Therefore, since in this chapter we focus on the general non-binary CSP uniform problem, we next deal with hypergraph-based approaches only.

2.2 Solving Acyclic Problems

One of the most important and deeply studied island of tractability is the class of *acyclic structures* [8,14,23,32,55,63,67,81,82].

Definition 2. A relational structure $\mathcal{S}$ is acyclic if and only if its hypergraph $H(\mathcal{S})$ is acyclic or, equivalently, if it has a join forest. A *join forest* for the hypergraph $H(\mathcal{S})$ is a forest G whose set of vertices V_G is the set *edges*$(H(\mathcal{S}))$ and such that, for each pair of hyperedges $h_1, h_2 \in V_G$ having variables in common (i.e., such that $h_1 \cap h_2 \neq \emptyset$), the following conditions hold:

1. h_1 and h_2 belong to the same connected component of G, and
2. all variables common to h_1 and h_2 occur in every vertex on the (unique) path in G from h_1 to h_2. This is known as the *connectedness condition* of join forests.

If G is a tree, then it is called a *join tree* for $H(\mathcal{S})$. For instance, Figure 4 shows a join tree of the hypergraph $\mathcal{H}(Q_a)$ associated with query Q_a in Example 3. □

Solving CSP instances whose left-hand sides have associated acyclic hypergraphs is feasible in polynomial time. In fact, this was firstly observed within the database theory community. Therefore, it is easier and more convenient for the presentation to describe techniques and results tailored for acyclic instances in terms of queries and databases, keeping in mind their equivalence with CSPs and homomorphisms.

Consider a conjunctive query

$$Q: \quad ans(\mathbf{u_0}) \leftarrow r_1(\mathbf{u_1}) \wedge \cdots \wedge r_n(\mathbf{u_n})$$

on a database instance **db**. Recall that a *join* operation between a pair of atoms $r_i(\mathbf{u_i})$ and $r_j(\mathbf{u_j})$ with respect to **db**, denoted by $r_i(\mathbf{u_i}) \bowtie r_j(\mathbf{u_j})$, gives a new database relation r_{ij} containing the set of tuples $\{t_i \cdot t_j \mid t_i \in r_i, t_j \in r_j,$ and t_i matches $t_j\}$, where $\cdot$ is the concatenation operator, and two tuples are matching if they have the same values in correspondence to any variable $X \in \mathbf{u_i} \cap \mathbf{u_j}$ they have in common. In other words, r_{ij} contains all homomorphisms from the relational structure associated with such pair of atoms to the relational structures associated with the two relations r_i and r_j in **db**. Note that $r_i(\mathbf{u_i}) \bowtie r_j(\mathbf{u_j})$ can be computed in $O(|r_i(\mathbf{u_i})| \log(|r_i(\mathbf{u_i})|) + |r_j(\mathbf{u_j})| \log(|r_j(\mathbf{u_j})|) + |r_{ij}(\mathbf{u_i}, \mathbf{u_j})|)$, by using sorting and merging techniques. Of course this cost in general is quadratic, because the size of the output relation r_{ij} can be $O(|r_i(\mathbf{u_i})||r_j(\mathbf{u_j})|)$, if all pair of tuples from the two relations are matching. Moreover, recall that the projection $\prod_{\mathbf{v}} r(\mathbf{u})$ with respect to **db** of an atom $r(\mathbf{u})$ over a set of variables $\mathbf{v}$ is a new database relation r' containing the set of tuples obtained by restricting the tuples of r to (the values corresponding to) those variables that $\mathbf{u}$ and $\mathbf{v}$ have in common. This operation is feasible in linear time, plus the cost of deleting possible duplicates (in this chapter we deal with mathematical relations, which are sets, and not with the so called bag semantics, where duplicates are allowed). Finally, a *semijoin* operation between a pair of atoms $r_i(\mathbf{u_i})$ and $r_j(\mathbf{u_j})$ with respect to **db**, denoted by $r_i(\mathbf{u_i}) \ltimes r_j(\mathbf{u_j})$, modifies the database relation r_i by assigning to it the result of the relational expression $\prod_{\mathbf{u_i}}(r_i(\mathbf{u_i}) \bowtie r_j(\mathbf{u_j}))$. That is, the relation $r_i \in$ **db** is filtered by deleting those tuples that matches no tuple in r_j. This operation is feasible in $O(|r_i(\mathbf{u_i})| \log(|r_i(\mathbf{u_i})|) + |r_j(\mathbf{u_j})| \log(|r_j(\mathbf{u_j})|))$, by using sorting and merging techniques, and the size of the result is clearly bounded by $|r_i(\mathbf{u_i})|$.

It is well-known and easy to see that the *answer* of Q on a database instance **db** can be computed simply by performing the join of all query atoms and then computing the projection of the resulting final atom $r_Q(var(Q))$ over the output variables $\mathbf{u_0}$. The relation obtained from the projection is the desired answer-relation *ans*. Note that $r_Q(var(Q))$ encodes all homomorphisms between the relational structure associated with Q and the one associated with the database instance **db**. Of course, the decision problem does not need the final projection step, as it is sufficient to check whether $r_Q(var(Q))$ is empty or not. The above procedure is in fact used in real-world database applications, with the help of indices based on advanced data structures, join ordering techniques and so on. The problem is that, when the query involves many atoms (the typical CSP situation), the temporary atoms obtained after the join computations may grow exponentially, up to $O(|r_1| \cdot |r_2| \cdots |r_n|)$, where the number of tuples in any relation may be in the order of several thousands, with sizes of several hundreds of megabytes, in typical database applications. For the sake of presentation, we may consider the simple upper bound $O(|r_{max}|^n)$, where r_{max} is the relation of **db** having the largest size. It is worthwhile noting that a huge temporary relation may be computed even if the result is eventually empty, because it could be discovered at the very end, after the join of this temporary relation with the last query atom to be processed.

For acyclic instances, we can do much better. In fact, according to Yannakakis's algorithm [81], they can be evaluated by processing any of their join trees bottom-up, by performing upward semijoins, thus keeping the size of the intermediate relations small. At the end, if the relation associated with the root atom of the join tree is not empty, then the answer of the query is not empty, or, equivalently, there are some homomorphisms. Therefore, acyclic Boolean conjunctive queries can be evaluated in $O((n-1)|r_{max}|\log|r_{max}|)$, because we perform $n-1$ semijoins, one for each edge of the join tree, whose vertices are the n atoms in the body of the query.

Example 4. Consider again the Boolean query Q_a in Example 3 and its join tree in Figure 4. Then, we know that it can be evaluated very efficiently, in $O(6|r_{max}|\log|r_{max}|)$: we start computing the semijoin $p_3(X,Y,C,C',Z) \ltimes p_5(X,Z)$, then the semijoin of $p_3(X,Y,C,C',Z) \ltimes p_6(X,Z)$, and so on, following a topological order of the tree. Eventually, we compute the semijoin of $p_1(J,X,Y,X',Y')$ with its child, and we are done. The answer of the Boolean query Q_a is true if and only if the relation obtained after the last semijoin is not empty. ◁

Let us recall the highly desirable computational *properties of acyclic instances*:

1. Acyclicity is efficiently recognizable: deciding whether a hypergraph is acyclic is feasible in linear time [76] and belongs to the class L (deterministic logspace). The latter result is quite new: it follows from the fact that hypergraph acyclicity belongs to SL [37], and from the recent proof that SL is in fact equal to L [61].
2. Acyclic instances can be efficiently solved. Besides the polynomial time algorithm for Boolean acyclic queries, Yannakakis showed that the answer of a non-Boolean acyclic conjunctive query can be *computed* in input-output polynomial time [81]. Indeed, after the bottom-up step described above, one can perform the reverse top-down step, again based on semi-join operations but changing the order, this time filtering a child vertex of the join tree from those tuples that do not match with its parent tuples. The relations obtained after the top-down step contain only tuples whose values are part of some answer of the query, no useless tuples are kept. The new database consisting of these filtered relations is called *full reducer*. Moreover, we say that the new instance has the *global consistency* property: no tuple can be missed if we compute the join of all query atoms (with respect to the new database). Then, from the full reducer, we can easily compute all solutions with a backtrack-free procedure (i.e., with backtracks occurring only looking for further solutions, never for wrong choices). Note that this is in fact a polynomial-time output delay algorithm, because the cost of the first two preliminary phases is $O(n\log n)$, where n is the input size, and the delay between the computation of any two solutions in the final step is $O(n)$, and thus polynomial in the input.
3. *Pairwise Consistency entails Global Consistency* [6]. Recall that pairwise consistency holds if, for every pair of atoms p,q, no tuple is missed by computing the join $p \bowtie q$ with respect to the given database, that is, every tuple

in one relation has a corresponding matching tuple in the other one. Formally, both $p \ltimes q = p$ and $q \ltimes p = q$ hold. The acyclic instances that fulfil this property also fulfil the global consistency property, and thus we may easily compute all answers in polynomial-time output delay, e.g., trough the above mentioned backtrack-free procedure. Note that this equivalence between local and global consistency (as the other direction trivially holds) yields another simple algorithm for solving acyclic instances: Enforce pairwise consistency, by taking the semijoins between all pairs of atoms until a fixpoint is reached, or some database relation becomes empty. If the latter case occurs, then the query has no answer. Otherwise, we eventually get a pairwise consistent instance. In this case, we know that the query has some answer (the Boolean problem is solved), and we have obtained a full reducer, from which we may easily compute all answers, if desired. It is easy to see that the cost of this consistency enforcing procedure is $O(m^2 t|r_{max}| \log |r_{max}|)$, where t is the total number of tuples in the given database, which is higher than the $O((m-1)|r_{max}| \log |r_{max}|)$ we take by using Yannakakis's Algorithm. Of course, this is not surprisingly, because the latter one exploits the knowledge of a given join tree (computable in linear time), while the consistency-enforcing algorithm acts repeatedly on all pairs, in a brute-force like manner.

4. It has been shown that answering queries is highly parallelizable on acyclic queries, as this problem (actually, the decision problem of answering Boolean queries) is complete for the low complexity class LOGCFL [37]. Efficient parallel algorithms for Boolean and non-Boolean queries have been proposed in [37] and [35]. They run on parallel database machines that exploit the *inter-operation parallelism* [80], i.e., machines that execute different relational operations in parallel. These algorithms can be also employed for solving acyclic queries efficiently in a distributed environment.

2.3 Generalizing Acyclicity

We have seen our first example of an island of tractability: the class of acyclic relational structures. Many attempts have been made in the literature for extending the good results about this class to relevant classes of *nearly acyclic* structures. We call these techniques *structural decomposition methods*, because they are based on the acyclicization of cyclic (hyper)graphs. More precisely, each method specifies how appropriately transforming the hypergraph of a given instance into the join tree of an equivalent acyclic instance, by organizing its hyperedges into a polynomial number of clusters, and suitably arranging the clusters as a tree.

Example 5. Consider the following query: Q_0: $ans \leftarrow s_1(A,B,D) \wedge s_2(B,C,D) \wedge s_3(B,E) \wedge s_4(D,G) \wedge s_5(E,F,G) \wedge s_6(E,H) \wedge s_7(F,I) \wedge s_8(G,J)$. ◁

Figure 5 shows the *Query Decomposition* approach for Q_0: Each cluster contains a number of atoms; after performing the join of the atoms contained in each

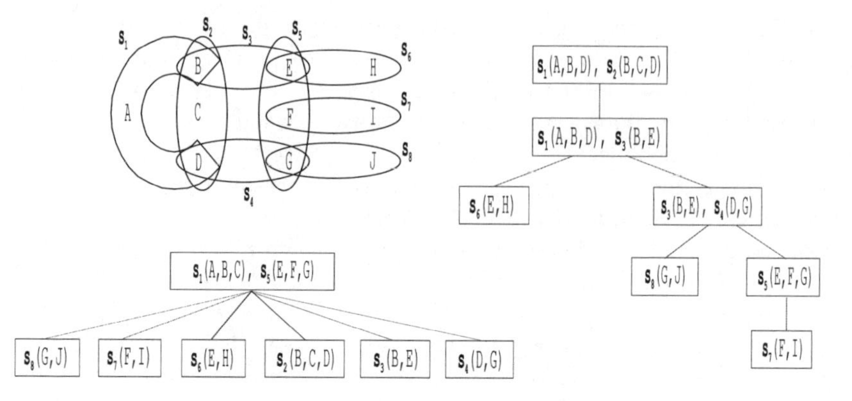

Fig. 5. Hypergraph $\mathcal{H}(Q_0)$ (left), two query decompositions of width 2 of $\mathcal{H}(Q_0)$ (right and bottom)

cluster, we obtain a join tree of an acyclic query which is equivalent to the original query. The resulting query can thus be answered efficiently (e.g., by using Yannakakis's algorithm). The tree produced by a structural query decomposition method on a given query Q is referred to as the *decomposition* of Q.

Thus, the efficiency of a structural decomposition method essentially depends on the maximum size of the produced clusters, measured (according to the chosen decomposition method) either in terms of the number of variables or in terms of the number of atoms. For a given decomposition, this size is referred-to as the *width* of the decomposition. For example, if we follow the query-decomposition approach and adopt the number of atoms, then the width of both decompositions shown in Figure 5 is 2. Intuitively, the complexity of transforming a given decomposition into an equivalent tree query is exponential in its width w. In fact, the evaluation cost of each of the (polynomially many) clusters is bounded by the cost of performing the (at most) w joins of its relations, which is in turn bounded by $O(|r_{max}|^w)$, where $|r_{max}|$ denotes the size of the largest relation r_{max} in the database.

In general, a rough upper bound for the cost of answering a given instance $(Q, \mathbf{db})$ according to any structural method D is given by $O(n^{w+1} \log n)$, where w is the D-width of $\mathcal{H}(Q)$ and n is the total size of the input problem, that is, the size of the query and of the database encoding [34]. Therefore, once we fix a bound k for such a width, the structural method D identifies a class of queries that can be answered in polynomial time exploiting D-decompositions, namely, the class of all queries having k-bounded D-width (i.e., D-width at most k).[1] More in general, if $\mathbf{A}$ is a class of relational structures whose hypergraphs have D-width bounded by some finite $k \geq 1$, and whose D-decompositions can be computed in P, then CSP($\mathbf{A}$, $\mathbf{All}$) $\in$ P. Moreover, if the membership in $\mathbf{A}$ is decidable in P, then $\mathbf{A}$ is an island of tractability.

[1] Intuitively, the D-width of a query Q is the minimum width of the decompositions of Q obtainable by method D.

The main islands of tractability based on structural decomposition methods are based on the notions of Biconnected Components [27], Tree Decompositions [14,21,26,48,55,65], Hinge Decompositions [49], Spread-cuts [17] [2], and Hypertree Decompositions [33,37,39,69].

We refer the interested reader to [34] for a detailed comparison of these decomposition methods (but the more recent Spread-cuts), and to [45] for further results about graph-based techniques when relational structures are represented according to different graph representations (primal graph, dual graph, hidden-variable encoding). See also [17] for a nice unifying view of such structural decomposition methods in terms of acyclic guarded covers of hypergraphs. Moreover, a survey of most of these techniques is currently available in the Wikipedia (look for "decomposition method", at `http://www.wikipedia.org`). We next focus on hypertree decompositions and their extensions, which give the largest classes of tractable instances.

3 Hypertree Decompositions

In this section, we describe the structural decomposition method based on hypertree decompositions. For more details on this notion, see [33,40].

3.1 Embedding Hypergraphs in (Hyper)Trees

A *hypertree for a hypergraph* $\mathcal{H}$ is a triple $\langle T, \chi, \lambda\rangle$, where $T = (N, E)$ is a rooted tree, and χ and λ are labeling functions that associate with each vertex $p \in N$ two sets $\chi(p) \subseteq var(\mathcal{H})$ and $\lambda(p) \subseteq edges(\mathcal{H})$. The *width* of a hypertree is the cardinality of its largest λ label, i.e., $max_{p\in N}|\lambda(p)|$.

We denote the set of vertices of any rooted tree T by $vertices(T)$, and its root by $root(T)$. Moreover, for any $p \in vertices(T)$, T_p denotes the subtree of T rooted at p. If T' is a subtree of T, we define $\chi(T') = \bigcup_{v\in vertices(T')} \chi(v)$.

Definition 3. [39] A *generalized hypertree decomposition* of a hypergraph $\mathcal{H}$ is a hypertree $HD = \langle T, \chi, \lambda\rangle$ for $\mathcal{H}$ that satisfies the following conditions:

1. *For each edge* $h \in edges(\mathcal{H})$, *all of its variables occur together in some vertex of the decomposition tree*, that is, there exists $p \in vertices(T)$ such that $h \subseteq \chi(p)$ (we say that *p covers h*).
2. *Connectedness Condition*: for each variable $Y \in var(\mathcal{H})$, the set $\{p \in vertices(T) \mid Y \in \chi(p)\}$ induces a (connected) subtree of T.
3. For each vertex $p \in vertices(T)$, all *variables in the χ labeling should belong to edges in the λ labeling*, that is, $\chi(p) \subseteq var(\lambda(p))$.

A *hypertree decomposition* is a generalized hypertree decomposition that satisfies the following additional condition:

[2] Actually, a first version of spread-cut decomposition has been presented in [16]. However, it turned out that deciding whether a hypergraph has spread-cut width at most k is NP-hard (even for $k = 4$). When we speak of islands of tractability based on spread-cuts, we thus consider the more recent notion defined in [17].

4. *Descendant Condition*: for each $p \in vertices(T)$, $var(\lambda(p)) \cap \chi(T_p) \subseteq \chi(p)$.

The HYPERTREE *width* $hw(\mathcal{H})$ (resp., generalized hypertree width $ghw(\mathcal{H})$) of $\mathcal{H}$ is the minimum width over all its hypertree decompositions (resp., generalized hypertree decompositions).

An edge $h \in edges(\mathcal{H})$ is *strongly covered* in *HD* if there exists $p \in vertices(T)$ such that $var(h) \subseteq \chi(p)$ and $h \in \lambda(p)$. In this case, we say that p strongly covers h. A decomposition *HD* of hypergraph $\mathcal{H}$ is a *complete decomposition* of $\mathcal{H}$ if every edge of $\mathcal{H}$ is strongly covered in *HD*. From any (generalized) hypertree decomposition *HD* of $\mathcal{H}$, we can easily compute a complete (generalized) hypertree decomposition of $\mathcal{H}$ having the same width. □

Note that the notions of hypertree width and generalized hypertree width are true generalizations of acyclicity, as the acyclic hypergraphs are precisely those hypergraphs having hypertree width and generalized hypertree width one. In particular, as we will see in the next section, the classes of conjunctive queries having bounded (generalized) hypertree width have the same desirable computational properties as acyclic queries [33].

At a first glance, a generalized hypertree decomposition of a hypergraph may simply be viewed as a clustering of the hyperedges (i.e., query atoms) where the classical connectedness condition of join trees holds. However, a generalized hypertree decomposition may deviate in two ways from this principle: **(1)** A hyperedge already used in some cluster may be reused in some other cluster; **(2)** Some variables occurring in reused hyperedges are not required to fulfill any condition.

For a better understanding of this notion, let us focus on the two labels associated with each vertex p: the set of hyperedges $\lambda(p)$, and the set of *effective* variables $\chi(p)$, which are subject to the connectedness condition (2). Note that all variables that appear in the hyperedges of $\lambda(p)$ but that are not included in $\chi(p)$ are "ineffective" for v and do not count w.r.t. the connectedness condition. Thus, the χ labeling plays the crucial role of providing a join-tree like re-arranging of all connections among variables. Besides the connectedness condition, this re-arranging should fulfill the fundamental Condition 1: every hyperedge (i.e., query atom, in our context) has to be properly considered in the decomposition, as for graph edges in tree-decompositions and for hyperedges in join trees (where this condition is actually even stronger, as hyperedges are in a one-to-one correspondence with vertices of the tree). Since the only relevant variables are those contained in the χ labels of vertices in the decomposition tree, the λ labels are "just" in charge of covering such relevant variables (Condition 3) with as few hyperedges as possible. Indeed, the width of the decomposition is determined by the largest λ label in the tree. This is the most important novelty of this approach, and comes from the specific properties of hypergraph-based problems, where hyperedges often play a predominant role. For instance, think of our database framework: the cost of evaluating a join operation with k atoms (read: k hyperedges) is $O(n^k)$, no matter of the number of variables occurring in the query.

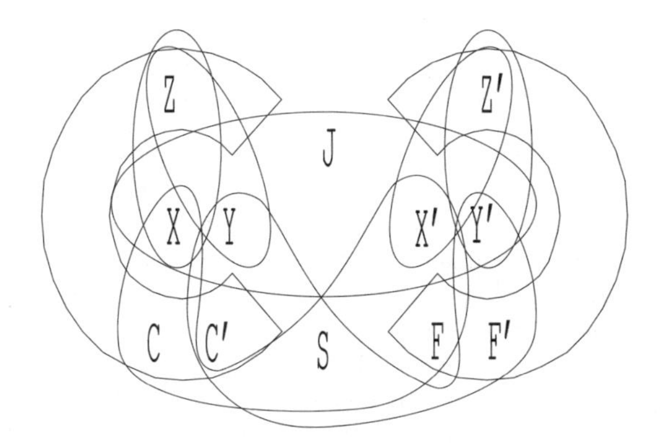

Fig. 6. Hypergraph $\mathcal{H}_1$ associated with query Q_1 in Example 6

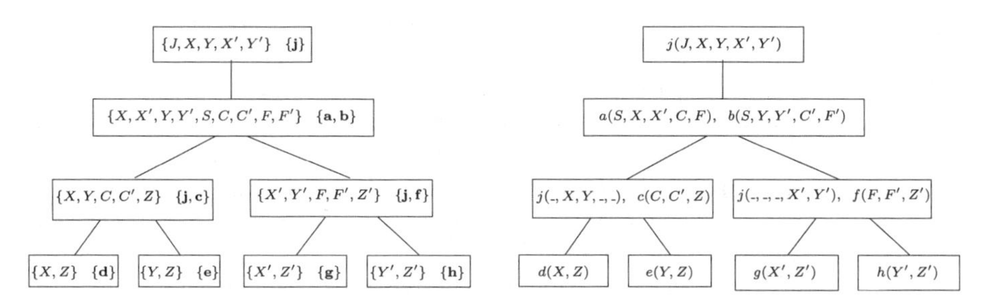

Fig. 7. A 2-width hypertree decomposition of hypergraph $\mathcal{H}_1$ in Example 6 (left), and atom representation (right)

Example 6. Consider the following conjunctive query Q_1:

$$\begin{aligned} ans \leftarrow\ & a(S, X, X', C, F) \ \wedge\ b(S, Y, Y', C', F') \ \wedge\ c(C, C', Z) \ \wedge\ d(X, Z) \ \wedge \\ & \wedge\ e(Y, Z) \ \wedge\ f(F, F', Z') \ \wedge\ g(X', Z') \ \wedge\ h(Y', Z') \ \wedge\ j(J, X, Y, X', Y'). \end{aligned}$$

Figure 6 shows the hypergraph $\mathcal{H}_1$ associated with Q_1. As you can see, this hypergraphs looks quite intricate, in fact $\mathcal{H}_1$ is cyclic, and thus $hw(\mathcal{H}_1) > 1$ holds. However, surprisingly, it turns out that it is quasi-acyclic. Indeed, Figure 7 shows a (complete) hypertree decomposition HD_1 of $\mathcal{H}_1$ having width 2, hence $hw(\mathcal{H}_1) = 2$.

In order to help the intuition, the figure also shows an alternative representation of this decomposition, called *atom* (or *hyperedge*) *representation* [33]: each node p in the tree is labeled by a set of atoms representing $\lambda(p)$; $\chi(p)$ is the set of all variables, distinct from '_', appearing in these hyperedges. Thus, in this representation, possible occurrences of the anonymous variable '_' take the place of variables in $var(\lambda(p)) - \chi(p)$.

Another example is depicted in Figure 5, which shows two hypertree decompositions of query Q_0. Indeed, we presented them as query decompositions, but every query decomposition is in fact a hypertree decomposition with

$var(\lambda(p)) = \chi(p)$ for each vertex p of the tree.[3] Both decompositions have width two and are complete hypertree decompositions of Q_0. ◁

Let $k \geq 1$ be a fixed positive integer. We say that a relational structure I has k-bounded (generalized) hypertree width if $(g)hw(\mathcal{H}(I)) \leq k$. A class of relational structures has k-bounded (generalized) hypertree width if all instances in the class have k-bounded (generalized) hypertree width. Also, we say that a class **A** has bounded (generalized) hypertree width if there is a integer $k \geq 1$ such that **A** has k-bounded (generalized) hypertree width.

Clearly enough, choosing a tree and a clever combination of χ and λ labeling for its vertices in order to get a decomposition below a fixed threshold-width k is not that easy, and is definitely more difficult than computing a simple tree decomposition, where only variables are associated with each vertex. In fact, it has been shown very recently that generalized hypertree-width is an intractable notion, as deciding whether a hypergraph has generalized hypertree width at most k is an NP-complete problem, for any fixed $k \geq 3$ [41].

It is thus very nice and somehow surprising that dealing with the hypertree width is a very easy task. More precisely, for any fixed $k \geq 1$, deciding whether a given hypergraph has hypertree width at most k is in LOGCFL, and thus it is a tractable and highly parallelizable problem. Correspondingly, the search problem of computing a k-bounded hypertree decomposition belongs to the functional version of LOGCFL, which is $\mathrm{L}^{\mathrm{LOGCFL}}$ [33].

Let us briefly discuss the only difference of hypertree decompositions with respect to generalized hypertree decompositions, that is, the *descendant condition* (Condition 4 in Definition 3). Consider a vertex p of a hypertree decomposition and a hyperedge $h \in \lambda(p)$ such that some variables $\bar{X} \subseteq h$ occur in the χ labeling of some vertex in the subtree T_p rooted at p. Then, according to this condition, these variables must occur in $\chi(p)$, too. This means, intuitively, that we have to deal with variables in $\bar{X}$ at this point of the decomposition tree, if we want to put h in $\lambda(p)$. For instance, as a consequence of this condition, for the root r of any hypertree decomposition we always have $\chi(r) = var(\lambda(r))$. However, once a hyperedge has been covered by some vertex of the decomposition tree, any subset of its variables can be used freely in order to decompose the remaining cycles in the hypergraph.

To shed more light on this restriction, consider what happens in the related (but intractable) hypergraph-based notions: in query decompositions [14], all variables are relevant; at the opposite side, in generalized hypertree decompositions, we can choose as relevant variables any subset of variables occurring in λ, without any limitation; in hypertree decompositions, we can choose any subset of relevant variables as long as the above descendant condition is satisfied. Therefore, the notion of hypertree width is clearly more powerful than the notion

[3] Actually, the tree labels in the original notion of query decompositions may contain both variables and atoms [14]. However, for the sake of presentation, we consider here the so called *pure* query decompositions, where the labels contain only atoms, since they are equivalent to the original ones [37].

of query width, but less general than the notion of generalized hypertree width, which is the most liberal notion.

For instance, consider again Figure 7: the variables in the hyperedge corresponding to atom j in $\mathcal{H}_1$ are jointly included only in the root of the decomposition, while we exploit two different subsets of this hyperedge in the rest of the decomposition tree. Note that the descendant condition is satisfied. Take the vertex at level 2, on the left: the variables J, X' and Y' are not in the χ label of this vertex (they are replaced by the anonymous variable '_'), but they do not occur anymore in the subtree rooted at this vertex. On the other hand, if we were forced to take all the variables occurring in every atom in the decomposition tree, it would not be possible to find a decomposition of width 2. Indeed, j is the only atom containing both pairs X, Y and X', Y', and it cannot be used again entirely, for its variable J cannot occur below the vertex labeled by a and b, otherwise it would violate the connectedness condition (i.e., Condition 2 of Definition 3). In fact, every query decomposition of this hypergraph has width 3, while the hypertree width is 2. In this case the generalized hypertree width is 2, as well, but in general it may be less than the hypertree width.

3.2 Characterizations of Hypertree Decompositions

Though the formal definition of hypertree width is rather involved, it is worthwhile noting that this notion has very natural characterizations in terms of games and logics [39]:

- **The robber and marshals game (R&Ms game).** It is played by one robber and a number of marshals on a hypergraph. The robber moves on variables, while marshals move on hyperedges. At each step, any marshal controls an entire hyperedge. During a move of the marshals from the set of hyperedges E to the set of hyperedges E', the robber cannot pass through the vertices in $B = (\cup E) \cap (\cup E')$, where, for a set of hyperedges F, $\cup F$ denotes the union of all hyperedges in F. Intuitively, the vertices in B are those not released by the marshals during their move. As in the monotonic robber and cops game defined for treewidth [71], it is required that the marshals capture the robber by monotonically shrinking the moving space of the robber. The game is won by the marshals if they corner the robber somewhere in the hypergraph. A hypergraph $\mathcal{H}$ has k-bounded hypertree width if and only if k marshals win the R&Ms game on $\mathcal{H}$.
- **Logical characterization of hypertree width.** Let L denote the existential conjunctive fragment of positive first order logic (FO). Then, the class of queries having k-bounded hypertree width is equivalent to the k-guarded fragment of L, denoted by $\mathrm{GF}_k(\mathrm{L})$. Roughly, we say that a formula Φ belongs to $\mathrm{GF}_k(\mathrm{L})$ if, for any subformula ϕ of Φ, there is a conjunction of up to k atoms jointly acting as a guard, that is, covering the free variables of ϕ. Note that this notion is related to the *loosely guarded fragment* as defined (in the context of full FO) by Van Benthem [7], where an arbitrary number of atoms may jointly act as guards (see also [43]).

3.3 Exploiting (Generalized) Hypertree Decompositions

We next describe how to use a given generalized hypertree decomposition for answering a conjunctive query. Of course, the same applies to hypertree decompositions, because *they are* generalized hypertree decompositions (satisfying an additional requirement).

Let $k \geq 1$ be a fixed constant, Q a conjunctive query over a database **db**, and $HD = \langle T, \chi, \lambda \rangle$ a generalized hypertree decomposition of Q of width $w \leq k$. Then, we can answer Q in two steps:

1. For each vertex $p \in vertices(T)$, compute the join operations among the atoms occurring together in $\lambda(p)$ (w.r.t. the given database **db**), and project onto the variables in $\chi(p)$. At the end of this phase, the conjunction of these intermediate results forms an acyclic conjunctive query, say Q', equivalent to Q. Moreover, the decomposition tree T represents a join tree of Q'.
2. Answer Q', and hence Q, by using any algorithm for acyclic queries, e.g. Yannakakis's algorithm.

For instance, Figure 4 shows the tree $\mathcal{JT}_a$ obtained after Step 1 above, from the query Q_1 in Example 6 and the generalized hypertree decomposition in Figure 7. E.g. observe how the vertex of $\mathcal{JT}_a$ labeled by atom $p_3(X, Y, C, C', Z)$ is built. Its database relation p_3 comes from the join of atoms $j(J, X, Y, X', Y')$ and $c(C, C', Z)$ (occurring in its corresponding vertex in Figure 7) w.r.t. **db**, and from the subsequent projection onto the variables X, Y, C, C', and Z (belonging to the χ label of that vertex). By construction, $\mathcal{JT}_a$ satisfies the connectedness condition, because the variables occurring in its vertices are precisely those occurring in the χ labeling of a hypertree decompositions, which fulfils the connectedness condition. Therefore, $\mathcal{JT}_a$ is a join tree. More precisely, it is the join tree of the acyclic query consisting of (the conjunction of) its atoms. In our example it is the join tree of the acyclic query Q_a in Example 4. Moreover, it is easy to see that Q_a (on the database obtained from **db** by performing the above mentioned join operations) has the same answer as Q_1 (on **db**) [33].

As we observed in our introduction to structural methods for the similar query decompositions, Step 1 is feasible in $O(m|r_{max}|^w)$ time, plus the (typically dominated) cost of the projections, where m is the number of vertices of T, and r_{max} is the relation of **db** having the largest size. For Boolean queries, Yannakakis's algorithm in Step 2 takes $O((m-1)|r_{max}|^w \log |r_{max}|)$ time. For non-Boolean queries, Yannakakis's algorithm works in input-output polynomial time (also, in polynomial-time output delay), and thus we should add to the above cost a term that depends on the answer of the given query (which may be exponential w.r.t. the input size). For instance, if we consider query Q_1, the above upper bound is $O(6|r_{max}|^2 \log |r_{max}| + 7|r_{max}|^2)$, which sounds very nice, compared with typical (non-structural) query answering algorithms, that would take $O(|r_{max}|^7)$ time, in the worst case.

It has been observed that, according to Definition 3, a hypergraph may have some (usually) undesirable hypertree decompositions [33], possibly with a large number m of vertices in the decomposition tree. For instance, a decomposition may contain two vertices with exactly the same labels. Therefore, a *normal form* for hypertree decompositions has been defined in [33], and then strengthened in [70], in order to avoid such kind of redundancies. The number m of vertices of these decompositions cannot exceed the number of variables v in $\mathcal{H}$, and is typically much smaller. It follows that, having such a decomposition of $\mathcal{H}$, an upper bound for query evaluation is $O((v-1)|r_{max}|^w \log |r_{max}| + v|r_{max}|^w)$ Moreover, $\mathcal{H}$ has a hypertree decomposition of width w if and only if it has a normal-form hypertree decomposition of the same width w. These same results can be proved for a suitable irredundant restriction of generalized hypertree decompositions. Therefore, the above upper bound on the query evaluation cost holds when such (irredundant) generalized hypertree decompositions are given, as well.

3.4 Islands of Tractability Based on Hypertree-Width

Let $k \geq 1$ be a finite integer, and let HW$[k]$ be the class of all relational structures whose associated hypergraphs have hypertree width at most k. For instance, the class of acyclic structures is in fact HW[1], because it is precisely the class of all structures having hypertree width 1. As we have seen in the previous section, for any fixed k, all these instances may be evaluated in polynomial time, once a suitable decomposition is given. Moreover, they can be easily recognized, as well.

Proposition 1 ([33]). *Let $k \geq 1$ be a fixed constant, and let* HW$[k]$ *be the class of all relational structures whose associated hypergraphs have hypertree width at most k. Then,* HW$[k]$ *is an island of tractability.*

In fact, both computing a width-k hypertree decomposition for a given structure, and solving the associated problem instance is feasible in (the functional version of) LOGCFL, and thus can be efficiently solved by parallel algorithms.

Recall that the class-membership tractability property does not hold for generalized hypertree decompositions, because deciding whether a hypergraph has generalized hypertree-width at most k is NP-complete, even if k is a fixed constant [41]. It follows that, unlike hypertree decompositions, generalized hypertree decompositions are not useful for identifying islands of tractability. However, as better described in Section 4.2, it is known that the difference between hypertree width and generalized hypertree width is within a small constant factor. It follows that a class of hypergraphs has k-bounded generalized hypertree width if and only if it has k'-bounded hypertree width, where k and k' are two positive integer constants. Therefore, the two notions identify the same set of tractable classes, because every class of CSPs that is tractable according to generalized hypertree width is tractable according to hypertree width, as well.

Moreover, since the width of the acyclic guarded covers in [17] is precisely the generalized hypertree width, the above result can be applied immediately to the islands of tractability based on every notion D in the unified view described in [17], because all of them are special kinds of acyclic guarded covers. That is, if k is a (finite) constant and **A** is a class of structures having k-bounded D-width, then **A** has k'-bounded hypertree width, where k' is a (finite) constant, possibly smaller than k, and never greater than $3k+1$ [4]. In this sense, hypertree decompositions allow us to identify the largest islands of tractability among all these generalizations of acyclicity. In Section 4.3, we will see a different kind of structural method that is incomparable with hypertree decompositions.

However, observe that the above important theoretical result does not mean that, in practice, hypertree decompositions is always the best choice. In fact, one has to choose the right decomposition method looking at the particular problem at hand. For instance, as shown in [17], there are some classes of hypergraphs where the spread-cut width is less than the hypertree width—of course, the maximum difference is within a constant factor, given the above result. In these cases, such a choice may be preferable, because the cost of evaluating CSP instances or queries is exponential in the width, whence even small improvements in the width may lead to a significant speed-up of the evaluation time. For completeness, note that the other side of this relationship between hypertree decompositions and spread-cut decompositions is not known: in principle, there may exist a class of CSPs having k-bounded hypertree width, but unbounded spread-cut width (hence such a class would be intractable, according to the spread-cut method).

Moreover, in order to achieve the minimum possible width, sometimes may be convenient to look for the more powerful generalized hypertree decompositions. Indeed, it is worthwhile noting that the cost of computing such decompositions depends only on the size of the given left-hand structure. In the database framework, this means that this cost depends only on the size of the query, while the benefits of having a small width involves the whole evaluation time, which heavily depends on the (typically) huge database size. Therefore, investing some time for computing a generalized hypertree decomposition of minimum width may well pay off. Instead, in the Constraint Satisfaction framework such left-hand structures are not small, as we usually have many constraints. In these cases, generalized hypertree decompositions are usually computed by resorting to heuristics, as described next.

3.5 Algorithms

Because of the practical relevance of the notion of (generalized) hypertree decomposition, several efforts have been spent to devise algorithms for its computation.

The first proposal in the literature appeared in [33], where it has been shown that, for any fixed $k \geq 1$, the search problem of computing a k-bounded hypertree decomposition is a tractable and highly parallelizable problem. Indeed, to establish this tractability result, an algorithm called k-`decomp` has been presented constructing a (normal-form) hypertree decomposition of minimal width less than or equal to k (if such a decomposition exists). However, k-`decomp` is an alternating algorithm that "runs" on alternating Turing machines using logarithmic workspace, and hence is not designed for real-world applications.

A more practical algorithm to compute hypertree decompositions, named `opt-`k`-decomp`, has been presented in [36]. In fact, `opt-`k`-decomp` is obtained by "uprolling" k-`decomp` in a sequential bottom-up fashion. As for many other decomposition methods, the running time of this algorithm to find a hypergraph decomposition is exponential in the parameter k. More precisely, opt-k-`decomp` runs in $O(m^{2k}v^2)$ time, where m and v are the number of edges and the number of vertices of the hypergraph, respectively. This algorithm has been improved subsequently in [50], where some techniques for limiting redundant computations have been discussed, which actually do not improve on the asymptotic worst-case bounds of the original algorithm.

Very recently, a different approach for computing hypertree decompositions has been discussed in [73]. Its basic idea is to exploit a backtracking procedure that stops as soon as it discovers a decomposition of width at most k (differently from opt-k-`decomp`, which implicitly builds a structure from which it is possible to enumerate easily all possible normal-form decompositions of width at most k). The time complexity of this approach is $O(v^{3k+3})$. Moreover, a technique to identify isomorphic components of a hypergraph in order to speed-up the computation is also presented, and integrated into the basic algorithm. Interestingly, this work also discusses a generalization of the backtracking procedure that results in two new tractable decomposition methods: HyperSpread and Connected Hypertree.

The main problem with the exact approaches discussed above is that, although polynomial, they needs a huge amount of time (and sometimes of memory) to deal with large hypergraphs. In fact, it is worthwhile noting that, unlike treewidth, recognizing k-bounded hypertree width is *not fixed-parameter tractable*, with respect to its natural parameter k [40]. Therefore, unless some unlikely collapse occurs in fixed-parameter complexity theory, a bad exponential dependency of the form $O(f_1(n)^{f_2(k)})$ is unavoidable in the running time of sound and complete algorithms.

Motivated by these bad news, most recent research focuses on heuristic approaches for the construction of (generalized) hypertree decompositions of "small" but not necessarily minimal width.

A successful algorithm of this kind has been presented in [59], where it has been observed that generalized hypertree decompositions can be constructed starting from tree decompositions, and subsequently covering the the variables at each vertex label by a small number of hyperedges. This latter condition can be straightforwardly implemented by set covering (heuristics), so that it is possible to use tree decomposition heuristics for the (heuristic) construction of generalized hypertree decompositions. In particular, in [59], *Bucket Elimination* is used in combination with several variable ordering heuristics.

Another successful technique has been discussed in [68], which shows how to use the *branch-decomposition* approach for ordinary graphs [18] for the heuristic construction of generalized hypertree decompositions (based on the fact that every branch decomposition of width k can be transformed into a tree decomposition of width at most $3k/2$).

In [56], it has been investigated the idea of generating generalized hypertree decompositions based on the tree decompositions of the primal and of the dual graph. To generate tree decompositions, Bucket Elimination is coupled with three heuristics for finding good vertex orderings. Moreover, a further method based on recursive partitioning of the hypergraph in weakly connected subgraphs is discussed.

The use of tabu search for computing (generalized) hypertree decompositions has been considered in [60].

More recently, [42] considered a combination of exact and heuristic approaches in the sense that the search space is restricted by a fixed upper bound k, and some heuristics are used to accelerate the search for a generalized hypertree decomposition of width at most k (but not necessarily the minimal one). The resultant algorithm `det-`k`-decomp` is based on backtracking, and can also be implemented for parallel executions.

See the Hypertree Decomposition HomePage [69], for available implementations of algorithms for computing hypertree decompositions, and further links to heuristics and other papers on this subject.

4 Extending Hypertree Decompositions

In this section, we describe some recent approaches to generalize the notion of hypertree decompositions. The first one allows us to express preferences, for those problems where "any" decomposition (even below a desired width) is not enough, and we rather want the best one, according to a given criterion. The second approach aims at identifying decomposition methods that are more powerful than hypertree decompositions but still polynomially computable, whose degree of generality is somewhere between hypertree decompositions and generalized hypertree decompositions. The third one, based on the notion of fractional edge cover, is instead more powerful than generalized hypertree decompositions, but likely intractable.

4.1 Weighted Hypertree Decompositions

In this section, we consider hypertree decompositions with an associated weight, which encodes our preferences, and allows us to take into account further requirements, besides the width. We will see how to answer queries more efficiently, by looking for their best decompositions.

Formally, given a hypergraph $\mathcal{H}$, a *hypertree weighting function* (short: `HWF`) $\omega_{\mathcal{H}}$ is any polynomial-time function that maps each generalized hypertree decomposition $HD = \langle T, \chi, \lambda \rangle$ of $\mathcal{H}$ to a real number, called the *weight* of *HD*.

For instance, a very simple `HWF` is the function $\omega_{\mathcal{H}}^{w}(HD) = \max_{p \in vertices(T)} |\lambda(p)|$, that weights a decomposition *HD* just on the basis of its worse vertex, that is the vertex with the largest λ label, which also determines the width of the decomposition.

In many applications, finding such a decomposition having the minimum width is not the best we can do. We can think of minimizing the number of vertices having the largest width w and, for decompositions having the same numbers of such vertices, minimizing the number of vertices having width $w-1$, and continuing so on, in a lexicographical way. To this end, we can define the HWF $\omega_{\mathcal{H}}^{lex}(HD) = \sum_{i=1}^{w} |\{p \in N \text{ such that } |\lambda(p)| = i\}| \times B^{i-1}$, where $N = vertices(T)$, $B = |edges(\mathcal{H})| + 1$, and w is the width of HD. Note that any output of this function can be represented in a compact way as a radix B number of length w, which is clearly bounded by the number of edges in $\mathcal{H}$. Consider again the query Q_0, and the hypertree decomposition, say HD', of $\mathcal{H}(Q_0)$ shown in Figure 5, on the right. It is easy to see that HD' is not the best decomposition w.r.t. $\omega_{\mathcal{H}}^{lex}$ and the class of hypertree decompositions in normal form. Indeed, $\omega_{\mathcal{H}}^{lex}(HD') = 4 \times 9^0 + 3 \times 9^1$, and thus the decomposition HD'' shown on the bottom of Figure 5 is better than HD', as $\omega_{\mathcal{H}}^{lex}(HD'') = 6 \times 9^0 + 1 \times 9^1$.

Let $k > 0$ be a fixed integer and $\mathcal{H}$ a hypergraph. We define the class $kHD_{\mathcal{H}}$ (resp., $kNFD_{\mathcal{H}}$) as the set of all hypertree decompositions (resp., normal-form hypertree decompositions) of $\mathcal{H}$ having width at most k.

Definition 4. *[70]*Let $\mathcal{H}$ be a hypergraph, $\omega_{\mathcal{H}}$ a weighting function, and $\mathcal{C}_{\mathcal{H}}$ a class of generalized hypertree decompositions of $\mathcal{H}$. Then, a decomposition $HD \in \mathcal{C}_{\mathcal{H}}$ is *minimal* w.r.t. $\omega_{\mathcal{H}}$ and $\mathcal{C}_{\mathcal{H}}$, denoted by $[\omega_{\mathcal{H}}, \mathcal{C}_{\mathcal{H}}]$-*minimal*, if there is no $HD' \in \mathcal{C}_{\mathcal{H}}$ such that $\omega_{\mathcal{H}}(HD') < \omega_{\mathcal{H}}(HD)$. □

For instance, the $[\omega_{\mathcal{H}}^{w}, kHD_{\mathcal{H}}]$-*minimal* decompositions are exactly the k-bounded hypertree decompositions having the minimum possible width, while the $[\omega_{\mathcal{H}}^{lex}, kHD_{\mathcal{H}}]$-*minimal* hypertree decompositions are a subset of them, corresponding to the lexicographically minimal decompositions described above.

It is not difficult to show that, for general weighting functions, the computation of minimal decompositions is a difficult problem even if we consider just bounded hypertree decompositions [70]. We thus restrict our attention to simpler HWFs.

Let $\langle \mathbb{R}^+, \oplus, \min, \bot, +\infty \rangle$ be a *semiring*, that is, $\oplus$ is a commutative, associative, and closed binary operator, $\bot$ is the neuter element for $\oplus$ (e.g., 0 for $+$, 1 for $\times$, etc.) and the absorbing element for min, and min distributes over $\oplus$.[4] Given a function g and a set of elements $S = \{p_1, ..., p_n\}$, we denote by $\bigoplus_{p_i \in S} g(p_i)$ the value $g(p_1) \oplus \ldots \oplus g(p_n)$.

Definition 5. *[70]*Let $\mathcal{H}$ be a hypergraph. Then, a *tree aggregation function* (short: TAF) is any hypertree weighting function of the form

$$\mathrm{F}_{\mathcal{H}}^{\oplus,v,e}(HD) = \bigoplus_{p \in N} \Big(v_{\mathcal{H}}(p) \ \oplus \bigoplus_{(p,p') \in E} e_{\mathcal{H}}(p,p') \Big),$$

associating a value from $\mathbb{R}^+$ with the hypertree decomposition $HD = \langle (N, E), \chi, \lambda \rangle$, where $v_{\mathcal{H}} : N \mapsto \mathbb{R}^+$ and $e_{\mathcal{H}} : N \times N \mapsto \mathbb{R}^+$ are two polynomial functions evaluating vertices and edges of hypertrees, respectively. □

---

[4] For the sake of presentation, we refer to min and hence to minimal hypertree decompositions. However, it is easy to see that all the results can be generalized to any semiring, possibly changing min, $\mathbb{R}^+$, and $+\infty$.

We next focus on a tree aggregation function that is useful for query optimization. We refer the interested reader to [70] for further examples and applications.

Given a query Q over a database **db**, let $HD = \langle T, \chi, \lambda \rangle$ be a hypertree decomposition in normal form for $\mathcal{H}(Q)$. For any vertex p of T, let $E(p)$ denote the relational expression $E(p) = \bowtie_{h \in \lambda(p)} \prod_{\chi(p)} rel(h)$, i.e., the join of all relations in **db** corresponding to hyperedges in $\lambda(p)$, suitably projected onto the variables in $\chi(p)$. Given also an incoming node p' of p in the decomposition HD, we define $v^*_{\mathcal{H}(Q)}(p)$ and $e^*_{\mathcal{H}(Q)}(p, p')$ as follows:

- $v^*_{\mathcal{H}(Q)}(p)$ is the estimate of the cost of evaluating the expression $E(p)$, and
- $e^*_{\mathcal{H}(Q)}(p, p')$ is the estimate of the cost of evaluating the semi-join $E(p) \ltimes E(p')$.

Let $cost_{\mathcal{H}(Q)}$ be the TAF $\mathtt{F}^{+,v^*,e^*}_{\mathcal{H}(Q)}(HD)$, determined by the above functions. Intuitively, $cost_{\mathcal{H}(Q)}$ weights the hypertree decompositions of the query hypergraph $\mathcal{H}(Q)$ in such a way that minimal hypertree decompositions correspond to "optimal" query evaluation plans for Q over **db**. Note that any method for computing the estimates for the evaluation of relational algebra operations from the quantitative information on **db** (relations sizes, attributes selectivity, and so on) may be employed for v^* and e^*. See, for instance, the standard techniques described in [30,51].

Clearly, all these powerful weighting functions would be of limited practical applicability, without a polynomial time algorithm for the computation of minimal decompositions. Surprisingly, it turns out that, unlike the traditional (non-weighted) framework, working with normal-form hypertree decompositions, rather than with any kind of bounded-width hypertree decomposition, does matter. Indeed, computing such minimal hypertree decompositions with respect to any tree aggregation function is a tractable problem, while it has been proved that the problem is still NP-hard if the whole class of bounded-width hypertree decomposition is considered. A polynomial time algorithm for this problem, called `cost-`k`-decomp`, is presented in [70] and subsequently improved in [31].

An example of the practical advantages of this approach is reported in Figure 8. It shows the execution times for evaluating query Q_1 in Example 6 and two further queries Q_2 (consisting of 8 atoms and 9 distinct variables) and Q_3 (consisting of 9 atoms, 12 distinct variables, and 4 output variables), on a well-known commercial DBMS system (name omitted for software license restrictions), by using two configurations of its query engine: in the first one, labeled by CommDB, the DBMS internal optimizer is used to find the query plans; in the second one, labeled by Cost-k-decomp, the DBMS query engine is feed with the query plan determined by the hypertree decomposition computed by `cost-`k`-decomp`. In particular, Figure 8 reports, on the left, the results for query Q_1 for different values of the parameter k. Observe that higher values of k clearly mean more power for the hypertree-based optimizer, that can possibly choose even decompositions with a non-minimum width, if it is convenient according to the database statistics available and the query-evaluation estimates. In fact, all these queries have hypertree width 2, and thus any $k \geq 2$ would work, but we get the

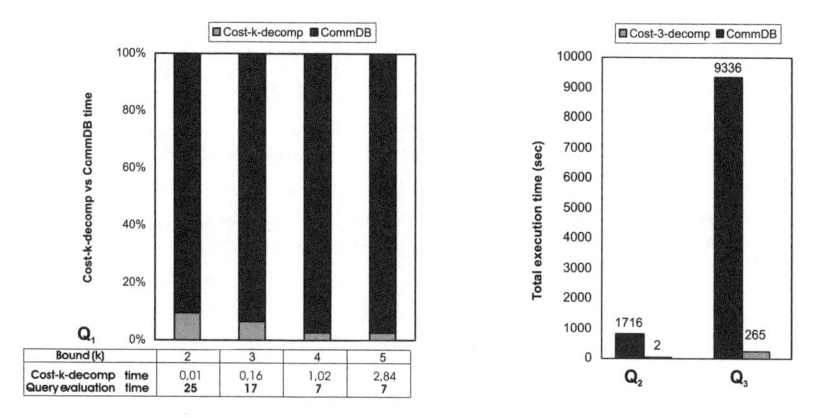

Fig. 8. Example performances for `cost-`k`-decomp`

best results since $k = 4$. The execution cost of `cost-`k`-decomp` is also reported. Note that its computation times are negligible, if compared with the CommDB evaluation times on the considered queries. On the right, Figure 8 shows the evaluation times for the other two queries, when the parameter $k = 3$ is used for `cost-`k`-decomp`. More details on these (and further) experimental results are available in [31].

4.2 Approximating Generalized Hypertree Decompositions

Recall from Section 3.1 that a generalized hypertree decomposition (GHD) is a decomposition that satisfies the first three conditions of Definition 3, but not necessarily Condition 4, the *Descendant Condition*. Given that GHDs are more liberally defined than hypertree decompositions (HDs), there are hypergraphs $\mathcal{H}$ whose general hypertree width $ghw(\mathcal{H})$ is smaller than their hypertree width $hw(\mathcal{H})$. Concrete examples of such hypergraphs can be found in [3,41]. Since GHDs—once computed—have the same favorable properties for problem solving as HDs, it would actually be better to use GHDs instead of HDs because of the smaller width of the former, which would lead to better solution algorithms. However, there is a problem with GHDs, that was already mentioned in Section 3. As shown in [41], checking whether for a constant k, $ghw(\mathcal{H}) \leq k$ is NP-hard (even for $k = 3$).

The unfavorable complexity results related to generalized hypertree decompositions motivate the search for somewhat weaker hypergraph decomposition methods that in some sense approximate GHDs, and that fulfill the criteria of polynomial query evaluation and polynomial recognizability. We can thus look for decomposition methods M which associate with each hypergraph $\mathcal{H}$ a set $M(\mathcal{H})$ of generalized hypertree decompositions of $\mathcal{H}$, and search for a generalized hypertree decomposition in $M(\mathcal{H})$ of minimal width. Intuitively, we thus consider methods which "approximate generalized hypertree width from above", and are still better than the notion of hypertree width. The width $MW(\mathcal{H})$ of a hypergraph $\mathcal{H}$, according to some decomposition method M, is the minimum generalized hypertree width of a decomposition in $M(\mathcal{H})$. For two methods M and N we write $M \leq N$ iff for each hypergraph H, $MW(H) \leq NW(H)$. If $M \leq N$ and there is some hypergraph H such that $MW(H) < NW(H)$, then we write $M < N$.

An example of a decomposition that approximates GHD is the *spread cut decomposition* (SCD) with the associated notion of *spread cut width (scw)* [17]. In a similar way as HDs, SCDs explicitly restrict the sets $\chi(p)$ that may appear at a decomposition node p. There are many other possible decompositions one can imagine that are weaker than GHDs but stronger than HDs and SCDs. A key question is, how much we can gain through such decompositions, in other terms, how do their associated widths compare with hypertree width.

In [4] the following was shown:

Proposition 2 ([4]). *For each hypergraph $\mathcal{H}$, $ghw(\mathcal{H}) \leq hw(\mathcal{H}) \leq 3 \times ghw(\mathcal{H}) + 1$.*

This means that hypertree width approximates generalized hypertree width by a factor of three. It is currently open whether this factor can be improved. Of course, all concepts of decomposition widths that are between hypertree width and generalized hypertree width also approximate generalized hypertree width by (at most) the same factor.

Subedge-based decomposition methods. Motivated by the goal to improve hypertree decompositions, and to get a clearer picture of the "grey area" of the islands of tractability lying between HDs and GHDs, the concept of *subedge-based decomposition methods* was defined in [41]. A *subedge* of a hypergraph $\mathcal{H} = (V, E)$ is a subset of some edge of $\mathcal{H}$. A subedge-based decomposition method M relies on a *subedge function.* This is a function f which associates to each integer $k > 0$ and each hypergraph $\mathcal{H}$ a set $f(\mathcal{H}, k)$ of subedges of $\mathcal{H}$. Moreover, the set of k-width M-decompositions can be obtained as follows: (1) obtain a hypertree decomposition D of $\mathcal{H}' = (V, E \cup f(\mathcal{H}, k))$, and (2) convert D into a generalized hypertree decomposition of $\mathcal{H}$ by replacing each subedge $e \in \lambda(p)$, for each decomposition node p, by some edge e' of $\mathcal{H}$ such that $e \subseteq e'$. We call such a decomposition method M *subedge-based.* The following result was derived in [41]:

Proposition 3 ([41]). *For each polynomially recognizable decomposition method $M \leq GHD$, there exists a polynomially recognizable subedge-based decomposition method M' such that $M' \leq M$.*

The above result is useful from a methodological point of view. In fact, it tells us that, when searching for some new decomposition method M such that $GHD < M < HD$, then we may concentrate on subedge-based decomposition, and thus study appropriate subedge functions.

Component Hypertree decompositions. In order to improve at the same time HDs and SCDs, in [41] a particular subedge function f^C was defined, whose definition is based on structural properties of the input hypergraph $\mathcal{H} = (V, E)$. In particular, each subedge in $f^C(\mathcal{H}, k)$ is obtained from a full edge $e \in E$ and some candidate decomposition block M of at most k edges containing e, by eliminating from e all vertices that are edge-connected to some induced component of $V \setminus \mathit{vertices}(M)$, or all vertices that are not edge-connected to any component of $V \setminus M$, or all vertices from $e \setminus \cup(M \setminus \{e\})$ that are edge-connected to some component of $V \setminus \mathit{vertices}(M)$. The new subedge based decomposition method defined through this subedge function f^C is called *component hypertree decomposition (CHD)* and its associated width is referred to as *component hypertree width.* In [41] it is shown that Component hypertree decompositions fulfill both criteria, polynomial query answering and polynomial recognizability, and thus define an island of tractability. Moreover, it was shown that CHD < HD and CHD < SCD.

The method of component hypertree decompositions is currently the most general known polynomially recognizable hypergraph decomposition method, i.e., the largest island of tractability. But it is easy to see that the subedge-based paradigm allows one to define even larger tractable islands.

The only published method we are aware of, that does not fit into the "subedge-based" framework is the fractional hypertree-decomposition method, described next.

4.3 Fractional Hypertree Decompositions

In [47], it was observed that, for solving problems in polynomial time using (sub)edge based decompositions, a key requirement is that the cluster of atoms at each vertex p of the decomposition tree corresponds to a subproblem with a polynomial number of solutions. Of course, this is guaranteed if each cluster $\lambda(p)$ (using the notation of hypertree decompositions) contains at most k atoms, for a fixed constant k. Indeed, we have seen that such a p identifies a subquery, whose answer is given by the join of the atoms in $\lambda(p)$ (possibly projected over some subset of their variables). Then, the size of this answer is $O(n^k)$, where n is the size of the largest relation associated with the atoms in $\lambda(p)$. A very interesting result by Grohe and Marx [47] shows that there is a more general property that makes the size of a query answer polynomial (w.r.t. the input size).

Definition 6. Let $\mathcal{H} = (V, E)$ be a hypergraph. A fractional edge cover is a non-negative weight assignment f to the hyperedges in E such that every vertex $X \in V$ is covered by total weight at least 1, that is, $\sum_{h \in E | X \in h} f(h) \geq 1$. The weight of the cover f is $\sum_{h \in E} f(h)$. Then, the fractional edge cover number (FCN) of $\mathcal{H}$, denoted by $\rho^*(\mathcal{H})$, is the smallest weight over all its fractional edge covers. □

By exploiting a combinatorial lemma known as Shearer's Lemma [15], it can be shown that, for every instance of the query problem $I = (Q, \mathbf{db})$, the size of the answer $Q(\mathbf{db})$ is $O(|I|^{\rho^*(\mathcal{H}(Q))})$. Thus, if a class of structures has k-bounded FCN, for some finite integer $k \geq 1$, $|Q(\mathbf{db})|$ is polynomial in the input size, and it can be seen that $Q(\mathbf{db})$ can be computed in polynomial time. Moreover, the FCN of a hypergraph can be computed in polynomial time by linear programming. It follows that this notion can be used to identify classes of tractable instances.

Proposition 4 ([47]). *Let $k \geq 1$ be a fixed constant, and let FCN[k] be the class of all structures whose associated hypergraphs have fractional edge cover number at most k. Then, FCN[k] is an island of tractability.*

It is worthwhile noting that this island of tractability is incomparable with those based on hypertree width and its extensions: it is known that there are classes of hypergraphs having bounded FCN, but unbounded (generalized) hypertree width. On the other hand, there are classes of very simple hypergraphs that have unbounded FCN, but bounded (generalized) hypertree width. E.g., the class of acyclic instances, having hypertree width 1, has no finite bound for the FCN. In fact, recall that the size of the answer of an acyclic query may be exponential in the input (even if it can be computed in polynomial-time output delay). In order to overcome these drawbacks, a new more powerful decomposition method has been defined.

A *fractional hypertree for a hypergraph* $\mathcal{H}$ is a triple $\langle T, \chi, \gamma \rangle$, where $T = (N, E)$ is a rooted tree, and χ and γ are labeling functions that associate with each vertex $p \in N$ a set $\chi_p \subseteq var(\mathcal{H})$ and a mapping $\gamma_p : edges(\mathcal{H}) \rightarrow [0, \infty)$. The *width* of a hypertree is the maximum weight over the vertices of *HD*, where the weight of a vertex p is $\sum_{e \in edges(\mathcal{H})} \gamma_p(e)$.

Definition 7. [47] A *fractional hypertree decomposition* of a hypergraph $\mathcal{H}$ is a fractional hypertree $HD = \langle T, \chi, \gamma \rangle$ for $\mathcal{H}$ that satisfies the following conditions:

1. *For each edge $h \in edges(\mathcal{H})$, all of its variables occur together in some vertex of the decomposition tree*, that is, there exists $p \in vertices(T)$ such that $h \subseteq \chi_p$.
2. *Connectedness Condition*: for each variable $Y \in var(\mathcal{H})$, the set $\{p \in vertices(T) \mid Y \in \chi_p\}$ induces a (connected) subtree of T.
3. For each vertex $p \in vertices(T)$, all variables in the χ_p label should be fractionally edge covered according to γ_p, that is, $\chi_p \subseteq \{X \in var(\mathcal{H}) \mid \sum_{h \in edges(\mathcal{H}) | X \in h} \gamma_p(h) \geq 1\}$.

The *fractional hypertree width (fhw)* of $\mathcal{H}$ is the minimum width over all its fractional hypertree decompositions. □

Note that the only difference with a generalized hypertree decomposition is the fractional covering condition 3, which replaces the analogous condition of standard GHDs, where covering of variables in any χ label is ensured by the union of the hyperedges in its corresponding λ label. Of course, the notion of width is consequently based on the weight of these fractional edge covers.

Let $k \geq 1$, and let FHW[k] be the class of all structures whose associated hypergraphs have k-bounded fractional hypertree width. From what we have seen in the previous sections, it is easy to see that every query Q whose structure is in FHW[k] is tractable (on every database **db**) if a fractional hypertree decomposition HD of $\mathcal{H}(Q)$ is given. Indeed, the only difference with the evaluation exploiting k-bounded GHDs is that here the hyperedges associated with each vertex p of the decomposition (i.e., those covering the variables in χ_p) may be many more than k. However, the FCN of the sub-query corresponding to p is at most k, and thus the polynomial-time evaluation of this sub-query is guaranteed by the FCN approach.

Let us call FHD the decomposition method based on fractional hypertree width. It was shown in [47] that $FHD < GHD$, but the computational properties of FHD are unexplored. It was conjectured in [41] that fractional hypertree-width is not polynomially recognizable unless P = NP.

We remark that the above results show how the general setting considered in this chapter, where relational structures do not have a fixed arity, is very different and more complex than the setting with fixed arity. Note that, in the latter case, most research problems are now solved, because everything works as for binary instances, where the bounded treewidth property is necessary and sufficient for tractability [46].

5 Structural Methods without Structural Decompositions

A new group of interesting results about structural tractability provided a number of polynomial algorithms for solving the homomorphism problem, with the guarantee that they give the correct output if the input instances belong to certain classes.

5.1 Promise Problems and Tractability

When the result of an algorithm is conditioned to some property of the input, possibly not easily checkable, following [46], we may give a different definition of tractability:

Definition 8 (Tractability, promise version). CSP($\mathbf{A}, \mathbf{All}$) is in *promise* polynomial-time for an arbitrary class $\mathbf{A}$ of problems if there is a polynomial time algorithm such that,

- if its input consists of (the encoding of) two structures $\mathcal{S} \in \mathbf{A}$ and $\mathcal{R}$, then it correctly decides whether there is a homomorphism from $\mathcal{S}$ to $\mathcal{R}$;
- if the input is not of this form, then the answer of the algorithm may be arbitrary. □

Note that every island of tractability is a tractable class, according to the above promise version of tractability. However, the converse does not hold, because checking the membership in the class is not required to be feasible in polynomial time, in Definition 8.

Another important notion for both CSPs and databases is the core of a relational structure.

Definition 9. Let A be a relational structure. The *core* of A is a substructure C of A such that $A \mapsto C$, and no proper substructure of C has this property, i.e., C is minimal. □

It is well known that the core of a relational structure is unique (up to isomorphism). Moreover, if (A, B) is a CSP instance and C is the core of A, then (A, B) is equivalent to the instance (C, B). That is, using the core is the same as using the original structure A.

For instance, in the case of a conjunctive query Q, its core is the minimal query Q' with $atoms(Q') \subseteq atoms(Q)$ such that there exists a homomorphism h such that $\forall r(\bar{X}) \in atoms(Q)$, $r(h(\bar{X})) \in atoms(Q)$. Therefore, in order to answer the query Q efficiently, we may think of computing its reduced equivalent version Q' and then computing the answer of Q'. However, as shown below, dealing with the core is an intractable problem. Moreover, note that, if all atoms of a query are over different relation symbols, then the query coincides with its core. Since in conjunctive queries there are typically only a few atoms over the same database relation, in practice exploiting the core is not very useful for query answering. Instead, in the constraint satisfaction framework, it is more frequent to find many constraint scopes with the same constraint relations.

Example 7. Recall the 3-coloring CSP in Example 1. The constraint scopes are the edges of the graph G to be colored, and for all constraints the relation is $r_{\neq}$, containing all pairs of different colors. Then, $r_{\neq}$ is just the triangle graph K_3, and in fact it is well known that G is 3-colorable if and only if $G \mapsto K_3$. Moreover, if G contains a triangle, this entails that G is 3-colorable if and only if K_3 is the core of G. ◁

From the above example, we clearly see that, given a pair of relational structures A and B, deciding whether B is the core of A is an NP-hard problem, where the hardness holds even if B is fixed, and thus is a constant structure not part of the input of the problem. More precisely, it has been shown that the precise complexity of this problem is DP-complete [24].

The following result, proved exploiting pebble games on relational structures, provides a large class of tractable instances, which extends—but in the promise version—the well-known island of tractability of the bounded treewidth structures.

Proposition 5 ([19]). *Let $w \geq 1$ be a fixed constant. Moreover, let* **A** *be a class of structures such that the core of each structure in* **A** *has tree-width at most w. Then* CSP(**A**, **All**) *is in promise polynomial-time.*

Interestingly, it was later proved that the above property is also necessary for tractability, as far as relational structures with fixed arity are concerned, and assuming that a (very unlikely) collapse does not occur in the theory of fixed-parameter tractability [22].

Proposition 6 ([46]). *Assume that* FPT $\neq W[1]$. *Then, for every recursively enumerable class* **A** *of structures of bounded arity,* CSP(**A**, **All**) *is in polynomial time only if the cores of all structures in* **A** *have bounded treewidth.*

This is a remarkable theoretical result. However, exploiting the above property in practice may be rather difficult, because of the hardness of dealing with the cores of relational structures, while it can be useful if we have some guarantee that our instances belong to the class **A**.

It is worthwhile noting that, if we do not have such a guarantee, (in general) the polynomial time algorithms for CSP(**A**, **All**) cannot give a checkable justification for their answers, or a way to compute a solution (unless P = NP). To see that, consider the following simple example, where checking the membership in the class is in fact "the" problem: Let 3COL be the class of all (structures encoding the) 3-colorable graphs that contains a triangle. Therefore, $\forall G \in$ 3COL the core of G is the triangle K_3, because K_3 is a subgraph of G, and $G \mapsto K_3$, from the 3-colorability of G. Moreover, the treewidth of K_3 is 2, and thus the core of all structures in 3COL is 2. From Proposition 5, CSP($3COL$, **All**) is in polynomial time for any right-hand class of structures and thus, in particular, for the class **B** consisting of the one relational structure $B = \langle \{red, green, blue\}, r_{\neq} \rangle$. Let ALG be any polynomial-time algorithm that solves the 3-coloring problem CSP($3COL$, **B**) (trivial, as restricted to the "yes" instances of 3COL) according to Definition 8. Now, let (G, B) be an input instance of this problem such that G is *any graph containing a triangle*, and assume that ALG outputs "yes" on (G, B). Then, this output may be

- **correct,** if G in fact belongs to the class 3COL, and hence is 3-colorable, by definition of the class, or
- **wrong,** if $G \notin$ 3COL, and thus the output of ALG may be arbitrary.

Therefore, a positive answer gives no information on the problem instance at hand (while a possible negative answer is always correct). However, if ALG gave some polynomial-time checkable justification for its answers, rather than a mere "yes" or "no", we would be able to solve in polynomial-time the NP-hard 3-colorability problem: given any graph G, just add a (disconnected) triangle to it, feed ALG with the resulting graph, get its output, and decide whether the result is correct by checking in polynomial-time its justification (or directly, for negative answers).

It follows that the usual approach of calling a Boolean procedure many times, in order to build a solution node-by-node by adding "something" to the initial instance at each call, does not work here. Why that? Intuitively, because such a "something" in general may lead us out of the required class **A**. In fact, we may also loose a possible initial guarantee of membership in **A**. Of course, such Boolean procedures may instead be useful if the class **A** has the property to be closed under adding such a "something."

5.2 No-Promise Solutions

More useful results for the applications are about *no-promise tractability*, where the output of algorithms is always reliable, and may be "don't know" only if the input instance does not belong to the desired class of structures.

Definition 10 (Tractability, no-promise version). $\mathrm{CSP}(\mathbf{A}, \mathbf{All})$ is in *no-promise* polynomial-time for an arbitrary class $\mathbf{A}$ of problems if there is a polynomial time algorithm such that, given (the encoding of) two structures $\mathcal{S}$ and $\mathcal{R}$,

- says "yes" only if $(\mathcal{S}, \mathcal{R}) \in \mathrm{CSP}(\mathbf{A}, \mathbf{All})$, that is only if there is a homomorphism from $\mathcal{S}$ to $\mathcal{R}$;
- says "no" only if $(\mathcal{S}, \mathcal{R}) \notin \mathrm{CSP}(\mathbf{A}, \mathbf{All})$;
- says "don't know" only if $\mathcal{S} \notin \mathbf{A}$. □

A nice result in this respect was recently proven by Chen and Dalmau [13]. It can be seen as a generalization of Proposition 5 to structures of unbounded arity, since the hypergraph-based notion of generalized hypertree-width is considered, instead of the graph-based notion of treewidth. It is shown that instances having bounded generalized hypertree-width are efficiently solvable, even without computing a decomposition. This is a remarkable result, as computing such a decomposition is NP-hard [41]. Moreover, the tractability result is extended to structures homomorphically equivalent to structures having this property, where A is homomorphically equivalent to A' if $A \mapsto A'$ and $A' \mapsto A$. Note that this is slightly more general than considering the core of the structure, because A' is not required to be a substructure of A.

Proposition 7 ([13]). *Let $w \geq 1$ be a fixed constant, and let $\mathbf{A}$ be a class of structures homomorphically equivalent to structures having generalized hypertree-width at most w. Then $\mathrm{CSP}(\mathbf{A}, \mathbf{All})$ is in no-promise polynomial-time. Moreover, "no" instances are always recognized (we never get a "don't know" answer for them).*

5.3 A Database View

We give a database-theory reading of Proposition 7 and of its proof. We may generalize the property of acyclic instances that pairwise consistency entails global consistency to the class of bounded generalized hypertree-width instances, where pairwise consistency is replaced by the following notion of local consistency.

Definition 11. Let Q be a conjunctive query over a database $\mathbf{db}$. We say that Q is ℓ-consistent w.r.t. $\mathbf{db}$ if, for every set of ℓ atoms $S = \{r_1(\bar{X}_1), r_2(\bar{X}_2), \ldots, r_\ell(\bar{X}_\ell)\} \subseteq atoms(Q)$, no tuple of any relation in S is missed by taking the join of all these ℓ atoms. That is, for every atom $r_i(\bar{X}_i) \in S$, we have $\prod_{\bar{X}_i}(r_1(\bar{X}_1) \bowtie r_2(\bar{X}_2) \bowtie \cdots \bowtie r_\ell(\bar{X}_\ell)) = r_i$. □

Note that this definition is a simple extension of the classical one: pairwise consistency is just 2-consistency. Moreover, as for pairwise consistency, for any fixed ℓ, it can be enforced in polynomial time. Indeed, Figure 9 shows a procedure that, given a query Q over a database $\mathbf{db}$, removes from the relations in $\mathbf{db}$ those tuples that do not fulfil the desired local consistency property, and outputs a new (smaller) database $\mathbf{db}'$ such that Q is ℓ-consistent w.r.t. $\mathbf{db}'$. If some relations in $\mathbf{db}'$ is empty, then we know that Q has no answer on $\mathbf{db}'$, as well as on the original database **db** (as we only performs operations that represent subqueries of the full—more constrained—query Q). Let t be the number of tuples in **db**, that is the sum the cardinalities of all relations in **db**, a the number of atoms in the body of Q, and r_{max} the relation having the largest size t_{max} over the relations in **db**. Then, it is easy to check that the cost of this procedure is $O(\binom{a}{\ell}t_{max}^{\ell}t)$, which is clearly polynomial in the input, if ℓ is a constant, because $\binom{a}{\ell}$ is $O(a^{\ell})$.

The following result shows that, for (sufficiently) low-width instances, this local consistency property entails global consistency.

Input: A conjunctive query Q over a database **db**.
Output: A database $\mathbf{db}' \subseteq \mathbf{db}$ such that Q is ℓ-consistent w.r.t. $\mathbf{db}'$, or *failure*, only if Q has no answer on **db**.

Repeat
  For each $S = \{r_1(\bar{X}_1), r_2(\bar{X}_2), \ldots, r_\ell(\bar{X}_\ell)\} \subseteq atoms(Q)$ **Do**
    Compute $r_S = r_1(\bar{X}_1) \bowtie r_2(\bar{X}_2) \bowtie \cdots \bowtie r_\ell(\bar{X}_\ell)$;
    For each $r_i \in S$ **Do**
      Compute $r_i = \prod_{\bar{X}_i}(r_1(\bar{X}_1) \bowtie r_2(\bar{X}_2) \bowtie \cdots \bowtie r_\ell(\bar{X}_\ell))$
Until a fixpoint is reached or some database relation becomes empty.
Let $\mathbf{db}'$ be the database obtained after the previous steps.
If some relation in $\mathbf{db}'$ is empty **Then Output** *failure*,
Else Output $\mathbf{db}'$.

Fig. 9. Enforcing ℓ-consistency

Proposition 8. *Let k be any positive integer, let Q_f be a conjunctive query over a database* **db***, and let Q be a query homomorphically equivalent to Q_f (that is, whose associated relational structures are homomorphically equivalent) such that $ghw(\mathcal{H}(Q)) \leq k$. Then, $2k$-consistency entails global consistency for Q w.r.t.* **db**.

Proof. From the hypothesis, there exists a width-k generalized hypertree-decomposition $HD = \langle T, \chi, \lambda \rangle$ of the hypergraph $\mathcal{H}(Q)$. From Section 3, we know that there is an equivalent acyclic query Q_a on a database $\mathbf{db}_a$ that we can build by computing, for each vertex p of T, the join of the (at most k) atoms in $\lambda(p)$, and then projecting the resulting relation on the variables in $\chi(p)$. Moreover, assume that Q is $2k$-consistent w.r.t. **db**. Then, we know that computing the join of any set of $2k$ atoms of Q we do not miss any tuple from any involved database relation. Of course this property trivially holds for any set of k' atoms, with $k' \leq 2k$. Now, consider the join $r_i(\bar{X}_i) \bowtie r_j(\bar{X}_j)$ of any pair of atoms $r_i(\bar{X}_i)$, $r_j(\bar{X}_j)$ of the acyclic query Q_a. Since either relation comes from the join of at most k atoms, this relational expression involves the join of at most $2k$ atoms of **db**. From the $2k$-consistency of Q w.r.t. **db**, no tuple can be missed in any relation by taking the join of such a set of atoms. It easily follows that $r_i(\bar{X}_i) \ltimes r_j(\bar{X}_j) = r_i$, as well as $r_j(\bar{X}_j) \ltimes r_i(\bar{X}_i) = r_j$. Therefore, the acyclic query Q_a is pairwise consistent and thus globally consistent w.r.t. $\mathbf{db}_a$. Then, its equivalent query Q (and hence the original query Q_f) is globally consistent w.r.t. **db**, too. □

Let $\ell = 2k$. It is worthwhile noting that, after Proposition 8, the simple procedure shown in Figure 9 is able to decide in polynomial time whether a given query Q has some answer on a database **db** without computing any decomposition of its hypergraph $\mathcal{H}(Q)$. If the procedure outputs *failure*, then $Q(\mathbf{db}) = \emptyset$; otherwise we have the guarantee that, if $ghw(\mathcal{H}(Q)) \leq k$, then Q has some answers on **db**. In practice, in this case we try to compute such an answer from the reduced database $\mathbf{db}'$. If we fail, we infer that $ghw(\mathcal{H}(Q)) > k$, but we cannot say anything about the answer of Q; otherwise, if we actually get an answer (all atoms/constraints are satisfied), we clearly know that $Q(\mathbf{db}) \neq \emptyset$, but we cannot say anything on the generalized hypertree width of the hypergraph—however, this is not really important, at this point!

After these interesting no-promise tractability results, one may wonder whether decompositions are still important to compute or not. Observe that, as for pairwise consistency in acyclic queries, consistency enforcing techniques are somehow brute-force: by enforcing the consistency on any group of $2k$ atoms, we also enforce pairwise consistency on any pair of groups of k atoms belonging to any possible k-width generalized hypertree decomposition of the query. However, while in the case of acyclic instances computing a join tree is feasible in linear time, in the case of bounded-width instances computing a generalized hypertree decomposition is NP-hard. It follows that there is not a clear answer here, it depends on the applications.

To be more precise, let us have a look at the cost of evaluation procedures. Let a be the number of atoms in the body of the given query Q and v the number of its variables. Applying the no-promise approach takes $O(\binom{a}{2k}t|r_{max}|^{2k})$, while by exploiting a given (generalized) hypertree decomposition HD of Q in normal form, we take $O((m-1)|r_{max}|^k \log|r_{max}| + m|r_{max}|^k)$, where m is the number of vertices of HD, which is at most the number of variables v for decompositions in normal form (actually, $m << v$ in typical applications). In the latter case, it is also easier to compute all solutions of the given problem, but we have to compute such a width-k decomposition HD. Of course, if the database/constraint relations are large, it is clearly preferable to spend some time in computing a good decomposition, since its computation time does not depend on the database size (see Section 3.5). Also, this task may be easier if we have some information on the structure of our problem instances that can profitably be exploited. On the other hand, if we miss any (useful) structural information on the problem and the relations sizes are very small, the no-promise approach may be useful, because the larger factor $\binom{a}{2k}$ and the larger exponent $2k$ may be balanced by the decomposition time.

For completeness, we observe that the no-promise approach can be also implemented working on groups of k atoms, and performing semi-join operations between all pairs of these groups, until a fixpoint is reached. This variation of the consistency enforcing procedure, which is similar to the procedure described in [13], takes $O(\binom{a}{k}^2 t|r_{max}|^k \log|r_{max}|)$. Thus, this latter approach may be preferable to the procedure shown in Figure 9 if we want to enforce consistency but there are some large relations in the database, because the database exponent becomes k instead of $2k$. However, note that $\binom{a}{k}^2$ may be much larger than $\binom{a}{2k}$. For instance, if $a = 50$ and $k = 3$, the former is about $384 \cdot 10^6$, while the latter one is about $16 \cdot 10^6$. In any case, compared with the hypertree-based approach (without considering the cost of computing the decomposition), if we assume $v = 100$ variables, the consistency enforcing procedure is about 3 million times slower in this example.

Open problem. It remains to understand whether the property of a structure to have k-bounded generalized hypertree width is also necessary for a correct use of the local consistency approach. We know that it holds for $k = 1$, because 2-consistency entails global consistency with respect to every right-hand structure if and only if (the core of) a structure has generalized hypertree width 1 (i.e., it is acyclic) [6]. This means that, if we consider a conjunctive query Q whose core is not acyclic, there exists a database **db** such that the consistency enforcing approach fails, that is, Q is pairwise consistent w.r.t. **db** but not globally consistent.

However, this problem is open for an arbitrary bound k. Let $\mathcal{S}$ be a structure and $\mathcal{S}'$ its core: is it the case that $2k$-consistency entails global consistency with respect to every right-hand (database) structure if and only if $ghw(\mathcal{H}(\mathcal{S}')) \leq k$?

It is worthwhile noting that this problem has been recently closed for structures having a fixed arity bound, where variables, graphs, and the notion of treewidth play the roles of hyperedges, hypergraphs, and generalized hypertree width, respectively. In this framework, one can define a variable oriented variant of the notion of k-consistency, by considering all partial homomorphisms whose domain contain at most k variables. As observed in [55], this is equivalent to play the existential k-pebble game on the given pair of structures. Let $a \geq 1$ be a fixed constant, let $\mathcal{S}$ be a relational structure whose relations have arity at most a, and let $\mathcal{S}'$ be its core. Then, the variable variant of k-consistency entails global consistency with respect to every right-hand structure if and only if the treewidth of (the graph associated with) $\mathcal{S}'$ is at most k [5].

References

1. Abiteboul, S., Hull, R., Vianu, V.: Foundations of Databases. Addison-Wesley, Reading (1995)
2. Abiteboul, S., Duschka, O.M.: Complexity of Answering Queries Using Materialized Views. In: Proc. of the PODS 1998, Seattle, Washington, pp. 254–263 (1998)
3. Adler, I.: Marshals, Monotone Marshals, and Hypertree-Width. Journal of Graph Theory 47(4), 275–296 (2004)
4. Adler, I., Gottlob, G., Grohe, M.: Hypertree-Width and Related Hypergraph Invariants. In: Proc. of EuroComb 2005, Berlin (2005)
5. Atserias, A., Bulatov, A., Dalmau, V.: On the Power of k-Consistency. In: Arge, L., Cachin, C., Jurdziński, T., Tarlecki, A. (eds.) ICALP 2007. LNCS, vol. 4596, pp. 279–290. Springer, Heidelberg (2007)

6. Beeri, C., Fagin, R., Maier, D., Yannakakis, M.: On the desirability of acyclic database schemes. Journal of the ACM 30(3), 479–513 (1983)
7. Van Benthem, J.: Dynamic Bits and Pieces. ILLC Research Report, University of Amsterdam (1997)
8. Bernstein, P.A., Goodman, N.: The power of natural semijoins. SIAM Journal on Computing 10(4), 751–771 (1981)
9. Bodlaender, H.L., Fomin, F.V.: Tree decompositions with small cost. Discrete Applied Mathematics 145(2), 143–154 (2005)
10. Cai, J., Chakaravarthy, V.T., Kaushik, R., Naughton, J.F.: On the Complexity of Join Predicates. In: Proc. of PODS 2001 (2001)
11. Chandra, A.K., Kozen, D.C., Stockmeyer, L.J.: Alternation. Journal of the ACM 26, 114–133 (1981)
12. Chandra, A.K., Merlin, P.M.: Optimal Implementation of Conjunctive Queries in relational Databases. In: Proc. of STOC 1977, pp. 77–90 (1977)
13. Chen, H., Dalmau, V.: Beyond hypertree width: Decomposition methods without decompositions. In: van Beek, P. (ed.) CP 2005. LNCS, vol. 3709, pp. 167–181. Springer, Heidelberg (2005)
14. Chekuri, C., Rajaraman, A.: Conjunctive query containment revisited. Theoretical Computer Science 239(2), 211–229 (2000)
15. Chung, F., Frank, P., Graham, R., Shearer, J.: Some intersection theorems for ordered sets and graphs. Journal of Combinatorial Theory, Series A 43, 23–37 (1986)
16. Cohen, D.A., Jeavons, P.G., Gyssens, M.: A Unified Theory of Structural Tractability for Constraint Satisfaction and Spread Cut Decomposition. In: Proc. IJCAI 2005, Edinburgh, UK, pp. 72–77 (2005)
17. Cohen, D.A., Jeavons, P.G., Gyssens, M.: A unified theory of structural tractability for constraint satisfaction problems. Journal of Computer and System Sciences 74(5), 721–743 (2008)
18. Cook, W., Seymour, P.: Tour merging via branch-decomposition. INFORMS Journal on Computing 15(3), 233–248 (2003)
19. Dalmau, V., Kolaitis, P.G., Vardi, M.Y.: Constraint Satisfaction, Bounded Treewidth, and Finite-Variable Logics. In: Van Hentenryck, P. (ed.) CP 2002. LNCS, vol. 2470, pp. 310–326. Springer, Heidelberg (2002)
20. Dechter, R.: Constraint networks. In: Shapiro, S.C. (ed.) Encyclopedia of Artificial Intelligence, 2nd edn., vol. 1, pp. 276–285. Wiley, Chichester (1992)
21. Dechter, R.: Constraint Processing. Morgan Kaufmann, San Francisco (2003)
22. Downey, R., Fellows, M.: Parameterized Complexity. Springer, Heidelberg (1999)
23. Fagin, R., Mendelzon, A.O., Ullman, J.D.: A simplified universal relation assumption and its properties. ACM Transactions on Database Systems 7(3), 343–360 (1982)
24. Fagin, R., Kolaitis, P.G., Popa, L.: Data exchange: getting to the core. ACM Trans. Database Syst. 30(1), 174–210 (2005)
25. Feder, T., Vardi, M.Y.: The computational structure of monotone monadic SNP and constraint satisfaction: a study through Datalog and group theory. SIAM Journal of Computing 28, 57–104 (1998)
26. Flum, J., Frick, M., Grohe, M.: Query evaluation via tree-decompositions. Journal of the ACM 49(6), 716–752 (2002)
27. Freuder, E.C.: A sufficient condition for backtrack-bounded search. Journal of ACM 32(4), 755–761 (1985)
28. Frick, M., Grohe, M.: Deciding first-order properties of locally tree-decomposable structures. Journal of the ACM 48(6), 1184–1206 (2001)

29. Garey, M.R., Johnson, D.S.: Computers and Intractability: A Guide to the Theory of NP-Completeness. Freeman, New York (1979)
30. Garcia-Molina, H., Ullman, J., Widom, J.: Database system implementation. Prentice Hall, Englewood Cliffs (2000)
31. Ghionna, L., Granata, L., Greco, G., Scarcello, F.: Hypertree Decompositions for Query Optimization. In: Proc. of ICDE 2007, pp. 36–45 (2007)
32. Goodman, N., Shmueli, O.: Tree queries: a simple class of relational queries. ACM Transactions on Database Systems 7(4), 653–6773 (1982)
33. Gottlob, G., Leone, N., Scarcello, F.: Hypertree decompositions and tractable queries. Journal of Computer and System Sciences 64(3), 579–627 (2002)
34. Gottlob, G., Leone, N., Scarcello, F.: A comparison of structural CSP decomposition methods. Artificial Intelligence 124(2), 243–282 (2000)
35. Gottlob, G., Leone, N., Scarcello, F.: Advanced parallel algorithms for processing acyclic conjunctive queries, rules, and constraints. In: Proceedings of the 2000 Conference on Software Engineering and Knowledge Engineering (SEKE 2000), Chicago, pp. 167–176 (2000)
36. Gottlob, G., Leone, N., Scarcello, F.: On tractable queries and constraints. In: Bench-Capon, T.J.M., Soda, G., Tjoa, A.M. (eds.) DEXA 1999. LNCS, vol. 1677, pp. 1–15. Springer, Heidelberg (1999)
37. Gottlob, G., Leone, N., Scarcello, F.: The complexity of acyclic conjunctive queries. Journal of the ACM 48(3), 431–498 (2001)
38. Gottlob, G., Leone, N., Scarcello, F.: Computing LOGCFL Certificates. Theoretical Computer Science 270(1-2), 761–777 (2002)
39. Gottlob, G., Leone, N., Scarcello, F.: Robbers, marshals, and guards: game theoretic and logical characterizations of hypertree width. Journal of Computer and System Sciences 66(4), 775–808 (2003)
40. Gottlob, G., Grohe, M., Musliu, N., Samer, M., Scarcello, F.: Hypertree decompositions: Structure, algorithms, and applications. In: Kratsch, D. (ed.) WG 2005. LNCS, vol. 3787, pp. 1–15. Springer, Heidelberg (2005)
41. Gottlob, G., Miklos, Z., Schwentick, T.: Generalized hypertree decompositions: NP-Hardness and Tractable Variants. In: Proc. of PODS 2007, pp. 13–22 (2007)
42. Gottlob, G., Samer, M.: A Backtracking-Based Algorithm for Computing Hypertree-Decompositions. arXiv:cs/0701083 (2007)
43. Grädel, E.: On the Restraining Power of Guards. Journal of Symbolic Logic 64, 1719–1742 (1999)
44. Greibach, S.H.: The Hardest Context-Free Language. SIAM Journal on Computing 2(4), 304–310 (1973)
45. Greco, G., Scarcello, F.: Non-Binary Constraints and Optimal Dual-Graph Representations. In: Proc. of IJCAI 2003, pp. 227–232 (2003)
46. Grohe, M.: The Complexity of Homomorphism and Constraint Satisfaction Problems Seen from the Other Side. Journal of the ACM 54(1) (2007)
47. Grohe, M., Marx, D.: Constraint solving via fractional edge covers. In: Proc. of SODA 2006, Miami, Florida, USA, pp. 289–298 (2006)
48. Grohe, M., Schwentick, T., Segoufin, L.: When is the evaluation of conjunctive queries tractable? In: Proc. of STOC 2001, Heraklion, Crete, Greece, pp. 657–666 (2001)
49. Gyssens, M., Jeavons, P.G., Cohen, D.A.: Decomposing constraint satisfaction problems using database techniques. Journal of Algorithms 66, 57–89 (1994)
50. Harvey, P., Ghose, A.: Reducing Redundancy in the Hypertree Decomposition Scheme. In: Proc. of 5th IEEE International Conference on Tools with Artificial Intelligence, pp. 474–481 (2003)

51. Ioannidis, Y.E.: Query Optimization. The Computer Science and Engineering Handbook, pp. 1038–1057 (1997)
52. Ioannidis, Y.E.: The History of Histograms (abridged). In: Proc. of VLDB 2003, Berlin, Germany, pp. 19–30 (2003)
53. Johnson, D.S.: A Catalog of Complexity Classes. In: Handbook of Theoretical Computer Science, Volume A: Algorithms and Complexity, pp. 67–161 (1990)
54. Kolaitis, P.G.: Constraint Satisfaction, Databases, and Logic. In: Proc. of IJCAI 2003, Acapulco, Mexico, pp. 1587–1595 (2003)
55. Kolaitis, P.G., Vardi, M.Y.: Conjunctive-query containment and constraint satisfaction. Journal of Computer and System Sciences 61(2), 302–332 (2000)
56. Korimort, T.: Constraint Satisfaction Problems – Heuristic Decomposition. Ph.D thesis, Vienna University of Technology (April 2003)
57. Kun, G.: Constraints, MMSNP and expander relational structures (2007) ArXiv.org, `http://www.citebase.org/abstract?id=oai:arXiv.org:0706.1701`
58. Maier, D.: The Theory of Relational Databases. Computer Science Press (1986)
59. McMahan, B.: Bucket eliminiation and hypertree decompositions. Implementation report, Institute of Information Systems (DBAI), TU Vienna (2004)
60. Musliu, N.: Tabu Search for Generalized Hypertree Decompositions. In: Proc. of MIC 2007 (2007)
61. Reingold, O.: Undirected ST-connectivity in log-space. In: Proc. of STOC 2005, Baltimore, MD, USA, pp. 376–385 (2005)
62. McMahan, B.J., Pan, G., Porter, P., Vardi, M.Y.: Projection Pushing Revisited. In: Bertino, E., Christodoulakis, S., Plexousakis, D., Christophides, V., Koubarakis, M., Böhm, K., Ferrari, E. (eds.) EDBT 2004. LNCS, vol. 2992, pp. 441–458. Springer, Heidelberg (2004)
63. Papadimitriou, C.H., Yannakakis, M.: On the complexity of database queries. In: Proc. of PODS 1997, Tucson, Arizona, pp. 12–19 (1997)
64. Pearson, J., Jeavons, P.G.: A Survey of Tractable Constraint Satisfaction Problems, CSD-TR-97-15, Royal Holloway, Univ. of London (1997)
65. Robertson, N., Seymour, P.D.: Graph minors. II. Algorithmic aspects of tree width. Journal of Algorithms 7, 309–322 (1986)
66. Ruzzo, W.L.: Tree-size bounded alternation. Journal of Cumputer and System Sciences 21, 218–235 (1980)
67. Saccà, D.: Closures of database hypergraphs. Journal of the ACM 32(4), 774–803 (1985)
68. Samer, M.: Hypertree-decomposition via Branch-decomposition. In: Proceeding of IJCAI 2005, pp. 1535–1536 (2005)
69. Scarcello, F.: The Hypertree Decompositions HomePage (2002), `http://www.deis.unical.it/scarcello/Hypertrees/`, `http://www.dbai.tuwien.ac.at/proj/hypertree/` maintained by N. Musliu
70. Scarcello, F., Greco, G., Leone, N.: Weighted Hypertree Decompositions and Optimal Query Plans. In: Proc. of PODS 2004, pp. 210–221 (2004)
71. Seymour, P.D., Thomas, R.: Graph Searching and a Min-Max Theorem for Tree-Width. Journal of Combinatorial Theory, Series B 58, 22–33 (1993)
72. Seidel, R.: A new method for solving constraint satisfaction problems. In: Proc. of IJCAI 1981 (1981)
73. Subbarayan, S., Andersen, H.R.: Backtracking Procedures for Hypertree, HyperSpread and Connected Hypertree Decomposition of CSPs. In: Proc. of IJCAI 2007, pp. 180–185 (2007)
74. Skyum, S., Valiant, L.G.: A complexity theory based on Boolean algebra. Journal of the ACM 32, 484–502 (1985)

75. Sudborough, I.H.: Time and Tape Bounded Auxiliary Pushdown Automata. In: Gruska, J. (ed.) MFCS 1977. LNCS, vol. 53, pp. 493–503. Springer, Heidelberg (1977)
76. Tarjan, R.E., Yannakakis, M.: Simple linear-time algorithms to test chordality of graphs, test acyclicity of hypergraphs, and selectively reduce acyclic hypergraphs. SIAM Journal on Computing 13(3), 566–579 (1984)
77. Ullman, J.D.: Principles of Database and Knowledge Base Systems. Computer Science Press (1989)
78. Vardi, M.: Complexity of relational query languages. In: Proc. of STOC 1982, San Francisco, California, United States, pp. 137–146 (1982)
79. Vardi, M.: Constraint Satisfaction and Database Theory. In: Tutorial at the 19th ACM Symposium on Principles of Database Systems, PODS 2000 (2000)
80. Wilschut, A.N., Flokstra, J., Apers, P.M.G.: Parallel evaluation of multi-join queries. In: Proceedings of SIGMOD 1995, San Jose, CA, USA, pp. 115–126 (1995)
81. Yannakakis, M.: Algorithms for acyclic database schemes. In: Proc. of VLDB 1981, Cannes, France, pp. 82–94 (1981)
82. Yu, C.T., Özsoyoğlu, M.Z.: On determining tree-query membership of a distributed query. Infor. 22(3), 261–282 (1984)
83. Yu, C.T., Özsoyoğlu, M.Z., Lam, K.: Optimization of Distributed Tree Queries. Journal of Computer and System Sciences 29(3), 409–445 (1984)

Constraint Satisfaction Problems with Infinite Templates

Manuel Bodirsky

École polytechnique, Laboratoire d'informatique (LIX), France
bodirsky@informatik.hu-berlin.de

Abstract. Allowing templates with infinite domains greatly expands the range of problems that can be formulated as a non-uniform constraint satisfaction problem. It turns out that many CSPs over infinite templates can be formulated with templates that are *ω-categorical.* We survey examples of such problems in temporal and spatial reasoning, infinite-dimensional algebra, acyclic colorings in graph theory, artificial intelligence, phylogenetic reconstruction in computational biology, and tree descriptions in computational linguistics.

We then give an introduction to the universal-algebraic approach to infinite-domain constraint satisfaction, and discuss how cores, polymorphism clones, and pseudo-varieties can be used to study the computational complexity of CSPs with ω-categorical templates. The theoretical results will be illustrated by examples from the mentioned application areas. We close with a series of open problems and promising directions of future research.

1 Introduction

Some of the oldest results in constraint satisfaction concern constraint satisfaction problems over *infinite* domains. Examples are the constraint satisfaction problem (CSP) for Allen's interval algebra and its fragments, and the CSPs for many other temporal reasoning formalisms. Also in spatial reasoning, some of the earliest computational problems can be formulated as CSPs, e.g., the CSP for the region connection calculus.

In the 1990's, several systematic results about the computational complexity of CSPs in spatial and temporal reasoning were obtained, mostly for binary constraint languages. An important step was Nebel and Bürckert's discovery of Ord-Horn, a tractable fragment of Allen's interval algebra. In a computer-assisted proof, they showed that Ord-Horn is a *largest tractable* fragment of Allen's interval algebra, i.e., adding any (binary) relation from Allen's interval algebra to Ord-Horn results in an NP-complete constraint language. Another important step in complexity classification was the complete classification of all tractable fragments of the region connection calculus RCC-5 in spatial reasoning by Jonsson and Drakengren [44]. The corresponding result for Allen's interval algebra was obtained later by Krokhin, Jeavons, and Jonsson [49]. Similar classifications for the cardinal direction calculus, for constraint languages for

N. Creignou et al. (Eds.): Complexity of Constraints, LNCS 5250, pp. 196–228, 2008.

branching time, and constraint languages for partially ordered time were mostly obtained by a case-by-case study and ad-hoc methods. There are some common themes [29] and tools (e.g., local consistency techniques), but so far there is no general theory of tractabilility and hardness for such problems.

A typical property of constraint satisfaction problems is that if we add constraints to an unsatisfiable instance of the CSP, the instance stays unsatisfiable. Moreover, if we form the disjoint union of two satisfiable instances of the CSP, the resulting instance is again satisfiable. In fact, all computational problems (i.e., all sets of relational structures) that share these two properties can be formulated as a homomorphism problem for a fixed infinite relational structure Γ (in other words, as a non-uniform CSP with an infinite template). Many examples of computational problems that have been studied in the literature have a natural formulation as a homomorphism problem for a fixed infinite structure Γ, and we will mention examples from graph theory, artificial intelligence, phylogenetic reconstruction, finite model theory, and tree description logics in computational linguistics.

This survey is about techniques to study the computational complexity of CSPs with infinite templates. We focus mostly on the border between CSPs that can be solved by a polynomial-time algorithm, and CSPs that are known to be NP-hard. This border has been of central interest in most of the mentioned work on temporal and spatial reasoning. Our main theme will be the question which of the powerful universal-algebraic techniques that are employed to study the complexity of the finite domain CSP (see e.g. [22]) can be generalized to infinite domain templates. It turns out that the universal-algebraic approach applies if the template satisfies an important and central property in model theory, called *ω-categoricity.*

The article is divided into two parts: in the first part we demonstrate that the class of computational problems that can be formulated with an ω-categorical template is very large (actually, it contains all the infinite-domain CSPs we have mentioned so far). In the second part, we set out to develop the universal-algebraic theory for ω-categorical templates. We recall the definitions of polymorphisms, algebras, and (pseudo-) varieties, and show that these concepts are useful not only for finite but also for infinite ω-categorical structures. Some of the tractability results presented here do not have a finite domain counterpart. We only give proofs of the statements in this text if they are are instructive, previously unpublished, or more conceptual than the existing proofs in the literature, and give references to the literature in all other cases.

2 Infinite Templates

The notation used in this text mostly follows Hodges' text book [42]. A *signature* τ is a set of relation and function symbols, each equipped with an arity. A *τ-structure* $\boldsymbol{A}$ is a set A (the *domain* of $\boldsymbol{A}$) together with a relation $R^{\boldsymbol{A}} \subseteq A^k$ for each k-ary relation symbol in τ and a function $f^{\boldsymbol{A}} : A^k \to A$ for each k-ary function symbol in τ. When there is no danger of confusion, we use the same symbol for a function and its function symbol, and for a relation and its relation symbol. By convention, $A, B, C, \ldots$ denotes the domain of the structures $\boldsymbol{A}$, $\boldsymbol{B}$, $\boldsymbol{C}$, ..., respectively. We sometimes write $(A, R_1^{\boldsymbol{A}}, R_2^{\boldsymbol{A}}, \ldots, f_1^{\boldsymbol{A}}, f_2^{\boldsymbol{A}}, \ldots)$ for the relational structure $\mathbf{A}$ with relations $R_1^{\boldsymbol{A}}, R_2^{\boldsymbol{A}}, \ldots$ and functions $f_1^{\boldsymbol{A}}, f_2^{\boldsymbol{A}}, \ldots$ We say that a structure is infinite if its domain is infinite. The most important special cases of structures that appear in this paper are *relational structures*, that is, structures with a purely relational signature, and *algebras*, that is, structures with a purely functional signature. Algebras with domain $A, B, C, \ldots$ are denoted by $\mathbb{A}, \mathbb{B}, \mathbb{C}, \ldots$

Homomorphisms. A *homomorphism* h from a structure $\boldsymbol{A}$ to a structure $\boldsymbol{B}$ with the same signature τ is a mapping from A to B that *preserves* each function and each relation for the symbols in τ; that is,

- if $(a_1, \ldots, a_k)$ is in $R^{\boldsymbol{A}}$, then $(h(a_1), \ldots, h(a_k))$ must be in $R^{\boldsymbol{B}}$;
- if $f^{\boldsymbol{A}}(a_1, \ldots, a_k) = a_0$, then $f^{\boldsymbol{B}}(h(a_1), \ldots, h(a_k)) = h(a_0)$.

In this article, a *(non-uniform) constraint satisfaction problem (CSP)* is a computational problem that is specified by a single structure with a finite relational signature, called the *template* of the CSP. Relational structures that denote templates for CSPs will be denoted by capital greek letters Γ, Δ (and their domain by $D(\Gamma)$, $D(\Delta)$, respectively).

Definition 1. *Let Γ be a (possible infinite) relational structure with a finite relational signature τ. Then CSP(Γ) is the computational problem to decide whether a given finite τ-structure homomorphically maps to Γ.*

We sometimes also write $\mathrm{CSP}(D, R_1, \ldots, R_l)$ instead of $\mathrm{CSP}((D, R_1, \ldots, R_l))$. Note that due to the assumption that the signature τ is finite, we can fix any representation of the relation symbols in τ to represent the input structure. $\mathrm{CSP}(\Gamma)$ can also be considered to be a set – the set of all finite structures that homomorphically map to Γ.

Logic. As for finite domain CSPs, there is another way of looking at the constraint satisfaction problem for Γ that uses terminology from logic. A first-order τ-formula is called *primitive positive* if it is of the form

$$\exists x_1, \ldots, x_m.\, \psi_1 \wedge \cdots \wedge \psi_l$$

where $\psi_1, \ldots, \psi_l$ are atomic τ-formulas.

It is straightforward to verify (in the same way as for finite domain CSPs) that $\mathrm{CSP}(\Gamma)$ is polynomial-time equivalent to the following computational problem (in fact, the two problems can be considered to be the same computational problem, up to formalization). The input consists of a primitive positive τ-sentence Φ (i.e., a primitive positive τ-formula without free variables), and the question is whether Φ is holds true in Γ. The conjuncts in the primitive positive sentence are then called the *constraints* of Φ. From this formulation of the constraint satisfaction problem it is obvious that $\mathrm{CSP}(\Gamma)$ is fully determined by the *first-order theory of* Γ (i.e., the set of first-order sentences that are valid in Γ).

We would like to remark that the logic perspective on CSP(Γ) is closely related to the evaluation problem for conjunctive queries studied in database theory. We freely switch between the relational homomorphism and the logic perspective whenever this is convenient. We also say that $\boldsymbol{A}$ is *satisfiable* (with respect to CSP(Γ) if $\boldsymbol{A}$ homomorphically maps to Γ, and the homomorphism from $\boldsymbol{A}$ to Γ is called a *solution* for $\boldsymbol{A}$.

Example 1. Consider the problem CSP($\mathbb{N}, =, \neq$). An instance of this problem consists of a set of variables, some linked by equality, some by disequality constraints. Such an instance is unsatisfiable if and only if there is a path $x_1, \ldots, x_n$ from a variable x_1 to a variable x_n that uses only equality edges, i.e., '$x_i = x_{i+1}$' is a constraint in the instance for each $1 \leq i \leq n-1$, and additionally '$x_1 \neq x_n$' is a constraint in the instance. Clearly, it can be tested in linear time in the size of the input instance whether the instance contains such a path.

Example 2. Next, consider the problem CSP($\mathbb{Q}, <$). Here, $<$ denotes the strict linear order on the relational numbers $\mathbb{Q}$ (i.e., $<$ is a binary relation). An instance of this problem can be viewed as a directed graph (potentially with loops), where there is an arc between the vertices x and y if there is the constraint '$x < y$' in the instance. It is easy to see that an instance homomorphically maps to ($\mathbb{Q}, <$) if and only if there is no directed cycle in the graph. Again, this can be tested in linear time, e.g., by depth-first search.

Example 3. The so-called *betweenness problem* is CSP($\mathbb{Q}$, *Betw*) where *Betw* is the ternary relation $\{(x, y, z) \in \mathbb{Q}^3 \mid x < y < z \vee z < y < x\}$. This problem is an NP-complete problem from the famous book of Garey and Johnson [37].

We have defined the constraint satisfaction problem only for relational structures, but the generalization to structures that also include function symbols is straightforward, and has been studied for finite domains [47]. If we allow function symbols *and* infinite domains, we can for instance formulate the famous unification problem (see, e.g., [56]) as CSP(Γ) for an appropriate infinite structure Γ. In this article we focus on relational templates only.

Having structures with function symbols will be convenient in Section 8 where we use algebras to study the computational complexity of CSPs. Moreover, several templates that we use to illustrate the theory have a convenient definition by means of structures that contain functions, e.g., in Subsections 5.3 and 5.2.

The next lemma is a convenient tool to determine whether a computational problem can be formulated as CSP(Γ) for an infinite relational structure Γ. The *disjoint union* of a set of τ-structures $\mathcal{C}$ is a τ-structure $\boldsymbol{B}$ whose domain is the disjoint union of the domains of the structures in $\mathcal{C}$. The relations in $\boldsymbol{B}$ are the union of the corresponding relations in the structures in $\mathcal{C}$.

Definition 2. *We say that a set $\mathcal{C}$ of relational structures is* closed under (finite) disjoint unions *iff whenever* $\mathbf{A}, \mathbf{B} \in \mathcal{C}$ *then the disjoint union of* $\mathbf{A}$ *and* $\mathbf{B}$ *is also in* $\mathcal{C}$. *We say that* $\mathcal{C}$ *is* closed under inverse homomorphisms *iff whenever* $\mathbf{B} \in \mathcal{C}$ *and* $\mathbf{A}$ *homomorphically maps to* $\mathbf{B}$ *then* $\mathbf{A} \in \mathcal{C}$.

The following is a fundamental lemma for constraint satisfaction problems.

Lemma 1. *Let $\mathcal{C}$ be a set of finite τ-structures, for a finite relational signature τ. Then $\mathcal{C} = CSP(\Gamma)$ for some relational structure Γ over an infinite domain if and only if $\mathcal{C}$ is closed under disjoint unions and inverse homomorphisms.*

Proof. Clearly, the disjoint union of two instances of CSP(Γ) that homomorphically map to Γ also homomorphically maps to Γ. Moreover, if $\boldsymbol{A}$ does not homomorphically map to Γ, and there is a homomorphism from $\boldsymbol{A}$ to $\boldsymbol{B}$, then $\boldsymbol{B}$ does not homomorphically map to Γ either.

For the other direction, suppose that $\mathcal{C}$ is a set of relational structures that is closed under disjoint unions and inverse homomorphisms. Let Γ be the (infinite) disjoint union of all structures in $\mathcal{C}$. Clearly, every structure in $\mathcal{C}$ homomorphically maps to Γ. Now, let $\boldsymbol{A}$ be a finite structure with a homomorphism h to Γ. By construction of Γ, the set $h(A)$ is contained in the disjoint union $\boldsymbol{B}$ of a finite set of structures from $\mathcal{C}$. Since $\mathcal{C}$ is closed under disjoint unions, $\boldsymbol{B}$ is in $\mathcal{C}$. Clearly, $\boldsymbol{A}$ homomorphically maps to $\boldsymbol{B}$, and because $\mathcal{C}$ is closed under inverse homomorphisms, $\boldsymbol{A}$ is in $\mathcal{C}$ as well. □

To apply the techniques presented in this paper it will be important that the templates of the CSPs are ω-categorical (ω-categoricity will be defined in Section 3). The structures produced by Lemma 1 are usually not ω-categorical.

Let us conclude this section with remarks concerning the computational complexity of CSP(Γ). So far, we have seen CSPs in Examples 1, 2, and 3 that are in P or NP-complete. However, it is not hard to come up with undecidable CSPs. For example, Hilbert's 10'th problem, the problem to decide whether a given diophantine equation has a solution, is undecidable [54], and is computationally equivalent to CSP$(\mathbb{Z}, R^+, R^*)$ where $R^+ = \{(x,y,z) \in \mathbb{Z}^3 \mid x+y+z=1\}$ and $R^* = \{(x,y,z) \in \mathbb{Z}^3 \mid xy = z\}$.

Bauslaugh [6] has constructed for every recursive function f an infinite digraph Γ such that CSP(Γ) is decidable, but has time complexity at least f. In this paper, we focus on CSPs that are in NP.

3 Good Templates: ω-Categorical Structures

Many important infinite-domain constraint satisfaction problems can be formulated with templates that are *ω-categorical.* The concept of ω-categoricity is of central importance in model theory.

Definition 3. *A countably infinite structure Γ is called ω-categorical if all countable models of its first-order theory are isomorphic to Γ.*

One of the first structures that were found to be ω-categorical (by Kantor) is the linear order of the rational numbers $(\mathbb{Q}, <)$. We will see many more examples of ω-categorical structures in this section and in Section 5. One of the standard approaches to verify that a structure is ω-categorical is via a so-called *back-and-forth* argument. We sketch the back-and-forth argument that shows that $(\mathbb{Q}, <)$

is ω-categorical; much more detail about this important concept in model theory can be found in [42, 57]. Let $\boldsymbol{A}$ be a countable model of the first-order theory of $(\mathbb{Q}, <)$. An isomorphism i between $\boldsymbol{A}$ and $(\mathbb{Q}, <)$ can be defined inductively as follows. Suppose that we have already defined i on a finite subset S of $\mathbb{Q}$ and that f is an embedding of the structure $\boldsymbol{S}$ induced by S in $(\mathbb{Q}, <)$ into $\boldsymbol{A}$. Since $<^{\boldsymbol{A}}$ is dense and unbounded, we can extend f to any other element of $\mathbb{Q}$ such that the extension is still an embedding from a substructure of $\mathbb{Q}$ into $\boldsymbol{A}$ (*going forth*). Symmetrically, for every element v of $\boldsymbol{A}$ we can find an element $u \in \mathbb{Q}$ such that the extension of f that maps u to v is also an embedding (*going back*). We now alternate between going forth and going back; when going forth, we extend the domain of f by the *next* element of $\mathbb{Q}$, according to some fixed enumeration of the elements in $\mathbb{Q}$. When going back, we extend f such that the image of A contains the *next* element of $\boldsymbol{A}$, according to some fixed enumeration of the elements of $\boldsymbol{A}$. If we continue in this way, we have defined the value of f on all elements of $\mathbb{Q}$. Moreover, f will be surjective, and an embedding, and hence an isomorphism between $(\mathbb{Q}, <)$ and $\boldsymbol{A}$.

There are many equivalent characterizations of ω-categoricity. Let G be a permutation group acting on a set X. For $n \geq 1$ the *orbit* of $(t_1, \ldots, t_n) \in X^n$ under G is the set $\{(\alpha(t_1), \ldots, \alpha(t_n)) \mid \alpha \in G\}$. We say that a permutation group is *oligomorphic* if for each $n \geq 1$ there are finitely many orbits of n-tuples. An accessible proof of the following theorem can be found in Hodges' book (Theorem 6.3.1 in [42]).

Theorem 1 (Engeler, Ryll-Nardzewski, Svenonius). *The following are equivalent:*

1. *Γ is ω-categorical;*
2. *The automorphism group of Γ is oligomorphic;*
3. *for each $n \geq 1$, there are finitely many inequivalent formulas with n free variables over Γ;*
4. *for each $n \geq 1$, the orbits of n-tuples are first-order definable in Γ.*

The second condition in Theorem 1 provides another possibility to verify that a structure is ω-categorical. We again illustrate this with the structure $(\mathbb{Q}, <)$. It is not difficult but a good exercise to verify that the orbit of an n-tuple $(t_1, \ldots, t_n)$ from $\mathbb{Q}^n$ in the automorphism group of $(\mathbb{Q}, <)$ is determined by the weak linear order that is induced by $t_1, \ldots, t_n$ in $(\mathbb{Q}, <)$ (we write *weak linear order*, and not *linear order*, because some of the elements $t_1, \ldots, t_n$ might be equal). Hence, there is a finite number of orbits of n-tuples, for all $n \geq 1$.

Lemma 2 below states a useful fact that ω-categorical structures have in common with finite structures. A homomorphism h from $\boldsymbol{A}$ to $\boldsymbol{B}$ is called a *strong homomorphism* if it also preserves the complements of the relations from $\boldsymbol{A}$. Note that an embedding is an injective strong homomorphism.

Lemma 2. *Let Γ be a finite or infinite ω-categorical structure with relational signature τ, and let* $\mathbf{A}$ *be a countable relational structure with the same signature τ. If there is no homomorphism (embedding) from* $\mathbf{A}$ *to Γ, then there is a finite substructure of* $\mathbf{A}$ *that does not homomorphically map (embed) to Γ.*

The lemma is an easy consequence of Königs tree lemma; a proof for homomorphisms can be found in [11], and the proof for embeddings (and, similarly, for strong homomorphisms and for injective homomorphisms) is analogous.

3.1 Fraïssé Amalgamation – ω-Categorical Structures for Everyone

A versatile tool to construct ω-categorical templates is Fraïssé-amalgamation. We present it here for structures that might contain relations *and* functions; this will sometimes be convenient even if we are only interested in relational structures (see Subsections 5.2 and 5.3).

If $\boldsymbol{A}$ and $\boldsymbol{B}$ are structures, $A \subseteq B$, and the inclusion map from A to B is an embedding, then we say that $\boldsymbol{A}$ is a *substructure* of $\boldsymbol{B}$. Note that for every subset S of B there is a unique smallest substructure $\boldsymbol{A}$ of $\boldsymbol{B}$ such that A contains S. We call $\boldsymbol{A}$ the *substructure of* $\mathbf{B}$ *generated by* S, also denoted by $\boldsymbol{B}[S]$. We say that $\boldsymbol{A}$ is *finitely generated* if S is a finite set of elements.

Let $\mathcal{C}$ be a set of finitely generated structures. We say that $\mathcal{C}$ has the

- HP *Hereditary property* if whenever $\boldsymbol{A} \in \mathcal{C}$ and $\boldsymbol{B}$ is a finitely generated substructure of $\boldsymbol{A}$ then $\boldsymbol{B}$ is isomorphic to a structure in $\mathcal{C}$.
- JEP *Joint embedding property* if whenever $\boldsymbol{A}, \boldsymbol{B} \in \mathcal{C}$ then there is $\boldsymbol{C} \in \mathcal{C}$ such that both $\boldsymbol{A}$ and $\boldsymbol{B}$ embed into $\boldsymbol{C}$.
- AP *Amalgamation property* if whenever $\boldsymbol{A}, \boldsymbol{B}_1, \boldsymbol{B}_2 \in \mathcal{C}$ and $e_1 : \boldsymbol{A} \to \boldsymbol{B}_1$ and $e_2 : \boldsymbol{A} \to \boldsymbol{B}_2$ are embeddings there exists $\boldsymbol{C} \in \mathcal{C}$ and embeddings $f_1 : \boldsymbol{B}_1 \to \boldsymbol{C}$ and $f_2 : \boldsymbol{B}_2 \to \boldsymbol{C}$ such that $f_1 e_1 = f_2 e_2$.

A structure Γ is *homogeneous* (sometimes also called *ultra-homogeneous* [42]) if every isomorphism between finitely generated substructures of Γ can be extended to an automorphism of Γ. It is well-known that a homogeneous structure Γ with a signature that contains finitely many relation symbols of arity k, for each k, is ω-categorical (this is not explicitly mentioned in Hodges' book [42], but can easily be shown with the background from there). Moreover, Γ admits *quantifier elimination*, i.e., for every first-order τ-formula there exists an equivalent quantifier-free τ-formula [42]. In fact, an ω-categorical structure is homogeneous if and only if it admits quantifier elimination (Statement 2.22 in [25]).

Theorem 2 (Fraïssé [36]; see [42]). *Let τ be a countable signature and let $\mathcal{C}$ be a non-empty finite or countable set of finitely generated τ-structures which has HP, JEP, and AP. Then there is a homogeneous and at most countable τ-structure Γ such that a structure is a finitely generated substructure of Γ if and only if it is isomorphic to a structure in $\mathcal{C}$. The structure Γ is unique up to isomorphism, and called the* Fraïssé-limit *of $\mathcal{C}$.*

With Theorem 2 we have seen a third possibility how to verify the ω-categoricity of our running example $(\mathbb{Q}, <)$: it suffices to verify that the class of all finite weak linear orders is an amalgamation class.

Remark 1. It is sometimes convenient to define an ω-categorical τ-structure Γ by defining an amalgamation class $\mathcal{C}$ with a signature that is larger than τ such that

Γ is a reduct[1] of the Fraïssé-limit of $\mathfrak{C}$. It is an easy consequence of Theorem 1 that Γ must then be ω-categorical.

3.2 New ω-Categorical Structures from Old

Many ω-categorical structures can be derived from other ω-categorical structures via first-order interpretations. Our definition of first-order interpretations essentially follows [42].

If $\delta(x_1, \ldots, x_k)$ is a first-order formula with the k free variables $x_1, \ldots, x_k$, and Γ is a structure, we we write $\delta(\boldsymbol{A}^k)$ for the k-ary relation that is defined by δ on $\boldsymbol{A}$.

Definition 4. *A relational σ-structure* $\mathbf{B}$ *has a* (first-order) interpretation *in a τ-structure* $\mathbf{A}$ *if there exists a natural number d, called the* dimension *of the interpretation, and*

- *a τ-formula $\delta(x_1, \ldots, x_d)$ – called* domain formula,
- *for each k-ary relation symbol R in σ a τ-formula $\phi_R(\overline{x}_1, \ldots, \overline{x}_k)$ where the $\overline{x}_i$ denote disjoint d-tuples of distinct variables – called the* defining formulae,
- *a τ-formula $\phi_=(x_1, \ldots, x_d, y_1, \ldots, y_d)$, and*
- *a surjective map $h : \delta(\mathbf{A}^d) \rightarrow B$ – called* coordinate map,

such that for all relations R in $\mathbf{B}$ *and all tuples $\overline{a}_i \in \delta(\mathbf{A}^d)$*

$$(h(\overline{a}_1), \ldots, h(\overline{a}_k)) \in R^{\mathbf{B}} \Leftrightarrow \mathbf{A} \models \phi_R(\overline{a}_1, \ldots, \overline{a}_k) \; , \; \textit{and}$$
$$h(\overline{a}_1) = h(\overline{a}_2) \Leftrightarrow \mathbf{A} \models \phi_=(\overline{a}_1, \overline{a}_2) \; .$$

If the formulas δ, ϕ_R, and $\phi_=$ are all primitive positive, we say that $\boldsymbol{B}$ has a primitive positive interpretation in $\boldsymbol{A}$. We say that $\boldsymbol{B}$ is *interpretable in Γ with finitely many parameters* if there are $c_1, \ldots, c_n \in A$ such that $\boldsymbol{B}$ is interpretable in the expansion of $\boldsymbol{A}$ by the singleton relations $\{c_i\}$ for all $1 \leq i \leq n$. First-order *definitions* are a special case of interpretations: a structure $\boldsymbol{B}$ is *(first-order) definable* in $\boldsymbol{A}$ if $\boldsymbol{B}$ has an interpretation in $\boldsymbol{B}$ of dimension one where the domain formula is logically equivalent to true.

Lemma 3 (see e.g. [42]). *If Γ is an ω-categorical structure, then every structure Δ that is first-order interpretable in Γ with finitely many parameters is ω-categorical as well.*

Suppose that $\boldsymbol{A}$ and $\boldsymbol{B}$ are finite or ω-categorical structures. In the remainder of the section we show that then the *disjoint union* $\boldsymbol{A} \uplus \boldsymbol{B}$ (which was defined in Section 2) and the *composition* $\boldsymbol{A} \circ \boldsymbol{B}$ (Definition 5 below) of $\boldsymbol{A}$ and $\boldsymbol{B}$ are finite or ω-categorical as well. Again, this will be convenient to specify templates of constraint satisfaction problems.

The composition of $\boldsymbol{A}$ and $\boldsymbol{B}$ is, roughly speaking, the structure obtained from $\boldsymbol{A}$ by replacing each element in $\boldsymbol{A}$ by a copy of $\boldsymbol{B}$.

[1] A *reduct* of a structure $\boldsymbol{A}$ is a structure that is obtained from $\boldsymbol{A}$ by removing relations and/or functions from $\boldsymbol{A}$.

Definition 5. *If* $\mathbf{A}, \mathbf{B}$ *are* τ*-structures, then* $\mathbf{A} \circ \mathbf{B}$ *is the* τ*-structure with domain* $A \times B$*, where* $((a_1, b_1), \ldots, (a_k, b_k)) \in R^{\mathbf{A} \circ \mathbf{B}}$ *for a* k*-ary relation symbol* $R \in \tau$ *iff* $(a_1, \ldots, a_k) \in R^{\mathbf{A}}$ *or* $(a_1 = \cdots = a_k$ *and* $(b_1, \ldots, b_k) \in \mathbf{B})$*.*

To verify ω-categoricity of disjoint union and composition it is most convenient to use the second condition from Theorem 1 (i.e., we will count the orbits of k-tuples in the resulting structure). For an ω-categorical structure $\boldsymbol{A}$, let $f_n(\boldsymbol{A})$ be the number of orbits of n-tuples in the automorphism group of $\boldsymbol{A}$, and let $F(\boldsymbol{A}, x)$ be the exponential generating function in one formal variable x, defined by $\sum_{n \geq 0} f_n(\boldsymbol{A}) x^n / n!$. The following reflects well-known facts in enumerative combinatorics (for a presentation in the context of oligomorphic permutation groups, see [25]).

Lemma 4. *Let* $\mathbf{A}, \mathbf{B}$ *be finite or* ω*-categorical structures. Then*

$$F(\mathbf{A} \uplus \mathbf{B}, x) = F(\mathbf{A}, x) F(\mathbf{B}, x)$$
$$F(\mathbf{A} \circ \mathbf{B}, x) = F(\mathbf{A}, F(\mathbf{B}, x) - 1)$$

In particular, for all $n \geq 1$ *we have that* $f_n(\mathbf{A} \uplus \mathbf{B})$ *and* $f_n(\mathbf{A} \circ \mathbf{B})$ *are finite, and therefore* $\mathbf{A} \uplus \mathbf{B}$ *and* $\mathbf{A} \circ \mathbf{B}$ *is finite or* ω*-categorical.*

We would like to remark that the automorphism group of $\boldsymbol{A} \circ \boldsymbol{B}$ is also known as the *wreath product* of the automorphism group of $\boldsymbol{B}$ with the automorphism group of $\boldsymbol{A}$.

4 Better Templates: Model-Complete Cores

It might be that the same CSP can be formulated with different ω-categorical templates. For example, if T_3 is the transitive tournament on three vertices, then CSP$(\mathbb{Q}, <)$ and CSP$(T_3 \circ (\mathbb{Q}, <))$ (for the definition of the operation $\circ$, see Subsection 3.2) are the same computational problem, namely graph acyclicity.

For finite and for ω-categorical templates we have a very elegant characterization of those pairs of templates that have the same CSP. Call two structures Γ and Δ *homomorphically equivalent* if there is a homomorphism from Γ to Δ and a homomorphism from Δ to Γ (this is in fact an equivalence relation). The following is an easy consequence of Lemma 2.

Lemma 5. *Let* Γ, Δ *be finite or* ω*-categorical. Then CSP(*Γ*) equals CSP(*Δ*) if and only if* Γ *and* Δ *are homomorphically equivalent.*

Lemma 5 is false for general relational structures. Consider for example the structure $(\mathbb{Z}, \{(x, y) \mid y = x + 1\})$ — the 'infinite line', and the structure $(\mathbb{N}, \{(x, y) \mid y = x + 1\})$ — the 'infinite ray'. Clearly, these two structures give rise to the same CSP, but there is no homomorphism from the line to the ray.

It turns out that every equivalence class of finite or ω-categorical templates (with respect to homomorphic equivalence) has a member with very good properties for the universal-algebraic approach. Moreover, this member is unique up to isomorphism.

Definition 6. *An ω-categorical structure Γ is called* model-complete *if every embedding of Γ into Γ is* elementary, *i.e., preserves all first-order definable relations. An ω-categorical structure is called a* core *if every endomorphism is an embedding.*

The structure $(\mathbb{Q}, <)$, for example, is easily seen to be a model-complete core, in contrast to $T_3 \circ (\mathbb{Q}, <)$, which is not a core.

Note that for every finite structure Γ every embedding of Γ into Γ is elementary (because it must be an automorphism of Γ).

Theorem 3 (of [7]). *Every ω-categorical relational structure Γ is homomorphically equivalent to a model-complete core Δ, which is unique up to isomorphism, and ω-categorical or finite. For all $k \geq 1$, the orbits of k-tuples in Δ are primitive positive definable.*

Hence, if we want to classify the complexity of CSP(Γ) for an ω-categorical structure Γ, we can without loss of generality assume that Γ is a core. We state an important consequence of Theorem 3, which is well-known (and also non-trivial) for finite templates. The consequence will be used in Theorem 10.

Corollary 1 (of [7]). *Let Γ be an ω-categorical model-complete core, and let Δ be the expansion of Γ by finitely many singleton relations, i.e., relations of the form $\{a\}$ for $a \in D(\Gamma)$. Then CSP(Γ) and CSP(Δ) are polynomial-time equivalent.*

5 Examples

Most of the examples of CSPs that we present in this Section can be solved in polynomial-time, and we come back to most of these problems in Section 10 when discussing tractability criteria for CSPs. Coming up with NP-hard CSPs is much easier: each CSP in this section becomes NP-hard if we expand the template by the relation $\{(x, y, z) \mid x = y \vee y = z\}$. More on NP-hardness criteria of CSPs follows in later sections, culminating in Section 9.

5.1 Temporal Reasoning

Temporal reasoning is an important sub-discipline in Artificial Intelligence. One of the most basic temporal reasoning problems is the constraint satisfaction problem of the so-called *point algebra*. Here, the variables denote time points, and the constraints are of the form $x = y$, $x < y$, $x \leq y$, and $x \neq y$. In this CSP, it does not matter whether we use as the domain the natural, the rational, or the real numbers. The problem can e.g. be formalized as CSP($\mathbb{Q}, =, <, \leq, \neq$), and can be solved in linear time in the size of the input instances.

If we do not restrict our attention to binary constraints, it is natural to study expansions of the point-algebra by higher-ary relations with a first-order definition in $(\mathbb{Q}, <)$. For instance, consider the relation

$$R^{\min} = \{(x, y, z) \mid x > y \vee x > z\} .$$

The relation $R^{\min}(x, y, z)$ specifies that x is larger than the minimum of y and z. The constraint satisfaction problem CSP($\mathbb{Q}, R^{\min}$) will be discussed in Section 10.

Another famous temporal reasoning problem is the CSP for *Allen's Interval Algebra.* It is easiest to describe the template for this CSP by an interpretation in $(\mathbb{Q}, <)$. The dimension of this interpretation is two, and the domain formula $\delta(x, y)$ is $x < y$. Hence, the variables of the CSP denote non-empty time intervals. The template contains for each inequivalent $\{<\}$-formula with four variables ϕ a binary relation R such that (a_1, a_2, a_3, a_4) satisfies ϕ if and only if $((a_1, a_2), (a_3, a_4)) \in R$. In particular, we have relations for containment of intervals, disjointness of intervals, and so forth. By Lemma 3, this template is ω-categorical. Its CSP is NP-complete [5]. The computational complexity of the CSP for all fragments of Allen's inteval algebra has been determined in [55,49].

5.2 Spatial Reasoning

One of the most fundamental spatial reasoning formalisms is the RCC-5 calculus (also known as the containment algebra in the theory of relation algebras [31]). In this formalism, the variables denote non-empty regions, and the basic relations in the calculus express containment of regions, disjointness of regions, etc; formal definitions will be provided below. The constraint satisfaction problem for RCC-5 is NP-complete; the computational complexity of its fragments was classified in [58,44].

To formulate the CSPs for RCC-5 and its fragments with ω-categorical templates, let $\mathbb{B}$ be the countable atomless boolean ring without an identity element. That is, $\mathbb{B}$ is an algebra with an operation $+$ for addition and an operation $\cdot$ for multiplication satisfying the usual axioms for rings. A ring is *boolean* if it satisfies $x \cdot x = x$ for all elements x. A ring is *without (multiplicative) identity element* if there is no element x_1 such that $x_1 \cdot y = y$ for all elements y. A ring without identity element is *atomless* if the partial order $\leq$ defined by $x \leq y$ if $x \cdot y = x$ does not have minimal elements. Every countable atomless boolean ring without an identity element is unique up to isomorphism [1], homogeneous (this can be shown by Theorem 2), and hence ω-categorical. We can interpret the elements of this boolean ring as non-empty sets (regions), where $x + y$ denotes the symmetric difference, and $x \cdot y$ the intersection of x and y.

Now, consider the relational structure that has the same domain B as the boolean ring $\mathbb{B}$, and that contains all binary relations with a first-order definition in $\mathbb{B}$. This is exactly the template for the constraint satisfaction problem for RCC-5 (several equivalent definitions can be found in the literature). Ordered by inclusion, there are five minimal non-empty binary relations with a first-order definition in $\mathbb{B}$, and they are known as the *basic relations* of RCC-5. Traditionally, these relations are denoted by DR,PO,PP,PPI,EQ, and they have the following first-order definitions in $\mathbb{B}$ (with their intuitive meaning in braces).

$\mathrm{DR}(x, y)$ iff $(x + y)x = x$	'x and y are disjoint'
$\mathrm{PP}(x, y)$ iff $xy = x \wedge x \neq y$	'y properly contains x'
$\mathrm{PPI}(x, y)$ iff $xy = y \wedge x \neq y$	'x properly contains y'
$\mathrm{EQ}(x, y)$ iff $x = y$	'x equals y'
$\mathrm{PO}(x, y)$ iff $\neg(\mathrm{DR}(x, y) \vee \mathrm{PP}(x, y) \vee \mathrm{PPI}(x, y) \vee \mathrm{EQ}(x, y))$	'x and y properly overlap'

It is known that $\mathrm{CSP}(B, \mathrm{DR}, \mathrm{PP}, \mathrm{PPI}, \mathrm{EQ}, \mathrm{PO})$ is in P [58]. A larger tractable fragment of RCC-5 is discussed in Section 10.

5.3 Vector Space CSPs

Next, we discuss a natural algebraic constraint satisfaction problem.

Vector Space CSP
INSTANCE: A set of equalities of the form $x + y = z$ and disequalities of the form $x + y \neq z$ over a set of variables V.
QUESTION: Can we assign d-dimensional boolean vectors to the variables such that all the equalities and disequalities are satisfied, for some d?

It is easy to observe that if there exists a solution, then there is a solution where $d = |V|$. However, note that $|V|$ grows with the input size, and it is easy to see that this problem cannot be formulated with a finite template.

The vector space CSP has a natural formulation as a CSP with an ω-categorical template. Let $\mathbb{V}$ be the infinite-dimensional vector space over the 2-element field $\mathbb{F}_2$. Infinite-dimensional vector spaces over finite fields are known to be homogeneous and ω-categorical [33], and they play an important role in the classification of so-called *strictly minimal sets* in ω-categorical structures [61, 26, 27]. They are also examples of *totally categorical structures*, i.e., structures with a first-order theory that has one model of *every* infinite cardinality, up to isomorphism.

The signature of $\mathbb{V}$ is operational, with function symbols for $+$ and a unary function for scalar multiplication for each element of the field. Then the template $\boldsymbol{V}$ of the above CSP is the relational structure with the same domain as $\mathbb{V}$, and with the ternary relation $\{(x, y, z) \mid x + y = z\}$ and the ternary relation $\{(x, y, z) \mid x + y \neq z\}$. Since $\boldsymbol{V}$ clearly has a first-order definition in $\mathbb{V}$, it is ω-categorical as well. It is not hard to see that $\boldsymbol{V}$ is a model-complete core. An polynomial time algorithm for $\mathrm{CSP}(\boldsymbol{V})$ will be discussed in Section 10.

5.4 Graph Coloring Problems

The $\boldsymbol{H}$-coloring problem in graph theory is the special case of the CSP where the template $\boldsymbol{H}$ has a single binary relation, and therefore can be viewed as a digraph (directed graph). If we allow infinite digraphs $\boldsymbol{H}$, many more graph coloring problems that have been studied in the literature can be formulated as $\boldsymbol{H}$-coloring problems.

The complexity of the following class of infinite $\boldsymbol{H}$-coloring problems has been classified completely by Feder, Hell, and Mohar [34].

Acyclic $\boldsymbol{H}$-coloring
INSTANCE: A finite digraph $\boldsymbol{G}$.
QUESTION: Is there a mapping c from G to H such that for every edge xy in $\boldsymbol{G}$ the digraph $\boldsymbol{H}$ contains the arc $c(x)c(y)$, and such that the pre-image of every vertex is acyclic?

For every digraph $\boldsymbol{H}$, we can formulate the acyclic $\boldsymbol{H}$-coloring problem as a constraint satisfaction problem with an ω-categorical template. The acyclic $\boldsymbol{H}$-coloring problem is CSP($\boldsymbol{H} \circ (\mathbb{Q}, <)$).

5.5 Phylogenetic Reconstruction

Modern biology holds the view that the species in the evolution of life on earth developed in a mostly tree-like fashion: at certain time periods, species separated into sub-species. The goal of phylogenetic reconstruction is to determine the evolutionary tree from given partial information about the tree. This motivates the computational problem of *rooted triple consistency*, defined below. In 1981, Aho, Sagiv, Szymanski, and Ullman [4] presented a quadratic time algorithm to this problem, motivated independently from computational biology by questions in database theory.

We fix some terminology concerning rooted trees. Let $\boldsymbol{T}$ be a tree with a distinguished vertex r, the *root* of $\boldsymbol{T}$. For $u, v \in T$, we say that u *lies below* v if the path from u to r passes through v. We say that u *lies strictly below* v if u lies below v and $u \neq v$. The *youngest common ancestor (yca)* of two vertices $u, v \in T$ is the node w such that both u and v liew below w and w has maximal distance from r.

Rooted-Triple-Consistency
INSTANCE: A finite relational structure (V, R^{yca}), where R^{yca} is a ternary relation.
QUESTION: Is there a rooted tee $\boldsymbol{T}$ with leaves X and a mapping $\alpha : V \to X$ such that for every triple $(x, y, z) \in R^{yca}$ the yca of $\alpha(x)$ and $\alpha(y)$ lies strictly below the yca of $\alpha(x)$ and $\alpha(z)$ in $\boldsymbol{T}$?

The rooted-triple-consistency problem can be formulated as CSP(Γ) for an ω-categorical template Γ. To define Γ, we first consider the following structure Δ with domain $\mathbb{N} \to \mathbb{Q}$, i.e., the set of all functions from the natural numbers to the rational numbers (hence, Δ is uncountable). For two elements f, g of Δ, let $k_{f,g}$ be the largest natural number such that $f(i) = g(i)$ for all $i < k_{f,g}$. The ternary relation $fg|h$ in Δ holds on elements f, g, h of Δ if they are pairwise distinct and either $k_{f,g} > k_{f,h}$ or ($k_{f,g} = k_{f,h}$ and $f(k_{f,g}) < h(k_{f,h})$). It is known that the first-order theory of Δ is ω-categorical [30]. It follows from the theorem of Löwenheim-Skolem (see [42]) that the first-order theory of Δ also has a countable model Γ, and it is straightforward to verify that CSP(Γ) is the rooted triple consistency problem.

5.6 Reasoning over Branching-Time

Another important model in temporal reasoning is *branching time*, where for every time point the past is linearly ordered, but the future is only partially ordered.

Branching-Time-Consistency
INSTANCE: A finite relational structure $I = (V, \leq, \|, \not\equiv)$ where $\leq$, $\|$, and $\not\equiv$ are binary relations
QUESTION: Is there a rooted tree $\boldsymbol{T}$ and a mapping $\alpha : V \to T$ such that in $\boldsymbol{T}$ the following is satisfied:
a) If $x \leq y$, then $\alpha(x)$ lies above $\alpha(y)$;
b) If $x \| y$, then neither $\alpha(x)$ lies strictly above $\alpha(y)$ nor $\alpha(y)$ lies strictly above $\alpha(x)$;
c) If $x \not\equiv y$, then $\alpha(x) \neq \alpha(y)$.

This problem can be formulated as CSP(Γ) for an ω-categorical structure $(D, \leq, \|, \not\equiv)$, which has been studied intensively in the theory of infinite permutation groups [30, 25]. The reduct $(D, \leq)$ is a partial order with the property mentioned above: for all $x \in D$, the set $\{y \mid y \leq x\}$ is linearly ordered.

The first polynomial-time algorithm for this problem is due to Hirsch [41], and has a worst-case running time in $O(n^5)$. This was later improved by Broxvall and Jonsson [19], who presented an algorithm running in $O(n^{3.376})$ (this algorithm uses an $O(n^{2.376})$ algorithm for fast integer matrix multiplication). A simpler algorithm which does not use fast matrix multiplication and runs in $O(nm)$ can be found in [15]. It is an easy exercise to efficiently reduce the rooted-triple-consisteny problem to the branching-time-consistency problem.

5.7 CSPs without an ω-Categorical Template

Of course, there are constraint satisfaction problems that can not be formulated with an ω-categorical template. Since it might not be obvious whether a CSP has this property, we demonstrate with a simple example how one can show that a CSP can *not* be formulated with an ω-categorical template.

Consider the CSP for the relational structure Γ with domain $\mathbb{Q}$ and the three relations $\{(x, y, z) \mid x + y = z\}$, $\neq$, $\{0\}$, and $\{1\}$. To specify constraints for instances of CSP(Γ) we write $x + y = z$, $x = 0$, and $x = 1$, to keep expressions simple.

Proposition 1. *CSP(Γ) cannot be formulated as CSP(Γ') for an ω-categorical template Γ'.*

Proof. Suppose for contradiction that there is such an ω-categorical template Δ. We construct an infinite sequence $p_0, p_1, p_2 \ldots$ of elements from pairwise distinct orbits of the automorphism group of Δ, which will be a contradiction by Theorem 1. Let z_0 and z_1 be elements in Δ such that $z_0 = 0$ and $z_1 = 1$ hold in Δ. Note that if there was an element z_0' distinct from z_0 such that $z_0' = 0$ holds in Δ, then $x = 0 \wedge x \neq y \wedge y = 0$ would be a satisfiable instance of CSP(Δ), but unsatisfiable in CSP(Γ), which is impossible by assumption. Hence, z_0 is uniquely defined. Let p_0 be z_0. Assume inductively that p_n is defined uniquely in Γ by a primitive positive formula $\phi(x)$. Consider the following instance of CSP(Γ), which is specified by the primitive positive formula ψ:

$$\phi(s) \wedge (s + x = y) \wedge (y + x = l) \wedge l = 1 .$$

Clearly, if we set s to p_n, x to $(1 - p_n)/2$, and y to $(1 + p_n)/2$, all conjuncts of ψ are satisfied in Γ. By assumption, ψ is also satisfiable over Δ, i.e., there is an assignment $h : \{s, x, y, l\} \to D(\Delta)$ to the free variables that satisfies all conjuncts in ψ. Let p_{n+1} be $h(y)$. We claim that p_{n+1} is uniquely determined: if there was solution with another value for y, then $\psi(s, x, y, l) \wedge \psi(s, x', y', l) \wedge x' \neq x$ would be satisfiable over Δ, but not over Γ, a contradiction. We also claim that p_{n+1} lies in an orbit that is distinct from the orbits of $p_1, \ldots, p_n$, respectively: otherwise, if there was an automorphism α of Γ that maps p_i to p_{n+1}, for some $i \leq n$, then p_{n+1} would not have been unique, since $\alpha(h)$ is an assignment that also satisfies all conjuncts but assigns a different value to y. □

6 CSPs and Existential Second-Order Logic

One of the motivating questions in the landmark paper of Feder and Vardi [35] is

Question 1. Which natural subclasses of NP do exhibit a *complexity dichotomy*, i.e., only contain problems that are either NP-complete or in P?

Whether the class FCSP of finite domain constraint satisfaction problems has such a complexity dichotomy is one of the greatest open problems in constraint satisfaction theory.

Feder and Vardi approached Question 1 from two sides. One the one hand, they conjectured that the class FCSP exhibits a dichotomy and made important contributions to this end. On the other hand, they investigated larger classes of computational problems that provably do not exhibit a complexity dichotomy. These larger classes are defined with syntactic restrictions of existential second order logic (ESO).

By Fagin's famous theorem, a set $\mathcal{C}$ of finite τ-structures is in NP if and only if there is a sentence Φ in existential second-order logic such that Φ is true on a τ-structure $\boldsymbol{A}$ if and only if $\boldsymbol{A}$ is in $\mathcal{C}$. An important syntactic restriction of existential second-order logic is SNP (for *strict NP*): here we require that the sentence is of the form

$$\exists R_1, \ldots, R_k . \forall x_1, \ldots, x_l . \, \psi$$

where $R_1, \ldots, R_k$ are (existentially quantified) relation symbols, $x_1, \ldots, x_l$ are (universally quantified) first-order variables, and ψ is a quantifier-free formula. Furthermore, we say that an SNP-formula is *monadic* if all second-order variables are unary. Let MSNP be the class of all problems that can be described by a monadic SNP sentence.

Theorem 4 (of [35]). *Every problem in NP has a polynomial-time equivalent problem in MSNP.*

Theorem 4 shows that there is no complexity dichotomy for all problems in MSNP, because if $P \neq NP$ there are problems in NP that are neither in P nor NP-complete [52].

However, Feder and Vardi also showed that if we restrict our attention to (infinite domain) constraint satisfaction problems in MSNP, we obtain a class that is 'computationally equivalent' to FCSP. Let CSP be the class of all constraint satisfaction problems (with domains of arbitrary cardinality).

Theorem 5 (of [35], [50]). *Every problem in CSP $\cap$ MSNP is polynomial-time equivalent to an FCSP (and FCSP is contained in CSP $\cap$ MSNP).*

Hence, a complexity dichotomy for FCSP is equivalent to a complexity dichotomy for CSP $\cap$ MSNP. The dichotomy conjecture for FCSP is wide open, and not undisputed. Instead of disproving the dichotomy conjecture for FCSP, it might be easier to disprove it for CSPs in SNP that can be formulated with an ω-categorical template.

Question 2. Is every problem in NP polynomial-time equivalent to a CSP with an ω-categorical template?

The logic of SNP and MSNP are natural logics in the context of constraint satisfaction with ω-categorical templates, as witnessed by the following theorem.

Theorem 6 (of [11]). *Every CSP in MSNP can be formulated as CSP(Γ) for an ω-categorical structure Γ.*

So far, we have not touched the question how infinite templates Γ might be finitely represented. For the definition of CSP(Γ) we don't have to fix a representation. But for several algorithmic "meta-questions" concerning CSP(Γ) such a representation is necessary.

Most natural CSPs with ω-categorical templates are in SNP. Thus, SNP sentences can be used to specify constraint satisfaction problems for ω-categorical templates (up to homomorphic equivalence, see Lemma 5). However, the following questions remain open.

Question 3. Is every CSP in SNP a CSP over an ω-categorical template?

7 Preservation Theorems

Primitive positive definability plays an important role in hardness proofs for the CSP. In this section, we discuss a model-theoretic preservation theorem that characterizes primitive positive definability. We then use this result to establish that the computational complexity of a CSP with a finite or ω-categorical template is fully described by an algebra that can be associated to a template.

Let $\boldsymbol{A}$ and $\boldsymbol{B}$ be two structures with the same signature τ. Then the *(direct) product* $\boldsymbol{C} = \boldsymbol{A} \times \boldsymbol{B}$ of $\boldsymbol{A}$ and $\boldsymbol{B}$ is the τ-structure with domain $A \times B$, and for each k-ary $R \in \tau$ the structure $\boldsymbol{C}$ has the relation that contains a tuple $((a_1, b_1), \ldots, (a_k, b_k))$ if and only if $R(a_1, \ldots, a_k)$ holds in $\boldsymbol{A}$ and $R(b_1, \ldots, b_k)$ holds in $\boldsymbol{B}$. For each k-ary $f \in \tau$ the structure $\boldsymbol{C}$ has the operation that maps $((a_1, b_1), \ldots, (a_k, b_k))$ to $(f(a_1, \ldots, a_k), f(b_1, \ldots, b_k))$. We write $\boldsymbol{A}^k$ for $\boldsymbol{A} \times \boldsymbol{A}^{k-1}$ if $k > 1$, and $\boldsymbol{A}^1$ for $\boldsymbol{A}$.

Definition 7. *Let* $\mathbf{A}$ *be a structure. Then* $f : A^k \to A$ *is called a* polymorphism *of* $\mathbf{A}$ *if* f *is a homomorphism from* $\mathbf{A}^k$ *to* $\mathbf{A}$.

If R is a relation over D, and f is a polymorphism of (D, R), we also say that f *preserves* R. A relation R has a primitive positive definition in a *finite* structure if and only if R is preserved by all polymorphisms of this structure. This was discovered by Geiger [38] and by Bodnarcuk et al. [18], and is of central importance in universal algebra.

The following generalization of this theorem to ω-categorical structures was shown in [17]. We give a new proof here, which derives the theorem from the well-known homomorphism preservation theorem in model-theory. The proof given in [17] is neither long nor very complicated; however, we believe that the proof given below is also interesting, because it demonstrates how universal-algebraic preservation theorems can be derived from classical preservation theorems in model theory.

Theorem 7 (of [17]). *Let* $\mathbf{A}$ *be an* ω*-categorical or a finite structure. A relation* R *has a primitive-positive definition in* $\mathbf{A}$ *if and only if* R *is preserved by all polymorphisms of* $\mathbf{A}$.

The corresponding preservation theorem in model theory is as follows (even though we have not been able to find a reference for this fact, it should be considered to be known). We say that a first-order formula ϕ *is equivalent to* ψ *modulo* a first-order theory T if $\forall \overline{x}. \phi(\overline{x}) \leftrightarrow \psi(\overline{x})$ holds in all models of T. We also say that $\phi(x_1, \ldots, x_l)$ is *preserved* by a homomorphism f from a direct product of models $\Gamma_1, \ldots, \Gamma_k$ of T to a model Γ of T if whenever ϕ holds on l-tuples $a_1, \ldots, a_l$ in $\Gamma_1, \ldots, \Gamma_k$, respectively, then ϕ also holds on $f((a_1^1, \ldots, a_l^1), \ldots, (a_1^k, \ldots, a_l^k))$ in Γ.

Theorem 8. *Let* T *be a first-order theory. A first-order formula* ϕ *is equivalent to a primitive positive formula modulo* T *if and only if* ϕ *is preserved by all homomorphisms from finite direct products of models of* T *to models of* T.

The theorem can be shown by a slight modification of the proofs in [46]. This approach relies on several concepts and basic results in model-theory, and is out of the scope of this paper. We recommend Hodges' book [42] as an accessible introduction to model-theoretic preservation theorems.

Proof (of Theorem 7). First, observe that every relation R with a primitive positive definition in Γ is preserved by all polymorphisms of Γ. This can be shown in a straightforward way by induction on the syntactic structure of primitive positive definitions. To prove the converse, let R be a k-ary relation that is preserved by all polymorphisms of Γ. In particular, R is preserved by all automorphisms of Γ, and Theorem 1 implies that R has a first-order definition ϕ in Γ.

Let T be the first-order theory of Γ. We claim that if R is preserved by all polymorphisms of Γ then ϕ is preserved by all homomorphisms from finite direct products of models $\Gamma_1, \ldots, \Gamma_l$ of T to a model Γ_0 of T. If we have shown the claim, Theorem 7 follows directly from Theorem 8.

So assume to the contrary that ϕ is not preserved by a homomorphism f from a finite direct product of models $\Gamma_1, \ldots, \Gamma_l$ of T to a model Γ_0 of T (note that $\Gamma_1, \ldots, \Gamma_l$, and Γ_0 might be uncountable). We prove by a standard application of the downward Löwenheim-Skolem theorem (see e.g. [42]) that ϕ is not preserved by some l-ary polymorphism of Γ, as follows. Let Δ be the disjoint union of $\Gamma_0, \Gamma_1, \ldots, \Gamma_l$. Additionally, Δ has a relation S^Δ that denotes the union of the relations defined by ϕ in $\Gamma_0, \Gamma_1, \ldots, \Gamma_l$, and has unary relations $P_0^\Delta, P_1^\Delta, \ldots, P_l^\Delta$, where P_i^Δ denotes the elements from Γ_i, for $1 \leq i \leq l$. Finally, Δ has an l-ary function g^Δ, which is interpreted as follows. If $v_1, \ldots, v_l$ are elements from Δ, where v_i is from Γ_i for $1 \leq i \leq l$, then $g^\Delta(v_1, \ldots, v_l)$ is the value of the homomorphism f from $\Gamma_1 \times \cdots \times \Gamma_l$ to Γ_0. Because ϕ is not preserved by f, there are k-tuples $t_1, \ldots, t_l$ from S^Δ such that $(f(t_1), \ldots, f(t_l))$ is not in S^Δ. For all other l-tuples from Δ, the interpretation of g^Δ is set arbitrarily to some fixed element from Δ. By downward Löwenheim-Skolem, the first-order theory of Δ has a countable model Δ'. Note that $P_i^{\Delta'}$ denotes a countable, but still necessarily an infinite set. It is also easy to see that the structures $\Gamma_1', \ldots, \Gamma_k'$ induced by the sets $P_1^{\Delta'}, \ldots, P_k^{\Delta'}$ in Δ' have the same first-order theory as Γ, and by ω-categoricity of Γ are isomorphic to Γ. The function $g^{\Delta'}$ maps tuples $v_1, \ldots, v_k$ where $v_i \in \Gamma_i'$ for $1 \leq i \leq k$ to Γ_0', and still the restriction of $g^{\Delta'}$ to those tuples violates the relation $S^{\Delta'}$. Hence, the restriction of $g^{\Delta'}$ to these tuples gives rise to a polymorphism of Γ that violates ϕ. □

8 The Algebra of a Template

In this section we discuss how the algebraic approach to constraint satisfaction might be extended to ω-categorical templates Γ. We have already seen in the previous section that the polymorphisms of Γ characterize primitive positive definability of relations in Γ. Therefore, the algebra defined on the domain D of Γ that contains as operations all the polymorphisms of Γ fully describes the computational complexity of CSP(Γ). The algebra that we just defined for an ω-categorical template Γ will be denoted by $\mathbb{A}_\Gamma$.

$\mathbb{A}_\Gamma$ has several important properties. First of all, its set of operations forms a *clone*, i.e., is closed under compositions and contains the projections [59]. Clones with infinite domains are a subject in its own right [39]. Clones that arise as the set of polymorphisms of a structure are always *locally closed*, a concept which only becomes relevant for infinite domains. We need a couple of definitions.

Let D be an infinite set, and let $\mathcal{O}^{(k)}$ be the set of operations from D^k to D, for $k \geq 1$. The symbol O denotes $\bigcup_{k=1}^{\infty} \mathcal{O}^{(k)}$. We say that an operation $f \in \mathcal{O}^{(k)}$ is *interpolated* by a set $\mathcal{F} \subseteq \mathcal{O}$ if for every finite subset A of D there is some operation $g \in \mathcal{F}$ such that $f(t) = g(t)$ for every $t \in A^k$. The set of all operations that are interpolated by $\mathcal{F}$ is denoted by $I(\mathcal{F})$. A clone is called a *local clone* (or *locally closed*) if $I(\mathcal{F}) = \mathcal{F}$. The smallest clone that contains $\mathcal{F}$ is the clone *generated* by $\mathcal{F}$, and denoted by $G(\mathcal{F})$. The smallest *local clone* that contains $\mathcal{F}$ is called the clone *locally generated* by F, and denoted by $L(\mathcal{F})$.

Proposition 2 (see e.g. [59]). *For all $\mathcal{F} \subseteq \mathcal{O}$ we have that $L(\mathcal{F}) = I(G(\mathcal{F}))$.*

We say that an algebra $\mathbb{A}$ is called *oligomorphic* if the unary bijective operations (i.e., the permutations) in $\mathbb{A}$ form an oligomorphic permutation group. Theorem 1 asserts that $\mathbb{A}_\Gamma$ is oligomorphic if Γ is ω-categorical. Conversely, it is not hard to see that for every oligomorphic algebra $\mathbb{A}$ whose operations form a local clone there is an ω-categorical structure Γ such that $\mathbb{A} = \mathbb{A}_\Gamma$. Such an ω-categorical structure Γ can be obtained by equipping the domain of $\mathbb{A}$ with all relations that are preserved by all operations in $\mathbb{A}$; we call this structure the *canonical structure* for $\mathbb{A}$.

An operation of an oligomorphic clone is called *elementary* if it is locally generated by its permutations. Clearly, for clones with a finite domain, the elementary operations are the operations that are composed of a projection with a permutation. Note that all endomorphisms of a model-complete ω-categorical core are elementary. Conversely, if all endomorphisms of an oligomorphic algebra $\mathbb{A}$ are elementary, then the canonical structure for $\mathbb{A}$ is a core [8].

We now define several other important properties of k-ary operations. A k-ary operation f is

- *idempotent* iff $f(x, \ldots, x) = x$;
- *essentially unary* iff there is a unary operation g such that $f(x_1, \ldots, x_k) = g(x_i)$ for some $i \in \{1, \ldots, k\}$;
- *essential* iff f is not essentially unary;
- *commutative* iff f is binary and $f(x, y) = f(y, x)$;
- a *quasi near-unanimity operation* (short, *qnu-operation*) iff $f(x, \ldots, x) = f(x, \ldots, x, y) = \cdots = f(x, \ldots, x, y, x, \ldots, x) = \cdots = f(y, x, \ldots, x)$;
- a *quasi Maltsev operation* iff $k = 3$ and $f(x, y, y) = f(y, y, x) = f(x, x, x)$;

An idempotent quasi near-unanimity and quasi Maltsev operation is known as near-unanimity and Maltsev operation, respectively.

9 The Pseudo-variety of a Template

A class $\mathcal{V}$ of algebras with the same signature τ is called a *pseudo-variety* if $\mathcal{V}$ contains all homomorphic images (Section 2), subalgebras (i.e., substructures as defined in Section 3), and finite direct products (Section 7) of algebras in $\mathcal{V}$. The difference between pseudo-varieties and varieties is that pseudo-varieties need not be closed under *direct products of arbitrary cardinality* (which we did not define here). The smallest pseudo-variety that contains an algebra $\mathbb{A}$ is called the pseudo-variety *generated* by $\mathbb{A}$.

The results in this section link the universal-algebraic concept of pseudo-varieties with the model-theoretic concept of primitive positive interpretations. We already mentioned that first-order interpretations are a convenient tool to construct ω-categorical structures from simpler ω-categorical structures. *Primitive positive* interpretations (see Section 3) can be used to study the computational complexity of constraint satisfaction problems.

Proposition 3. *Let Δ and Γ be a structures with finite relational signatures. If there is a primitive positive interpretation of Δ in Γ, then there is a polynomial-time reduction from CSP(Δ) to CSP(Γ).*

Proof. Let d be the dimension of the primitive positive interpretation of the τ-structure Δ in the σ-structure Γ, let $\delta(x_1, \ldots, x_d)$ be the domain formula, let $h : \delta(\Gamma^d) \to D(\Delta)$ be the coordinate map, and let $\phi_R(x_1, \ldots, x_{dk})$ be the formula for the k-ary relation R from Δ.

Let $\boldsymbol{A}$ be an instance of CSP(Δ), and let $A = \{a_1, \ldots, a_n\}$ be the elements of $\boldsymbol{A}$. We construct an instance $\boldsymbol{B}$ of CSP(Γ) as follows. The domain B of $\boldsymbol{B}$ consists of dn vertices $b_1^1, \ldots, b_n^d$. For all $1 \leq i \leq n$, we *impose* δ in $\boldsymbol{B}$ on $(b_i^1, \ldots, b_i^d)$: note that δ is a primitive positive σ-formula, and therefore can be simulated by a conjunction of constraints from τ, possibly with adding new vertices for the existentially quantified variables in the definition. If a tuple $(a_{i_1}, \ldots, a_{i_k})$ is contained in a k-ary relation R in $\boldsymbol{A}$, then we impose $\phi_R(b_{i_1}^1, \ldots, b_{i_1}^d, \ldots, b_{i_k}^1, \ldots, b_{i_k}^d)$ in $\boldsymbol{B}$. Clearly, the structure $\boldsymbol{B}$ is an instance of CSP(Γ) and can be constructed in polynomial time in the size of $\boldsymbol{A}$.

We claim that there is a homomorphism from $\boldsymbol{A}$ to Δ if and only if there is a homomorphism from $\boldsymbol{B}$ to Γ. Suppose $f : B \to D(\Gamma)$ is a homomorphism from $\boldsymbol{B}$ to Γ. By construction, if $R(a_{i_1}, \ldots, a_{i_k})$ holds in $\boldsymbol{A}$ then $\phi_R((f(b_{i_1}^1), \ldots, f(b_{i_1}^d)), \ldots, (f(b_{i_1}^1), \ldots, f(b_{i_1}^d)))$ holds in Γ. By the definition of interpretations, this is the case if and only if $R(h(f(b_{i_1}^1), \ldots, f(b_{i_1}^d)), \ldots, h(f(b_{i_k}^1), \ldots, f(b_{i_k}^d)))$ holds in Δ. Hence, the mapping $g : A \to D(\Delta)$ that maps a_i to $h(f(b_i^1), \ldots, f(b_i^d))$ is a homomorphism from $\boldsymbol{A}$ to Δ.

Now, suppose that f is a homomorphism from $\boldsymbol{A}$ to Δ. Since h is a surjective mapping from $\delta(\Gamma)^d$ to Δ, there are elements $e_i^1, \ldots, e_i^d$ in Γ such that $h(e_i^1, \ldots, e_i^d) = f(a_i)$, for all $i \in \{1, \ldots, n\}$. We claim that the mapping $g : B \to D(\Gamma)$ that maps b_i^j to e_i^j is a homomorphism from $\boldsymbol{B}$ to Γ. By construction, any constraint in $\boldsymbol{B}$ comes from a primitive positive formula $\phi_R(b_{i_1}^1, \ldots, b_{i_1}^d, \ldots, b_{i_k}^1, \ldots, b_{i_k}^d)$ that was introduced for a constraint $R(a_{i_1}, \ldots, a_{i_k})$ in $\boldsymbol{A}$. It therefore suffices to show that $\phi_R(g(b_{i_1}^1), \ldots, g(b_{i_1}^d), \ldots, g(b_{i_k}^1), \ldots, g(b_{i_k}^d))$ holds in Γ. Since f is a homomorphism from $\boldsymbol{A}$ to Δ, $R(f(a_{i_1}), \ldots, f(a_{i_k}))$ holds in Δ. By the choice of $e_1^1, \ldots, e_n^d$, this shows that $R(h(e_{i_1}^1, \ldots, e_{i_1}^d), \ldots, h(e_{i_k}^1, \ldots, e_{i_k}^d))$ holds in Γ. By the definition of interpretations, this is the case if and only if $\phi_R(e_{i_1}^1, \ldots, e_1^d, \ldots, e_{i_k}^1, \ldots, e_{i_k}^d)$ holds in Γ, which is what we had to show. □

Section 11 contains a hardness proof that uses this proposition. We now present the mentioned connection between primitive positive interpretations and pseudo-varieties. We have not been able to find the following theorem explicitely in the literature even in the case of finite algebras, and therefore present its proof in full detail.

Theorem 9. *Let Γ be a finite or ω-categorical relational structure. Then a structure Δ has a primitive positive interpretation in Γ if and only if there is an algebra $\mathbb{B}$ in the pseudo-variety generated by $\mathbb{A}_\Gamma$ such that all operations of $\mathbb{B}$ are polymorphisms of Δ.*

Proof. Let $\mathcal{V}$ be the pseudo-variety generated by $\mathbb{A}_\Gamma$. Similarly to the famous HSP theorem for varieties, every algebra in $\mathcal{V}$ is the homomorphic image of a subalgebra of a finite direct product of $\mathbb{A}_\Gamma$ (we can use the same proof as for the HSP theorem given in [24]).

First assume that there is an algebra $\mathbb{B}$ in $\mathcal{V}$ all of whose operations are polymorphisms of Δ. Then there exists a finite number $d \geq 1$, a subalgebra $\mathbb{C}$ of $(\mathbb{A}_\Gamma)^d$, and a surjective homomorphism h from $\mathbb{C}$ to $\mathbb{B}$.

We claim that Δ has a first-order interpretation with dimension d in Γ. All operations of $\mathbb{A}_\Gamma$ preserve C (viewed as a d-ary relation over $D(\Gamma)$), since $\mathbb{C}$ is a subalgebra of $(\mathbb{A}_\Gamma)^d$. By Theorem 7, this implies that C has a primitive positive definition $\delta(x_1, \dots, x_d)$ in Γ, which becomes the domain formula of our interpretation. As coordinate map we choose the mapping h.

If R is a k-ary relation in Δ, let R' the dk-ary relation over $D(\Gamma)$ that contains the dk-tuple $(a_1^1, \dots, a_1^d, \dots, a_k^1, \dots, a_k^d)$ whenever R contains the tuple $(h(a_1^1, \dots, a_1^d), \dots, h(a_k^1, \dots, a_k^d))$. Let f be any operation in $\mathbb{A}_\Gamma$. By assumption, the corresponding operation f in $\mathbb{B}$ preserves R. It is easy to verify that then the operation f in $\mathbb{A}_\Gamma$ preserves R'. Hence, all polymorphisms of Γ preserve R', and because Γ is ω-categorical, the relation R' has a primitive positive definition ϕ_R in Γ. The formula ϕ_R becomes the defining formula for R in the interpretation of Δ in Γ.

Finally, since h is an algebra homomorphismus, the kernel[2] K of h is a congruence[3] of $\mathbb{C}$. In other words, if $((a_1^1, \dots, a_2^d), (a_2^1, \dots, a_2^d)) \in K$ then $(f(a_1^1, \dots, a_2^d), f(a_2^1, \dots, a_2^d)) \in K$ for all operations f from $\mathbb{C}$, and hence $((f(a_1^1), \dots, f(a_2^d)), (f(a_2^1), \dots, f(a_2^d))) \in K$ for all operations f from $\mathbb{A}_\Gamma$. It follows that K, viewed as a $2d$-ary relation over $D(\Gamma)$, is preserved by all operations from $\mathbb{A}_\Gamma$. Theorem 7 implies that K has a primitive positive defintion in Γ. This definition becomes the formula $\phi_=$ in the interpretation of Δ in Γ. It is straightforward to verify that we have found a primitive positive interpretation of Δ in Γ.

To prove the opposite direction, suppose that Δ has a primitive positive interpretation in an ω-categorical τ-structure Γ. We have to show that $\mathcal{V}$ contains

---

[2] The *kernel* of a map $f : A \to B$ is the equivalence relation on A that contains all pairs (x, y) such that $f(x) = f(y)$.

[3] A *congruence* of an algebra $\mathbb{A}$ is an equivalence relation that is preserved by all operations in $\mathbb{A}$; equivalently, a congruence is the kernel of a homomorphic image of $\mathbb{A}$.

an algebra $\mathbb{B}$ such that all operations in $\mathbb{B}$ are polymorphisms of Δ. Let d be the dimension and δ be the domain formula of the interpretation. Clearly, the set $\delta(\Gamma^d)$ is preserved by all operations in $\mathbb{A}_\Gamma$, and therefore induces a subalgebra $\mathbb{C}$ of $(\mathbb{A}_\Gamma)^d$.

We first show that the kernel of the coordinate map h of the interpretation is a congruence ρ of $\mathbb{C}$. For all d-tuples $\overline{a}, \overline{b} \in C$ the $2d$-tuple $(\overline{a}, \overline{b})$ satisfies $\phi_=$ in Γ if and only if $h(\overline{a}) = h(\overline{b})$. Hence, the restriction of $\phi_=$ to C defines the kernel of h. Since $\phi_=$ is primitive positive definable in Γ, it is preserved by all polymorphisms of Γ, and therefore is a congruence of $(\mathbb{A}_\Gamma)^d$. By the second isomorphism theorem in universal algebra [24], the restriction of this congruence to C is a congruence ρ of $\mathbb{C}$.

Also by basic universal algebra [24] the *natural map* g that maps an element c from $\mathbb{C}$ to its congruence class c/ρ in the quotient algebra $\mathbb{C}/\rho$ is a surjective homomorphism from $\mathbb{C}$ to $\mathbb{C}/\rho$. By the first isomorphism theorem, the elements of $\mathbb{B} := \mathbb{C}/\rho$ are in bijective correspondence to the elements of Δ, and we assume without loss of generality that $\mathbb{B}$ and Δ have the same domain. We finally verify that every operation in $\mathbb{B}$ is a polymorphism of Δ: It suffices to prove that every relation R of Δ is preserved by all operations f in $\mathbb{B}$. Let ϕ_R be the defining τ-formula of R in the primitive positive interpretation of Δ in Γ. The operation f in $\mathbb{B}$ preserves R if and only if the operation f in $\mathbb{A}_\Gamma$ preserves the relation defined by ϕ_R. Since the operations in $\mathbb{A}_\Gamma$ are polymorphisms of Γ, and since ϕ_R is a primitive positive τ-formula, the operation f preserves the relation defined by ϕ_R. □

Theorem 10. *Let Γ be ω-categorical. Then CSP(Γ) is NP-hard if there is an expansion Δ of the model-complete core of Γ by finitely many singleton relations such that $\mathcal{V}(\mathbb{A}_\Delta)$ contains an 2-element algebra where all operations are projections.*

Proof. By Corollary 1 there is a polynomial-time reduction from CSP(Δ) to CSP(Γ). Combining Theorem 9 with Proposition 3, we can show that CSP(Δ) is NP-hard, by reduction from the well-known NP-complete problem positive 1-in-3-3SAT, which can be formulated as CSP($\{0,1\}, \{(1,0,0),(0,1,0),(0,0,1)\}$). □

All known ω-categorical templates with an NP-hard CSP satisfy the condition in Theorem 10. It is therefore natural to ask whether the sufficient condition for hardness given in Theorem 10 is also necessary. For finite templates, it has been conjectured that this is true [21].

Question 4. Does Theorem 10 describe exactly the NP-hard CSPs with an ω-categorical template?

10 Tractable Templates

For finite templates Γ, all known tractable problems CSP(Γ) can be solved by a Datalog program [35] or by an algorithm for templates with a Maltsev polymorphism [23, 43] (generalizing group-theoretic algorithms introduced in [35]), or combinations of those [20].

An ω-categorical model-complete core Γ can not have a Maltsev polymorphism, not even a quasi Maltsev polymorphism [25]. This does not imply that group-theoretic algorithms do not occur for ω-categorical templates, as we will see in Subsection 10.2.

Datalog continues to play an important role also for ω-categorical templates. All tractability results known for temporal and spatial reasoning can be formulated with Datalog programs. We will discuss in this section a result showing that every ω-categorical model-complete core with a *quasi near-unanimity operation* can be solved by a Datalog program. In fact, in this case we can use the Datalog program to compute a *globally consistent* instance. We also present examples of templates in temporal and spatial reasoning that have a quasi near-unanimity polymorphism.

For a large family of CSPs we can show tractability because the template has a certain binary bijective polymorphism. The powerful tractability criterion explains several results in the literature on temporal and spatial reasoning and will be introduced in Subsection 10.2.

We finally present an ω-categorical template whose CSP can be solved by a simple linear-time algorithm, but which can neither be solved by Datalog nor by algebraic algorithms related to Maltsev polymorphisms.

10.1 From Local to Global Consistency

An introduction to Datalog for constraint satisfaction with finite templates can be found in [Cite Bulatov,Krokhin,Larose in this book]. Datalog programs can also be used to solve CSPs with infinite templates in polynomial time: recall that every Datalog program derives on a given finite input structure only a polynomially bounded number of facts. Thus, any Datalog program can be evaluated in polynomial time. An example of a CSP with an infinite template that can be solved by a $(2,3)$-Datalog program is $\mathrm{CSP}(\mathbb{Q}, \neq, \leq)$ [60].

Datalog is particularly useful when the template is ω-categorical, since in this case there exist *canonical Datalog programs*, as in the case of finite templates.

Definition 8. *Let Γ be an ω-categorical structure. The* canonical (l,k)-Datalog program *for CSP(Γ) is a Datalog program Π that contains an IDB for every at most l-ary primitive positive definable relation in Γ (by ω-categoricity of Γ, there are only finitely many such relations). The empty 0-ary relation will be denoted by* false. *The EDBs are precisely the relation symbols in τ. The program Π contains a rule ψ :- $\psi_1, \ldots, \psi_j$ if the implication $\psi_1 \wedge \cdots \wedge \psi_j \Rightarrow \psi$ contains at most k variables, is valid in Γ, and ψ is of the form $R(y_1, \ldots, y_s)$ for an IDB R and $s \leq l$.*

The final stage in the evaluation of a canonical Datalog program on a given instance gives rise to another instance $\boldsymbol{A}$ of $\mathrm{CSP}(\Gamma')$, where Γ' is the expansion of Γ by all at most l-ary primitive positive definable relations, and where $\boldsymbol{A}$ contains all the derived tuples from these relations [32].

Proposition 4 (of [11]). *A constraint satisfaction problem CSP(Γ) with an ω-categorical template Γ can be solved with an (l,k)-Datalog program if and only if the canonical (l,k)-Datalog program solves CSP(Γ).*

Global consistency. Some templates Γ have the strong property that the instance computed by the canonical Datalog program is always *globally consistent*, see Definition 9 below.

Definition 9. *Let Γ be a structure with finite relational signature. An instance* $\mathbf{A}$ *of CSP(Γ) is called* strongly k-consistent *if for every subset $S = \{v_1, \ldots, v_l\}$ of A with $l \leq k$ and every homomorphism h from $\mathbf{A}[\{v_1, \ldots, v_{l-1}\}]$ to Γ there exists an extension of h that is a homomorphism from $\mathbf{A}[S]$ to Γ. An instance $\mathbf{A}$ of CSP(Γ) is called* globally consistent *if it is k-consistent for all $1 \leq k \leq |A|$.*

Note that if Γ has the property that every strongly k-consistent instance of CSP(Γ) is globally consistent, then CSP(Γ) can be solved in polynomial time. As an example, consider again CSP($\mathbb{Q}, \leq, \neq$). It is known that the canonical $(4,5)$-Datalog program computes globally consistent instances, but the canonical $(3,4)$-Datalog program does not [48]. We will see a universal-algebraic explanation of this fact in the next paragraph.

Quasi near-unanimity functions. We present a universal-algebraic characterization of those ω-categorical model-complete cores where the canonical (l,k)-Datalog program computes globally consistent instances. Recall the definition of *(quasi) near-unanimity functions* given in Section 8. As an example of a near-unanimity function, consider the structure $(\mathbb{Q}, \leq, <)$, and the operation *median*, which is the ternary function that returns the median of its three arguments. More precisely, for three elements x, y, z from $\mathbb{Q}$, suppose that $\{x, y, z\} = \{a, b, c\}$, where $a \leq b \leq c$. Then *median*(x, y, z) is defined to have value b. It is easy to verify that *median* is a ternary near-unanimity function, and that it is a polymorphism of $(\mathbb{Q}, \leq, <)$.

The structure $(\mathbb{N}, \neq)$ is an example of a structure that has no near-unanimity polymorphism (exercise!), but has a ternary quasi near-unanimity operation, defined as follows. Consider the structure $(\mathbb{N}, \neq)^3$, where for all $x, y \in \mathbb{N}$ the triples of the form (y, x, x), (x, y, x), (x, x, y), and (x, x, x) are identified. The resulting structure homomorphically maps to $(\mathbb{N}, \neq)$ (Lemma 2), and this homomorphism is a ternary polymorphism of $(\mathbb{N}, \neq)$ and a quasi near-unanimity operation.

Similarly, we can construct a 5-ary quasi near-unanimity polymorphism of $(\mathbb{N}, R)$ where $R = \{(x, y, u, v) \mid x \neq y \vee u \neq v\}$). To see that this structure has no 4-ary quasi near-unanimity, consider the tuples $t_1 = (1,0,0,0), t_2 = (0,1,0,0), t_3 = (0,0,1,0)$, and $t_4 = (0,0,0,1)$. Any 4-ary quasi near-unanimity operation applied to (t_1, t_2, t_3, t_4) yields a tuple (c, c, c, c) for some $c \in \mathbb{N}$. But note that $t_1, \ldots, t_4$ are in R, whereas c is not.

Theorem 11 (of [11]). *An ω-categorical model-complete core Γ has a k-ary quasi near-unanimity polymorphism if and only if every strongly k-consistent instance of CSP(Γ) is globally consistent.*

Examples. A 5-ary quasi near-unanimity polymorphism of $(\mathbb{Q}, \leq, \neq)$ was described in [9]. We give a different construction of such a polymorphism here. Let $f : \mathbb{Q}^{10} \rightarrow \mathbb{Q}$ be an injective operation that preserves $\leq$. Such an operation exists; we can simply take any operation such that $f(x_1, \dots, x_{10}) < f(y_1, \dots, y_{10})$ if and only if $(x_1, \dots, x_{10})$ is lexicographically smaller than $(y_1, \dots, y_{10})$.

Proposition 5. *Let $h(x, y, z, u, v)$ be the 5-ary operation*

$$f(median(x, y, z), median(x, y, u), median(y, z, v), ..., \\ median(y, u, v), median(z, u, v)\ )$$

(That is, we apply the median to all of the 10 three-element subsets of $\{x, y, z, u, v\}$.) Then h is a 5-ary quasi near-unanimity polymorphism of $(\mathbb{Q}, \leq, \neq)$.

It can also be shown that the template with the basic relations of the spatial reasoning formalisms RCC-5 (see Section 5) has a 5-ary quasi near-unanimity polymorphism [9].

We would like to remark that the concept of quasi near-unanimity operations is a natural concept also for finite domain constraint satisfaction problem. For example, it allows to formulate the well-known dichotomy of CSP(H) for undirected graphs H as follows: *CSP(H) is in P if and only if H has a quasi near-unanimity operation* (unless P=NP). This is easy to derive from the well-known theorem that CSP(H) is NP-hard if H is not bipartite [40].

10.2 Horn Tractability

In this section we describe a powerful tractability criterion, which in particular explains many results in temporal and spatial reasoning. But also CSPs with a group-theoretic algorithm will be treated here.

We say that a relation R has a *quantifier-free Horn definition* in a τ-structure Γ if R can be defined by a quantifier-free τ-formula in conjunctive normal form in which each clause contains at most one positive literal. If $\boldsymbol{A}$ is a structure, we denote by $\hat{\boldsymbol{A}}$ the expansion[4] of $\boldsymbol{A}$ that also contains the complement for each relation in $\boldsymbol{A}$.

Theorem 12 (of [10]). *Let Γ be an ω-categorical homogeneous structure, and let Δ be a structure with a first-order definition in Γ. If Δ has a polymorphism i which is a strong homomorphism from Γ^2 to Γ, then Δ has a quantifier-free Horn definition in Γ. Moreover, if CSP($\hat{\Gamma}$) is tractable, then CSP(Δ) is tractable as well.*

Note that if i is bijective (in other words, if i is an isomorphism between Γ^2 and Γ) then the binary polymorphism of Δ satisfies the equation

$$i(x, y) = a(i(y, x))$$

[4] An *expansion* of a structure $\boldsymbol{A}$ is a structure that is obtained from $\boldsymbol{A}$ by adding relations and/or functions to $\boldsymbol{A}$.

for some automorphism a of Δ [10]. In this case, every expansion of Δ by some relation with a first-order definition in Γ that is not equivalent to a quantifier-free Horn formula over Γ has an NP-hard CSP [10]; in this sense, the tractability result is strongest possible. We illustrate the use of this theorem by examples.

Solving Equations over Infinite Vector Spaces. We come back to the application on solving systems of equations over boolean vectors, described in Subsection 5.3. To recall, the template for this CSP had a first-order definition in the infinite-dimensional vector space $\mathbb{V}$ over the 2-element field $\mathbb{F}_2$. It is well-known that $\mathbb{V}^2$, the direct product of $\mathbb{V}$ with itself, is isomorphic to $\mathbb{V}$; similarly, we have that $\boldsymbol{V}^2$ (as defined in Subsection 5.3) is isomorphic to $\boldsymbol{V}$. Let i be the isomorphism. Note that $\hat{\boldsymbol{V}} = \boldsymbol{V}$, since $\boldsymbol{V}$ already contains the complements of all its relations. It was shown in [10] that CSP($\boldsymbol{V}$) can be solved in polynomial time.

Finally, let Δ be any relational structure where all relations have a first-order definition in Γ which is a quantifier-free Horn formula. It is straightforward to verify that such relations are preserved by i. Now we can apply Theorem 12, and obtain the result that CSP(Δ) can be solved in polynomial time as well.

Spatial Reasoning. Theorem 12 can also be applied to study tractable CSPs in spatial reasoning. The countable atomless boolean ring without identity element $\mathbb{B}$, as introduced in Subsection 5.2, is isomorphic to $\mathbb{B}^2$ (because $\mathbb{B}^2$ is again an atomless boolean ring without identity element, and all such boolean rings are isomorphic [1]); let $i : B^2 \rightarrow B$ denote this isomorphism. The constraint satisfaction problem for $(B, \mathrm{DR}, \mathrm{PP}, \neg\mathrm{DR}, \neg\mathrm{PP})$ is in P [58] (in fact, this structure has a 5-ary qnuf, as shown in [9]). Theorem 12 then implies that CSP(Γ) is in P if all relations of Γ have a first-order definition in $(B, \mathrm{DR}, \mathrm{PP})$ and are preserved by i. These are precisely the relations that have a first-order Horn definition in $(B, \mathrm{DR}, \mathrm{PP})$. In particular, we obtain a result by Renz and Nebel [58] who determined a largest tractable fragment of RCC-5. It follows from the result in [44] that this fragment is the unique largest tractable fragment that contains the basic relations DR and PP. However, note that Theorem 12 also applies to relational structures Δ with relations of arbitrary arity.

10.3 Other Tractable Templates

All known tractable problems of the form CSP(Γ) for templates Γ with a finite domain can be solved by Datalog, or by an algebraic algorithm, or combinations and extensions of these approaches. For infinite templates Γ, there are new kinds of algorithms to solve CSP(Γ) in polynomial time. We present a simple example of such an algorithm, and give some details here, because the algorithm is a prototype for several more advanced algorithms [16, 4].

In Section 5, we have already mentioned the problem CSP($\mathbb{Q}, R^{\min}$), which we call the *min-ordering problem*: in this problem we are given a set A of variables, and a set $R^{\min}$ of triples on these variables. We want to find a mapping $\alpha : A \rightarrow \mathbb{Q}$ such that for each triple (x, y, z) either $\alpha(x) > \alpha(y)$ or $\alpha(x) > \alpha(z)$. It can be shown [14] that there is no Datalog program that solves this problem. However, here is a simple linear-time algorithm for CSP($\mathbb{Q}, R^{\min}$).

The algorithm we present relies on the fact that $R^{\min}$ has the binary operation *min* as a polymorphism, where $\min(x, y)$ is by definition the minimum of x and y. The operation min is an example of a *semilatttice operation*, which is an associative, commutative, and idempotent operation. For finite structures it is known that if a template Γ has an semilattice polymorphism, then CSP(Γ) is tractable by a Datalog program. This result does not carry over to ω-categorical templates (we have already mentioned that $(\mathbb{Q}, R^{\min})$ has a semilattice polymorphism, but cannot be solved by a Datalog program).

Let $\boldsymbol{A}$ be an instance of the min-ordering problem that has a solution α. It will be easy to generalize our algorithm to all templates where all relations have a first-order definition in $(\mathbb{Q}, <)$ and are preserved by min. The algorithm was found in the course of the classification of the complexity of CSP(Γ) for all Γ with a first-order definition in $(\mathbb{Q}, <)$, and is one of the simplest cases in this classification [13, 14].

Definition 10. *A set S of variables in an instance $\mathbf{A}$ of the min-ordering problem is called* free *if the instance has a solution α where $\alpha(x) \leq \alpha(y)$ for all $y \in A$ and $x \in S$.*

If x is from a free set of variables S, then $R^{\min}$ cannot contain a triple (x, y, z) where x appears in the first entry, because then either $\alpha(y)$ or $\alpha(z)$ are strictly smaller than $\alpha(x)$, a contradiction. We say that a variable $x \in A$ is *blocked* if $R^{\min}$ contains a tuple (x, y, z) where x appears in the first entry (and *unblocked* otherwise).

Lemma 6. *If all variables in an instance of the min-ordering problem are blocked, then the instance is unsatisfiable.*

Proof. Suppose the instance has a solution, then there must exist a free set of variables. But as we have seen, free variables cannot be blocked. □

It turns out that if an instance of the min-ordering problem is satisfiable, then the set of *all* min-candidates is free. This follows from the correctness of the following algorithm.

Min-Ordering($\boldsymbol{A}$)
Input: A structure $(A, R^{\min})$ with a ternary relation $R^{\min}$.

If $A = \emptyset$ then **accept**
else
  Compute the set S of min-candidates;
  If $S = \emptyset$ then **reject**;
  else Min-Ordering($\boldsymbol{A}[A \setminus S]$)
end if

Fig. 1. The Min-Ordering algorithm

Proposition 6. *The Min-Ordering procedure given in Figure 1 correctly decides the min-ordering problem in linear time in the input size.*

Proof. Let $\boldsymbol{A}$ be an instance of the min-ordering problem. If at some point during the execution of the Min-Ordering procedure on $\boldsymbol{A}$ we recursively apply the min-ordering procedure to a substructure $\boldsymbol{B}$ of $\boldsymbol{A}$ and do not find an unblocked variable, then Lemma 6 implies that $\boldsymbol{B}$ and therefore also $\boldsymbol{A}$ does not have a solution.

Otherwise, it is clear that $\boldsymbol{A}$ has a solution if the set of variables is empty. So, suppose the algorithm finds a non-empty set S. Inductively assume that $\boldsymbol{A}[A \setminus S]$ has a solution α. Then the extension α' of α that maps all variables in S to a value smaller than the value of $\alpha(y)$ for all other variables $y \in A \setminus S$ is clearly a solution to $\boldsymbol{A}$. Hence, the algorithm is correct. It is not difficult to see that it can be implemented such that it has a linear worst-case running time. □

A similar, but more complicated algorithm based on finding a free set of variables can be used to solve the branching-time consistency problem (and therefore also the rooted triple consistency problem, as we have noticed before) in quadratic time [16].

11 Classification for Equality Constraint Languages

In this section, we show how to use the concepts we have seen so far to describe a full classification for a very restricted, but important class of ω-categorical templates. We study countably infinite relational structures that are first-order definable in $(\mathbb{N}, =)$. Equivalently, we study relational structures where all relations have a definition by a boolean combination of atomic formulas of the form $x = y$. Such templates are called *equality constraint languages* [12]. Example 1 in Section 2 is an example of an equality constraint language. Another example is $(\mathbb{N}, S)$ where

$$S = \{(x, y, z) \in \mathbb{N}^3 \mid x = y = z \vee x \neq y \neq z \neq x\} .$$

It is not hard to see that equality constraint languages are precisely those templates that are preserved by *all* permutations of the domain.

Theorem 13 (of [12]). *Let Γ be an equality constraint language. If Γ is preserved by a unary constant operation, or by a binary injective operation, then CSP(Γ) is in P. Otherwise, all polymorphisms of Γ are essentially unary and preserve $\neq$, and CSP(Γ) is NP-complete.*

We can answer Question 4 positively in case that Γ is an equality constraint language. In fact, Theorem 13 has the following reformulation using pseudo-varieties and equations.

Theorem 14. *Let Γ be an equality constraint language. The either $\mathcal{V}(\mathbb{A}_\Gamma)$ contains a 2-element algebra where all operations are projections, and CSP(Γ) is NP-complete, or $\mathbb{A}_\Gamma$ contains operations a, f satisfying $f(x, y) = a(f(y, x))$.*

Proof. If $\mathbb{A}_\Gamma$ contains operations a, f satisfying $f(x,y) = a(f(y,x))$ then this equation holds for all members of $\mathcal{V}(\mathbb{A}_\Gamma)$, and since the projections do not satisfy this equation, the two cases mentioned in the theorem are indeed disjoint.

Now, suppose that all polymorphisms of $\mathbb{A}_\Gamma$ are essentially unary and preserve $\neq$. Let $f : D(\Gamma)^2 \to \{0,1\}$ be such that (x,y) is mapped to 0 for $x = y$ and to 1 for $x \neq y$. It is straightforward to verify that f is a surjective algebra homomorphism from $\mathbb{A}_\Gamma^2$ to an algebra $\mathbb{B}$ where all operations are projections. Theorem 10 implies that CSP(Γ) is NP-complete.

If $\mathbb{A}_\Gamma$ contains an operation that does not preserve $\neq$, or is not essentially unary, then Theorem 13 shows that $\mathbb{A}_\Gamma$ contains a constant operation or a binary injective operation. It is easy to see that in both cases there is a permutation a and a binary operation f in $\mathbb{A}_\Gamma$ such that $f(x,y) = a(f(y,x))$. □

12 Outlook

In the opinion of the author, there are three important directions of future research on CSPs with infinite templates.

1. **Apply the universal-algebraic approach to unify and generalize existing CSP complexity results in the application areas.** Similarly as for the mentioned classification for equality constraint languages, it would be interesting to have a complete classification for templates for reasoning over time points, time intervals (i.e., the generalization of Allen's interval algebra to constraint languages with arbitrary, and not only binary constraints), for reasoning over partially ordered time, branching time, and spatial reasoning with RCC-5, and for other important qualitative reasoning formalisms that have been studied in the literature. I am optimistic that with the universal-algebraic approach a classification in the style of Theorem 13 can be obtained in all these areas. In particular, I expect that in this way we discover new and interesting tractable languages.
2. **Further develop the universal-algebraic theory for ω-categorical templates.**
   (a) When is it useful to work with varieties instead of pseudo-varieties? For finite templates Γ, the pseudo-variety and the variety of the Γ contain the same finite algebras. Birkhoff's theorem thus asserts that the computational complexity of CSP(Γ) is captured by the set of equations satisfied by the operations in $\mathbb{A}_\Gamma$. Indeed, also for ω-categorical templates several tractability criteria can be formulated by equations satisfied by the polymorphisms of the template (see Section 10).
   (b) Develop tame congruence theory for oligomorphic algebras $\mathbb{A}$. An interesting approach might be to apply classical tame congruence theory to the finite algebras in the pseudo-variety generated by $\mathbb{A}$, and to study the implications for the algebra $\mathbb{A}$ (recall the examples in Section 11).
   (c) Suppose we have classified the computational complexity for all templates with a first-order definition in another ω-categorical structure Γ.

Can we lift this classification to a classification for all structures with a first-order interpretation in Γ?

(d) Follow the successful lines in model-theory where ω-categorical structures were classified under certain additional assumptions, for example strict minimality [61, 26], finite homogeneity or stability [51, 28], or smooth approximability [45, 27]. These classifications are usually up to inter-definability, sometimes even bi–inter-pretability, so even if a classification is complete in the model-theoretic sense, quite some effort might be necessary to classify the complexity of the corresponding CSPs. Finally, we would like to mention that many of the ω-categorical templates in Artificial Intelligence have an automorphism group that is a *Jordan group* [53, 25], for which strong classification results are known [2, 3].

3. **Clarify which constraint satisfaction problems in NP can be formulated with an ω-categorical template.** We have seen an example that does not have such a formulation in Section 5. But we have also seen that all CSPs in MSNP do have a formulation with an ω-categorical template. Is the same true for all problems in SNP? Which constraint satisfaction problems in NP are in SNP?

Acknowledgements. I would like to thank the referee for his critical and helpful comments, all my CSP co-authors for the cooperation: Hubie Chen, Victor Dalmau, Jan Kára, Martin Kutz, Jaroslav Nešetřil, Michael Pinsker, Timo von Oertzen.

References

1. Abian, A.: Categoricity of denumerable atomless boolean rings. Studia Logica 30(1), 63–67 (1972)
2. Adeleke, S., Neumann, P.M.: Structure of partially ordered sets with transitive automorphism groups. AMS Memoir 57(334) (1985)
3. Adeleke, S.A., Macpherson, D.: Classification of infinite primitive jordan permutation groups. Proceedings of the London Mathematical Society s3-72(1), 63–123 (1996)
4. Aho, A., Sagiv, Y., Szymanski, T., Ullman, J.: Inferring a tree from lowest common ancestors with an application to the optimization of relational expressions. SIAM Journal on Computing 10(3), 405–421 (1981)
5. Allen, J.F.: Maintaining knowledge about temporal intervals. Communications of the ACM 26(11), 832–843 (1983)
6. Bauslaugh, B.L.: The complexity of infinite h-coloring. J. Comb. Theory, Ser. B 61(2), 141–154 (1994)
7. Bodirsky, M.: Cores of countably categorical structures. In: Logical Methods in Computer Science, LMCS (2007), doi:10.2168/LMCS-3(1:2)
8. Bodirsky, M., Chen, H.: Oligomorphic clones. Algebra Universalis 57(1), 109–125 (2007)
9. Bodirsky, M., Chen, H.: Qualitative temporal and spatial reasoning revisited. In: Duparc, J., Henzinger, T.A. (eds.) CSL 2007. LNCS, vol. 4646, pp. 194–207. Springer, Heidelberg (2007)

10. Bodirsky, M., Chen, H., Kara, J., von Oertzen, T.: Maximal infinite-valued constraint languages. In: Arge, L., Cachin, C., Jurdziński, T., Tarlecki, A. (eds.) ICALP 2007. LNCS, vol. 4596, pp. 546–557. Springer, Heidelberg (2007)
11. Bodirsky, M., Dalmau, V.: Datalog and constraint satisfaction with infinite templates. In: Durand, B., Thomas, W. (eds.) STACS 2006. LNCS, vol. 3884, pp. 646–659. Springer, Heidelberg (2006)
12. Bodirsky, M., Kára, J.: The complexity of equality constraint languages. In: Grigoriev, D., Harrison, J., Hirsch, E.A. (eds.) CSR 2006. LNCS, vol. 3967, pp. 114–126. Springer, Heidelberg (2006)
13. Bodirsky, M., Kára, J.: The complexity of temporal constraint satisfaction problems (preprint, 2007)
14. Bodirsky, M., Kára, J.: A fast algorithm and lower bound for temporal reasoning (preprint, 2007)
15. Bodirsky, M., Kutz, M.: Pure dominance constraints. In: Alt, H., Ferreira, A. (eds.) STACS 2002. LNCS, vol. 2285, pp. 287–298. Springer, Heidelberg (2002)
16. Bodirsky, M., Kutz, M.: Determining the consistency of partial tree descriptions. Artificial Intelligence 171, 185–196 (2007)
17. Bodirsky, M., Nešetřil, J.: Constraint satisfaction with countable homogeneous templates. In: Baaz, M., Makowsky, J.A. (eds.) CSL 2003. LNCS, vol. 2803, pp. 44–57. Springer, Heidelberg (2003)
18. Bodnarčuk, V.G., Kalužnin, L.A., Kotov, V.N., Romov, B.A.: Galois theory for post algebras, part I and II. Cybernetics 5, 243–539 (1969)
19. Broxvall, M., Jonsson, P.: Point algebras for temporal reasoning: Algorithms and complexity. Artif. Intell. 149(2), 179–220 (2003)
20. Bulatov, A.: A graph of a relational structure and constraint satisfaction problems. In: Proceedings of the 19th IEEE Annual Symposium on Logic in Computer Science (LICS 2004), Turku, Finland (2004)
21. Bulatov, A., Jeavons, P.: Algebraic structures in combinatorial problems. Technical report MATH-AL-4-2001, Technische Universitat Dresden. International Journal of Algebra and Computing (submitted, 2001)
22. Bulatov, A., Krokhin, A., Jeavons, P.G.: Classifying the complexity of constraints using finite algebras. SIAM Journal on Computing 34, 720–742 (2005)
23. Bulatov, A.A., Dalmau, V.: A simple algorithm for Mal'tsev constraints. SIAM J. Comput. 36(1), 16–27 (2006)
24. Burris, S., Sankappanavar, H.: A Course in Universal Algebra. Springer, Berlin (1981)
25. Cameron, P.J.: Oligomorphic Permutation Groups. Cambridge Univ. Press, Cambridge (1990)
26. Cherlin, G., Harrington, L., Lachlan, A.: $\aleph_0$-categorical, $\aleph_0$-stable structures. Annals of Pure and Applied Logic 28, 103–135 (1985)
27. Cherlin, G., Hrushovski, E.: Finite Structures with Few Types. Princeton University Press, Princeton (2003)
28. Cherlin, G., Lachlan, A.H.: Stable finitely homogeneous structures. TAMS 296, 815–850 (1986)
29. Cohen, D., Jeavons, P., Jonsson, P., Koubarakis, M.: Building tractable disjunctive constraints. Journal of the ACM 47(5), 826–853 (2000)
30. Droste, M.: Structure of partially ordered sets with transitive automorphism groups. AMS Memoir 57(334) (1985)
31. Düntsch, I.: Relation algebras and their application in temporal and spatial reasoning. Artificial Intelligence Review 23, 315–357 (2005)

32. Ebbinghaus, H.-D., Flum, J.: Finite Model Theory, 2nd edn. Springer, Heidelberg (1999)
33. Evans, D.: Examples of $\aleph_0$-categorical structures. In: Kaye, R., Macpherson, H.D. (eds.) Automorphisms of first-order structures, pp. 33–72. Oxford University Press, Oxford (1994)
34. Feder, T., Hell, P., Mohar, B.: Acyclic homomorphisms and circular colorings of digraphs. SIAM J. Discrete Math. 17(1), 161–169 (2003)
35. Feder, T., Vardi, M.: The computational structure of monotone monadic SNP and constraint satisfaction: A study through Datalog and group theory. SIAM Journal on Computing 28, 57–104 (1999)
36. Fraïssé, R.: Theory of Relations. North-Holland, Amsterdam (1986)
37. Garey, M., Johnson, D.: A guide to NP-completeness. CSLI Press (1978)
38. Geiger, D.: Closed systems of functions and predicates. Pacific Journal of Mathematics 27, 95–100 (1968)
39. Goldstern, M., Pinsker, M.: A survey of clones on infinite sets. arXiv:0701030 (preprint, 2007)
40. Hell, P., Nešetřil, J.: On the complexity of H-coloring. Journal of Combinatorial Theory, Series B 48, 92–110 (1990)
41. Hirsch, R.: Expressive power and complexity in algebraic logic. Journal of Logic and Computation 7(3), 309–351 (1997)
42. Hodges, W.: A shorter model theory. Cambridge University Press, Cambridge (1997)
43. Idziak, P.M., Markovic, P., McKenzie, R., Valeriote, M., Willard, R.: Tractability and learnability arising from algebras with few subpowers. In: LICS, pp. 213–224 (2007)
44. Jonsson, P., Drakengren, T.: A complete classification of tractability in RCC-5. J. Artif. Intell. Res. 6, 211–221 (1997)
45. Kantor, W.M., Macpherson, H.D., Liebeck, M.W.: $\aleph_0$-categorical structures smoothly approximated by finite substructures. Proc. London Math. Soc. 59, 439–463 (1989)
46. Keisler, J.: Reduced products and Horn classes. Trans. Amer. Math. Soc. 117, 307–328 (1965)
47. Klíma, O., Tesson, P., Thérien, D.: Dichotomies in the complexity of solving systems of equations over finite semigroups. Theory Comput. Syst. 40(3), 263–297 (2007)
48. Koubarakis, M.: Tractable disjunctions of linear constraints: Basic results and applications to temporal reasoning. Theoretical Computer Science 266, 311–339 (2001)
49. Krokhin, A.A., Jeavons, P., Jonsson, P.: Reasoning about temporal relations: The tractable subalgebras of Allen's interval algebra. JACM 50(5), 591–640 (2003)
50. Kun, G.: Constraints, mmsnp, and expander relational structures. arXiv:0706.1701 (2007)
51. Lachlan, A.H.: Stable finitely homogeneous structures: A survey. In: Algebraic Model Theory, NATO ASI Series, vol. 496, pp. 145–159 (1996)
52. Ladner, R.E.: On the structure of polynomial time reducibility. JACM 22(1), 155–171 (1975)
53. Macpherson, D.: A survey of jordan groups. In: Kaye, R., Macpherson, H.D. (eds.) Automorphisms of first-order structures, pp. 73–110. Oxford University Press, Oxford (1994)
54. Matiyasevich, Y.: Enumerable sets are diophantine. Doklady Akademii Nauk SSSR 191, 279–282 (1970)

55. Nebel, B., Bürckert, H.-J.: Reasoning about temporal relations: A maximal tractable subclass of Allen's interval algebra. JACM 42(1), 43–66 (1995)
56. Paterson, M.S., Wegman, M.N.: Linear unification. Journal of Computer and System Sciences 16, 158–167 (1978)
57. Poizat, B.: A Course in Model Theory: An Introduction to Contemporary Mathematical Logic. Springer, Heidelberg (2000)
58. Renz, J., Nebel, B.: On the complexity of qualitative spatial reasoning: A maximal tractable fragment of the region connection calculus. Artif. Intell. 108(1-2), 69–123 (1999)
59. Szendrei, A.: Clones in universal Algebra. Seminaire de mathematiques superieures. Les Presses de L'Universite de Montreal (1986)
60. van Beek, P., Cohen, R.: Exact and approximate reasoning about temporal relations. Computational Intelligence 6, 132–144 (1990)
61. Zilber, B.: Uncountable categorical theories, Tranlations of Mathematical Monographs, vol. 117. Amer. Math. Soc. (1993)

Partial Polymorphisms and Constraint Satisfaction Problems

Henning Schnoor[1] and Ilka Schnoor[2]

[1] Institut für Informatik, Christian-Albrechts-Universität Kiel, Christian-Albrechts-Platz 4, D-24118 Kiel, Germany
schnoor@ti.informatik.uni-kiel.de
[2] Institut für Theoretische Informatik, Universität Lübeck, Ratzeburger Allee 160, D-23538 Lübeck, Germany
schnoor@tcs.uni-luebeck.de

Abstract. The Galois connection between clones and and co-clones has received a lot of attention in the context of complexity considerations for constraint satisfaction problems. However, it fails if we are interested in a reduction giving equivalence instead of only satisfiability-equivalence. We show how a similar Galois connection involving weaker closure operators can be applied for these problems. As an example of the usefulness of our construction, we show how to obtain very short proofs of complexity classifications in this context.

Keywords: partial polymorphisms, clones, constraint satisfaction problems, computational complexity.

1 Introduction

Constraint satisfaction problems (CSPs) play an important role in computational complexity theory. In particular, the non-uniform version of the problem, $\mathsf{CSP}(\Gamma)$ has been studied. In this context, a set of relations Γ, called a constraint language, over a finite domain is fixed, and so-called Γ-formulas, i.e., conjunctions of applications of relations from Γ to variables, are studied. For the Boolean domain, these problems generalize many well-known restrictions of the propositional satisfiability problem, for example 2SAT, 3SAT, Horn-SAT, etc. For non-Boolean domains, CSPs can be used to express various problems related to graphs (like search, colorability problems, and others) as well as database queries and scheduling problems. Many combinatorial problems can be expressed in this context, and therefore constraint satisfaction problems can be seen as embodying the quintessence of the combinatorial aspects of complexity theory.

The study of constraint related problems in computational complexity started with Thomas Schaefer's seminal paper [Sch78]. He showed that for Boolean constraint languages Γ, the complexity of the satisfiability problem $\mathsf{CSP}(\Gamma)$ (the problem to determine if a given Γ-formula is satisfiable) is dichotomic: For

N. Creignou et al. (Eds.): Complexity of Constraints, LNCS 5250, pp. 229–254, 2008.

a given Γ, this problem can be solved in P, or is already complete for NP, hence avoiding the infinitely many complexity degrees between these two classes which are known to exist (if P $\neq$ NP) due to a classic result by Richard Ladner [Lad75]. Since Schaefer's work, many results have been obtained about the complexity of the problem $\mathsf{CSP}(\Gamma)$ for non-Boolean domains, leading to a proof of a similar dichotomy theorem for the three-element case by Andrei Bulatov [Bul06].

One of the most successful techniques to obtain results on the complexity of constraint-related problems has been the application of tools from Universal Algebra. In particular, a Galois connection between constraint languages and functions on finite domains has been studied. This Galois connection can be used to show that if two constraint languages Γ_1 and Γ_2 have the same algebraic closure properties, then the complexity of the problems $\mathsf{CSP}(\Gamma_1)$ and $\mathsf{CSP}(\Gamma_2)$ is the same [Jea98] (up to $\leq_m^{\log}$-reductions, [ABI$^+$05]). The Galois connection relates the *expressive power* of a constraint language to its set of *polymorphisms*, i.e., algebraic closure properties. By a work of Emil Post [Pos41], the structure of the polymorphism sets occurring here, so-called *clones*, is well known for the Boolean case. Hence the Galois connection can be used to transfer results from these well-known classes to the complexity of constraint satisfaction problems.

The Galois connection gives a procedure transforming Γ_1-formulas into equivalent Γ_2-formulas, where in the newly constructed Γ_2-formulas, additional existentially quantified variables and equality clauses may occur. It is easy to see that this leads to a transformation from Γ_1- to Γ_2-formulas preserving satisfiability, as in the satisfiability problem, new free variables in a formula can be regarded as existentially quantified, and the occurring equality clauses can be dealt with using variable identification. Similarly, in many cases where the newly introduced variables can be "hidden" in some way (for example, in many cases of problems involving quantified formulas), this transformation can easily be seen to preserve all relevant properties of the involved formulas. An introduction to this technique can be found in [BCRV04], and background over the algebraic properties of the involved structure is given in Dietlinde Lau's excellent survey [Lau06].

However, for most problems different from satisfiability, this transformation is not sufficient. In the case of problems like counting and enumeration of solutions for constraint formulas, it is evident that the newly introduced variables are problematic, as they can change the set and the number of solutions to a given formula. In [SS06], it was shown that this feature of the Galois connection makes it inapplicable for the enumeration problem over non-Boolean domains. Similarly, if the goal is to determine whether two formulas are equivalent, the introduction of equality clauses is an additional problem.

The identification of equality-constrained variables is also problematic when considering low complexity classes. In [ABI$^+$05], it was shown that the introduction of equality clauses makes it impossible to refine the logspace reduction given by the Galois connection to a reduction computable in AC^0.

From these considerations, it is evident that other algebraic tools are required when looking at questions in the constraint context different from determining the $\leq_m^{\log}$-degree of complexity of the satisfiability problem. A natural approach is to consider restricted closures, where the problematic features from the Galois connection mentioned above, i.e., introduction of new variables and equality clauses, are absent. This closure operator has been studied in mathematics already in the 1960s, and a similar relation to closure properties of the involved relations was proven. This refined Galois connection tells us that instead of looking at the set of polymorphisms, we need to consider the set of *partial polymorphisms*. These form a structure which is a refinement of the clone structure exhibited by Post for the Boolean case. In particular, we know that instead of the countably many classes arising in Post's classification, there is an uncountable number of *partial clones* [AV94]. In other words, the step from the usual Galois connection to the refinement comes with the price of additional complexity of the mathematical structures involved.

In light of these considerations, it is surprising to see that for many of the computational problems where the above-mentioned difficulties occur, a complexity classification, when achieved, often follows the lines of the well-studied closure operator. In this case, we say that the Galois connection *can be applied*. The question for which problems the Galois connection can be applied is an interesting question in the constraint context. While proofs for an "a priori"-application of the Galois connection are very often close to trivial, there are many cases where the fact that the Galois connection can be applied only follows from a complete complexity classification of the involved problem (for example enumeration [CH97] and equivalence [BHRV02] problems in the Boolean case).

The contribution of this paper is twofold: First, we exhibit, among the potentially uncountably many partial clones corresponding to each clone $\mathcal{C}$, a unique one which leads to the constraint problems with the lowest complexity among those with polymorphism set $\mathcal{C}$, and show how canonical constraint languages with this complexity behavior can be constructed. These "weak bases" can be used to show lower complexity bounds for all constraint languages sharing the same set of polymorphisms, even when the usual Galois connection cannot be applied to the problem at hand. Therefore, they address a common problem encountered when proving complexity classifications. Second, we give an answer to the above question: we give a method which can be used to determine if for a given problem, the usual Galois connection can be applied.

The paper is structured as follows: In Section 2, we give the necessary definitions. In Section 3, we state the refined Galois connection. In Section 4, we construct the "weak bases" mentioned above. While this work focuses on eliminating the need for existential quantification, Section 5 then considers the problem of required equality clauses. Finally, in Section 6, we demonstrate our technique with giving very short re-proofs of the above mentioned results about enumeration and equivalence for Boolean constraint languages, where much of the technical difficulty of the original proofs can be avoided using our algebraic results. We also prove a complexity classification for the implication problem for constraint formulas, which is closely related to the equivalence problem.

2 Preliminaries

We first briefly repeat the basic definitions in the constraint satisfaction context. In this paper, we only consider finitary relations over a finite domain. For a tuple v, the value $v[i]$ denotes its i-th component. A *constraint language* Γ is a finite set of non-empty, finitary relations over a finite domain D. For a (not necessarily finite) set of relations Γ, a Γ-formula is a conjunction of the form

$$\varphi = \bigwedge_{i=1}^{n} R_i(x_1^i, \ldots, x_{k_i}^i),$$

where each R_i is a k_i-ary relation from Γ, and the x_j^i are (not necessarily distinct) variables. For a single-element constraint language $\{R\}$, we often simply speak about R-formulas, etc. We denote the set of variables occurring in φ with $\text{VAR}(\varphi)$. An assignment $I\colon \text{VAR}(\varphi) \to D$ *satisfies* φ or is a *solution* of φ, if for all $i \in \{1, \ldots, n\}$, it holds that $(I(x_1^i), \ldots, I(x_{k_i}^i))$ is a tuple from R_i. We say that a constraint language is *Boolean*, if the domain D has cardinality 2. The unary relation D is also denoted as $\top$, since $\top(x)$ is a tautological clause. Finally, for any set D, let OP_D denote the total finitary functions on D.

There is a close relationship between formulas and relations, as each formula defines the relation of its satisfying assignments. We say that a formula *represents* or *expresses* the relation of its solutions.

We now define three closure operators for relations and constraint languages. The first one, $\langle . \rangle$, is the one usually considered in the constraint context, which has successfully been applied to classify the complexity of satisfiability problems. The other two are the refinements where we disallow the introduction of equality clauses and/or new variables.

Definition 2.1. *Let Γ be a set of relations.*

- $\langle \Gamma \rangle$ *is the set of relations which can be expressed as a $\Gamma \cup \{=\}$-formula with additional existentially quantified variables.*
- $\langle \Gamma \rangle_{\nexists}$ *is the set of relations which can be expressed as a $\Gamma \cup \{=\}$-formula.*
- $\langle \Gamma \rangle_{\nexists,\neq}$ *is the set of relations which can be expressed as a Γ-formula.*

In the notation above, $\neq$ does not mean that the closure operator is allowed to use the inequality predicate, but that the operator is not allowed to use the equality predicate (unless it is present in Γ), similarly $\nexists$ means that the introduction of existentially quantified variables is not allowed. It is evident that for any set of relations Γ, the inclusion $\langle \Gamma \rangle_{\nexists,\neq} \subseteq \langle \Gamma \rangle_{\nexists} \subseteq \langle \Gamma \rangle$ holds.

The operators defined above are closure operators in the usual mathematical sense, i.e., for an operator $C \in \{\langle . \rangle, \langle . \rangle_{\nexists}, \langle . \rangle_{\nexists,\neq}\}$ and sets of relations Γ_1 and Γ_2, it holds that $\Gamma_1 \subseteq C(\Gamma_1)$, if $\Gamma_1 \subseteq \Gamma_2$, then $C(\Gamma_1) \subseteq C(\Gamma_2)$ also holds, and finally $C(C(\Gamma_1)) = C(\Gamma_1)$. Sets Γ satisfying $C(\Gamma) = \Gamma$ are also called C-closed. The $\langle . \rangle$-closed sets are called *co-clones*, and the $\langle . \rangle_{\nexists}$-closed sets are called *weak systems*. We say that Γ is a *base* of the co-clone $\langle \Gamma \rangle$.

It is obvious that each closure operator $C \in \{\langle . \rangle, \langle . \rangle_{\nexists}, \langle . \rangle_{\nexists,\neq}\}$ gives rise to a formula transformation. Let $\Gamma_1 \subseteq C(\Gamma_2)$. Then a Γ_1-formula can be transformed into an equivalent Γ_2-formula by replacing a clause $R(x_1, \ldots, x_n)$ for some relation $R \in \Gamma_1$ with its Γ_2-implementation, possibly adding new existentially quantified variables and/or equality clauses, depending on the closure operator. To remove these features and produce an actual Γ_2-formula, we proceed as follows: In the case of equality clauses $(x = y)$, we simply replace every occurrence of the variable y with x, and remove the equality clause. For existentially quantified variables, we simply drop the quantifiers, leaving the variables free. The Γ_2-formula constructed in this way is not necessarily equivalent to the old one, but the transformation preserves many properties of the formulas which we are interested in. In particular, it preserves satisfiability, and hence is very useful when studying the complexity of satisfiability problems.

In the following, let Problem be a computational problem where the input instances are propositional formulas, and let Problem(Γ) be its restriction to Γ-formulas as inputs. As examples, we introduce the following: CSP is the problem to determine whether a formula is satisfiable, EQUIV is the problem to determine whether two input formulas are equivalent, and ENUM is the problem to enumerate the set of solutions for a formula. If P $\neq$ NP, then all these problems are intractable: the problems CSP and EQUIV are NP-complete resp. coNP-complete. The problem ENUM is not a decision problem and hence cannot easily be related to these standard complexity classes, but any efficient enumeration algorithm can be used to efficiently decide the satisfiability problem.

For the problem Problem and a complexity reduction $\leq$, we say that the operator C *can be* $\leq$*-applied a to* Problem, if for any two constraint languages Γ_1 and Γ_2 such that $\Gamma_1 \subseteq C(\Gamma_2)$, Problem (Γ_1) $\leq$-reduces to Problem (Γ_2). The enumeration problem does not fall into the category of counting and decision problems for which established notions of reductions exist. However, in the constraint context it turns out that the question if there is an efficient enumeration algorithm exists for a given constraint language Γ is again closely related to the co-clone generated by Γ. We therefore say that a closure operator C *can be applied to the enumeration problem* if for all constraint languages Γ_1 and Γ_2 such that $\Gamma_1 \subseteq C(\Gamma_2)$, it holds that if there is an efficient enumeration algorithm for Γ_2-formulas, then there also is one for Γ_1-formulas.

Such results can very often be obtained "for free," since in many cases transforming Γ_1-formulas into Γ_2-formulas in the way described above gives a reduction from the problem Problem (Γ_1) to Problem (Γ_2). For example, it was shown in [ABI+05] that the closure operator $\langle . \rangle$ leads to a $\leq_m^{\log}$-reduction, and $\langle . \rangle_{\nexists,\neq}$ leads to an $\leq_m^{\mathrm{AC}^0}$-reduction for the satisfiability problem. Similarly, it is obvious that the closure operator $\langle . \rangle_{\nexists,\neq}$ leads to a parsimonious reduction for the problem to count solutions for a constraint formula, and $\langle . \rangle_{\nexists}$ can be applied to the enumeration problem, since $\langle . \rangle_{\nexists}$ allows us to transform a Γ_1-formula into a Γ_2-formula in such a way that there is a canonical and easily computable relation between the solution sets of the formulas.

It is worth noting that, contrary to intuition, the operator $\langle . \rangle_{\nexists}$ provably cannot be applied to the counting problem. Consider the languages $\Gamma_1 = \{x, \overline{x}\}$ and $\Gamma_2 = \{x, \overline{x}, =\}$. Obviously, $\langle \Gamma_1 \rangle_{\nexists} = \langle \Gamma_2 \rangle_{\nexists}$, but Γ_1-formulas always have exactly

0 or exactly 1 solutions, while for every k, there is a Γ_2-formula with exactly 2^k solutions, namely $(x_1 = x_1) \wedge (x_2 = x_2) \wedge \cdots \wedge (x_k = x_k)$. Therefore, there is no parsimonious reduction between their respective counting problems.

Note that the formula transformations given by the closure operators do not only differ with respect to the question for which problems they can be applied, the complexity of the resulting transformation procedure is different as well. If we allow equality clauses in the closure operator, then a transformation to the final constraint language must be able to remove them, and will usually do this by identifying variables. Hence it needs to perform a search in an undirected graph, and therefore needs the computational power of LOGSPACE. When this problem does not arise, the formula transformation is simply a local replacement, and can be computed by an AC^0-reduction. The operator $\langle . \rangle_{\nexists,\neq}$ obviously allows for a transformation of formulas into equivalent ones. Therefore, this operator can be applied to the majority of problems in the constraint context. In particular, this holds for the equivalence problem, the counting problem, and all other problems mentioned. Since the formula transformation in this case just needs local replacement, it can be performed by AC^0-reductions. It is natural that the weakest of our closure operators gives us the strongest results: it is applicable to most problems, and its formula transformation needs the least computational resources.

From the above discussion, we conclude:

Proposition 2.2. – $\langle . \rangle$ *can be* $\leq_m^{\log}$*-applied to* CSP*,*
– $\langle . \rangle_{\nexists}$ *can be applied to* ENUM*,*
– $\langle . \rangle_{\nexists,\neq}$ *can be* $\leq_m^{\mathrm{AC}^0}$*-applied to* EQUIV*.*

It is obvious that if $\langle . \rangle$ can be applied to a problem, then the weaker closure operators can be applied as well, and similarly, if $\langle . \rangle_{\nexists}$ can be applied, then this also is true for $\langle . \rangle_{\nexists,\neq}$ In many cases, it cannot easily be seen that one of the above closure operators can be applied, but this result follows from a full complexity classification of the problem. This is true for the following problems:

Theorem 2.3. – $\langle . \rangle$ *can be applied to* ENUM *over Boolean domains [CH97],*
– $\langle . \rangle$ *can be* $\leq_m^p$*-applied to the* EQUIV *over Boolean domains [BHRV02].*

For both the enumeration and the equivalence problem, it is not clear that the complexity of the problem only depends on $\langle \Gamma \rangle$, since the introduction of new existentially quantified variables changes the set of solutions of a given formula significantly. Yet in the Boolean case, this was shown to be true. On the other hand, it was proven that $\langle . \rangle$ cannot be applied to ENUM over non-Boolean domains [SS06]. In the survey [CV08], Creignou and Vollmer give an extensive list of problems, and the known results about whether the $\langle . \rangle$ can be applied to them. We give a general criterion about applicability of this operator at the end of Section 4.

The study of the closure operators we defined above is closely related to algebraic properties of the involved relations. In particular, the notion of a polymorphism has proven to be very useful. We will now define a generalization, which we then use to state the refined Galois connection.

Definition 2.4. *Let D be a finite domain, and let $R \subseteq D^n$ be a relation. Let $A \subseteq D^m$, and let $f\colon A \to D$ be a function. We say that f is a* partial polymorphism *of R, $f \in \mathrm{pPol}(R)$, if the following holds: For any $u_1 = (u_1^1, \ldots, u_n^1), \ldots, u_m = (u_1^m, \ldots, u_n^m) \in R$, if $(u_i^1, \ldots, u_i^m) \in A$ for all $i \in \{1, \ldots, n\}$, then the coordinate-wise application*

$$f(u_1, \ldots, u_m) := (f(u_1^1, \ldots, u_1^m), \ldots, f(u_n^1, \ldots, u_n^m))$$

is in R. If $A = D^m$, we say that f is a polymorphism *of R, $f \in \mathrm{Pol}(R)$.*

For a set of relations Γ, the set $\mathrm{Pol}(\Gamma)$ ($\mathrm{pPol}(\Gamma)$) denotes the set of (partial) functions which are a polymorphism of every relation in Γ. We say that f in the definition above is a *partial function* on D if $A \neq D^m$, and f is *total* if $A = D^m$. The condition demanding that the vectors $(u_i^1, \ldots, u_i^m)$ must be elements of A ensures that the coordinate-wise application of the (possibly partial) function f gives a well-defined vector. It is obvious that for any relation R, it holds that $\mathrm{Pol}(R) = \mathrm{pPol}(R) \cap \mathrm{OP}_D$. For a set of functions B, the set $\mathrm{Inv}(B)$ denotes the set of relations R such that $B \subseteq \mathrm{pPol}(R)$. It is worth noting that although the operators $\mathrm{pPol}(.)$ and $\mathrm{Pol}(.)$ obviously differ, there is no need for two "versions" of the $\mathrm{Inv}(.)$-operator.

Polymorphisms have interesting algebraic properties: We say that a set $\mathcal{C}$ of (partial) functions on D is a *(partial) clone*, if $\mathcal{C}$ contains all projections and is closed under arbitrary composition. $\mathcal{C}$ is a *strong partial clone*, if in addition for every (partial) function $f \in B$, every further restriction of f (i.e., a partial function whose domain is a subset of f's domain, and that agrees with f for all values in its domain) again is a member of B. For a set B of (partial) functions, we denote with $[B]$ ($[B]_{\mathrm{p}}$) the (strong partial) clone generated by B, i.e., the smallest (strong partial) clone which is a superset of B. It is obvious that $[B]$ ($[B]_{\mathrm{p}}$) is the set of functions which can be generated from B by adding all projections and closing the set under arbitrary composition (and further restriction).

It is easy to see that the set $\mathrm{Pol}(\Gamma)$ is always a clone for a constraint language Γ, and the set $\mathrm{pPol}(\Gamma)$ always is a strong partial clone. For the Boolean case, Emil Post obtained a complete classification of all clones in [Pos41]. This classification is a powerful tool for the complexity classifications of Boolean constraint-related problems: Whenever it can be shown that $\langle . \rangle$ can be applied to a problem, then his list of clones gives, via the Galois connection, a complete list of cases to consider. Therefore in these cases, the Galois connection and the well-known complete list of Boolean clones often makes the complexity analysis of a problem where $\langle . \rangle$ can be applied a straight-forward task.

3 The Galois Connection

The following theorem summarizes the main properties of the standard Galois connection which has been used to prove complexity results in the constraint satisfaction context:

Theorem 3.1 ([Gei68, BKKR69]). *Let Γ be a constraint language over a finite domain D, and let R be a non-empty relation over D. Then $R \in \langle \Gamma \rangle$ if and only if* $\mathrm{Pol}(\Gamma) \subseteq \mathrm{Pol}(R)$.

In particular, this theorem can be used to show the following complexity result. It was first stated for polynomial-time many-one reductions in [Jea98], and refined to a logspace reduction in [ABI+05].

Theorem 3.2 ([Jea98, ABI+05]). *Let Γ_1 and Γ_2 be constraint languages over a finite domain D, such that* $\mathrm{Pol}(\Gamma_2) \subseteq \mathrm{Pol}(\Gamma_1)$. *Then* $\mathsf{CSP}(\Gamma_1) \leq_m^{\log} \mathsf{CSP}(\Gamma_2)$.

This theorem is useful because as mentioned before, the sets $\mathrm{Pol}(\Gamma)$ are always clones, and their structure is well-studied. The following theorem is the refinement of the Galois connection exhibited in Theorem 3.1 to the $\langle . \rangle_{\nexists}$ operator. It was proven by Boris Romov in 1981 [Rom81], but is already implicit in [Gei68].

Theorem 3.3 ([Rom81]). *Let Γ be a constraint language over a finite domain D, and let R be a non-empty relation over D. Then $R \in \langle \Gamma \rangle_{\nexists}$ if and only if* $\mathrm{pPol}(\Gamma) \subseteq \mathrm{pPol}(R)$.

The above Theorem 3.3 does not consider the closure operator $\langle . \rangle_{\nexists, \neq}$. There also is a further refinement of the Galois connection to this case, where instead of functions, we consider partial hyperfunctions. However, we will show in Section 5 that for many practical applications, this is not necessary.

It should be noted that the seemingly subtle step from polymorphisms to partial polymorphisms has unfortunate consequences. While, as mentioned, the lattice of clones for the Boolean case is well understood and completely classified due to Post's work [Pos41], the picture for partial clones is much less clear. Already over the Boolean domain, there is an uncountable number of partial clones [AV94], and the structure of the lattice of partial clones is not completely known. This is the key reason why results that the operator $\langle . \rangle$ can be applied are so useful in practice: For a given Boolean constraint language Γ, it is easy to determine its set of polymorphisms. In fact, this can be done automatically by an algorithm. Hence, if there is a classification theorem giving the complexity of $\mathsf{Problem}(\Gamma)$ as an easy function of $\langle \Gamma \rangle$, then the complexity of $\mathsf{Problem}(\Gamma)$ can be determined automatically.

In [Rom81], Theorem 3.1 was also proven to hold for infinite constraint languages. In the Boolean case, there are exactly 8 clones $\mathcal{C}$ for which there does not exist a finite constraint language Γ with $\langle \Gamma \rangle = \mathrm{Inv}(\mathcal{C})$, for larger finite domains, there exists a continuum of such co-clones [BKKR69].

4 Weak Bases

The Galois connection stated in Theorem 3.3 shows that in order to classify the complexity of problems where the standard Galois connection cannot be proven to be applicable in a straight-forward way, we need to study the sets of partial polymorphisms of a given constraint language. It is obvious that a given clone corresponds to many different partial clones. In fact, since there are uncountably many partial clones and in the Boolean case, there are only countably many clones, there are clones with uncountably many partial clones associated with them. We therefore need a more detailed examination of the partial clones occurring inside a given clone. As mentioned, the sets $\mathrm{pPol}\,(\Gamma)$ always form strong partial clones, and therefore we can restrict ourselves to those.

Definition 4.1. *For a clone $\mathcal{C}$ over a finite domain D, let*

$$\mathcal{I}(\mathcal{C}) =_{def} \{\mathcal{D} \mid \mathcal{D} \text{ is a strong partial clone and } \mathcal{D} \cap \mathrm{OP}_D = \mathcal{C}\}$$

denote the interval of $\mathcal{C}$. *Further, let*

$$\mathcal{I}_{\cup}(\mathcal{C}) =_{def} \bigcup_{\mathcal{D} \in \mathcal{I}(\mathcal{C})} \mathcal{D}$$

be the union of all partial clones in the interval $\mathcal{I}(\mathcal{C})$.

The set $\mathcal{I}(\mathcal{C})$ represents all strong partial clones which are "inside" the given clone $\mathcal{C}$. Hence the set of all partial functions which are in any of these partial clones form the largest set of partial functions which can be related naturally to $\mathcal{C}$. It turns out that this set has very important properties. We will soon see that the set $\mathcal{I}_{\cup}(\mathcal{C})$ forms a strong partial clone. It is the largest strong partial clone contained in $\mathcal{I}(\mathcal{C})$, and hence the interval $\mathcal{I}(\mathcal{C})$ contains exactly the strong partial clones $\mathcal{D}$ such that $[\mathcal{C}]_{\mathrm{p}} \subseteq \mathcal{D} \subseteq \mathcal{I}_{\cup}(\mathcal{C})$. Therefore, a constraint language Γ satisfying $\mathrm{pPol}\,(\Gamma) = \mathcal{I}_{\cup}(\mathcal{C})$ is in a certain way one of those which lead to the "easiest" problems among those languages with polymorphism set $\mathcal{C}$. It turns out that constraint languages satisfying this condition are central for our complexity considerations.

Definition 4.2. *Let $\mathcal{C}$ be a clone over a domain D. Then a constraint language Γ over D is called a* weak base *for* $\mathrm{Inv}\,(\mathcal{C})$ *if* $\mathrm{pPol}\,(\Gamma) = \mathcal{I}_{\cup}(\mathcal{C})$.

These weak bases are the main tool that we introduce to prove hardness results for constraint-related problems. One of the important properties of weak bases, as discussed informally above, is the following:

Corollary 4.3. *Let $\mathcal{C}$ be a clone over D and Γ a weak base for* $\mathrm{Inv}\,(\mathcal{C})$*. Then for any constraint language Γ' with $\langle\Gamma\rangle = \langle\Gamma'\rangle$, it follows that $\Gamma \subseteq \langle\Gamma'\rangle_{\nexists}$.*

Proof. Since Γ is a weak base for $\mathrm{Inv}\,(\mathcal{C})$, we know that $\mathrm{pPol}\,(\Gamma) = \mathcal{I}_{\cup}(\mathcal{C})$. Now let Γ' be a constraint language over D such that $\langle\Gamma'\rangle = \langle\Gamma\rangle$. From Theorem 3.1, it follows that $\mathrm{Pol}\,(\Gamma) = \mathrm{Pol}\,(\Gamma') = \mathcal{C}$. Therefore, $\mathrm{pPol}\,(\Gamma') \in \mathcal{I}(\mathcal{C})$ and hence $\mathrm{pPol}\,(\Gamma') \subseteq \mathcal{I}_{\cup}(\mathcal{C}) = \mathrm{pPol}\,(\Gamma)$. Therefore, from Theorem 3.3, it follows that $\langle\Gamma\rangle_{\nexists} \subseteq \langle\Gamma'\rangle_{\nexists}$, and since $\Gamma \subseteq \langle\Gamma\rangle_{\nexists}$ obviously holds, this concludes the proof. $\square$

For complexity considerations, Corollary 4.3 is of crucial importance. If we have a weak base Γ for some co-clone $\mathrm{Inv}(\mathcal{C})$, then we know that this constraint language is in the $\langle . \rangle_{\nexists}$-closure of all constraint languages Γ' with polymorphism set $\mathcal{C}$. This implies that for all computational problems from the constraint context where the $\langle . \rangle_{\nexists}$-operator can be applied, the problem for the constraint language Γ reduces to the problem for the constraint language Γ'. Therefore, the language Γ can be regarded as being the "easiest" among those with the same set of polymorphisms. As the name weak base suggests, such a Γ has the lowest expressive power with respect to the $\langle . \rangle_{\nexists}$-operator among those with the same set of polymorphisms.

Corollary 4.4. *Let $\leq$ be a complexity reduction, let* Problem *be a computational problem such that $\langle . \rangle_{\nexists}$ can be $\leq$-applied to* Problem, *let Γ' be a constraint language, and let Γ be a weak base of $\langle \Gamma' \rangle$. Then* Problem (Γ) *reduces to* Problem (Γ').

Proof. Let $\mathcal{C} =_{\mathrm{def}} \mathrm{Pol}(\Gamma')$. Since Γ a weak base of $\langle \Gamma' \rangle$, Corollary 4.3 implies that $\Gamma \subseteq \langle \Gamma' \rangle_{\nexists}$. Since $\langle . \rangle_{\nexists}$ can be $\leq$-applied to Problem, we know that Problem (Γ) reduces to Problem (Γ'). □

Corollary 4.4 states that weak bases are *exactly* the tools we need if we want to prove hardness results for all constraint languages generating a particular co-clone, if the closure operator $\langle . \rangle_{\nexists}$ can be applied to the given computational problem. It is therefore very desirable to be able to construct weak bases—from the definition, it is not clear that finite languages having these properties actually exist. Hence the rest of this section shows how weak bases (which in our case will always be single relations) can be constructed. We will show that in fact for all clones $\mathcal{C}$, such that the co-clone has a finite base in the usual sense (i.e., a finite constraint language Γ such that $\mathrm{Pol}(\Gamma) = \mathcal{C}$), a weak base exists. For other clones $\mathcal{C}$, a finite weak base cannot exist, because in particular, if $\mathrm{pPol}(\Gamma) = \mathcal{I}_{\cup}(\mathcal{C})$, then $\mathrm{Pol}(\Gamma) = \mathcal{C}$. Also, the clones with finite bases (weak or otherwise) are the ones which usually appear in practical applications, since often, we are concerned with a finite vocabulary.

We will give a characterization of $\mathcal{I}_{\cup}(\mathcal{C})$ and show that it again is a partial clone with total functions $\mathcal{C}$. In order to prove this result, we first introduce another way to relate a partial function to the clone $\mathcal{C}$:

Definition 4.5. *Let $\mathcal{C}$ be a clone over a finite domain D and $f \colon A \to D$ be a partial function, where $A \subseteq D^n$. Then f is $\mathcal{C}$-*total*, if for all functions $g_1, \ldots, g_n \in \mathcal{C}$ the function*

$$f(g_1(x_1^1, \ldots, x_{m_1}^1), \ldots, g_n(x_1^n, \ldots, x_{m_n}^n))$$

is either non-total or from $\mathcal{C}$, where the x_j^i are (not necessarily distinct) variables.

It is immediate that a total function is $\mathcal{C}$-total if and only if it is an element of $\mathcal{C}$. The first thing we need to note about $\mathcal{C}$-total functions is that they form a strong partial clone:

Theorem 4.6. *Let $\mathcal{C}$ be a clone over a finite domain D, and let $\mathcal{F}$ be a set of $\mathcal{C}$-total functions over D. Then every function in $[\mathcal{F}]_p$ is $\mathcal{C}$-total. In particular, the set of $\mathcal{C}$-total functions is a strong partial clone.*

Proof. It obviously suffices to prove the first part. Note that projections are trivially $\mathcal{C}$-total, hence it remains to show that the set of $\mathcal{C}$-total functions is closed under arbitrary composition and restriction of the domain of functions. It is obvious that a restriction of a $\mathcal{C}$-total function remains $\mathcal{C}$-total. Therefore let $f, h_1, \ldots, h_n$ be $\mathcal{C}$-total partial functions where f is of arity n and h_i is of arity m_i for all $i \in \{1, \ldots, n\}$. We need to prove that any function f' obtained from these by composition again is $\mathcal{C}$-total. Hence let f' be defined as $f' = f(h_1(x_1^1, \ldots, x_{m_1}^1), \ldots, h_n(x_1^n, \ldots, x_{m_n}^n))$. Let $g_1^1, \ldots, g_{m_n}^n$ be functions from $\mathcal{C}$, where each g_j^i is an $l_{i,j}$-ary function. We need to prove that the function

$$f'' = f\ (h_1(g_1^1(x_1^{1,1}, \ldots, x_{l_{1,1}}^{1,1}), \ldots, g_{m_1}^1(x_1^{1,m_1}, \ldots, x_{l_{1,m_1}}^{1,m_1})),$$
$$\vdots$$
$$h_n(g_1^n(x_1^{n,1}, \ldots, x_{l_{n,1}}^{n,1}), \ldots, g_{m_n}^n(x_1^{n,m_n}, \ldots, x_{l_{n,m_n}}^{n,m_n})))$$

is either non-total, or a function from $\mathcal{C}$. We consider two cases. First, assume that one of the terms $h_i(g_1^i(x_1^{i,1}, \ldots, x_{l_{i,1}}^{i,1}), \ldots, g_{m_i}^i(x_1^{i,m_i}, \ldots, x_{l_{i,m_i}}^{i,m_i}))$ represents a non-total function. Then obviously, the function f'' is non-total as well. Now assume that all of these terms represent total functions. Then, since the h_i are $\mathcal{C}$-total, and the functions g_j^i are from $\mathcal{C}$, we know that all of these terms represent functions from $\mathcal{C}$. Since f is $\mathcal{C}$-total, this implies that the function f'' is a function from $\mathcal{C}$ or non-total, as required. □

We now show that the $\mathcal{C}$-total functions are exactly those which are contained in the clones from $\mathcal{I}(\mathcal{C})$. One important consequence of this theorem is that, together with Theorem 4.6, it implies that the set $\mathcal{I}_\cup(\mathcal{C})$ forms a strong partial clone. This implies that the set $\mathcal{I}(\mathcal{C})$ has a largest element, which is not immediate. The fact that $\mathcal{I}_\cup(\mathcal{C})$ is a strong partial clone is a crucial requirement for weak bases to exist.

Theorem 4.7. *Let $\mathcal{C}$ be a clone on the finite domain D. Then $\mathcal{I}_\cup(\mathcal{C})$ is exactly the set of partial functions which are $\mathcal{C}$-total.*

Proof. First, let $f \in \mathcal{I}_\cup(\mathcal{C})$. Then there is a strong partial clone $\mathcal{D}$, such that $\mathcal{D} \cap \mathrm{OP}_D = \mathcal{C}$, and $f \in \mathcal{D}$. In particular, it follows that $\mathcal{C} \subseteq \mathcal{D}$. Thus, since $\mathcal{D}$ is a partial clone, the composition of f and functions from $\mathcal{C}$ still gives a function from $\mathcal{D}$, in particular a function which is either non-total, or a member of $\mathcal{C}$. Hence, f is $\mathcal{C}$-total.

For the other direction, let f be a $\mathcal{C}$-total function. Since every function from $\mathcal{C}$ is trivially $\mathcal{C}$-total, Theorem 4.6 implies that every function from $[\{f\} \cup \mathcal{C}]_\mathrm{p}$ is $\mathcal{C}$-total. Since every function from OP_D which is $\mathcal{C}$-total is a member of $\mathcal{C}$, it follows that $[\{f\} \cup \mathcal{C}]_\mathrm{p} \cap \mathrm{OP}_D = \mathcal{C}$. Therefore, by definition we have that $[\{f\} \cup \mathcal{C}]_\mathrm{p} \in \mathcal{I}(\mathcal{C})$, and therefore $f \in \mathcal{I}_\cup(\mathcal{C})$. □

We now construct weak bases, i.e., constraint languages whose partial polymorphisms are exactly $\mathcal{I}_{\cup}(\mathcal{C})$ for a given clone $\mathcal{C}$. In order to introduce the construction, we need some more notation. In the following, we consider relations as matrices, where the rows of the matrix correspond to the tuples of the relation. For this representation to be unique, we demand that the rows are ordered lexicographically. For example, the relation $\{(0,0,1,2),(2,1,0,1),(1,1,2,0)\}$ is represented by the matrix

$$\begin{pmatrix} 0\,0\,1\,2 \\ 1\,1\,2\,0 \\ 2\,1\,0\,1 \end{pmatrix}.$$

For a relation R, let $R(l,k)$ be the value at row l and column k in the matrix representation of R. By $R(l,-)$ we denote the l-th row vector and by $R(-,k)$ the k-th column vector of R. The relation D-COLS_s is defined to be the $|D|^s$-ary relation of cardinality s such that $D\text{-}\mathrm{COLS}_n(l,k) =_{\mathrm{def}} D^n(k,l)$. The main feature of this relation is that its columns contain every s-ary tuple over the domain D.

Example 4.8. The columns of the matrix representation of $\{0,1\}$-COLS_3 are exactly the binary numbers from 0 to $2^3-1=7$:

$$\{0,1\}\text{-}\mathrm{COLS}_3 = \begin{pmatrix} 0\,0\,0\,0\,1\,1\,1\,1 \\ 0\,0\,1\,1\,0\,0\,1\,1 \\ 0\,1\,0\,1\,0\,1\,0\,1 \end{pmatrix}$$

For a set of D-valued functions $\mathcal{F}$, the *$\mathcal{F}$-closure* of a relation R over D, denoted by $\mathcal{F}(R)$, is the relation $\bigcap_{S\in\mathrm{Inv}(\mathcal{F}),R\subseteq S} S$, i.e. the minimal superset of R that is closed under $\mathcal{F}$. This relation can be obtained from R by applying all functions from $\mathcal{F}$ repeatedly, and adding the result to R. It is immediate that $\mathcal{F}(R) = [\mathcal{F}](R)$. We say R is an *$\mathcal{F}$-core* of $\mathcal{F}(R)$. For a single partial function f we write $f(R)$ instead of $\{f\}(R)$ and speak of f-closures and f-cores.

Definition 4.9. *Let $\mathcal{C}$ be a clone. We say that* $\mathrm{Inv}(\mathcal{C})$ *has* core-size s *if there is a relation R such that $\langle R\rangle = \mathrm{Inv}(\mathcal{C})$ and R has a $\mathcal{C}$-core with cardinality s.*

Obviously, the core-size of a clone is not unique. As we will see in Theorem 4.11, the core-size of a co-clone is closely related to weak bases of $\mathrm{Inv}(\mathcal{C})$.

Example 4.10. Let D be the Boolean clone generated by $f(x,y,z) = x\overline{y} \vee x\overline{z} \vee (\overline{y}\wedge\overline{z})$. Then $\mathrm{Inv}(\mathrm{D})$ has core-size 1.

Proof. From [BRSV05], we know that for the Boolean relation $R=\{(0,1),(1,0)\}$ it holds that $\mathrm{Pol}(R) = \mathrm{D}$, and hence $\langle R\rangle = \mathrm{Inv}(\mathrm{D})$. Since $f(0,0,0)=1$ and $f(1,1,1)=0$, we know that the set $\{(0,1)\}$ is an f-core of R, and therefore a $\mathcal{C}$-core of R. Therefore we have shown that R has a $\mathcal{C}$-core of cardinality 1. □

Since every relation R is a $\mathrm{Pol}(R)$-core of itself, it is obvious that every co-clone $\langle R\rangle$ has core-size $|R|$. Therefore the results from [BRSV05], in which finite bases were given for each co-clone where they exist, directly give core-sizes of all Boolean co-clones with finite bases (it is obvious that every co-clone which has a finite base also has a base with just one relation, by taking the Cartesian product). The significance of $\mathcal{C}$-cores and their cardinalities is explained in the next theorem: The knowledge of the core-size of a co-clone is enough to exhibit a relation R which is a weak base of $\mathrm{Inv}(\mathcal{C})$. In the following, we often simply write $\mathcal{C}(\mathrm{COLS}_s)$ instead of $\mathcal{C}(D\text{-COLS}_s)$, as the domain should be clear from the context.

We will now show that the relations of the form $\mathcal{C}(\mathrm{COLS}_s)$ are weak bases, and therefore, as shown above, constitute the "easiest" bases with regard to all problems where the closure operator $\langle . \rangle_{\nexists}$ can be applied.

Theorem 4.11. *Let* $\mathcal{C}$ *be a clone and let* $\mathrm{Inv}(\mathcal{C})$ *have core-size* s*. Then the relation* $\mathcal{C}(\mathrm{COLS}_s)$ *is a weak base of* $\mathrm{Inv}(\mathcal{C})$*.*

Proof. We need to show that $\mathrm{pPol}(\mathcal{C}(\mathrm{COLS}_s)) = \mathcal{I}_{\cup}(\mathcal{C})$. By the prerequisites there exists a relation R such that $\mathrm{Pol}(R) = \mathcal{C}$, and which has a $\mathcal{C}$-core S of size s. Let $\mathcal{B} =_{\mathrm{def}} \mathrm{pPol}(\mathcal{C}(\mathrm{COLS}_s))$. First we show that $\mathcal{I}_{\cup}(\mathcal{C}) \subseteq \mathcal{B}$. Let $f \in \mathcal{I}_{\cup}(\mathcal{C})$ be an n-ary partial function and assume that $f \notin \mathcal{B}$. That means there are $t_1, \ldots, t_n \in \mathcal{C}(\mathrm{COLS}_s)$ such that $f(t_1, \ldots, t_n)$ is well-defined and not from $\mathcal{C}(\mathrm{COLS}_s)$. Since $D\text{-COLS}_s$ is a $\mathcal{C}$-core of $\mathcal{C}(\mathrm{COLS}_s)$, every $t \in \mathcal{C}(\mathrm{COLS}_s)$ can be derived by applying a function from $\mathcal{C}$ to the elements of $D\text{-COLS}_s$. Therefore there are s-ary functions $h_1, \ldots, h_n \in C$ such that $t_i = h_i(D\text{-COLS}_s(1, -), \ldots, D\text{-COLS}_s(s, -))$. We look at the partial function

$$g =_{\mathrm{def}} f \circ (h_1(x_1, \ldots, x_n), \ldots, h_n(x_1, \ldots, x_s)).$$

The following equation is true:

$$g(D\text{-COLS}_s(1, -), \ldots, D\text{-COLS}_s(s, -)) = f(t_1, \ldots, t_n) \notin D\text{-COLS}_s$$

Since f is defined for $(t_1, \ldots, t_n)$, it holds that g is defined on every column of $D\text{-COLS}_s$, and since $D\text{-COLS}_s$ contains of all possible s-tuples on the domain D, this means that g is a total function. Then $g \in \mathcal{C}$, because f is $\mathcal{C}$-total due to Theorem 4.7. It follows that $g \in \mathrm{Pol}(\mathcal{C})(D\text{-COLS}_s)$, but this is a contradiction. Thus, $f \in \mathcal{B}$.

We prove $\mathcal{B} \subseteq \mathcal{I}_{\cup}(\mathcal{C})$ by showing that $\mathcal{B} \cap \mathrm{OP}_D = \mathcal{C}$. The main idea is that we find R in $D\text{-COLS}_s$ and therefore these two relations have the same set of polymorphisms. By construction, it is obvious that $\mathcal{C} \subseteq \mathcal{B} \cap \mathrm{OP}_D$. To prove the other inclusion note that, since S and $D\text{-COLS}_s$ have the same size, all columns of S are columns of $D\text{-COLS}_s$ as well. Let k be the arity of S and $c_1, \ldots, c_k \in \{1, \ldots, |D|^s\}$ such that for all $i \in \{1, \ldots, k\}$ hold

$$S(-, i) = D\text{-COLS}_s(-, c_i).$$

Since S is a $\mathcal{C}$-core of R, it follows that, if we only look at the columns $c_1, \ldots, c_k$ of $\mathcal{C}(\mathrm{COLS}_s)$, we see R. More precisely, it holds

$$R = \left\{ (d_1, \ldots, d_k) \mid \begin{array}{l} \text{there is a } 1 \leq j \leq q \text{ such that} \\ d_i = \mathcal{C}(\mathrm{COLS}_s)(j, c_i) \text{ for all } 1 \leq i \leq k \end{array} \right\} \quad (1)$$

where q is the size of $\mathcal{C}(\mathrm{COLS}_s)$.

Now let $f \in \mathrm{OP}_D \setminus \mathcal{C}$ be an n-ary function. Since $f \notin \mathrm{Pol}(R)$, there are $t_1, \ldots, t_n \in R$ such that $f(t_1, \ldots, t_n) \notin R$. Let $r_1, \ldots, r_n \in \{1, \ldots, q\}$ such that

$$t_i = (\mathcal{C}\,(\mathrm{COLS}_s)\,(r_i, c_1), \ldots, \mathcal{C}\,(\mathrm{COLS}_s)\,(r_i, c_k)).$$

We take a look at $v =_{\mathrm{def}} f(\mathcal{C}\,(\mathrm{COLS}_s)\,(r_1, -), \ldots, \mathcal{C}\,(\mathrm{COLS}_s)\,(r_n, -))$. It holds that $(v[c_1], \ldots, v[c_k]) = f(t_1, \ldots, t_n) \notin R$ and using Equation 1 it follows $v \notin \mathcal{C}(\text{-}\mathrm{COLS}_s)$. This means $f \notin \mathcal{B}$ and therefore $\mathcal{B} \cap \mathrm{OP}_D \subseteq \mathcal{C}$.

Thus we showed $\mathcal{B} \cap \mathrm{OP}_D = \mathcal{C}$, which means $\mathcal{B} \subseteq \mathcal{I}_{\cup}\,(\mathcal{C})$. □

Theorem 4.11 can be used to construct weak bases for any co-clone which has a finite base at all, if we know a core-size of the co-clone. Therefore, an important question is how to determine a core-size of a co-clone. Moreover, we are interested in determining a core-size as small as possible, as the weak bases constructed above grow exponentially in the core-size. In particular, we want to be prove smaller core-sizes than the ones provided by the above-mentioned base-list in [BRSV05] for the Boolean domain. We first note that Theorem 4.11 already gives a way to characterize core-sizes, since its converse is also true:

Proposition 4.12. *Let $\mathcal{C}$ be a clone. Then* $\mathrm{Inv}\,(\mathcal{C})$ *has core-size s if and only if* $\mathrm{Pol}\,(\mathcal{C}\,(\mathrm{COLS}_s)) = \mathcal{C}$.

Proof. First assume that $\mathrm{Pol}\,(\mathcal{C}\,(\mathrm{COLS}_s)) = \mathcal{C}$. By definition of $\mathcal{C}\,(\mathrm{COLS}_s)$, we know that D-COLS_s is a $\mathcal{C}$-core of $\mathcal{C}\,(\mathrm{COLS}_s)$ which has size s. Hence, $\mathrm{Inv}\,(\mathcal{C})$ has core-size s. For the other direction, if $\mathrm{Inv}\,(\mathcal{C})$ has core-size s, then the result follows directly from Theorem 4.11. □

We now give a second criterion to determine if a number s is a core-size of a co-clone. In the following, ${\mathrm{OP}_D}^s$ denotes the set of s-ary total functions on D.

Theorem 4.13. *Let $\mathcal{C}$ be a clone. Then* $\mathrm{Inv}\,(\mathcal{C})$ *has core-size s if and only if for all functions f, if $[\{f\} \cup \mathcal{C}] \cap {\mathrm{OP}_D}^s \subseteq \mathcal{C}$, it follows that $f \in \mathcal{C}$.*

Proof. We prove both directions indirectly. First assume that $\mathrm{Inv}\,(\mathcal{C})$ does not have core-size s, then due to Proposition 4.12, there is some function $f \in \mathrm{Pol}\,(\mathcal{C}(D\text{-}\mathrm{COLS}_s))$ such that $f \notin \mathcal{C}$. Now let g be an arbitrary function from $[\{f\} \cup \mathcal{C}] \cap {\mathrm{OP}_D}^s$. We show that $g \in \mathcal{C}$. Since $f \in \mathrm{Pol}\,(\mathcal{C}(D\text{-}\mathrm{COLS}_s))$, and $\mathcal{C} \subseteq \mathrm{Pol}\,(\mathcal{C}(D\text{-}\mathrm{COLS}_s))$, and since $\mathrm{Pol}\,(\mathcal{C}(D\text{-}\mathrm{COLS}_s))$ is a clone, it follows that $g \in \mathrm{Pol}\,(\mathcal{C}(D\text{-}\mathrm{COLS}_s))$. Let $u_1, \ldots, u_s$ denote the rows of D-COLS_s. In particular it then follows that $g(u_1, \ldots, u_s) \in \mathcal{C}(D\text{-}\mathrm{COLS}_s)$. Due to the construction of this relation, it follows that there is a function $h \in \mathcal{C}$ such that $h(u_1, \ldots, u_s) = g(u_1, \ldots, u_s)$. Since g is an s-ary function, and all tuples from D^s occur in D-COLS_s, it follows that g is the same function as h, and in particular, $g \in \mathcal{C}$.

For the other direction, assume that f is an l-ary function with $[\{f\} \cup \mathcal{C}] \cap {\mathrm{OP}_D}^s \subseteq \mathcal{C}$ and $f \notin \mathcal{C}$. It suffices to prove that $f \in \mathrm{Pol}\,(\mathcal{C}(D\text{-}\mathrm{COLS}_s))$, the result then follows from Proposition 4.12. Again, let $u_1, \ldots, u_s$ denote the rows of D-COLS_s. In order to prove that f is a polymorphism of $\mathcal{C}(D\text{-}\mathrm{COLS}_s)$, let $t_1, \ldots, t_l \in \mathcal{C}(D\text{-}\mathrm{COLS}_s)$. Due to construction of the relation, this implies that there are functions $h_1, \ldots, h_l \in \mathcal{C}$ such that $t_i = h_i(u_1, \ldots, u_s)$ for all relevant i. Then $f(t_1, \ldots, t_l) = f(h_1(u_1, \ldots, u_s), \ldots, h_t(u_1, \ldots, u_s))$. Let g be defined as $g(x_1, \ldots, x_s) = f(h_1(x_1, \ldots, x_s), \ldots, h_t(x_1, \ldots, x_s))$. By construction, it follows that $g \in [\{f\} \cup \mathcal{C}] \cap \mathrm{OP}_D{}^s$, and hence $g \in \mathcal{C} \subseteq \mathrm{Pol}\,(\mathcal{C}(D\text{-COLS}_s))$. Obviously, $g(u_1, \ldots, u_s) = f(t_1, \ldots, t_l)$, and therefore $f(t_1, \ldots, t_l) \in \mathcal{C}(D\text{-COLS}_s)$. It follows that f is a polymorphism of $\mathcal{C}(D\text{-COLS}_s)$, as claimed. □

From the above Theorem, we immediately obtain the following corollary:

Corollary 4.14. *Let $\mathcal{C}$ be a clone on a finite domain. If s is a core-size of* $\mathrm{Inv}\,(\mathcal{C})$*, then $s+1$ is a core-size of* $\mathrm{Inv}\,(\mathcal{C})$ *as well.*

Proof. Let f be a function such that $[\{f\} \cup \mathcal{C}] \cap \mathrm{OP}_D{}^{s+1} \subseteq \mathcal{C}$. We need to show that $f \in \mathcal{C}$, the result then follows from Theorem 4.13. Since s is a core-size of $\mathrm{Inv}\,(\mathcal{C})$, from the same theorem it follows that it is sufficient to show that $[\{f\} \cup \mathcal{C}] \cap \mathrm{OP}_D{}^s \subseteq \mathcal{C}$. Hence, let g be an s-ary function from $[\{f\} \cup \mathcal{C}]$, and assume that $g \notin \mathcal{C}$. Let g' be the $s+1$-ary function defined by $g'(\alpha_1, \ldots, \alpha_{s+1}) := g(\alpha_1, \ldots, \alpha_s)$. Since $g \notin \mathcal{C}$, it is obvious that $g' \notin \mathcal{C}$, since $g \in [g']$. This is a contradiction, since $g' \in [\{f\} \cup \mathcal{C}] \cap \mathrm{OP}_D{}^{s+1} \subseteq \mathcal{C}$. □

We therefore have two possibilities to determine a core-size of a co-clone: Proposition 4.12 gives us a criterion which can easily be verified by an algorithm, and Theorem 4.13 gives a more theoretical answer to this question. Since we can determine the core-size of the co-clone, we have now established that we can easily construct relations which have exactly those partial polymorphisms that we are interested in, and hence allow us to apply the closure operator $\langle . \rangle_{\nexists}$.

In some situations, working directly with the relations $\mathcal{C}\,(\mathrm{COLS}_s)$ is inconvenient. Therefore, we introduce the following construction. For relations R and S with the same cardinality, where $R = \{r_1, \ldots, r_n\}$, $S = \{s_1, \ldots, s_n\}$, we denote with $R \circ S$ the relation containing the tuples $r_1 \circ s_1, \ldots, r_n \circ s_n$, where for tuples $a = (\alpha_1, \ldots, \alpha_l)$ and $b = (\beta_1, \ldots, \beta_k)$, the tuple $a \circ b$ is defined as $(\alpha_1, \ldots, \alpha_l, \beta_1, \ldots, \beta_k)$.

Definition 4.15. *Let R be a relation over the finite domain D. Then the* extension *of R, $R^{[ext]}$, is defined as* $\mathrm{Pol}\,(R)\,\big(R \circ D\text{-COLS}_{|R|}\big)$.

The extension of a relation is closely related to the weak bases of the form $\mathcal{C}\,(\mathrm{COLS}_s)$ constructed earlier. There are two main differences: For once, the expressibility result for the extension of R is weaker than that for the earlier weak bases. For the latter, we will see in Section 5 that it can be expressed by the relevant constraint languages without equality (possibly needing the $\top$-relation), while for the former we can only obtain this result for the less strict closure operator $\langle . \rangle_{\nexists}$. Another disadvantage of the extension is that the arity of this relation is usually huge: It is exponential in the cardinality of the original relation R. Therefore, it is not feasible to work with the extension for specific relations. The reason we introduce it is that the extension is a very helpful tool if we do not need to consider specific examples, but want to perform uniform proofs for arbitrary relations. The main feature of the extension is that it allows us to start with a relation R and obtain a relation which is closely related to R in such a way that it can easily simulate R-formulas, and on the other hand can be expressed by all constraint languages having the same set of polymorphisms as R. We will see an application of this idea in the proof of Theorem 6.3.

Proposition 4.16. *Let R be a relation. Then $R^{[ext]}$ is a weak base of $\langle R \rangle$, and $R^{[ext]} \in \langle \Gamma \rangle_{\nexists}$ for all constraint languages Γ with* $\mathrm{Pol}(\Gamma) = \mathrm{Pol}(R)$.

Proof. It suffices to prove $\mathrm{pPol}\left(R^{[\mathrm{ext}]}\right) = \mathcal{I}_{\cup}(\mathrm{Pol}(R))$, the second point then follows from Corollary 4.3. In order to prove this equality, from Theorem 4.11 we know that $\mathrm{pPol}\left(\mathcal{C}(D\text{-COLS}_{|R|})\right) = \mathcal{I}_{\cup}(\mathcal{C})$. Now note that $R^{[\mathrm{ext}]}$ can be obtained from $\mathcal{C}(D\text{-COLS}_{|R|})$ by reordering and doubling columns, an operation which obviously leaves the $\langle . \rangle_{\nexists}$-closure and therefore, due to Theorem 3.3, the set of partial polymorphisms invariant. □

We give an example for the extension.

Example 4.17. Let $R = \{(0,1),(1,0)\}$. Then $R^{[\mathrm{ext}]} = \begin{pmatrix} 0\,1\,0\,0\,1\,1 \\ 0\,1\,1\,0\,1\,0 \\ 1\,0\,0\,1\,0\,1 \\ 1\,0\,1\,1\,0\,0 \end{pmatrix}$.

Proof. We have already mentioned in Example 4.10, that $\mathrm{Pol}(R) = \mathrm{D}$, where D is the Boolean clone generated by the function $f(x,y,z) = x\overline{y} \vee x\overline{z} \vee (\overline{y} \wedge \overline{z})$. Since $|R| = 2$, and the domain D is the Boolean universe $\{0,1\}$, it follows that

$$D\text{-COLS}_{|R|} = \begin{pmatrix} 0\,0\,1\,1 \\ 0\,1\,0\,1 \end{pmatrix}.$$

Therefore the relation $R \circ D\text{-COLS}_{|R|}$, which is obtained by concatenating the tuples from R and those from $D\text{-COLS}_{|R|}$ is the relation

$$\begin{pmatrix} 0\,1\,0\,0\,1\,1 \\ 1\,0\,0\,1\,0\,1 \end{pmatrix}.$$

By definition, $R^{[\mathrm{ext}]}$ is the f-closure of this relation, and it can be verified that this is exactly the relation as claimed: Note that $f(x,x,x) = \overline{x}$, and therefore $R^{[\mathrm{ext}]}$ must contain all tuples from $R \circ D\text{-COLS}_{|R|}$ along with their negations. It can be verified that the set of these four tuples then is indeed closed under the function f, and hence under $\mathrm{D} = \mathrm{Pol}(R)$. □

From the earlier Example 4.10, we know that this is not an "optimal" weak base for the coclone $\mathrm{Inv}(\mathrm{D})$, since we there showed that this co-clone has core-size 1. Hence, from Theorem 4.11 it follows that $f(\{0,1\}\text{-COLS}_1)$ already is a weak base for $\mathrm{Inv}(\mathrm{D})$. Since $\{0,1\}\text{-COLS}_1$ is the relation $\{(0,1)\}$, it follows that the closure of this relation under f, which can be verified to be the relation R, is a weak base for $\mathrm{Inv}(\mathrm{D})$. Therefore, there is an $\left\{R^{[\mathrm{ext}]}, =\right\}$-formula expressing R. It can be verified that the formula $R^{[\mathrm{ext}]}(x,y,x,x,y,y)$ indeed expresses R. On the other hand, since $R^{[\mathrm{ext}]}$ is the extension of R, we know that $R^{[\mathrm{ext}]} \in \langle R \rangle_{\nexists}$, and therefore there must be an $\{R, =\}$-formula expressing $R^{[\mathrm{ext}]}$. It can be seen that $R^{[\mathrm{ext}]}(a,b,c,d,e,f)$ is equivalent to $R(a,b) \wedge (a=d) \wedge (b=e) \wedge R(c,f)$. The fact that the equality relation is not actually required to express R with the relation $R^{[\mathrm{ext}]}$ is not a coincidence. We will give some more details on this in Section 5.

We now return to the question commented on in the introduction: When can the $\langle . \rangle$-operator be applied? Our weak bases are in a certain way the "easiest" constraint languages among those with the same sets of polymorphisms. On the other hand, the entire co-clone $\mathrm{Inv}\,(\mathcal{C})$ certainly is a "hardest" set of relations with polymorphisms $\mathcal{C}$. For the Boolean case, we have another option of "hardest" bases: In [CKZ05], Creignou, Kolaitis, and Zanuttini constructed, for each Boolean clone $\mathcal{C}$, a set of relations $\Gamma_{\mathcal{C}}$ such that every constraint language Γ with $\mathrm{Pol}\,(\Gamma) = \mathcal{C}$ is contained in $\langle \Gamma_{\mathcal{C}} \rangle_{\nexists}$. These sets are in most cases not finite, but they are highly uniform, for example consisting of all relations OR^m for every possible arity m. In the following, assume that for each clone $\mathcal{C}$ over an arbitrary domain, the set $\Gamma_{\mathcal{C}}$ is some fixed set satisfying $\mathrm{Inv}\,(\mathcal{C}) = \langle \Gamma_{\mathcal{C}} \rangle_{\nexists}$ and $\mathrm{Pol}\,(\Gamma_{\mathcal{C}}) = \mathcal{C}$.

We are now in a position to state a characterization which tells us exactly when the usual closure operator can be applied to a given problem.

Theorem 4.18. *Let $\leq$ be a transitive complexity reduction, let* Problem *be a computational problem for constraint formulas such that $\langle . \rangle_{\nexists}$ can be applied to* Problem. *Then the following statements are equivalent:*

1. *$\langle . \rangle$ can be $\leq$-applied to* Problem
2. *For every constraint language Γ, every finite subset Γ' of $\Gamma_{\mathrm{Pol}(\Gamma)}$, and every weak base Γ_w of $\langle \Gamma \rangle$,* Problem (Γ') *$\leq$-reduces to* Problem (Γ_w).

Note that the second point of the theorem is equivalent to saying that there is *some* weak base Γ_w of $\langle \Gamma \rangle$ for which the condition holds: Since $\langle . \rangle_{\nexists}$ can be $\leq$-applied to Problem, all weak bases of $\langle \Gamma \rangle$ lead to the same complexity.

Proof. First assume that $\langle . \rangle$ can be $\leq$-applied to Problem, let Γ be a constraint language, let R be a weak base of $\langle \Gamma \rangle$, and let Γ' be a finite subset of $\Gamma_{\mathrm{Pol}(\Gamma)}$. In particular, this implies that $\mathrm{Pol}\,(\Gamma') \supseteq \mathrm{Pol}\,(\Gamma) = \mathrm{Pol}\,(R)$. Since $\langle . \rangle$ can be $\leq$-applied to Problem, this implies that Problem $(\Gamma') \leq$ Problem (R).

For the other direction, assume that for Γ' and R with the required properties, Problem (Γ') always reduces to Problem (R), and let Γ_1 and Γ_2 be constraint languages such that $\mathrm{Pol}\,(\Gamma_2) \subseteq \mathrm{Pol}\,(\Gamma_1)$. We need to show that Problem $(\Gamma_1) \leq$ Problem (Γ_2). For $i = 1, 2$, let $\mathcal{C}_i := \mathrm{Pol}\,(\Gamma_i)$. Since $\mathrm{Pol}\,(\Gamma_2)$ is a subset of $\mathrm{Pol}\,(\Gamma_1)$, we know that $\Gamma_1 \subseteq \langle \Gamma_2 \rangle \subseteq \langle \Gamma_{\mathcal{C}_2} \rangle_{\nexists}$. Since Γ_1 is finite, there exists a finite subset Γ' of $\Gamma_{\mathcal{C}_2}$ such that $\Gamma_1 \subseteq \langle \Gamma' \rangle_{\nexists}$. Since $\langle . \rangle_{\nexists}$ can be $\leq$-applied to Problem, this implies that Problem $(\Gamma_1) \leq$ Problem (Γ'). Let R be a weak base of $\mathrm{Inv}\,(\mathcal{C}_2)$. Due to the prerequisites, since Γ' is a finite subset of $\Gamma_{\mathcal{C}_2}$, we know that Problem $(\Gamma') \leq$ Problem (R). Since R is a weak base of $\mathrm{Inv}\,(\mathcal{C}_2)$, Corollary 4.4 implies that Problem $(R) \leq$ Problem (Γ_2). Therefore we have that Problem $(\Gamma_1) \leq$ Problem $(\Gamma') \leq$ Problem $(R) \leq$ Problem (Γ_2), and since $\leq$ is transitive, it follows that Problem $(\Gamma_1) \leq$ Problem (Γ_2), as required. □

For the Boolean case, when working with the sets $\Gamma_{\mathcal{C}}$ from [CKZ05], the finite subsets of $\Gamma_{\mathrm{Pol}(\Gamma)}$ appearing in the above constructions are of a very regular form: Since the sets $\Gamma_{\mathrm{Pol}(\Gamma)}$ are uniform sets of relations containing, for example, the relation OR of arbitrary arities, for the arbitrary finite subsets it suffices to consider subsets containing, for example, all OR-relations up to some finite arity.

5 The Equality Relation

Up to now, we only considered the closure operator $\langle . \rangle_{\nexists}$, and showed that our weak bases give rise to the problems of the lowest complexity among those generating the same co-clone. We now show how these techniques can be applied to the more restricted operator $\langle . \rangle_{\nexists,\neq}$. In order to do this, we need to consider a technical property of relations, which is closely related to the introduction of equality clauses in the closure operator $\langle . \rangle_{\nexists}$.

Definition 5.1. *Let R be an n-ary relation over the finite domain D. We say that R is* =-redundant, *if there are two identical columns in the matrix representation of R. We say that R is* $\top$-redundant *if there is $1 \leq i \leq n$ such that for all $\beta, \beta', \alpha_1, \ldots, \alpha_{i-1}, \alpha_{i+1}, \ldots, \alpha_n \in D$, it holds $(\alpha_1, \ldots, \alpha_{i-1}, \beta, \alpha_{i+1}, \ldots, \alpha_n) \in R$ if and only if $(\alpha_1, \ldots, \alpha_{i-1}, \beta', \alpha_{i+1}, \ldots, \alpha_n) \in R$. In this case we also say that the i-th column of R is $\top$-redundant. We say that R is* redundant *if it is =-redundant or $\top$-redundant.*

Redundant relations play a role in connection with the =-operator which is allowed in two of our closure operators: It is obvious that any non-empty relation which is defined with a formula containing a non-tautological equality clause is =-redundant. In particular, the equality relation itself is obviously redundant. The following proposition is obvious:

Proposition 5.2. *Let Γ be a constraint language, and let R be an =-irredundant relation such that $R \in \langle \Gamma \rangle_{\nexists}$. Then $R \in \langle \Gamma \cup \{\top\} \rangle_{\nexists,\neq}$.*

Proof. Since $R \in \langle \Gamma \rangle_{\nexists}$, we know that there is a $\Gamma \cup \{=\}$-formula which represents R, and hence there also is a $\Gamma \cup \{=, \top\}$-formula φ representing R with a minimal number of equality clauses. If no equality clause occurs in φ, then by definition $R \in \langle \Gamma \cup \{\top\} \rangle_{\nexists,\neq}$. Therefore assume that a clause $x_i = x_j$ occurs in φ. If $i = j$, then the clause is obviously tautological and can be replaced with $\top(x_i)$, a contradiction to the minimality of φ. Hence assume $i \neq j$. Then in every solution I of φ, $I(x_i)$ and $I(x_j)$ have the same value, and since φ represents R, we know that the i-th component and the j-th component of every tuple in R have the same value. Since $i \neq j$, then this is a contradiction, since R is irredundant. $\square$

Since obviously, D-COLS_s is =-irredundant for any domain D and any number s, it follows that all relations of the form $\mathcal{C}(\mathrm{COLS}_s)$ are =-irredundant. Therefore, we obtain the following result:

Corollary 5.3. *Let* $\mathcal{C}$ *be a clone, and let* $\mathrm{Inv}\,(\mathcal{C})$ *have core-size* s*. Then for every constraint language* Γ *with* $\mathrm{Pol}\,(\Gamma) = \mathcal{C}$*, it holds that* $\mathcal{C}\,(\mathrm{COLS}_s) \in \langle \Gamma \cup \top \rangle_{\nexists,\neq}$*.*

Proof. By Theorem 4.11, we know that $\mathrm{pPol}\,(\mathcal{C}\,(\mathrm{COLS}_s)) = \mathcal{I}_\cup\,(\mathcal{C})$. Therefore, from Corollary 4.3, we know that $\mathcal{C}\,(\mathrm{COLS}_s) \in \langle \Gamma \rangle_{\nexists}$. Since $\mathcal{C}\,(\mathrm{COLS}_s)$ is $=$-irredundant, Proposition 5.2 implies the result. □

It can easily be seen that there are co-clones where weak bases of the form $\mathcal{C}\,(\mathrm{COLS}_s)$ are not $\top$-irredundant (the co-clones corresponding to the Boolean clones containing all functions f satisfying $f(0,\ldots,0) = 0$, or all functions f with $f(1,\ldots,1)$ are two examples). There are several ways to deal with this problem. One possibility is to allow a constraint formula to formally depend on variables which do not appear in any clause, which is equivalent to allowing the $\top$-operator in every Γ-formula regardless of the constraint language. A more satisfying solution is to remove the redundancy from the relations.

Definition 5.4. *Let* R *be an* $=$*-irredundant relation over a finite domain* D*, where* R *is not of the form* D^n *for any* n*. Then the* irredundant core *of* R *is obtained from* R *by deleting all* $\top$*-redundant columns from its matrix representation.*

Example 5.5. Let R_0 be the Boolean clone generated by the set $\{\wedge, \oplus\}$. It can be verified that R_0 contains exactly those Boolean functions f for which $f(0,\ldots,0) = 0$ holds, and that the minimal core size of $\mathrm{Inv}\,(\mathrm{R}_0)$ is 1. The weak base $\mathrm{R}_0\,(\mathrm{COLS}_1)$ of $\mathrm{Inv}\,(\mathrm{R}_0)$ is the relation $R := \{(0,0),(0,1)\}$, and hence it is $\top$-redundant in the second column. Its irredundant core is therefore the relation $\{(0)\}$, which can be represented by the formula $R(x,x)$.

Since the relations of the form $\mathcal{C}\,(\mathrm{COLS}_s)$ are $=$-irredundant, their irredundant cores therefore neither need the equality nor the $\top$-relation to be represented. Therefore, they form the analog of weak bases for the $\langle . \rangle_{\nexists,\neq}$-operator:

Corollary 5.6. *Let* $\mathrm{OP}_D \neq \mathcal{C}$ *be a clone over a finite domain* D *and let* s *be a core-size of* $\mathrm{Inv}\,(\mathcal{C})$*. Let* R *be the irredundant core of* $\mathcal{C}\,(\mathrm{COLS}_s)$*. Then for all constraint languages* Γ *with* $\mathrm{Pol}\,(\Gamma) = \mathcal{C}$*, it holds that* $R \in \langle \Gamma \rangle_{\nexists,\neq}$*. Further,* $\mathrm{Pol}\,(R) = \mathcal{C}$*.*

Proof. Let Γ be a constraint language with $\mathrm{Pol}\,(\Gamma) = \mathcal{C}$. From Theorem 4.11, we know that $\mathcal{C}\,(\mathrm{COLS}_s)$ is a weak base of $\mathrm{Inv}\,(\mathcal{C})$. Since $\mathcal{C}$ does not contain all functions from OP_D, we know that $\mathcal{C}\,(\mathrm{COLS}_s)$ cannot be of the form D^n, since this relation clearly is invariant under all functions from OP_D. It is obvious that $\mathrm{pPol}\,(R) = \mathrm{pPol}\,(\mathcal{C}\,(\mathrm{COLS}_s))$, and hence R also is a weak base of $\mathrm{Inv}\,(\mathcal{C})$, in particular it follows that $\mathrm{Pol}\,(R) = \mathcal{C}$. From Corollary 4.3, we therefore know that $R \in \langle \Gamma \rangle_{\nexists}$. Since $\mathcal{C}\,(\mathrm{COLS}_s)$ is trivially $=$-irredundant, and R is the irredundant core of this relation, we know that R is $=$-irredundant. Therefore, from Proposition 5.2, we know that $R \in \langle \Gamma \cup \{\top\} \rangle_{\nexists,\neq}$. Therefore, there is a $\Gamma \cup \{\top\}$-formula φ expressing R. By construction, R is $\top$-irredundant, and hence all appearing $\top$-clauses in φ can be removed without changing the relation expressed by φ. Therefore, we can assume that φ is a Γ-formula, and hence $R \in \langle \Gamma \rangle_{\nexists,\neq}$, as claimed. □

Corollary 5.6 shows that the irredundant cores of our weak bases of the form $\mathcal{C}(\mathrm{COLS}_s)$ are the exact analog of weak bases for the $\langle.\rangle_{\nexists,\neq}$-operator: Just like the weak bases for the $\langle.\rangle_{\nexists}$-operator, they represent the "easiest" cases among the constraint languages with the same co-clone. The following summarizes the complexity behavior of interest to us:

Corollary 5.7. *Let* Problem *be a problem such that* $\langle.\rangle_{\nexists,\neq}$ *can be* $\leq$*-applied to* Problem. *Let* $\mathrm{OP}_D \neq \mathcal{C}$ *be a clone over the finite domain* D *such that* $\mathrm{Inv}(\mathcal{C})$ *has core-size* s*, and let* R *be the irredundant core of* $\mathcal{C}(\mathrm{COLS}_s)$*. Then for any constraint language* Γ *with* $\mathrm{Pol}(\Gamma) = \mathcal{C}$*, it holds that* $\mathsf{Problem}(R) \leq \mathsf{Problem}(\Gamma)$.

Finally, the analogous statement of Theorem 4.18 obviously also holds for the closure operator $\langle.\rangle_{\nexists,\neq}$ instead of $\langle.\rangle_{\nexists}$. The bases from [CKZ05] are also bases for the corresponding co-clone with regard to the closure operator $\langle.\rangle_{\nexists,\neq}$.

6 Applications

For problems where the complexity of the problem does not only depend on the set of total polymorphisms, like enumeration for the non-Boolean case [SS06], it is obvious that for a classification it is necessary to look into the partial clones, which constitute a refinement of the clone lattice. However, as we mentioned in Theorem 2.3 and the surrounding discussion, there are many cases where the closure operator $\langle.\rangle$ can be applied, but this only follows from a full complexity classification of the problem.

For the already-mentioned problems ENUM and EQUIV, we show how our methods can be used to obtain a full complexity classification. We will only cover the hardness proofs of the considered problems in detail, and just give a rough idea of how the polynomial time algorithms work, because the purpose of this paper is to explain how our techniques can be applied to these problems. For the enumeration problem, the $\langle.\rangle_{\nexists}$-operator can be applied, and hence we can work with the extension of a relation, $R^{[\mathrm{ext}]}$. For the equivalence problem, this is not obvious. Therefore we work with the $\langle.\rangle_{\nexists,\neq}$-closure and hence with irredundant weak bases of the form $\mathcal{C}(\mathrm{COLS}_s)$.

Since we consider these problems over the Boolean domain only, we give some background on the Boolean case, in which a full list of all clones is known [Pos41]. For a Boolean constraint language Γ, we say that Γ is *Schaefer* if Γ has a polymorphism which depends on at least two variables. Using this terminology, we can state a version of Schaefer's theorem. Note that the problem $\mathsf{CSP}(\Gamma \cup \{\{(0)\}, \{(1)\}\})$ is the problem to determine whether a given Γ-formula which also may contain clauses like $x = 0$ or $y = 1$ is satisfiable. This problem also is known as the "satisfiability problem with constants."

Theorem 6.1 ([Sch78]). *Let* Γ *be a constraint language over the Boolean domain. If* Γ *is Schaefer, then* $\mathsf{CSP}(\Gamma \cup \{\{(0)\}, \{(1)\}\})$ *can be solved in polynomial time. Otherwise, this problem is* NP*-complete under* $\leq_m^{\log}$*-reductions.*

This version of Schaefer's theorem covers almost all of the tractable cases of $\mathsf{CSP}(\Gamma)$, the only additional cases are the trivial ones: if the constant 0- or 1-function is a polymorphism of Γ, then every Γ-formula is trivially satisfiable by the all-0-vector or the all-1-vector.

The clone N contains all Boolean functions which are either constant or depend only on one variable, i.e., constants, projections, and negation. The clone N_2 contains all non-constant functions from N. The clone I contains both constants and the projections, I_1 (I_0) contains the projections and the constant 1 (the constant 0), and finally I_2 contains only the projections. It is easy to see that a constraint language Γ is not Schaefer if and only if its polymorphism set is one of the above-defined clones, and all of these clones are subsets of N. The Schaefer property is essential in the case of Boolean constraint languages, because for many problems, the tractable instances are exactly those which are Schaefer.

6.1 Enumeration

An *efficient enumeration algorithm* enumerates, given a formula φ, the set of solutions of φ such that the time between the printing of the first solution, the time between the printing of each two consecutive solutions, and between the printing of the final solution and the termination of the algorithm is bounded by a polynomial in the input. In order to obtain a complexity classification of this problem, we use CSP* as an intermediate problem, which is the problem to decide whether a propositional formula has a solution which is not constant. We first note that the $\langle . \rangle_{\nexists}$-operator can be applied to CSP*—this is trivial for the $\langle . \rangle_{\nexists,\neq}$-operator, but applying the stronger closure operator requires an additional argument.

Proposition 6.2. *$\langle . \rangle_{\nexists}$ can be $\leq_m^{\log}$-applied to* CSP*.

Proof. Let $\Gamma_1 \subseteq \langle \Gamma_2 \rangle_{\nexists}$. We show $\mathsf{CSP}^*(\Gamma_1) \leq_m^{\log} \mathsf{CSP}^*(\Gamma_2)$. If every Γ_2-formula only has constant solutions, there is an a such that all relations in Γ_2 contain only the $(a, \ldots, a)$-tuple. Since we can transform Γ_1-formulas into equivalent $(\Gamma_2 \cup \{=\})$-formulas, it follows that $\mathsf{CSP}^*(\Gamma_1) \in$ LOGSPACE. Hence assume that this case does not occur.

For a Γ_1-formula φ, we can compute an equivalent $(\Gamma_2 \cup \{=\})$-formula φ'. In φ', identify variables connected with equality clauses, and replace $(x = x)$ with $\top(x)$. Remove $\top(x)$ if x appears in a non-equality clause. Then φ has a non-constant solution if and only if φ' has one. If no clause $\top(x)$ occurs in φ', then φ' is the desired Γ_2-formula. Otherwise, x is irrelevant for φ, and thus φ has a non-constant solution if and only if it has a solution at all. Let ψ be a Γ_2-formula such that ψ does not only have constant solutions, and such that $\mathrm{VAR}(\psi) \cap \mathrm{VAR}(\varphi') = \emptyset$. Then φ has a solution if and only if φ' has a solution, if and only if $\varphi' \wedge \psi$ has a non-constant solution, completing the reduction. □

In [CH97], the following result was shown using a number of technical implementation results. We now show how this theorem can be proven in a much simpler way, using the tools introduced in Section 4.

Theorem 6.3. *Let Γ be a finite constraint language such that Γ is not Schaefer. Then* $\mathsf{CSP}^*(\Gamma)$ *is* NP*-complete under* $\leq_m^{\log}$*-reductions.*

Proof. It is obvious that the problem is in NP. For $\Gamma = \{R_1, \dots, R_n\}$ let $R = R_1 \times \cdots \times R_n$. Since $\mathsf{CSP}^*(R) \leq_m^{\log} \mathsf{CSP}^*(\Gamma)$, we can assume $\Gamma = \{R\}$ without loss of generality. We show $\mathsf{CSP}(\{R, \{(0)\}, \{(1)\}\}) \leq_m^{\log} \mathsf{CSP}^*\left(R^{[\text{ext}]}\right) \leq_m^{\log} \mathsf{CSP}^*(R)$. For the first reduction let

$$\varphi = \bigwedge_{i=1}^{k} R(x_1^i, \dots, x_r^i) \bigwedge_{i=1}^{l} y^i \bigwedge_{i=1}^{m} z^i$$

for not necessarily distinct variables x_j^i, y^i, z^i (the variables y^i and z^i are the ones appearing in the $\{(0)\}$ and $\{(1)\}$-constraints). We construct an $R^{[\text{ext}]}$-formula ψ. If $\{y^1, \dots, y^l\} \cap \{z^1, \dots, z^m\} \neq \emptyset$ we set $\psi = R^{[\text{ext}]}(x, \dots, x)$, otherwise we define

$$\psi = \bigwedge_{i=1}^{k} R^{[\text{ext}]}(x_1'^i, \dots, x_r'^i, f, w_1^i, \dots, w_{2^{|R|}-2}^i, t),$$

with $x_j'^i = \begin{cases} t & \text{if } x_j^i \in \{y^1, \dots, y^l\}, \\ f & \text{if } x_j^i \in \{z^1, \dots, z^m\}, \\ x_j^i & \text{otherwise.} \end{cases}$ for new distinctive variables t, f, and w_j^i for $1 \leq i \leq k$ and $1 \leq j \leq 2^{|R|} - 2$.

By construction of $R^{[\text{ext}]}$, for every $(v_1, \dots, v_r) \in R$ there are $t_1, \dots, t_{2^{|R|}-2} \in \{0,1\}$ such that $(v_1, \dots, v_r, 0, t_1, \dots, t_{2^{|R|}-2}, 1)$ is an element of $R^{[\text{ext}]}$. Therefore every truth assignment for φ can be extended to a non-constant solution for ψ.

Now let ψ have a non-constant solution I. Then there must be a clause in ψ such that I restricted to the variables appearing in the clause is non-constant (otherwise I would be constant because t appears in every clause). Since Γ is not Schaefer, $\mathrm{Pol}(\Gamma) = \mathrm{Pol}(R) \subseteq \mathrm{N}$.

Case: $\mathrm{Pol}(R) \in \{\mathrm{I}, \mathrm{I}_0, \mathrm{I}_1, \mathrm{I}_2\}$. If $R^{[\text{ext}]}$ has elements where the last $2^{|R|}$ values are all 0 or all 1, then that are the tuples $(0, \dots, 0)$ and $(1, \dots, 1)$. In all other tuples the $(r+1)$-st value, which in all clauses of ψ is addressed by f, is 0 and the last value, addressed by t, is 1. Since I is non-constant for some clause in ψ, that means that $I(f) = 0$ and $I(t) = 1$.

Case: $\mathrm{Pol}(R) \in \{\mathrm{N}, \mathrm{N}_2\}$. Again the last $2^{|R|}$ values of a tuple of $R^{[\text{ext}]}$ can be all equal only if all values of the tuple are equal. In all other elements the $(r+1)$-st and the last value are not equal. Therefore it holds that $I(f) \neq I(t)$. Since $\mathrm{Pol}\left(R^{[\text{ext}]}\right) = \mathrm{Pol}(R) \in \{\mathrm{N}, \mathrm{N}_2\}$, it holds that $R^{[\text{ext}]}$ is complementive, so we can assume $I(f) = 0$ and $I(t) = 1$ without loss of generality (otherwise consider the solution I' defined by $I'(x) = 1$ iff $I(x) = 0$).

So, the assignment J defined by $J(x) = \begin{cases} I(t) & \text{if } x \in \{y^1, \dots, y^l\} \\ I(f) & \text{if } x \in \{z^1, \dots, z^m\} \\ I(x) & \text{otherwise} \end{cases}$ is a solution for φ. Hence, φ is satisfiable if and only if $\psi \in \mathsf{CSP}^*\left(R^{[\text{ext}]}\right)$. The second reduction $\mathsf{CSP}^*\left(R^{[\mathrm{ext}]}\right) \leq_m^{\log} \mathsf{CSP}^*(R)$ is trivial since due to Proposition 4.16, $R^{[\mathrm{ext}]} \in \langle R \rangle_{\nexists}$, and $\langle . \rangle_{\nexists}$ can be $\leq_m^{\log}$-applied to CSP* due to Proposition 6.2. □

From this result, the complexity classification for the enumeration problem easily follows:

Corollary 6.4. *Let Γ be a Boolean constraint language. If Γ is Schaefer, then there exists an efficient algorithm computing for each Γ-formula its set of solutions. If Γ is not Schaefer, then such an algorithm does not exist, unless* $\mathrm{P} = \mathrm{NP}$.

Proof. First assume that Γ is Schaefer. Due to Theorem 6.1, it follows that for any Γ-formula φ with variables $x_1, \ldots, x_n$, the questions if $\varphi[x_1/0]$ or $\varphi[x_1/1]$ are satisfiable can be solved in polynomial time. This easily gives an efficient algorithm to enumerate all satisfying assignments for φ, by recursively enumerating the solutions for these two formulas, if they are satisfiable.

Now assume that Γ is not Schaefer. Due to Theorem 6.3, we know that the problem to determine for a given Γ-formula if it has a non-constant solution is NP-complete. Obviously, any efficient enumeration algorithm can be used to answer this question. Hence, such an algorithm cannot exist, unless $\mathrm{P} = \mathrm{NP}$. □

6.2 Equivalence and Implication

The equivalence problem, $\mathsf{EQUIV}(\Gamma)$, is the problem to decide whether two Γ-formulas have the same set of solutions. This problem is obviously in coNP, and can be shown to be truth-table reducible to the constraint satisfaction problem for Γ-formulas with constants ([BHRV02]). Hence, the equivalence problem is solvable in polynomial time if Γ is Schaefer due to Theorem 6.1. For the dichotomy, it remains to prove that the problem is coNP-hard in all other cases. The result was originally proven by Böhler, Hemaspaandra, Reith, and Vollmer in [BHRV02]. Again, we show how the proof can be simplified using our techniques (note that the proof from [BHRV02] also relies on the complexity classification of CSP^* from [CH97]).

Our proof for the result on equivalence also classifies the complexity of the implication problem for constraint formulas: Let $\mathsf{IMPL}(\Gamma)$ be the problem to decide, given Γ-formulas φ_1 and φ_2, if φ_1 implies φ_2. To our knowledge, this problem has not yet been studied in the constraint context.

Theorem 6.5. *Let Γ be a Boolean constraint language. If Γ is Schaefer, then* $\mathsf{IMPL}(\Gamma)$ *and* $\mathsf{EQUIV}(\Gamma)$ *can be solved in polynomial time, otherwise they are* coNP*-complete under $\leq_m^{\log}$-reductions.*

Proof. Obviously, $\mathsf{IMPL}(\Gamma) \leq_m^{\log} \mathsf{EQUIV}(\Gamma)$, since φ_1 implies φ_2 if and only if φ_1 is equivalent to $\varphi_1 \wedge \varphi_2$. Hence, it suffices to prove the polynomial time result for equivalence and the hardness result for implication. The former follows from the above-mentioned Turing reduction from $\mathsf{EQUIV}(\Gamma)$ to the constraint satisfaction problem for Γ-formulas with constants. For the latter, let R be the Cartesian product over all relations in Γ, then $\mathrm{Pol}(R) = \mathrm{Pol}(\Gamma)$, and since $\mathsf{IMPL}(R) \leq_m^p \mathsf{IMPL}(\Gamma)$, by Theorem 6.3, it suffices to show $\mathsf{CSP}^*(R) \leq_m^{\log} \overline{\mathsf{IMPL}(R)}$.

Let $R' := \{0,1\}\text{-COLS}_{|R|}$, and $R'' := \mathrm{Pol}(R)(R') = \mathrm{Pol}(R)\left(\mathrm{COLS}_{|R|}\right)$. By construction, the first column of R' contains only 0s, and the last column contains only 1s. Since Γ is not Schaefer, R is not either, and thus not closed under conjunction. Hence, R has at least two elements, and R' is at least 4-ary.

Let φ be an R-formula containing at least 2 variables, and let x be a variable from φ. For each variable $y \in \mathrm{VAR}(\varphi)$, introduce a clause $R''(x, y, \ldots, y, x)$, and let ψ denote the conjunction of all these clauses. Due to Corollary 5.3, $R'' \in \langle R \cup \top \rangle_{\nexists,\neq}$. Therefore, ψ can be written as an $\{R, \top\}$-formula. We will later see that ψ can in fact be written as an R-formula.

We show that φ has a non-constant solution if and only if φ does not imply ψ. Note that φ and ψ have the same constant solutions: The constant c-assignment is a solution for φ if and only if $(c, \ldots, c) \in R$ if and only if c is a polymorphism of R if and only if c is a polymorphism of R'' (since $\mathrm{Pol}(R) = \mathrm{Pol}(R'')$) if and only if the constant c-assignment is a solution for ψ.

We prove that all solutions of ψ are constant. By construction of R'', the only tuples in R'' where the first and the last components are identical are constant tuples: In the relation R', the first and the last components of each tuple are different. Since $\mathrm{Pol}(R)$ only contains injective functions and constants, tuples in which these values are identical are only generated by the application of constant polymorphisms. Therefore, every variable is assigned the same value as x in every solution for ψ, i.e., ψ only has constant solutions. In particular, since ψ has at least two variables, this implies that there can be no variable x which only appears in a clause of the form $\top(x)$. Therefore, all clauses of the form $\top(x)$ can be removed, and we can assume that ψ is an R-formula.

If φ has a non-constant solution, then φ obviously cannot imply ψ, since the latter only has constant solutions. On the other hand, if φ only has constant solutions, then φ implies ψ, since they have the same constant solutions. Therefore, the reduction is complete. □

7 Conclusion and Future Research

We have shown that the refinement of the usual Galois connection to partial polymorphisms can be applied to give easier proofs for complexity classifications over the Boolean domain. By constructing weak bases, we have exhibited natural relations such that it is sufficient to prove hardness results for these, thus addressing a common difficulty in many proofs of complexity classifications for constraint-related problems.

We believe that our techniques can be used for classifications of the complexity of many problems in the constraint context where it cannot be shown in a straightforward way that the $\langle . \rangle$-operator can be applied. The next interesting question in this line of research is the application of our tools to problems where the $\langle . \rangle$ operator cannot be applied at all. Here the weak bases can still be used to give hardness results, and Theorem 4.18 can be used to determine the clones which give rise to problems with different complexity. Among other problems, we believe that using these techniques, further results on the enumeration problem for non-Boolean domains can be achieved.

Acknowledgments

We thank Marcel Erné, Boris Romov, and Gustav Nordh for hints and discussions. We also thank the anonymous referee for helpful suggestions and improvements.

References

[ABI+05] Allender, E., Bauland, M., Immerman, N., Schnoor, H., Vollmer, H.: The complexity of satisfiability problems: Refining schaefer's theorem. In: Proceedings of the 30th International Symposium on Mathematical Foundations of Computer Science, pp. 71–82 (2005)

[AV94] Alekseev, V., Voronenko, A.: On some closed classes in partial two-valued logic. Discrete Mathematics and Applications 4(5), 401–419 (1994)

[BCRV04] Böhler, E., Creignou, N., Reith, S., Vollmer, H.: Playing with Boolean blocks, part II: Constraint satisfaction problems. SIGACT News 35(1), 22–35 (2004)

[BHRV02] Böhler, E., Hemaspaandra, E., Reith, S., Vollmer, H.: Equivalence and isomorphism for Boolean constraint satisfaction. In: Bradfield, J.C. (ed.) CSL 2002 and EACSL 2002. LNCS, vol. 2471, pp. 412–426. Springer, Heidelberg (2002)

[BKKR69] Bodrarchuk, V., Kaluzhnin, L., Kotov, V., Romov, B.: Galois theory for Post algebras i. Kibernetika 5(3), 1–10 (1969)

[BRSV05] Böhler, E., Reith, S., Schnoor, H., Vollmer, H.: Bases for Boolean co-clones. Information Processing Letters 96, 59–66 (2005)

[Bul06] Bulatov, A.: A dichotomy theorem for constraint satisfaction problems on a 3-element set. Journal of the ACM 53(1), 66–120 (2006)

[CH97] Creignou, N., Hébrard, J.: On generating all solutions of generalized satisfiability problems. Informatique Théorique et Applications/Theoretical Informatics and Applications 31(6), 499–511 (1997)

[CKZ05] Creignou, N., Kolaitis, P., Zanuttini, B.: Preferred representations of Boolean relations. Technical Report TR05-119, Electronic Colloquium on Computational Complexity, ECCC (2005)

[CV08] Creignou, N., Vollmer, H.: Boolean Constraint Satisfaction Problems. In: Creignou, N., Kolaitis, P.G., Vollmer, H. (eds.) Complexity of Constraints. LNCS, vol. 5250. Springer, Heidelberg (2008)

[Gei68] Geiger, D.: Closed systems of functions and predicates. Pac. J. Math. 27(1), 95–100 (1968)

[Jea98] Jeavons, P.: On the algebraic structure of combinatorial problems. Theoretical Computer Science 200, 185–204 (1998)

[Lad75] Ladner, R.: On the structure of polynomial-time reducibility. Journal of the ACM 22, 155–171 (1975)

[Lau06] Lau, D.: Function Algebras on Finite Sets: Basic Course on Many-Valued Logic and Clone Theory (Springer Monographs in Mathematics). Springer, New York (2006)

[Pos41] Post, E.: The two-valued iterative systems of mathematical logic. Annals of Mathematical Studies 5, 1–122 (1941)

[Rom81] Romov, B.: The algebras of partial functions and their invariants. Cybernetics and Systems Analysis 17(2), 157–167 (1981)

[Sch78] Schaefer, T.: The complexity of satisfiability problems. In: Proceedings 10th Symposium on Theory of Computing, pp. 216–226. ACM Press, New York (1978)

[SS06] Schnoor, H., Schnoor, I.: Enumerating all solutions for constraint satisfaction problems. In: Creignou, N., Kolaitis, P., Vollmer, H. (eds.) Complexity of Constraints, number 06401 in Dagstuhl Seminar Proceedings. Internationales Begegnungs- und Forschungszentrum fuer Informatik (IBFI), Schloss Dagstuhl, Germany (2006)

Introduction to the Maximum Solution Problem

Peter Jonsson[1] and Gustav Nordh[2]

[1] Department of Computer and Information Science, Linköpings Universitet
S-581 83 Linköping, Sweden
petej@ida.liu.se
[2] Laboratoire d'Informatique, École Polytechnique
F-91128 Palaiseau, France
nordh@lix.polytechnique.fr

Abstract. This paper surveys complexity and approximability results for the Maximum Solution (Max Sol) problem. Max Sol is an optimisation variant of the constraint satisfaction problem. Many important and well-known combinatorial optimisation problems are instances of Max Sol: for example, Max Sol restricted to the domain $\{0, 1\}$ is exactly the Max Ones problem (which, in turn, captures problems such as Independent Set and 0/1 Integer Programming). By using this relationship, many different problems in logic, graph theory, integer programming, and algebra can be given a uniform treatment. This opens up for new ways of analysing and solving combinatorial optimisation problems.

1 Introduction

A large number of natural optimisation problems can be viewed as instances of the Maximum Solution (Max Sol) problem. In this survey, we will describe the problem and present complexity and approximability results. This section provides the formal definition together with some background information. Some additional material on approximability and universal algebra can be found in Section 2. All results for Max Sol on Boolean domains are collected in Section 3. In order to familiarise the reader with the basic tools, Section 3 also contains a derivation of new, sharper inapproximability bounds for the Boolean Max Sol problem. Sections 4–9 contain results for the Max Sol problem on non-Boolean domains; the examples come from many different areas such as logic, graph theory, and algebra. To make the survey more readable, we devote the first of these sections to two general tractability results. The Max Sol problem is compared to other optimisation formalisms in Section 10, and some open questions are posed in Section 11.

1.1 Background and Basic Assumptions

We introduce Max Sol by first considering the well-known Max Ones problem [35]: an instance of Max Ones consists of constraints applied to a number

N. Creignou et al. (Eds.): Complexity of Constraints, LNCS 5250, pp. 255–282, 2008.

of Boolean variables, and the goal is to find an assignment that satisfies all constraints while maximising the number of variables set to 1. The only difference between MAX ONES and MAX SOL is that we do not require the domain of the variables to be Boolean—the domain of the variables in MAX SOL are allowed to be any subset of the natural numbers, and the objective is to find an assignment that satisfy all the constraints and that maximise the sum of the variables. We parameterise the problem according to the set of allowed constraint types, i.e. for any set Γ of relations, MAX SOL(Γ) denotes the set of problems where the constraint types are restricted to Γ. The main goal of this survey is to present complexity and approximability results for MAX SOL(Γ) under different choices of Γ.

Let us now take a look at the maximisation problem INTEGER PROGRAMMING:

Instance: $m \times n$ matrix A of rationals, m-vector b of rationals, and n-vector c of rationals.
Solution: An n-vector x of integers such that $Ax \geq b$ and $x \geq 0$;
Measure: $c^T x$.

Obviously, one can view the integer programming problem as a MAX SOL(Γ) problem for a certain constraint language Γ over the integers. However, we will restrict ourselves to a certain subclass of MAX SOL problems in this survey; in fact, this class do *not* contain INTEGER PROGRAMMING. The restrictions we encompass are two finiteness conditions: we only consider finite domains and finite constraint languages. This enables us to use algebraic techniques for studying MAX SOL.

The finiteness conditions do not prevent us from capturing interesting problems, though. It is easy to see that MAX SOL over the domain $\{0, 1\}$ captures, for instance, MAX INDEPENDENT SET (problem GT23 in [2]), and certain variants of MAX 0/1 PROGRAMMING (problem MP2 in [2]). There are also many interesting non-Boolean MAX SOL problems: examples include problems in integer programming [25], multiple-valued logic [33], and equation solving over Abelian groups [36].

Only considering finite constraint languages may seem quite restrictive but, in many cases, it is not. Consider for instance integer programming over the bounded domain $\{0, \ldots, d-1\}$, i.e., the size of the domain is bounded by a constant but the length of the inequalities are not. Each row in the constraint matrix can be viewed as an inequality

$$a_1x_1 + a_2x_2 + \ldots + a_kx_k \geq b.$$

Obviously, such an inequality is equivalent to the following three inequalities

$$\begin{aligned} a_1x_1 + a_2x_2 + \ldots + a_{\lfloor k/2 \rfloor}x_{\lfloor k/2 \rfloor} - z &\geq 0 \\ -a_1x_1 - a_2x_2 - \ldots - a_{\lfloor k/2 \rfloor}x_{\lfloor k/2 \rfloor} + z &\geq 0 \\ z + a_{\lfloor k/2 \rfloor+1} + \ldots + a_kx_k &\geq b \end{aligned}$$

where z denotes a fresh variable that is given the weight 0 in the objective function. By repeating this process, one ends up with a set of inequalities where each inequality contains at most three variables, and the optimal solution to this instance have the same measure as the original instance. There are at most $2^d + 2^{d^2} + 2^{d^3}$ different relations (and thus inequalities) of length ≤ 3 on a d element domain. Since the size of the domain is constant, we have reduced the problem to one with a finite constraint language. Finally, this reduction is polynomial-time: each inequality of length k in the original instance give rise to at most $3^{\lceil \log_2 k \rceil} = O(k^2)$ inequalities and at most $O(k^2)$ new variables.

The restriction to finite domains appears to be more problematic since it provably excludes certain prominent problems (such as unbounded INTEGER PROGRAMMING). There has been some efforts lately in order to make the algebraic framework applicable to infinite-domain problems [4]. To the best of our knowledge, such extended methods have not been applied to the MAX SOL problem.

1.2 Formal Definition

Let us now formally define MAX SOL: let $D \subset \mathbb{N}$ (*the domain*) be a finite set. The set of all n-tuples of elements from D is denoted by D^n. Any subset of D^n is called an n-ary relation on D. The set of all finitary relations over D is denoted by R_D. A *constraint language* over a finite set, D, is a finite set $\Gamma \subseteq R_D$. Constraint languages are the way in which we specify restrictions on our problems. The constraint satisfaction problem over the constraint language Γ, denoted CSP(Γ), is defined to be the decision problem with instance (V, D, C), where

- V is a set of variables,
- D is a fixed finite set of values (sometimes called a domain), and
- C is a set of constraints $\{C_1, \ldots, C_q\}$, in which each constraint C_i is a pair (s_i, R_i) where s_i is a list of variables of length m_i, called the constraint scope, and R_i is an m_i-ary relation over the set D, belonging to Γ, called the constraint relation.

The question is whether there exists a solution to (V, D, C) or not, that is, a function from V to D such that, for each constraint in C, the image of the constraint scope is a member of the constraint relation. To exemplify this definition, let NAE be the following ternary relation on $\{0, 1\}$: $NAE = \{0, 1\}^3 \setminus \{(0, 0, 0), (1, 1, 1)\}$. It is easy to see that the well-known **NP**-complete problem NOT-ALL-EQUAL 3-SAT can be expressed as CSP($\{NAE\}$).

The optimisation problem that we are going to study, WEIGHTED MAXIMUM SOLUTION(Γ) (which we abbreviate MAX SOL(Γ)) is defined as follows:

Instance: Tuple (V, D, C, w), where D is a fixed finite subset of $\mathbb{N}$, (V, D, C) is a CSP(Γ) instance, and $w : V \to \mathbb{N}$ is a weight function.
Solution: An assignment $f : V \to D$ to the variables such that all constraints are satisfied.
Measure: $\sum_{v \in V} w(v) \cdot f(v)$

Note that in the definition of the Max Sol(Γ)) problem, the domain D is part of the input even though it is implicitly defined by the constraint language Γ. From a complexity point of view, this does not matter since the domain is defined to be a fixed finite set and the size of the domain is constant.

We illustrate this definition with a simple example:

Example 1. Consider the domain $D = \{0, 1\}$ and the binary relation $R = \{(0,0),(1,0),(0,1)\}$. Then, Max Sol($\{R\}$) is exactly the weighted Max Independent Set problem.

Several related problems can be defined along the same lines, e.g. Min Sol where the objective is to minimise $\sum_{v \in V} w(v) \cdot f(v)$ and Max AW Sol where we allow the weight function w to be a function from V to the integers $\mathbb{Z}$. The Boolean variants of these problems have been studied earlier [30,35]. Note that for Boolean constraint languages Γ, the Max Sol(Γ) problem is usually denoted Max Ones(Γ).

There are several aspects of the definition of Max Sol that can be discussed. Below, we consider two points that have been questioned and/or criticised in the past.

I. Choice of measure function. Note that our choice of measure function in the definition of Max Sol(Γ) is just one of several reasonable choices. Another reasonable alternative, used in [36], would be to let the domain D be any finite set and introduce an additional function $g : D \to \mathbb{N}$ mapping elements from the domain to natural numbers. The measure could then be defined as $\sum_{v \in V} w(v) \cdot g(f(v))$. This would result in a parameterised problem Max Sol(Γ, g) where the goal is to classify the complexity of Max Sol(Γ, g) for all combinations of constraint languages Γ and functions g. Note that our definition of Max Sol(Γ) is equivalent to the definition of Max Sol(Γ, g) if in addition g is required to be injective. Our main motivation for the choice of measure function is to stay close to integer programming and Max Ones. However, we will use the alternative definition when studying equations over Abelian groups in Section 7.

II. Weighted vs. unweighted problems. We remark that we do not deal explicitly with the unweighted version of the problem (in the unweighted version, all variables have weight 1). The correspondence, in terms of approximability, between the weighted and unweighted versions of the problem has already been discussed in depth [13,35]. In summary, Khanna *et al.* [35] prove that if Max Sol(Γ) is in **poly-APX**, then hardness results (for the weighted version) implies the corresponding hardness results for the unweighted version. The basic idea of the proof is to simulate weights by replication of variables.

Moreover, since all tractability results for Max Sol(Γ) (that we are aware of) hold for the weighted version of the problem, it should be clear that the classifications we report here also hold for the unweighted version of the Max Sol(Γ) problem. But, in general, it is still open whether tractability for the unweighted version implies tractability of the weighted version of the Max Sol(Γ) problem. The correspondence between unweighted and weighted problems does not hold if negative weights are allowed [30].

1.3 Methods

The complexity and approximability of MAX ONES(Γ) are completely known for all choices of Γ [35]. For any Boolean constraint language Γ, MAX ONES(Γ) is either in **PO** or is **APX**-complete or **poly-APX**-complete or finding a solution of non-zero value is **NP**-hard or finding any solution is **NP**-hard. The exact borderlines between the different cases are given in [35]. For larger domains, it seems significantly harder to obtain an exact characterisation of approximability than in the Boolean case. Such a characterisation would, for instance, show whether the dichotomy conjecture for constraint satisfaction problems is true or not – a famous open question which is believed to be difficult [16]. Hence, it seems reasonable to study restricted (but as general as possible) families of constraint languages where the complexity and approximability can be determined. In doing so, the *algebraic approach* appears to be indispensable. When the algebraic approach is applicable to a certain problem, there is an equivalence relation on the constraint languages such that two constraint languages which are equivalent under this relation have the same complexity. More specifically, two constraint languages are in the same equivalence class if they generate the same *relational clone*. The relational clone generated by Γ, captures the expressive power of Γ and is denoted by $\langle \Gamma \rangle$. Hence, instead of studying every possible finite set of relations it is enough to study the relational clones.

The algebraic approach is known to be applicable to MAX SOL. In fact, it is known that constraint languages Γ_1 and Γ_2 such that $\langle \Gamma_1 \rangle = \langle \Gamma_2 \rangle$, then MAX SOL($\Gamma_1$) S-reduces to MAX SOL(Γ_2), and vice-versa. An *S-reduction* is a certain strong approximation-preserving reduction: if $\langle \Gamma_1 \rangle = \langle \Gamma_2 \rangle$, then Γ_1 and Γ_2 are very similar with respect to approximability. For instance, if MAX SOL(Γ_1) is **NP**-hard to approximate within some constant c, then MAX SOL(Γ_2) is **NP**-hard to approximate within c, too. We note that the clone-theoretic approach was not used in the original classification of MAX ONES.

2 Preliminaries

The purpose of this section is to provide a brief overview of approximability and the algebraic approach. We refer the reader to [2] for a deeper treatment of approximability and to [7,41] for a deeper treatment of the algebraic approach.

2.1 Approximability, Reductions, and Completeness

A *combinatorial optimisation problem* is defined over a set of *instances* (admissible input data); each instance I has a finite set $\mathsf{sol}(I)$ of *feasible solutions* associated with it. Given an instance I and a feasible solution s of I, $m(I, s)$ denotes the positive integer *measure* of s. The objective is, given an instance I, to find a feasible solution of *optimum* value with respect to the measure m. The optimal value is the largest one for *maximisation* problems and the smallest one for *minimisation* problems. A combinatorial optimisation problem is said to be an **NPO** problem if its instances and solutions can be recognised in polynomial time, the solutions are polynomially bounded in the input size, and the objective function can be computed in polynomial time (see, e.g., [2]).

We say that a solution $s \in \mathsf{sol}(I)$ to an instance I of an **NPO** problem Π is *r-approximate* if it is satisfying

$$\max\left\{\frac{m(I,s)}{\text{OPT}(I)}, \frac{\text{OPT}(I)}{m(I,s)}\right\} \leq r,$$

where $\text{OPT}(I)$ is the optimal value for a solution to I. An approximation algorithm for an **NPO** problem Π has *performance ratio* $\mathcal{R}(n)$ if, given any instance I of Π with $|I| = n$, it outputs an $\mathcal{R}(n)$-approximate solution.

We define **PO** to be the class of **NPO** problems that can be solved (to optimality) in polynomial time. An **NPO** problem Π is in the class **APX** if there is a polynomial-time approximation algorithm for Π whose performance ratio is bounded by a constant. Similarly, Π is in the class **poly-APX** if there is a polynomial-time approximation algorithm for Π whose performance ratio is bounded by a polynomial in the size of the input. Completeness in **APX** and **poly-APX** is defined using appropriate reductions, called AP-reductions and A-reductions, respectively [12,35]. AP-reductions are more sensitive than A-reductions and every AP-reduction is also an A-reduction [35]. In this paper we will not need the added flexibility of A-reductions for proving our **poly-APX**-completeness results. Hence, we only need the definition of AP-reductions.

Definition 2. *An* **NPO** *problem Π_1 is said to be AP-*reducible *to an* **NPO** *problem Π_2 if two polynomial-time computable functions F and G and a constant α exist such that*

(a) for any instance I of Π_1, $F(I)$ is an instance of Π_2;
(b) for any instance I of Π_1, and any feasible solution s' of $F(I)$, $G(I, s')$ is a feasible solution of I;
(c) for any instance I of Π_1, and any $r \geq 1$, if s' is an r-approximate solution of $F(I)$ then $G(I, s')$ is an $(1 + (r-1)\alpha + o(1))$-approximate solution of I where the o-notation is with respect to $|I|$.

In some cases we will use another kind of reduction, S-reductions. They are defined as follows:

Definition 3. *An* **NPO** *problem Π_1 is said to be S-*reducible *to an* **NPO** *problem Π_2 if two polynomial-time computable functions F and G exist such that*

(a) given any instance I of Π_1, algorithm F produces an instance $I' = F(I)$ of Π_2, such that the measure of an optimal solution for I', $\text{OPT}(I')$, is exactly $\text{OPT}(I)$.
(b) given $I' = F(I)$, and any solution s' to I', algorithm G produces a solution s to I such that $m_1(I, G(s')) = m_2(I', s')$, where m_1 is the measure for Π_1 and m_2 is the measure for Π_2.

Obviously, the existence of an S-reduction from Π_1 to Π_2 implies the existence of an AP-reduction from Π_1 to Π_2. The reason why we need S-reductions is that AP-reductions do not (generally) preserve membership in **PO** [35]. We also note that S-reduction preserve approximation thresholds exactly for problems in **APX**: let Π_1, Π_2 be problems in **APX**, assume that it is **NP**-hard to approximate Π_1 within c, and that there exists an S-reduction from Π_1 to Π_2. Then, it is **NP**-hard to approximate Π_2 within c, too.

2.2 Algebraic Approach

We begin by giving a number of basic definitions. An *operation* on a finite set D (the domain) is an arbitrary function $f : D^k \to D$. Any operation on D can be extended in a standard way to an operation on tuples over D, as follows: let f be a k-ary operation on D and let R be an n-ary relation over D. For any collection of k tuples, $\boldsymbol{t_1}, \boldsymbol{t_2}, \ldots, \boldsymbol{t_k} \in R$, the n-tuple $f(\boldsymbol{t_1}, \boldsymbol{t_2}, \ldots, \boldsymbol{t_k})$ is defined as follows: $f(\boldsymbol{t_1}, \boldsymbol{t_2}, \ldots, \boldsymbol{t_k}) = (f(\boldsymbol{t_1}[1], \boldsymbol{t_2}[1], \ldots, \boldsymbol{t_k}[1]), f(\boldsymbol{t_1}[2], \boldsymbol{t_2}[2], \ldots, \boldsymbol{t_k}[2]), \ldots, f(\boldsymbol{t_1}[n], \boldsymbol{t_2}[n], \ldots, \boldsymbol{t_k}[n]))$, where $\boldsymbol{t_j}[i]$ is the i-th component in tuple $\boldsymbol{t_j}$. A technique that has been shown to be useful in determining the computational complexity of CSP(Γ) is that of investigating whether the constraint language Γ is invariant under certain families of operations [28].

Now, let $R_i \in \Gamma$. If f is an operation such that for all $\boldsymbol{t_1}, \boldsymbol{t_2}, \ldots, \boldsymbol{t_k} \in R_i$ $f(\boldsymbol{t_1}, \boldsymbol{t_2}, \ldots, \boldsymbol{t_k}) \in R_i$, then R_i is *invariant* (or, in other words, closed) under f. If all constraint relations in Γ are invariant under f then Γ is invariant under f. An operation f such that Γ is invariant under f is called a *polymorphism* of Γ. The set of all polymorphisms of Γ is denoted $Pol(\Gamma)$. Given a set of operations F, the set of all relations that are invariant under all the operations in F is denoted $Inv(F)$. Whenever there is only one operation under consideration, we write $Inv(f)$ instead of $Inv(\{f\})$.

We will need a number of operations in the sequel: an operation f over D is said to be

- a *constant* operation if f is unary and $f(a) = c$ for all $a \in D$ and some $c \in D$;
- a *majority* operation if f is ternary and $f(a, a, b) = f(a, b, a) = f(b, a, a) = a$ for all $a, b \in D$;
- a *binary commutative idempotent* operation if f is binary, $f(a, a) = a$ for all $a \in D$, and $f(a, b) = f(b, a)$ for all $a, b \in D$;
- an *affine* operation if f is ternary and $f(a, b, c) = a - b + c$ for all $a, b, c \in D$ where $+$ and $-$ are the binary operations of an Abelian group $(D, +, -)$.

Example 4. Let $D = \{0, 1, 2\}$ and let f be the majority operation on D where $f(a, b, c) = a$ if a, b and c are all distinct. Furthermore, let

$$R = \{(0, 0, 1), (1, 0, 0), (2, 1, 1), (2, 0, 1), (1, 0, 1)\}.$$

It is easy to verify that for every triple of tuples, $\boldsymbol{x}, \boldsymbol{y}, \boldsymbol{z} \in R$, we have $f(\boldsymbol{x}, \boldsymbol{y}, \boldsymbol{z}) \in R$. For example, if $\boldsymbol{x} = (0, 0, 1), \boldsymbol{y} = (2, 1, 1)$, and $\boldsymbol{z} = (1, 0, 1)$, then

$$f(\boldsymbol{x}, \boldsymbol{y}, \boldsymbol{z}) = \Big(f(\boldsymbol{x}[1], \boldsymbol{y}[1], \boldsymbol{z}[1]), f(\boldsymbol{x}[2], \boldsymbol{y}[2], \boldsymbol{z}[2]), f(\boldsymbol{x}[3], \boldsymbol{y}[3], \boldsymbol{z}[3]) \Big) =$$

$$\big(f(0, 2, 1), f(0, 1, 0), f(1, 1, 1)\big) = (0, 0, 1) \in R.$$

We can conclude that R is invariant under f or, equivalently, that f is a polymorphism of R.

We sometimes need to define relations in terms of other relations, using certain logical formulas. In doing so, the algebraic approach provides a convenient tool. A first-order formula ϕ over a constraint language Γ is said to be *primitive positive* (or *pp-formula* for short) if it is of the form

$$\exists \mathbf{x} : (R_1(\mathbf{x}_1) \wedge \ldots \wedge R_k(\mathbf{x}_k))$$

where $R_1, \ldots, R_k \in \Gamma$ and $\mathbf{x}_1, \ldots, \mathbf{x}_k$ are vectors of variables such that the arity of R_i equals the length of the vector $\mathbf{x}_i$ for all i. Note that a pp-formula ϕ with m free variables defines an m-ary relation $R \subseteq D^m$, denoted $R \equiv_{pp} \phi$; the relation R is the set of all m-tuples satisfying the formula ϕ.

We continue by defining a closure operation $\langle \cdot \rangle$ on sets of relations: for any set $\Gamma \subseteq R_D$ the set $\langle \Gamma \rangle$ consists of all relations that can be expressed using relations from $\Gamma \cup \{=_D\}$ ($=_D$ is the equality relation on D), conjunction, and existential quantification, i.e. $\langle \Gamma \rangle$ is the set of all relations that can be expressed via pp-formulas over Γ. The sets of relations of the form $\langle \Gamma \rangle$ are referred to as *relational clones*. There is a very strong connection between a relational clone $\langle \Gamma \rangle$ and the operators that Γ is invariant under:

Theorem 5 ([37]). *For every set $\Gamma \subseteq R_D$, $\langle \Gamma \rangle = Inv(Pol(\Gamma))$.*

The following result shows that the algebraic approach is applicable when studying the approximability of MAX SOL(Γ):

Theorem 6 ([33]). *Let Γ be a constraint language and $\Gamma' \subseteq \langle \Gamma \rangle$ finite. Then,* MAX SOL(Γ') *is S-reducible to* MAX SOL(Γ).

The concept of *cores* has been important when studying the complexity of CSP; a constraint language is a core if every unary polymorphism is injective (i.e. a permutation). We will use a related concept (that was introduced in [33]) when studying MAX SOL.

Definition 7. *A constraint language Γ is a* max-core *if and only if there is no non-injective unary operation f in $Pol(\Gamma)$ such that $f(d) \geq d$ for all $d \in D$. A constraint language Γ' is a max-core of Γ if and only if Γ' is a max-core and $\Gamma' = f(\Gamma)$ for some unary operation $f \in Pol(\Gamma)$ such that $f(d) \geq d$ for all $d \in D$.*

The constraint language $\{R\}$ where $R = \{(0,0), (1,1), (2,1), (1,2)\}$ is *not* a max-core since it admits a unary polymorphism $f : \{0,1,2\} \to \{0,1,2\}$ defined such that $f(0) = 1$, $f(1) = 1$, and $f(2) = 2$. It is easy to see that $\{R'\}$ is a max-core of $\{R\}$ where $R' = \{(1,1), (2,1), (1,2)\}$.

Lemma 8 ([33]). *If Γ' is a max-core of Γ, then* MAX SOL(Γ) *and* MAX SOL(Γ') *are polynomial-time equivalent.*

3 Boolean Domain

The MAX SOL(Γ) problem over Boolean constraint languages Γ goes under the name MAX ONES(Γ) and the approximability for every finite Boolean constraint language has been classified by Khanna *et al.* in [35]. Before stating this result, we recall the following standard restrictions on Boolean constraint languages (see, e.g. [12]).

Definition 9

- *Γ is* 0-valid *if* $(0, 0, \ldots, 0) \in R$ *for every relation* R *in* Γ,
- *Γ is* 1-valid *if* $(1, 1, \ldots, 1) \in R$ *for every relation* R *in* Γ,
- *Γ is* Horn *if every relation R in Γ is the set of models of a CNF formula having at most one unnegated variable in each clause,*
- *Γ is* dual Horn *if every relation R in Γ is the set of models of a CNF formula having at most one negated variable in each clause,*
- *Γ is* bijunctive *if every relation R in Γ is the set of models of a CNF formula having at most two literals in each clause,*
- *Γ is* affine *if every relation R in Γ is the set of models of a system of linear equations over $GF(2)$, the field with two elements,*
- *Γ is* width-2 affine *if every relation R in Γ is the set of models of a system of linear equations over $GF(2)$ in which each equation has at most 2 variables.*

Theorem 10 ([35]). *Given a constraint language Γ over the Boolean domain* $\{0, 1\}$,

- *if Γ is 1-valid or dual-Horn or width-2 affine, then* MAX SOL(Γ) *is in* **PO**;
- *otherwise if Γ is affine, then* MAX SOL(Γ) *is* **APX**-*complete;*
- *otherwise if Γ is Horn or bijunctive, then* MAX SOL(Γ) *is* **poly-APX**-*complete;*
- *otherwise if Γ is 0-valid, then finding a solution of positive measure is* **NP**-*hard;*
- *otherwise finding a feasible solution to* MAX SOL(Γ) *is* **NP**-*hard.*

We remark that Khanna *et al.*'s proof of the preceding theorem does not make use of the algebraic approach via relational clones. The bulk of their proof is spent on showing numerous explicit implementations/reductions between different constraint languages. The advantage of using the algebraic approach is that there is no need to deal with these implementations/reductions explicitly.

In fact, it is quite easy to strengthen Khanna *et al.*'s result and give tight approximability thresholds for MAX SOL(Γ) over Boolean constraint languages.

Theorem 11. *Given a constraint language Γ over* $\{0, 1\}$,

- *if Γ is 1-valid or dual-Horn or width-2 affine, then* MAX SOL(Γ) *is in* **PO**;
- *otherwise if Γ is affine, then* MAX SOL(Γ) *is 2-approximable but not approximable to within $2 - \varepsilon$ for any $\varepsilon > 0$ unless* **P**=**NP**;

- *otherwise if Γ is Horn or bijunctive, then* Max Sol(Γ) *is approximable to within $O(|V|)$ but not to within $O(|V|^{1-\varepsilon})$ for any $\varepsilon > 0$ unless* **P** = **NP**;
- *otherwise if Γ is 0-valid, then finding a solution of positive measure is* **NP**-*hard;*
- *otherwise finding a feasible solution to* Max Sol(Γ) *is* **NP**-*hard.*

All the bounds except the $2 - \varepsilon$ hardness for affine constraint languages (that are not 1-valid) and the $O(|V|^{1-\varepsilon})$-hardness for Horn and bijunctive constraint languages are proved in (or easily follows from) [35]. The $2-\varepsilon$ hardness for affine constraint languages consisting of relations expressed by equations on the form

$$\{x_1 + \cdots + x_k = c \ (\mathrm{mod}\ 2) \mid k \text{ even}, \ c \in \{0, 1\}\}$$

follows from a more general result due to Kuivinen [36]. The proof in [36] consists of a gap preserving reduction from the problem Max-E3-Lin-2. Max-E3-Lin-2 is the following problem: given a set of equations over $\mathbb{Z}_2$ with exactly three variables per equation, satisfy as many equations as possible. It is proved in [26] that it is **NP**-hard to approximate Max-E3-Lin-2 within $2 - \varepsilon$ for any $\varepsilon > 0$.

The remaining affine constraint languages that need to be classified consist of relations expressed by equations on the form

$$\{x_1 + \cdots + x_k = 0 \ (\mathrm{mod}\ 2) \mid k \in \mathbb{N}\}.$$

The $2 - \varepsilon$ hardness for these constraint languages can be proved by modifying the corresponding reduction in [36]. For more details, please consult [32].

The $O(|V|^{1-\varepsilon})$-hardness for Horn and bijunctive constraint languages (that are neither 1-valid, nor dual-Horn, nor affine) may seem trivial at first sight since Theorem 6 together with the inclusion structure among Boolean relational clones tells us that there is an S-reduction from Max Independent Set to Max Sol over any such constraint language. But please keep in mind that S-reductions do not (in general) preserve instance size. For example, if Max Sol(Γ_1) is **NP**-hard to approximate within $O(|V|)$ and there is an S-reduction from Max Sol(Γ_1) to Max Sol(Γ_2) which blows up the instance size by a quadratic factor. Then, this only implies that Max Sol(Γ_2) is **NP**-hard to approximate within $O(|V|^2)$. Hence, we cannot use the S-reduction implied by Theorem 6 directly since it may increase instance sizes by a polynomial factor. Fortunately, it is possible with some extra work to give an S-reduction from Max Independent Set to Max Sol(Γ), which increase instance sizes by at most a factor 2, for Horn and bijunctive constraint languages (that are neither 1-valid, nor dual-Horn, nor affine). In fact, in the Horn case it is possible to give an S-reduction which only introduces two extra variables. Hence, the $O(|V|^{1-\varepsilon})$-hardness result follows from the corresponding hardness result for Max Independent Set [26,44].

4 Tractability Results

In this section, we present tractability results for two classes of constraint languages: *injective* constraint languages and *generalised max-closed* constraint languages. These two classes includes almost all of the known tractable classes of constraint languages presented in the literature for this problem. In particular, they can be seen as substantial and nontrivial generalisations of the tractable classes known for the corresponding MAX ONES problem over the Boolean domain. We have already (in Section 3) pointed out that there are only three tractable classes of constraint languages over the Boolean domain: width-2 affine, 1-valid, and dual-Horn [35]. Width-2 affine constraint languages are examples of injective constraint languages and the classes of 1-valid and dual-Horn constraint languages are examples of generalised max-closed constraint languages. The monotone constraints which are, for instance, studied by Hochbaum *et al.* [24,25] (in relation with integer programming) and Woeginger [42] (in relation with constraint satisfaction) are also related to generalised max-closed constraints. Hochbaum & Naor [25] show that monotone constraints can be characterised as those constraints that are simultaneously invariant under the max and min operators. Hence, monotone constraints are also generalised max-closed constraints as long as the underlying domain is finite.

4.1 Injective Relations

We begin by formally defining *injective relations.*

Definition 12. *A binary relation, $R \in R_D$, is called* injective *if there exists a subset $D' \subseteq D$ and an injective function $\pi : D' \to D$ such that*

$$R = \{(x, \pi(x)) \mid x \in D'\}.$$

It is important to note that the function π is *not* assumed to be total on D. Let I^D denote the set of all injective relations on the domain D and let $\Gamma_I^D = \langle I^D \rangle$. We say that a constraint language Γ is injective if $\Gamma \subseteq \Gamma_I^D$.

Example 13. Let $D = \{0, 1\}$ and let $R = \{(x, y) \mid x, y \in D,\ x + y \equiv 1 \pmod 2\}$. R is injective because the function $f : D \to D$ defined as $f(0) = 1$ and $f(1) = 0$ is injective. More generally, let $G = (D', +, -)$ be an arbitrary Abelian group and let $c \in D'$ be an arbitrary group element. It is easy to see that the relation $\{(x, y) \mid x, y \in D',\ x + y = c\}$ is injective.

R is an example of a relation which is invariant under an affine operation. Such relations have previously been studied in relation with the MAX ONES problem in [36]. We will give some additional results for such constraints in Section 7. With the terminology used in [36], R is said to be *width*-2 *affine.* The relations which can be expressed as the set of solutions to an equation with two variables over an Abelian group are exactly the width-2 affine relations in [36], so the injective relations are a superset of the width-2 affine relations. The following result is a direct consequence of [10, Sec. 4.4].

Theorem 14. *If Γ is injective, then* MAX SOL(Γ) *is in* **PO**.

An alternative way of proving Theorem 14 goes like this: MAX SOL(Γ_I^D) is in **PO** if and only if MAX SOL(I^D) is in **PO** (by Theorem 6) so we can concentrate on I^D. Given an instance of MAX SOL(I^D), consider the graph having the variables as vertices and edges between the vertices/variables occurring together in the same constraint. Each connected component of this graph represents an independent subproblem that can be solved in separately. If a value is assigned to a variable/vertex, all variables/vertices in the same component will be forced to take a value by propagating this assignment. Hence, each connected component have at most $|D|$ different solutions (that can be easily enumerated) and an optimal one can be found in polynomial time.

Injective relations can also be defined via a polymorphism: define the *discriminator* $t : D^3 \to D$ such that

$$t(a, b, c) = \begin{cases} c & \text{if } a = b, \\ a & \text{otherwise.} \end{cases}$$

It is known that $\Gamma_I^D = Inv(t)$ [41, Theorem 4.2].

4.2 Generalised Max-Closed Relations

We begin by giving the basic definition.

Definition 15. *A constraint language Γ over a domain $D \subset \mathbb{N}$ is* generalised max-closed *if and only if there exists a binary operation $f \in Pol(\Gamma)$ such that f satisfies the following two conditions:*

1. *for all $a, b \in D$ such that $a \neq b$ it holds that if $f(a, b) \leq \min(a, b)$, then $f(b, a) > \max(a, b)$; and*
2. *for all $a \in D$ it holds that $f(a, a) \geq a$.*

The following two examples will clarify the definition above.

Example 16. Assume that the domain D is $\{0, 1, 2, 3\}$. As an example of a generalised max-closed relation consider $R = \{(0, 0), (1, 0), (0, 2), (1, 2)\}$. R is invariant under max and is therefore generalised max-closed as max satisfies the properties of Definition 15. Now, consider the relation Q defined as

$$Q = \{(0, 1), (1, 0), (2, 1), (2, 2), (2, 3)\}.$$

Q is not invariant under max because

$$\max((0, 1), (1, 0)) = (\max(0, 1), \max(1, 0)) = (1, 1) \notin Q.$$

Let the operation $\circ : D^2 \to D$ be defined by the following Cayley table (note that we write $x \circ y$ instead of $\circ(x, y)$):

$\circ$	0	1	2	3
0	0	2	2	3
1	2	1	2	2
2	2	2	2	3
3	3	2	3	3

Now, it is easy to verify that $Inv(\circ)$ is a set of generalised max-closed relations and that $Q \in Inv(\circ)$.

Example 17. Consider the relations R_1 and R_2 defined as,

$$R_1 = \{(1,1,1),(1,0,0),(0,0,1),(1,0,1)\}$$

and $R_2 = R_1 \setminus \{(1,1,1)\}$. The relation R_1 is 1-valid because the tuple consisting only of ones is in R_1, i.e., $(1,1,1) \in R_1$. The relation R_2, on the other hand, is not 1-valid but is dual-Horn positive because it is invariant under max. Note that both R_1 and R_2 are generalised max-closed since R_1 is invariant under $f(x,y) = 1$ and R_2 is invariant under $f(x,y) = \max(x,y)$. It is in fact the case that every dual-Horn relation is invariant under max so the 1-valid and dual-Horn relations are subsets of the generalised max-closed relations.

We are now ready to explain the tractability of generalised max-closed constraint languages.

Theorem 18 ([32]). *If Γ is generalised max-closed, then* MAX SOL(Γ) *is in* **PO**.

The tractability of generalised max-closed constraint languages crucially depends on the following property. If Γ is generalised max-closed, then all relations

$$R = \{(d_{11}, d_{12}, \ldots, d_{1m}), \ldots, (d_{t1}, d_{t2}, \ldots, d_{tm})\}$$

in Γ have the property that the tuple

$$\boldsymbol{t}_{\max} = (\max\{d_{11}, \ldots, d_{t1}\}, \ldots, \max\{d_{1m}, \ldots, d_{tm}\})$$

is in R, too.

This property is the basis for the simple consistency based algorithm for CSPs over max-closed constraint languages, from [29]. We use the same algorithms to solve MAX SOL(Γ) when Γ is generalised max-closed. The algorithm for CSPs over max-closed constraint languages from [29], gives us a solution (if one exists) which has the property that the value assigned to each variable is the maximum value this variable is allowed to take in any solution. Hence, this solution is also the optimum solution to the corresponding MAX SOL problem. Since the only property of max-closed constraint languages that is exploited by this algorithm is that the tuple $\boldsymbol{t}_{\max}$ is in every relation, it follows that the same algorithm solves (to optimum) MAX SOL(Γ) for generalised max-closed constraint languages.

5 Clausal Constraints

We will now introduce our first example of a non-Boolean MAX SOL problem. We consider a framework for expressing constraint languages based on regular signed logic [23] over totally-ordered sets. This approach was introduced by Creignou *et al.* [11]. The set of relations that we consider is based on *regular signed logic* [23], where the underlying domain is a (possibly infinite) totally-ordered set of integers $\{0, 1, \ldots\}$. This logic provides us with convenient concepts for defining a class of relations with strong modelling capabilities. Jeavons and

Cooper [29] have proved that any constraint can be expressed as the conjunction of expressions over this class of relations. A disadvantage with their approach is that the resulting set of constraints may by exponentially large (in the number of tuples in the constraint to be expressed). An improved algorithm solving the same problem has been suggested by Gil *et al.* [17]. It takes a constraint/relation represented by the set of all assignments/tuples that satisfies it and outputs in polynomial time (in the number of tuples) an expression that is equivalent to the original constraint.

Let V be a set of variables. For $x \in V$ and $a \in D$, the inequalities $x \geq a$ and $x \leq a$ are called positive and negative literals, respectively. A *clause* is a disjunction of literals. A *clausal pattern* is a multiset of the form $P = (+a_1, \ldots, +a_p, -b_1, \ldots, -b_q)$ where $p, q \in \mathbb{N}$ and $a_i, b_i \in D$ for all i. The pattern P is said to be *negative* if $p = 0$ and *positive* if $q = 0$. The sum $p+q$, also denoted $|P|$, is the *length* of the pattern.

A *clausal language* L is a set of clausal patterns. Given a clausal language L, an L-*clause* is a pair $(P, \mathbf{x})$, where $P \in L$ is a pattern and $\mathbf{x}$ is a vector of not necessarily distinct variables from V such that $|P| = |\mathbf{x}|$. A pair $(P, \mathbf{x})$ with a pattern $P = (+a_1, \ldots, +a_p, -b_1, \ldots, -b_q)$ and variables $\mathbf{x} = (x_1, \ldots, x_{p+q})$ represents the clause

$$(x_1 \geq a_1 \vee \ldots \vee x_p \geq a_p \vee x_{p+1} \leq b_1 \vee \ldots \vee x_{p+q} \leq b_q),$$

where $\vee$ is the disjunction operator. An L-formula ϕ is a conjunction of a finite number of L-clauses. An *assignment* is a mapping $I : V \rightarrow D$ assigning a domain element $I(x)$ to each variable $x \in V$. An assignment I *satisfies* an L-formula ϕ if and only if

$$(I(x_1) \geq a_1 \vee \ldots \vee I(x_p) \geq a_p \vee I(x_{p+1}) \leq b_1 \vee \ldots \vee I(x_{p+q}) \leq b_q)$$

holds for every clause in ϕ. It can be easily seen that the literals $+0$ and $-d$ are superfluous since the inequalities $x \geq 0$ and $x \leq d$ vacuously hold. Without loss of generality, it is sufficient to only consider patterns and clausal languages without such literals. We see that clausal patterns are nothing more than a convenient way of specifying certain relations — consequently, we can use them for defining constraint languages. Thus, we make the following definitions: given a clausal language L and a clausal pattern $P = (+a_1, \ldots, +a_p, -b_1, \ldots, -b_q)$, we let $Rel(P)$ denote the corresponding relation, i.e. $Rel(P) = \{\mathbf{x} \in D^{p+q} \mid (P, \mathbf{x}) \text{ hold}\}$ and $\Gamma_L = \{Rel(P) \mid P \in L\}$.

It is easy to see that several well-studied optimisation problems are captured by this framework.

Example 19. Let the domain D be $\{0, 1\}$. The problem Independent Set (where the objective is to find an independent set of maximum weight in an undirected graph) can be viewed as the Max Sol($\Gamma_{(-0,-0)}$) problem. Similarly, Max Sol($\Gamma_{(-0,\ldots,-0)}$) (with k literals) is the Max k-Hypergraph Stable Set problem.

We can now present sufficient conditions for when MAX SOL over clausal languages is tractable and prove that it is **APX**-hard otherwise. To do so, we use a family of operations $\max_u : D^2 \to D$, $u \in D$, defined such that

$$\max_u(a,b) = \begin{cases} u & \text{if } \max(a,b) \leq u \\ \max(a,b) & \text{otherwise} \end{cases}$$

Theorem 20 ([33]). MAX SOL(Γ_L) *is tractable if* Γ_L *is invariant under* $\max_u$ *for some* $u \in D$*. Otherwise,* MAX SOL(Γ_L) *is* **APX**-*hard.*

Note that $Inv(\max_u)$ is generalised max-closed so the tractability part of Theorem 20 follows immediately from Theorem 18. The **APX**-hardness is proved by reductions from MAX INDEPENDENT SET and MAX-E3SAT.

The MIN SOL and MAX AW SOL problems for clausal constraints have also been studied: define a new family of operations $\min_u : D^2 \to D$, $u \in D$, such that

$$\min_u(a,b) = \begin{cases} u & \text{if } \min(a,b) \geq u \\ \min(a,b) & \text{otherwise} \end{cases}$$

Theorem 21 ([33]). MIN SOL(Γ_L) *is tractable if* Γ_L *is invariant under* $\min_u$ *for some* $u \in D$*. Otherwise,* MIN SOL(Γ_L) *is* **APX**-*hard.* MAX AW SOL(Γ_L) *is tractable if* Γ_L *is simultaneously invariant under* max *and* min*. Otherwise,* MAX AW SOL(Γ_L) *does not admit polynomial-time approximation schemes.*

The tractability proof for MIN SOL is analogous to the proof of Theorem 18 while the tractability proof for MAX AW SOL is based on supermodular maximisation [27,39]. Note that we do not prove **APX**-hardness for MAX AW SOL; the reason is that we are now forced to handle instances with negative optimal measure and **APX**-hardness and *AP*-reductions are only defined for problems with positive measure. However, it is still possible to rule out the existence of polynomial-time approximation schemes. For ordinary approximation problems, we say that a solution s is r-approximate if

$$\frac{\text{OPT}(I)}{r} \leq m(I,s) \leq \text{OPT}(I) \cdot r.$$

This does not work for problems with negative optima since in this case $\frac{\text{OPT}(I)}{r} \geq \text{OPT}(I) \cdot r$. With this in mind, we say that Π admits a PTAS if there exists an algorithm A satisfying the following property: for any instance I of Π and any rational value $r > 1$, $A(I,r)$ returns a solution s (in time polynomial in $|I|$) such that $m(I,s) \in [\text{OPT}(I)/r, r \cdot \text{OPT}(I)]$.

6 Binary Symmetric Relations

The complexity of CSP(R) is known for every binary and symmetric relation R due to Hell and Nešetřil's celebrated result: CSP(R) is **NP**-complete unless R is bipartite or contains a loop (and the problem is easily solvable in polynomial time). Such a dichotomy is not known for the MAX SOL problem but there are some preliminary results by Jonsson *et al.* [34]. These results are presented next.

From now on, we view binary symmetric relations as undirected graphs in the obvious way where vertices denote domain elements and an edge (a, b) is present if and only if the tuples $(a, b), (b, a)$ are members of the relation. We do not restrict ourselves to reflexive or irreflexive graphs, i.e. each vertex may or may not have a loop. We begin by showing that we can concentrate on connected graphs.

Let $\mathcal{H} = \{H_1, \ldots, H_n\}$ be a set of connected graphs and let H be the disjoint union of these graphs. We are interested in the complexity of MAX SOL(H), given the complexities of the individual problems. Let $\mathcal{H}_i = \mathcal{H} \setminus \{H_i\}$. We say that H_i *extends* the set $\mathcal{H}_i$ if there exists an instance $I = (V, D, C, w)$ of MAX SOL(H_i) such that for all $1 \leq j \leq n, j \neq i$ it holds that $\text{OPT}(I) > \text{OPT}(I_j)$ where $I_j = (V, D_j, \{H_j(x, y) \mid H_i(x, y) \in C\}, w)$. We call I a *witness* to the extension.

Assume that for some $1 \leq i \leq n$, it holds that H_i does not extend $\mathcal{H}_i$. It is clear that for any connected instance $I = (V, D, C, w)$ of MAX SOL(H), we have $\text{OPT}(I) = \text{OPT}(I_j)$ for some j, where $I_j = (V, D_j, \{H_j(x, y) \mid H(x, y) \in C\}, w)$. Furthermore, since H_i does not extend $\mathcal{H}_i$, we know that we can choose this $j \neq i$. Let H' be the disjoint union of the graphs in $\mathcal{H}_i$. Then, $\text{OPT}(I) = \text{OPT}(I')$, where $I' = (V, D, \{H'(x, y) \mid H(x, y) \in C\}, w)$ is an instance of MAX SOL(H'). For this reason, we may assume that every $H_i \in \mathcal{H}$ extends every graph in $\mathcal{H}_i$.

Lemma 22 ([34]). *Let $H_1, \ldots, H_n$ be graphs and H their disjoint union. If the problems* MAX SOL(H_i)*, $1 \leq i \leq n$ are all tractable, then* MAX SOL(H) *is tractable. If* MAX SOL(H_i) *is* **NP***-hard and H_i extends the set $\{H_1, \ldots, H_{i-1}, H_{i+1}, \ldots, H_n\}$ for some i, then* MAX SOL(H) *is* **NP***-hard.*

Next, we need a couple of algorithmic results. Let $F = \{I_1, \ldots, I_k\}$ be a family of intervals on the real line. A graph G with $V(G) = F$ and $(I_i, I_j) \in E(G)$ if and only if $I_i \cap I_j \neq \emptyset$ is called an *interval graph.* If the intervals are chosen to be inclusion-free, G is called a *proper interval graph.*

Let $F_1 = \{I_1, \ldots, I_k\}$ and $F_2 = \{J_1, \ldots, J_l\}$ be two families of intervals on the real line. A graph G with $V(G) = F_1 \cup F_2$ and $(I_i, J_j) \in E(G)$ if and only if $I_i \cap J_j \neq \emptyset$ is called an *interval bigraph.* If the intervals in each family are chosen to be inclusion-free, G is called a *proper interval bigraph.*

Lemma 23 ([34]). *If H is a connected graph which is a proper interval graph or a proper interval bigraph, then* MAX SOL(H) *is polynomial time solvable.*

We will now present some complexity results for MAX SOL(H). Let H be an irreflexive graph such that $deg(H) \leq 2$. It is easy to see that H is the disjoint union of paths and cycles. By Lemma 22, we can without loss of generality assume that H contains only one connected component. Every irreflexive path is a proper interval bigraph so Lemma 23 immediately gives a tractability result.

Proposition 24 ([34]). *If H is an irreflexive path, then* MAX SOL(H) *is in* **PO**.

Assume now that H is a cycle. If H is an odd cycle, then CSP(H) and MAX SOL(H) are **NP**-complete by Hell and Nešetřil's [22] result. If H is isomorphic to C_4, then H cannot be a max-core. One can see that the max-core of H is an irreflexive path so MAX SOL(H) is in **PO** by Proposition 24 and Lemma 8. More generally, every irreflexive even cycle that is not a max-core has a max-core that is an irreflexive path. Thus, we can additionally assume that H is a max-core.

Proposition 25 ([34]). *Let H be a max-core which is isomorphic to an even cycle, i.e., C_{2k}. Assume a bipartition $V(H) = \{d_1, \ldots, d_k\} \cup \{d'_1, \ldots, d'_k\}$ of H with $d_1 < d_2 < \cdots < d_k$ and $d'_1 < d'_2 < \cdots < d'_k$ and, without loss of generality, assume that $d_k > d'_k$. We denote by $Pol_1(H)$ the set of unary polymorphisms of H. Let $F = \{\pi \in Pol_1(H) \mid \exists j \neq k : \pi(d_j) \neq d_j \vee \pi(d'_j) \neq d'_j\}$. If there exist non-negative constants $a_1, \ldots, a_{k-1}, a'_1, \ldots, a'_{k-1}$ such that for each $\pi \in Pol_1(H) \setminus F$, it is true that*

$$\sum_{i=1}^{k-1} \left(a_i \cdot d_i + a'_i \cdot d'_i\right) > \sum_{i=1}^{k-1} \left(a_i \cdot \pi(d_i) + a'_i \cdot \pi(d'_i)\right).$$

Then, MAX SOL(H) *is* **NP**-*hard and, otherwise,* MAX SOL(H) *is in* **PO**.

The proof of Proposition 25 largely builds on constructing unary relations via pp-formulas and exploiting complexity results for the *retraction* problem [15].

We now turn our attention to graphs H such that $|V(H)| \leq 4$. For $|V(H)| = 2$ there are only two (types) of max-cores, H_1 and H_2 in Figure 1. MAX SOL(H_1) is essentially the problem of finding the largest (heaviest) bipartition in a bipartite graph, which is in **PO**. MAX SOL(H_2) is closely related to the MAX INDEPENDENT SET problem (in fact, it is exactly MAX INDEPENDENT SET for $d_1 = 0$ and $d_2 = 1$) and it is **NP**-hard (actually, **poly-APX**-complete when $d_1 = 0$, and **APX**-complete otherwise).

The complexity for 3-element graphs is completely determined in the next theorem. The main difficulty in proving the theorem is the tractability part where the CRITICAL INDEPENDENT SET problem [1,43] plays an important role. The hardness results can be obtained quite comfortably by reductions from the retraction problem and the MAX INDEPENDENT SET problem.

Theorem 26 ([34]). *There are six (types of) max-cores over $\{d_1, d_2, d_3\}$ where $d_1 < d_2 < d_3$, denoted $H_1, \ldots, H_6$ and shown in Figure 2.* MAX SOL(H) *is* **NP**-*hard for all of these except for H_5.* MAX SOL(H_5) *is in* **PO** *if $d_3 + d_1 \leq 2d_2$ and* **NP**-*hard otherwise.*

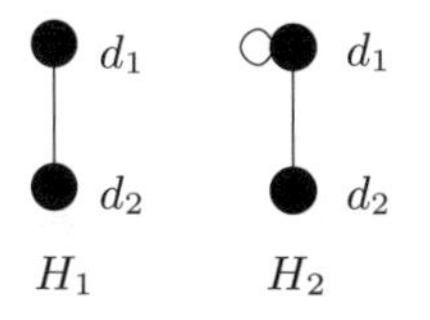

Fig. 1. Max-cores H_1 and H_2 for $V(H) = \{d_1, d_2\}$ where $d_1 < d_2$

We consider the graph H_5 closer since it highlights one of the difficulties with obtaining complexity classifications for MAX SOL(Γ). Consider the graph $H = \{(2,0),(0,2),(0,0),(0,1),(1,0),(1,1)\}$ in Figure 3 and note that it is isomorphic to H_5 by setting $d_1 = 0$, $d_2 = 1$, and $d_3 = 2$. One consequence of Theorem 26 is that MAX SOL(H) is in **PO**. However, it is easy to see that H is a max-core but not an injective relation nor a relation that is invariant under a generalised max operation. The relation H is thus an example of a relation whose tractability cannot be explained in terms of the general results in Section 4. In fact, MAX SOL(H) is essentially the CRITICAL INDEPENDENT SET problem, which was shown to be in **PO** in [1,43] by a rather clever algorithm.

If we instead consider the graph $H' = \{(3,0),(0,3),(0,0),(0,1),(1,0),(1,1)\}$ in Figure 4 (which is also isomorphic to H_5), then Theorem 26 tells us that MAX SOL(H') is **NP**-hard. Hence, despite the striking similarity of the graphs H and H', they have different complexity with respect to the MAX SOL problem. Moreover, it follows from Theorem 6 that the difference in complexity between MAX SOL(H) and MAX SOL(H') can be explained by analysing the set of polymorphisms of H and H', i.e., $Pol(H)$ and $Pol(H')$. In our opinion, this example, and

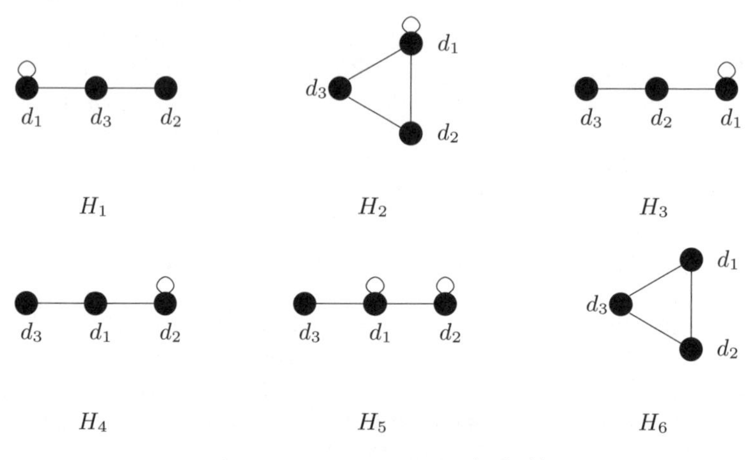

Fig. 2. Max-cores H_i for $|V(H)| = 3$

Fig. 3. A graph H for which MAX SOL is in **PO**

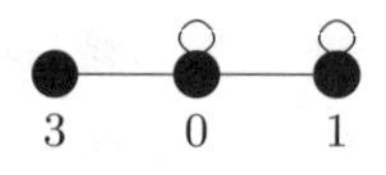

Fig. 4. A graph H' for which MAX SOL is **NP**-hard

in particular the fact that we have not (yet) been able to explain the tractability of MAX SOL(H) in terms of properties of $Pol(H)$ indicate that it might be quite challenging to classify the complexity of MAX SOL(Γ) over finite domains.

After this short digression we move on to the $|V| = 4$ case. The complexity of the $|V| = 4$ case is not completely known, but if we restrict ourselves to the vertex set $\{0, 1, 2, 3\}$, then we have the following result:

Theorem 27 ([34]). *Let H be a max-core on $D = \{0, 1, 2, 3\}$. Then,* MAX SOL(H) *is in* **PO** *if H is an irreflexive path, and otherwise,* MAX SOL(H) *is* **NP***-hard.*

The proof of Theorem 27 builds on the same ideas as the proof of Theorem 26.

7 Equations over Groups

The complexity of solving equations over different algebraic structures is a very well-studied topic; we refrain from giving a long list of examples but simply remind the reader that solving linear equations is an instance of this problem. If we assign natural numbers to the elements of the structure, then it is obvious that we can also view such a problem as an instance of the MAX SOL problem. In this section, we consider the MAX SOL problem over group equations and this motivates the next definition.

Definition 28. *Let $\mathcal{G} = (G; +, -, 0_G)$ be a group (we use $+$ for the binary group operation, $-$ for inversion and 0_G for the identity element) and $g : G \to \mathbf{N}$ a function. The* MAX SOL *problem over group equations is denoted by* MAX SOL EQN($\mathcal{G}$, g)*. An instance of* MAX SOL EQN($\mathcal{G}$, g) *is defined to be a triple (V, E, w) where,*

- *V is a set of variables,*
- *E is a set of equations of the form $w_1 + \ldots + w_k = 0_G$, where each w_i is either a variable, an inverted variable or a group constant, and*
- *w is a weight function $w : V \to \mathbf{N}$.*

The objective is to find an assignment $f : V \to G$ to the variables such that all equations are satisfied and the sum

$$\sum_{v \in V} w(v) \cdot g(f(v))$$

is maximised.

Note that the function g and the group $\mathcal{G}$ are not parts of the input so MAX SOL EQN($\mathcal{G}$, g) is parameterised by $\mathcal{G}$ and g.

Goldmann and Russell [18] have shown that solving systems of equations over non-Abelian groups is **NP**-hard. Thus, it is **NP**-hard to find feasible solutions to MAX SOL EQN($\mathcal{H}$, g) if $\mathcal{H}$ is non-Abelian, too. It is therefore sufficient to study MAX SOL EQN($\mathcal{H}$, g) where $\mathcal{H}$ is Abelian.

The main result is for groups of the form $\mathbf{Z}_p$ where p is prime. For a function $g : \mathbf{Z}_p \to \mathbf{N}$ we define the following two quantities,

$$g_{\max} = \max_{x \in \mathbf{Z}_p} g(x) \qquad \text{and} \qquad g_{\text{sum}} = \sum_{x \in \mathbf{Z}_p} g(x).$$

We are now ready to state the result.

Theorem 29 ([36]). *For every prime p and every function $g : \mathbf{Z}_p \to \mathbf{N}$,* Max Sol Eqn$(\mathbf{Z}_p, g)$ *is approximable within α where*

$$\alpha = \frac{p \cdot g_{\max}}{g_{\text{sum}}}.$$

Furthermore, for every prime p and every non-constant function $g : \mathbf{Z}_p \to \mathbf{N}$ Max Sol Eqn$(\mathbf{Z}_p, g)$ *is not approximable within $\alpha - \epsilon$ for any $\epsilon > 0$, unless* $\mathbf{P} = \mathbf{NP}$.

Note that if g is a constant function then every solution has the same measure. Obviously, an optimal can be found in polynomial time in this case. The inapproximability proof builds on the inapproximability of Max-Ek-Lin-G [26] combined with a series of reductions. The approximability result is obtained by an application of random sampling. This gives a randomised algorithm, which in expectation produces solutions of the required quality. A straightforward derandomisation of this algorithm is possible.

8 Maximal Constraint Languages

A *maximal constraint language* Γ is a constraint language such that $\langle \Gamma \rangle \subset R_D$, and if $R \notin \langle \Gamma \rangle$, then $\langle \Gamma \cup \{R\} \rangle = R_D$. That is, the maximal constraint languages are the largest constraint languages that are not able to express all finitary relations over D. This implies, among other things, that there exists an operation f such that $\langle \Gamma \rangle = Inv(f)$ whenever Γ is a maximal constraint language [38]. The complexity of the Csp(Γ) problem for all maximal constraint languages on domains $|D| \leq 3$ was determined in [8]. Moreover, it was shown in [8] that the only case that remained to be classified in order to extend the classification to all maximal constraint languages over a finite domain was the case where $\langle \Gamma \rangle = Inv(f)$ for binary commutative idempotent operations f. These constraint languages were finally classified by Bulatov in [5].

Theorem 30 ([5,8]). *Let Γ be a maximal constraint language on an arbitrary finite domain D. Then,* Csp(Γ) *is in* $\mathbf{P}$ *if $\langle \Gamma \rangle = Inv(f)$ where f is a constant operation, a majority operation, a binary commutative idempotent operation, or an affine operation. Otherwise,* Csp(Γ) *is* $\mathbf{NP}$*-complete.*

We now present an approximability classification, from [32], of Max Sol(Γ) for all maximal constraint languages Γ over $|D| \leq 4$. Moreover, it is proved in [32] that the only cases that remain to be classified, in order to extend the classification to all maximal constraint languages over finite domains, are constraint languages Γ such that $\langle\Gamma\rangle = Inv(f)$ for a binary commutative idempotent operation f. It is also proved in [32] that if a certain conjecture regarding minimal clones generated by binary operations, due to Szczepara [40], holds, then the classification can be extended to capture also these last cases.

Theorem 31. *Let Γ be maximal constraint language on a finite domain D, with $|D| \leq 4$, and $\langle\Gamma\rangle = Inv(f)$.*

1. *If Γ is generalised max-closed or an injective constraint language, then* MAX SOL(Γ) *is in* **PO***;*
2. *else if f is an affine operation, a constant operation different from the constant 0 operation, or a binary commutative idempotent operation satisfying $f(0, b) > 0$ for all $b \in D \setminus \{0\}$ (assuming $0 \in D$); or if $0 \notin D$ and f is a binary commutative idempotent operation or a majority operation, then* MAX SOL(Γ) *is* **APX***-complete;*
3. *else if f is a binary commutative idempotent operation or a majority operation, then* MAX SOL(Γ) *is* **poly-APX***-complete;*
4. *else if f is the constant 0 operation, then finding a solution with non-zero measure is* **NP***-hard;*
5. *otherwise, finding a feasible solution is* **NP***-hard.*

Moreover, if Conjecture 131 from [40] holds, then the results above hold for arbitrary finite domains D.

The proof of the preceding theorem consists of a careful analysis of the approximability of MAX SOL(Γ) for all maximal constraint languages Γ such that $\langle\Gamma\rangle = Inv(f)$, where f is one of the types of operations in Theorem 30.

9 Homogeneous Constraint Languages

In this section, we describe a classification result from [32] on the complexity of MAX SOL when the constraint language is *homogeneous.* A constraint language is called homogeneous if every *permutation relation* is contained in the language.

Definition 32. *A binary relation R is a* permutation relation *if there is a permutation $\pi : D \to D$ such that*

$$R = \{(x, \pi(x)) \mid x \in D\}.$$

Let Q denote the set of all permutation relations on D. The complexity classification of MAX SOL(Γ) when $Q \subseteq \Gamma$ from [32] provide the exact borderlines between tractability, **APX**-completeness, **poly-APX**-completeness, and **NP**-hardness of finding a feasible solution. Due to space constraints we only describe the borderline between tractability (i.e., membership in **PO**) and **NP**-hardness. This classification is described in Theorem 33.

Dalmau completely classified the complexity of CSP(Γ) when Γ is a homogeneous constraint language [14], and this classification relies heavily on the structure of homogeneous algebras. An algebra is called *homogeneous* if and only if every permutation on its universe is an automorphism of the algebra. For a formal definition and further information on the properties of homogeneous algebras we refer the reader to [41].

The approximability classification of MAX SOL(Γ) when Γ is a homogeneous constraint language uses the same approach as in [14], namely, the inclusion structure of homogeneous algebras is exploited. The (only) tractable class of MAX SOL(Γ), over homogenous constraint languages, can be characterised by the presence in $Pol(\Gamma)$ of a discriminator operation t (as defined in Section 4.1).

Theorem 33 ([32]). *Let Γ be a homogeneous constraint language. If a discriminator operation t is in $Pol(\Gamma)$, then* MAX SOL(Γ) *is in* **PO**. *Otherwise,* MAX SOL(Γ) *is* **NP**-*hard.*

Since a constraint language Γ is injective if and only if $\Gamma \subseteq Inv(t)$, the tractability part of the theorem follows from Theorem 14. The hardness part of the theorem is proved by reductions from variants of the MAX INDEPENDENT SET problem and the MAX SOL EQN(G, g) problem, as defined in Section 7.

As a direct consequence of Theorem 33 we get that the class of injective relations, as defined in Section 4.1, is a maximal tractable class for MAX SOL(Γ). That is, if we add a single relation which is not an injective relation to the class of all injective relations, then the problem is no longer in **PO** (unless **P** = **NP**).

10 Outlook

In this section, we describe the relationship between the MAX SOL problem and some other frameworks for optimisation problems.

10.1 Relation to Valued CSPs

We begin by giving a simplified account of the VCSP (valued CSP) framework studied in, e.g., [9,10]. The VCSP framework involves two types of constraints: crisp constraints (which must be satisfied) and soft constraints (which are satisfied by any assignment, but different assignments may generate different costs). More formally, a soft constraint (or valued constraint) is a pair (σ, φ) where $\sigma = (x_1, \ldots, x_k)$ is a k-tuple of variables (the constraint scope) and φ is a k-ary cost function from D to $\mathbb{Z}$. The cost of the assignment $(a_1, \ldots, a_k)$ is given by $\varphi(a_1, \ldots, a_k)$. The objective is to find a solution (satisfying all crisp constraints) that minimise (or maximise) the total cost of all soft constraints.

Just as for the ordinary CSP problem, much effort has been put into studying the complexity of the VCSP problem for various constraint languages. The constraint language can be viewed as consisting of two parts: the crisp constraint language Γ (a set of relations) and the valued constraint language Γ_V (a set of cost functions). The VCSP(Γ, Γ_V) problem can now be defined as follows:

Definition 34. *A* VCSP(Γ, Γ_V) *instance is a tuple (V, C, C_V, D), where V is a set of variables, C is a set of crisp constraints (whose constraint relations are in Γ), C_V is a set of valued constraints (whose cost functions are in Γ_V), and*

D is the domain. The objective is to find an assignment $f : V \to D$ such that all constraints in C are satisfied and the sum

$$\sum_{(\sigma,\varphi)\in C_V} \varphi(f(\sigma[1]),\ldots,f(\sigma[i]))$$

is minimised.

We denote the corresponding problem where the objective is to find the solution that maximises the sum above by MAX-VCSP(Γ, Γ_V). Note that the definitions of VCSP problems in [9,10] are much more sophisticated and elegant than our simplified account above. We also remark that, in order to comply with the existing literature on VCSP problems, we do not assume that the domain and the constraint languages Γ and Γ_V are finite.

It is easy to see that instances of MAX SOL(Γ), MIN SOL(Γ), and MAX AW SOL(Γ) problems can be seen as instances of VCSP problems.

Proposition 35.

- *Instances of* MAX SOL(Γ) *can be seen as instances of* MAX-VCSP(Γ, Γ_V) *where Γ_V consists of a single unary cost function φ, defined such that $\varphi(d) = d$ for all $d \in D$.*
- *Instances of* MIN SOL(Γ) *can be seen as instances of* VCSP(Γ, Γ_V) *where Γ_V consists of a single unary cost function φ, defined such that $\varphi(d) = d$ for all $d \in D$.*
- *Instances of* MAX AW SOL(Γ) *can be seen as instances of* MAX-VCSP(Γ, Γ_V) *where Γ_V consists of two unary cost functions φ and $\overline{\varphi}$, such that $\varphi(d) = d$ for all $d \in D$, and $\overline{\varphi}(d) = -d$ for all $d \in D$.*

Variable weights can be handled in the VCSP *setting by repeated applications of the unary cost function φ (and $\overline{\varphi}$ in the case of negative weights).*

10.2 Relation to Minimum Cost Homomorphism

The MINIMUM COST HOMOMORPHISM problem (denoted MIN HOM(H)) has recently received a lot of attention [19,20,21]. The problem MIN HOM(H) can be defined as follows:

Definition 36. *An instance of* MIN HOM(H) *is a graph G together with a weight function $w(x,y) : V(G) \times V(H) \to \mathbb{N}$. The objective is to find a homomorphism $h : V(G) \to V(H)$ such that the sum*

$$\sum_{v\in V(G)} w(v,h(v))$$

is minimised.

An instance of the MIN HOM(H) problem can also be seen as an instance of the VCSP(Γ, Γ_V) problem, where Γ consists of the single binary relation H and Γ_V is a set of unary cost functions, defined such that $\varphi_v(d) = w(v,d)$ for all $v \in V(G)$ and $d \in V(H)$. Note that Γ_V, as defined here, is not finite. We note in passing that an analogous result to Theorem 6 holds for MIN HOM(H).

Proposition 37. *Let H be a graph and $\Gamma' \subseteq \langle H \rangle$ finite. Then,* Min Hom(Γ') *is S-reducible to* Min Hom(H).

A dichotomy is known for the complexity of Min Hom(H) in the case when H is an undirected graph.

Theorem 38 ([19]). *If each component of H is a proper interval graph or a proper interval bigraph, then the problem* Min Hom(H) *is in* **PO**. *In all other cases,* Min Hom(H) *is* **NP***-hard.*

A list extension of the Max AW Sol(H) problem (denoted List Max AW Sol(H)) is studied in [34]. The Max AW Sol(H) problem is extended by introducing lists, $\{L(v) \subseteq V(H) \,|\, v \in V(G)\}$, and requiring that any solution must assign to v one of the vertices in $L(v)$. It is not hard to realise that the List Max AW Sol(H) problem is a restriction of the Min Hom(H) problem. One of the results of [34] is a complexity classification of List Max AW Sol(H) for undirected graphs H.

Theorem 39 ([34]). *Let H be an undirected graph with loops allowed. Then* List Max AW Sol(H) *is solvable in polynomial time if all components of H are proper interval graphs or proper interval bigraphs. Otherwise* List Max AW Sol(H) *is* **NP***-hard.*

The preceding result, together with the complexity classification of Min Hom(H) from [19], gives us the following corollary.

Corollary 40. *Let H be an undirected graph with loops allowed. Then,* List Max AW Sol(H) *is polynomial-time equivalent to* Min Hom(H).

11 Open Questions

The long-term goal for this line of research is, of course, to completely classify the approximability of Max Sol for all finite constraint languages. However, this is probably a hard problem since not even a complete classification for the corresponding decision problem Csp is known. A more manageable task would be to completely classify Max Sol for constraint languages over small domains (say, of size 3 or 4). For size 3, this has already been accomplished for Csp [6] and Max Csp [31]. Another obvious open problem is to classify the complexity of Max Sol(Γ) for the remaining maximal constraint languages Γ described in Section 8. The known results for the complexity of Max Sol suggest the following conjecture:

Conjecture 41. For every finite constraint language Γ, one of the following holds:

1. Max Sol(Γ) is in **PO**;
2. Max Sol(Γ) is **APX**-complete;
3. Max Sol(Γ) **poly-APX**-complete;
4. it is **NP**-hard to find a non-zero solution to Max Sol(Γ); or
5. it is **NP**-hard to find any solution to Max Sol(Γ).

If this conjecture is true, then there does not exist any constraint language Γ such that MAX SOL(Γ) has a polynomial-time approximation scheme (PTAS) but MAX SOL(Γ) is not in **PO**. However, if one impose simultaneous restrictions on the allowed constraint types and the way constraints are applied to variables (instead of only restricting the allowed constraint types), then the situation changes. Consider for example MAX INDEPENDENT SET (and, equivalently, MAX ONES($\{(0,0),(1,0),(0,1)\}$)): the unrestricted problem is **poly-APX**-complete and not approximable within $O(n^{1-\epsilon})$, $\epsilon > 0$ (unless **P**=**NP**) [44], but the problem restricted to planar instances admits a PTAS [3]. One may ask several questions in connection with this: For which constraint languages does MAX SOL admit a PTAS on planar instances? Or more generally: under what restrictions on variable scopes does MAX SOL(Γ) admit a PTAS?

The investigation of the complexity of MAX SOL(Γ) when Γ is a graph [34] indicates that giving a complete complexity classification of MAX SOL(Γ) for every fixed constraint language Γ is probably harder than first anticipated. In particular, the tractable class for the MAX SOL problem identified in [34] (and described in Section 6) depends very subtly on the values of the domain elements and no characterisation of this tractable class in terms of polymorphisms is known. Hence, this tractable class seems to be of a different flavour compared to the other tractable classes for the MAX SOL problem [32,33,35]. On the other hand, there is a complete classification for the complexity of the arbitrary weighted list version of the problem, LIST MAX AW SOL(Γ), in the important case where Γ is an undirected graph.

Interestingly, for undirected graphs H the borderline between tractability and **NP**-hardness for LIST MAX AW SOL(H) coincide exactly with Gutin *et al.*'s [19] recent complexity classification of MIN HOM(H). This is surprising, since the MIN HOM(H) problem is much more expressive than the LIST MAX AW SOL(H) problem, and we were expecting graphs H such that MIN HOM(H) were **NP**-hard and LIST MAX AW SOL(H) were in **PO**. Moreover, it is not hard to prove that MIN HOM(Γ) over Boolean constraint languages Γ is in **PO** if Γ is width-2 affine, or if Γ is Horn and dual-Horn; and that MIN HOM(Γ) is **NP**-hard for all other Boolean constraint languages Γ. This borderline between **PO** and **NP**-hardness coincides with the borderline between **PO** and **NP**-hardness for MAX AW SOL(Γ) over Boolean constraint languages Γ proved in [30]. The obvious question raised by these results is how far can we extend the correspondence in complexity between LIST MAX AW SOL(Γ) and MIN HOM(Γ)? More specifically, is it the case that LIST MAX AW SOL(Γ) is in **PO** (**NP**-hard) if and only if MIN HOM(Γ) is in **PO** (**NP**-hard) for arbitrary finite constraint languages Γ?

Acknowledgments

The authors thank Fredrik Kuivinen and Johan Thapper for making important contributions to this survey. We also thank the anonymous referee for providing some very useful comments. Peter Jonsson is partially supported by the *Center for Industrial Information Technology* (CENIIT) under grant 04.01, and by the *Swedish Research Council* (VR) under grant 621-2003-3421. Gustav Nordh is partially supported by the *Swedish-French Foundation*, and by the *National Graduate School in Computer Science* (CUGS), Sweden.

References

1. Ageev, A.: On finding critical independent and vertex sets. SIAM J. Discrete Math. 7(2), 293–295 (1994)
2. Ausiello, G., Crescenzi, P., Gambosi, G., Kann, V., Marchetti Spaccamela, A., Protasi, M.: Complexity and approximation: Combinatorial optimization problems and their approximability properties. Springer, Heidelberg (1999)
3. Baker, B.: Approximation algorithms for NP-complete problems on planar graphs. Journal of the ACM 41, 153–180 (1994)
4. Bodirsky, M., Nešetřil, J.: Constraint satisfaction with countable homogeneous templates. Journal of Logic and Computation 16(3), 359–373 (2006)
5. Bulatov, A.: A graph of a relational structure and constraint satisfaction problems. In: Proceedings of the 19th IEEE Symposium on Logic in Computer Science (LICS 2004), pp. 448–457 (2004)
6. Bulatov, A.: A dichotomy theorem for constraint satisfaction problems on a 3-element set. Journal of the ACM 53(1), 66–120 (2006)
7. Bulatov, A., Jeavons, P., Krokhin, A.: Classifying the complexity of constraints using finite algebras. SIAM J. Comput. 34(3), 720–742 (2005)
8. Bulatov, A., Krokhin, A., Jeavons, P.: The complexity of maximal constraint languages. In: Proceedings of the 33rd ACM Symposium on Theory of Computing (STOC 2001), pp. 667–674 (2001)
9. Cohen, D., Cooper, M., Jeavons, P.: An algebraic characterisation of complexity for valued constraint. In: Benhamou, F. (ed.) CP 2006. LNCS, vol. 4204, pp. 107–121. Springer, Heidelberg (2006)
10. Cohen, D., Cooper, M., Jeavons, P., Krokhin, A.: The complexity of soft constraint satisfaction. Artificial Intelligence 170(11), 983–1016 (2006)
11. Creignou, N., Hermann, M., Krokhin, A., Salzer, G.: Complexity of clausal constraints over chains. Theory Comput. Syst. 42(2), 239–255 (2008)
12. Creignou, N., Khanna, S., Sudan, M.: Complexity classifications of Boolean constraint satisfaction problems. SIAM, Philadelphia (2001)
13. Crescenzi, P., Silvestri, R., Trevisan, L.: On weighted vs unweighted versions of combinatorial optimization problems. Inf. Comput. 167(1), 10–26 (2001)
14. Dalmau, V.: A new tractable class of constraint satisfaction problems. Annals of Mathematics and Artificial Intelligence 44(1-2), 61–85 (2005)
15. Feder, T., Hell, P., Huang, J.: List homomorphisms and circular arc graphs. Combinatorica 19, 487–505 (1999)
16. Feder, T., Vardi, M.Y.: The computational structure of monotone monadic SNP and constraint satisfaction: A study through datalog and group theory. SIAM J. Comput. 28(1), 57–104 (1999)
17. Gil, À., Hermann, M., Salzer, G., Zanuttini, B.: Efficient algorithms for constraint description problems over finite totally ordered domains. In: Basin, D., Rusinowitch, M. (eds.) IJCAR 2004. LNCS, vol. 3097, pp. 244–258. Springer, Heidelberg (2004)
18. Goldmann, M., Russell, A.: The complexity of solving equations over finite groups. In: IEEE Conference on Computational Complexity, pp. 80–86 (1999)

19. Gutin, G., Hell, P., Rafiey, A., Yeo, A.: A dichotomy for minimum cost graph homomorphisms. European J. Combin. 29(4), 900–911 (2008)
20. Gutin, G., Rafiey, A., Yeo, A.: Minimum cost and list homomorphisms to semicomplete digraphs. Discrete Applied Mathematics 154(6), 890–897 (2006)
21. Gutin, G., Rafiey, A., Yeo, A., Tso, M.: Level of repair analysis and minimum cost homomorphisms of graphs. Discrete Applied Mathematics 154(6), 881–889 (2006)
22. Hell, P., Nešetřil, J.: On the complexity of H-colouring. Journal of Combinatorial Theory B 48, 92–110 (1990)
23. Hähnle, R.: Complexity of many-valued logics. In: Proceedings of the 31st IEEE International Symposium on Multiple-valued Logic (ISMVL 2001), pp. 137–148 (2001)
24. Hochbaum, D., Megiddo, N., Naor, J., Tamir, A.: Tight bounds and 2-approximation algorithms for integer programs with two variables per inequality. Mathematical Programming 62, 69–84 (1993)
25. Hochbaum, D., Naor, J.: Simple and fast algorithms for linear and integer programs with two variables per inequality. SIAM J. Comput. 23(6), 1179–1192 (1994)
26. Håstad, J.: Some optimal inapproximability results. Journal of the ACM 48(4), 798–859 (2001)
27. Iwata, S., Fleischer, L., Fujishige, S.: A combinatorial strongly polynomial algorithm for minimizing submodular functions. Journal of the ACM 48(4), 761–777 (2001)
28. Jeavons, P., Cohen, D., Gyssens, M.: Closure properties of constraints. Journal of the ACM 44, 527–548 (1997)
29. Jeavons, P., Cooper, M.: Tractable constraints on ordered domains. Artificial Intelligence 79, 327–339 (1996)
30. Jonsson, P.: Boolean constraint satisfaction: complexity results for optimization problems with arbitrary weights. Theoretical Computer Science 244(1-2), 189–203 (2000)
31. Jonsson, P., Klasson, M., Krokhin, A.: The approximability of three-valued Max CSP. SIAM J. Comput. 35(3), 1329–1349 (2006)
32. Jonsson, P., Kuivinen, F., Nordh, G.: Max Ones generalised to larger domains. SIAM J. Comput. 38(1), 329–365 (2008)
33. Jonsson, P., Nordh, G.: Generalised integer programming based on logically defined relations. In: Královič, R., Urzyczyn, P. (eds.) MFCS 2006. LNCS, vol. 4162, pp. 549–560. Springer, Heidelberg (2006)
34. Jonsson, P., Nordh, G., Thapper, J.: The maximum solution problem on graphs. In: Kučera, L., Kučera, A. (eds.) MFCS 2007. LNCS, vol. 4708, pp. 228–239. Springer, Heidelberg (2007)
35. Khanna, S., Sudan, M., Trevisan, L., Williamson, D.: The approximability of constraint satisfaction problems. SIAM J. Comput. 30(6), 1863–1920 (2001)
36. Kuivinen, F.: Tight approximability results for the maximum solution equation problem over Z_p. In: Jedrzejowicz, J., Szepietowski, A. (eds.) MFCS 2005. LNCS, vol. 3618, pp. 628–639. Springer, Heidelberg (2005)
37. Pöschel, R., Kaluznin, L.: Funktionen- und Relationenalgebren. DVW, Berlin (1979)
38. Rosenberg, I.: Minimal clones I: the five types. In: Szabó, L., Szendrei, Á. (eds.) Lectures in Universal Algebra. North-Holland, Amsterdam (1986)
39. Schrijver, A.: A combinatorial algorithm for minimizing submodular functions in polynomial time. Journal of Combinatorial Theory B 80, 346–355 (2000)
40. Szczepara, B.: Minimal clones generated by groupoids. PhD thesis, Université de Montréal (1996)

41. Szendrei, Á.: Clones in Universal Algebra. In: Séminaires de Mathématiques Supérieures, University of Montreal, vol. 99 (1986)
42. Woeginger, G.: An efficient algorithm for a class of constraint satisfaction problems. Operations Research Letters 30(1), 9–16 (2002)
43. Zhang, C.: Finding critical independent sets and critical vertex subsets are polynomial problems. SIAM J. Discrete Math. 3(3), 431–438 (1990)
44. Zuckerman, D.: Linear degree extractors and the inapproximability of max clique and chromatic number. In: Proceedings of the 38th ACM Symposium on Theory of Computing (STOC 2006), pp. 681–690 (2006)

Present and Future of Practical SAT Solving

Oliver Kullmann[1,*]

Computer Science Department, Swansea University
Swansea, SA2 8PP, UK
O.Kullmann@Swansea.ac.uk
http://cs.swan.ac.uk/~csoliver

Abstract. We review current SAT solving, concentrating on the two paradigms of *conflict-driven* and *look-ahead* solvers, and with a view towards the unification of these two paradigms. A general "modern" scheme for DPLL algorithms is presented, which allows natural representations for "modern" solvers of these two types.

1 Introduction

The number as well as the breadth of applications of SAT solving, like verification of hardware and software or solving difficult concrete combinatorial problem instances, has steadily increased over the last 10 years. Two main paradigms for backtracking solvers have emerged, the "conflict-driven solver" and the "look-ahead solver", the former better suited for verification problems, the latter better for difficult problems. Both paradigms now have reached some form of plateau, and the purpose of this article is to present these two different plateaus (kind of "fixed points"), and to discuss several ideas towards possible combinations of these two approaches, to overcome the current relative stagnation (regarding the core algorithms). We focus on "practical" algorithms which "work" (at least reasonably often), as represented by the SAT conference and the SAT competition.

Two basic approaches for SAT solving in general can be identified: Local search (for satisfying assignments), and backtracking. Local search has still a stronghold for satisfiable random formulas, and in recent years the theoretically very interesting "survey propagation" algorithm was developed, but outside the domain of (satisfiable) random formulas local search lost influence, and in this article we will not consider it (unquestionably there is a lot of potential, but it needs further development). Steady improvements of look-ahead backtracking solvers are noticeable, but the strongest development took place w.r.t. conflict-driven backtracking solvers, and accordingly we will put emphasise in this article on the notion of "clause learning". Motivated by the success of SAT, extensions of SAT (like pseudo-boolean formulas, quantified boolean formulas or "SAT modulo theory") become increasingly popular, but yet no clear pattern emerged here. For the whole area there are many beliefs, many observations, but no proofs;

* Partially supported by grant EPSRC GR/S58393/01.

N. Creignou et al. (Eds.): Complexity of Constraints, LNCS 5250, pp. 283–319, 2008.

nevertheless it seems that the subjects of this article are mature enough that a more systematic treatment might be possible.

Yet, in SAT no "theoretical idea" had impact on the "practice" of SAT solving: Although there have been many attempts, they never went far enough, and we do not understand the practical applications. I believe

- practice needs a *dedicated* effort, much more details and care in some areas, and more looseness in other areas,
- but there is much more to discover than the current "trivial" solvers!

So in this article I want somehow to present the "practical world" — hopefully we can learn from their observations and ideas. The observable stagnation regarding the core algorithms can be overcome in my opinion by unifying yet separate development lines:

1. The three main paradigms for SAT solving, "conflict-driven", "look-ahead" and "local search" should be combined (in a new, intelligent way).
2. SAT (with its focus on global structure) and CSP (with its focus on local structure) need to be unified. Here with "global structure" I allude at the fact that SAT problems in CNF representation come "chopped into little pieces", and the solution process considers statistical properties arising from pieces potentially belonging to very different parts of the problem instance. In contrast, traditionally the field of constraint satisfaction studies extensively "intelligent" problem representations, but at the cost of more global (and less predictable) structures.

In this article we focus on *look-ahead versus conflict-driven* (both are resolution-based, and thus closer together), trying to bring out the (quite different) underlying ideas, and to discuss how potentially those two approaches could be brought together (and what (considerable) problems have to be overcome for such a unification).[1] An outline of this article follows:

1. Conjunctive normal forms seem to be at the heart of "SAT", and they are discussed in Section 2.
2. An overview on polynomial time methods which seem relatively close to "practice" is given in Section 3.
3. The gist of this article is given by the new general scheme $\mathfrak{G}$ for DPLL-algorithm presented in Subsection 4.1, unifying the look-ahead and the conflict-driven paradigms. Specialisations yield a general look-ahead scheme `la` in Subsection 5.1, and a general conflict-driven scheme `cd` in Subsection 6.1.3.
4. DPLL in general is discussed in Section 4, while look-ahead solvers are presented in Section 5, and the main features of conflict-driven solvers are the subject of Section 6.
5. Approaches for understanding and extending clause-learning are outlined in Section 7.
6. Finally some conclusions are drawn in Section 8.

[1] The new `OKsolver`, tentatively called "OKsolver2009" and developed in the framework of the `OKlibrary` (`http://www.ok-sat-library.org`), an open-source library for generalised SAT solving (embracing CSP), aims at unifying all three paradigms.

Now before going into more details, I will try to "outline in a nutshell" fundamental concepts and ideas. The basic notion for a backtracking solver (of any kind) is that of a "partial assignment" φ, fixing some variables to values determined previously (according to the current path in the search tree from the root to the current node), while leaving other variables open, either to be decided later, or to be fixed by reasoning, or left open since they do not play a role. Since it as at the core of (current) SAT solving, in this article we concentrate on problems represented by conjunctive normal forms, or, more combinatorially, represented by "clause-sets", where each clause C can be seen as a negated partial assignment φ, a constraint forbidding all (total) assignments which extend φ. We will represent this via $C = C_{\varphi}$ in Subsection 2.2.2. Furthermore we only consider backtracking approaches, due to its current dominance for practical applications.

The two basic paradigms for backtracking SAT solvers (also "DPLL solvers") are "look-ahead" and "conflict-driven". The look-ahead paradigm, based on stronger polynomial-time reductions and stronger heuristics, is similar to CSP solvers using appropriate constraint propagators, only that here now the emphasise is on the use of partial assignments. Look-ahead solvers are easily parallelisable, and thus might become more important again in the future, at this time however conflict-driven solvers are in the foreground. Again, the basic approach is known from constraint programming, based on "clause-learning", which is no-good learning of clauses C_{φ} for the current path φ, but using several mechanisms to strengthen the learning effort (this is easier in this setting, since the problem representation (via CNF) is just what is needed to represent the no-goods).

Conflict-driven solvers are derived from DPLL-solvers, however it seems appropriate to describe their behaviour no longer in the usual tree-based (recursive) fashion, but as a simple iterative approach (more similar to dynamic programming, as put by David Mitchell). The basic idea is for a problem instance P to guess a satisfying partial assignment φ, guided by only the most basic look-ahead, namely unit-clause propagation, and once a contradiction was realised, then this "conflict" is analysed for its "real causes", and added via clause-learning to the clause database. The whole solution process then might completely re-start from scratch again (but using the modified problem instance P', including now the learned clauses), though in practice only a part of the current path is undone, just to the point where the freshly learned information has obvious consequences.

SAT solving based on DPLL-approaches is close to the resolution calculus, and from the early beginnings by Martin Davis and Hilary Putnam these connections have been exploited. Resolution is just the logical consequence relation restricted to the clause language and only two premises — it is easy to see that then the conclusion is either trivial or the (unique) "resolvent" of the two parent clauses. Now a "resolution-based" solver (as common in ATP at the level of first-order logic) starts with the premises, and tries to derive the empty clause from it, while a DPLL-solver in effect reverts this process: The goal becomes the input clause-set F (the root of the tree), for which backtracking seeks to find the premises from which F can be shown unsatisfiable. Conflict-driven solvers can be understood as breaking the tree-like structure, and so in a sense they combine the resolution-based approaches with the backtracking approaches. A classic references for the resolution calculus is [46], while a recent overview on proof systems is [42]. The earliest paper on the connection of backtracking (in the form of decision trees) and resolution is [36], while a fuller account, also applicable to more general problem representations, can be found in [26].

2 Conjunctive Normal Forms

One peculiar aspect of "SAT solving" is the focus on a very specific form of constraint satisfaction problems:

- only boolean variables are considered;
- only (disjunctive) clauses are allowed as constraints (but involving arbitrarily many variables).

In other words, only boolean CNF's are considered for core SAT solving, and recent extensions build on top of this. Due to the tight coupling of this representation with the algorithms and data structures, the role of (boolean) CNF's goes far beyond sheer problem representation, and seems actually to a certain degree essential for the success of SAT solving for applications. In this section we will discuss the various aspects of "boolean CNF's" and their role for SAT solving.

In this article we put the emphasise on boolean clause-sets, since they (seem) to embody the "secret" of SAT, but it is natural to ask to what extend the special properties of boolean clause-sets can be generalised to CSP's (with unrestricted constraint scopes). A natural settings is given by "signed CNF", which allow for non-boolean variables v with domains D_v, and where literals are of the form "$v \notin S$" for some "sign" $S \subseteq D_v$; this is the most general form of clause-sets (as CNFs) where literals contain only a single variable, and many logical properties can be generalised (for a recent entry point to the literature see [1]). However for the more combinatorial properties of boolean clause-sets, signed clause-sets seem a vast generalisation, and the notion of "clause-sets with non-boolean variables" should better be reserved in my opinion to clauses containing only literals of the form "$v \neq \varepsilon$" for $\varepsilon \in D_v$; such "sets of no-goods" are thoroughly studied in [27].

2.1 Clause-Sets and Partial Assignments

The theoretical foundations of SAT are best framed as follows:

- We have **variables** with boolean domain $\{0, 1\}$; the set of all variables is $\mathcal{VA}$.
- From variables we build **positive literals** and **negative literals**, stating that the variables must become true resp. false. Identifying positive literals with variables, and denoting negative literals with underlying variable v by "$\overline{v}$", we obtain the set $\mathcal{LIT} = \mathcal{VA} \,\dot{\cup}\, \overline{\mathcal{VA}}$ of all literals, where now complementation becomes an involution (a self-inverse bijection) of $\mathcal{LIT}$ onto itself. The underlying variable of literal x is denoted by $\mathrm{var}(x)$.

- Two literals x, y **clash** if they have the same underlying variable, but different "polarities" (or "signs"), that is, iff $x = \overline{y}$.
- **Clauses** are finite and clash-free sets of literals, understood as disjunctions; the set of all clauses is denoted by $\mathcal{CL}$. A special clause is the empty clause $\bot := \emptyset \in \mathcal{CL}$, and $\mathrm{var}(C) := \{\mathrm{var}(x) : x \in C\}$ for $C \in \mathcal{CL}$.
- **Clause-sets** are finite sets of clauses, understood as conjunctions; the set of all clause-sets is denoted by $\mathcal{CLS}$. A special clause-set is the empty clause-set $\top := \emptyset \in \mathcal{CLS}$, and $\mathrm{var}(F) := \bigcup_{C \in F} \mathrm{var}(C)$ for $F \in \mathcal{CLS}$.
- A **partial assignment** is a map $\varphi : V \to \{0, 1\}$ for some finite set $V \subseteq \mathcal{VA}$ of variables, the set of all partial assignments is $\mathcal{PASS}$, and we use $\mathrm{var}(\varphi) := V$. $\varphi(x)$ for literals x is defined (in the obvious way) if $\mathrm{var}(x) \in \mathrm{var}(\varphi)$; a term $\langle x_1 \to \varepsilon_1, \ldots, x_m \to \varepsilon_m \rangle$ for literals x_i (with different underlying variables) and $\varepsilon_i \in \{0, 1\}$ denotes the partial assignment φ with $\mathrm{var}(\varphi) = \mathrm{var}(\{x_1, \ldots, x_m\})$ and $\varphi(x_i) = \varepsilon_i$.
- The most important operation for SAT is the operation

$$* : \mathcal{PASS} \times \mathcal{CLS} \to \mathcal{CLS}$$

  of partial assignments on clause-sets, called "application", where $\varphi * F$ for $\varphi \in \mathcal{PASS}$ and $F \in \mathcal{CL}$ is obtained from F by removing all satisfied clauses (those $C \in F$ containing $x \in C$ with $\varphi(x) = 1$), and removing all falsified literals (i.e., literals x with $\varphi(x) = 0$) from the remaining clauses.
- Finally the set $\mathcal{SAT}$ of **satisfiable clause-sets** is defined as the set of $F \in \mathcal{CLS}$ such that there exists $\varphi \in \mathcal{PASS}$ with $\varphi * F = \top$, while $\mathcal{USAT} := \mathcal{CLS} \setminus \mathcal{SAT}$ denotes the set of **unsatisfiable clause-sets**.

Several special properties of this setting need to be pointed out:

1. A fundamental fact, justifying the use of *partial* assignments, is that if a partial assignment φ satisfies a clause-set F then every **extension** of φ also satisfies F (this is just guaranteed by the process of applying partial assignments), where an "extension" of φ is just a partial assignment ψ with $\varphi \subseteq \psi$ (using the definition of partial assignments as maps, that is, as sets of ordered pairs).
2. Every clause is falsifiable, and thus the property that a partial assignment φ satisfies a clause-set F, i.e., $\varphi * F = \top$, is equivalent to the property that every extension $\psi \supseteq \varphi$ of φ can be further extended to satisfy F. But that φ **falsifies** F, i.e., $\bot \in \varphi * F$, says much less (it's trivial to falsify, but hard to satisfy) and is in general only the final visible expression (during the search process) of unsatisfiability, for example F might be unsatisfiable right away. To obtain more symmetric conditions, one could say that φ "allows satisfaction" of F if $\varphi * F$ is satisfiable, that is, there exists an extension of φ which satisfies F, while φ "disallows satisfaction" of F if $\varphi * F$ is unsatisfiable. Then one could characterise look-ahead solvers as solvers which try to give good indications that the current partial assignment (under construction) allows satisfaction, while conflict-driven solvers could be characterised as solvers trying to find better and better reasons that the current partial assignment disallows satisfaction.

3. If a variable v is set to a value ε, then we can apply the partial assignment $\langle v \to \varepsilon \rangle$ to a clause-set F and obtain a new clause-set $\langle v \to \varepsilon \rangle * F$ which does not contain the variable anymore. This allows us to replace iterated application of partial assignments as in $\psi * (\varphi * F)$ by a single application of the **composition** $\psi \circ \varphi$ of both partial assignments via $\psi * (\varphi * F) = (\psi \circ \varphi) * F$, where the construction of the composition is obvious for variables v where ψ, φ do not clash, while in case of a clash only φ is relevant, since after application of φ the variable v has been eliminated and the value of ψ on v doesn't matter.

Theoretically clause-*sets* are a convenient framework — are they also used in practice ? Let us consider the corresponding conditions:

- Associativity of disjunction and conjunction is always implemented (typically by using lists to represent clauses and clause-sets).
- Commutativity:
  - Commutativity of disjunction (in a clause) may be implemented by ordering the literals in a clause. Often this is not done, but it seems not very relevant.
  - Commutativity of conjunction is never implemented, and especially for industrial benchmarks the order of clauses is quite important (successful conflict-driven solvers employ a lot of "dirty tricks").
- Input-clauses containing clashes are removed, and are also never introduced.
- Idempotency:
  - Literals are not repeated in clauses, achieved by preprocessing the input, while maintenance is basically trivial (for the type of solvers considered, where resolution operations are restricted to the preprocessing phase).
  - However repeated clauses are only removed during pre-processing (if at all, and then applying the stronger reduction of subsumption-elimination (see Subsection 3.4.1)), while they may be created during solving.

To summarise:

1. "Clauses in practice" can be adequately understood as we defined them.
2. "Clause-sets in practice" are actually lists of clauses (order is important, and the effort of removing duplicated clauses is too high).

The notion of literals, clause and clause-sets install several normalisation conditions when regarding literals, clauses and clause-sets as boolean functions (or constraints):

- Literals are never constant true or constant false.
- A clause is never constant true, while $\bot$ is the unique clause which is constant false.
- The unique clause-set which is constant true is $\top$.
- Exactly the unsatisfiable clause-sets are constant false. The clause-sets which are falsified by every partial assignment are those containing $\bot$, which is reduced to the unique form $\{\bot\}$ by reduction r_0 (see Subsection 3.1).

2.2 Properties

Partial assignments and boolean clause-sets have many special properties, compared to the situation for constraint satisfaction, and in this subsection the properties which seem most outstanding for practical SAT solving are discussed:

1. If we have a unit-clause $\{x\} \in F$, then the assignment $\langle x \to 1 \rangle$ is enforced, and this process of "unit-clause elimination" is considered in Subsection 2.2.1 (while unit-clause propagation is the subject of Subsection 3.1.1).
2. The close relation between clauses and partial assignment, the basis for clause-learning, is considered in Subsection 2.2.2.
3. The input-problem doesn't need to be given in a special form, but we can apply the logic of clause-sets and partial assignments under very general conditions, which are discussed in Subsection 2.2.3.

2.2.1 Unit-Clause Elimination

Arguably the most important aspect of clauses for SAT solving is:

Once all literals are falsified except of one, then
the remaining variable gets an enforced value.

This is based on three properties of clauses and literals:

1. falsification of clauses only by giving every variable the wrong value;
2. easy satisfaction of a clause by giving just one variable not the wrong value;
3. since there are only two values, there is no choice for a right value.

The first two properties are still maintained by generalised clauses, allowing non-boolean variables v with domain D_v and literals "$v \neq \varepsilon$" for some $\varepsilon \in D_v$, but the third property requires boolean variables, and thus the strong form of unit-clause elimination is characteristic for *boolean* clause-sets. Repeated elimination of unit-clauses is called "unit-clause propagation", and this most basic process for SAT solving is considered in more details in Subsection 3.1.1).

2.2.2 Correspondence between Clauses and Partial Assignments

At least second in importance to unit-clause elimination is the 1-1 correspondence between clauses and partial assignments:

For every partial assignment φ there is exactly one clause C_φ, such that the falsifying assignments for C_φ are exactly the extensions of φ.
And conversely, for every clause C there is exactly one partial assignment φ_C such that the clauses falsified by φ_C are exactly the sub-clauses of C.

Obviously C_φ consists exactly of the literals falsified by φ, while φ_C sets exactly the literals in C to false, and these two formations are inverse to each other:

1. This correspondence establishes the close relation between the **search trees** of backtracking algorithms and **resolution refutations**, as further explained in Subsection 7.1.
2. Clauses C_φ for falsifying assignments φ are also called "no-goods", and are the essence of "learning" as explored in Subsection 6.1.

2.2.3 An Axiomatic Approach

A generalisation of "satisfiability" by allowing arbitrary problem representations, on which partial assignments "operate", has been introduced and studied in [26]. The key observation here is, when considering only partial *assignments*, where assigned variables get a unique value[2], then a sequence $\psi * (\varphi * F)$ of applications of partial assignments can be elegantly represented by a unique application of the composition $(\psi \circ \varphi) * F$, and the essence of basic satisfiability considerations can be captured by a simple algebraic framework, where problem instances are left unspecified, and only via the application of partial assignments can we "query" them. Clauses and resolution then appear as a meta-structure on top of the given domain of problem instances. In Subsection 7.3 we will use this theory, which allows generalised resolution "modulo oracles", for a "compressed" form of learning.

2.3 Data Structures

Especially the handling of variables can be somewhat complicated from a software engineering point of view, when maximal generality is the goal, however from the purely algorithmic point of view there are no real complications involved:

- Variables are often implemented as unsigned positive integers; literals are then signed integers (other than zero). If variables are not already positive integers themselves, then they need to be associated with an index, so that we can establish constant time access to properties of variables.
- An alternative is to implement variables and literals by pointers to associated data structures with relevant meta-data (like occurrence numbers etc.).
- In any case variables and literals need to be light-weight objects which can be easily copied (note the difference between a literal and a literal-occurrence: given n variables, there are $2n$ literals, while the number of literal-occurrences is the sum of the clause-lengths).

Regarding the implementation of clauses and clause-sets, the basic decision is whether to use "lazy" or "eager" data structures; this is further discussed in Subsection 4.2, and here it suffices to say that conflict-driven solvers are lazy (avoiding to do much work at each node, since they might backtrack soon anyway), while look-ahead solvers are more eager (since the look-ahead at each node needs better support):

1. In the lazy case it is sufficient to implement clauses as vectors (fixed after reading the input).
2. While for the eager case clauses are dynamically changed, and are implemented as doubly-linked lists of literal-occurrences.
3. The list of all clauses is not of great importance (since one should avoid to run through all clauses), but clauses are accessed through the "clause-literal graph" discussed below.
4. In the eager as well as the lazy case, clauses must enable quick access to associated statistical data.

[2] While multivalued assignments can allow a set of values.

Clause-sets are

- generalisations of hypergraphs (adding signs to vertices), as well as
- special cases of hypergraphs (with literals as vertices).

Hypergraphs can be represented by bipartite graphs. For clause-sets we obtain the bipartite *clause-literal graph*, which is of fundamental importance:

- the nodes are the literals on one side, and the clauses on the other side;
- edges indicate membership of literals in clauses.

The *clause-variable graph*, connecting now clauses with variables, is also called "incidence graph". Using the standard adjacency-list representation of digraphs and representing graphs by symmetric digraphs, we obtain a basic implementation of clause-sets through the representation of the clause-literal graph, allowing quick access to the literal-occurrences in a clause as well as to the clauses in which a literal occurs. This representation can be considered as fundamental for the lazy as well as for the eager approach, where the former saves certain elements, while the latter adds further structure. Some remarks on the clause-literal graph:

1. More correct is to speak of a 3-partite graph, where the clause-literal graph is augmented with an additional layer for variables.
2. Literal-occurrence correspond to edges between clauses and literals.
3. I consider the graph and hypergraph concepts as a good conceptual framework, however it is used only implicitly by solver implementations.[3]
4. The technique of "watched literals" together with the "lazy datastructure" for clause-sets can be considered as removing certain (directed) edges from literals to clauses in the clause-literal graph: From a clause we can still reach all contained literals, but a clause is reachable only from two "watched" literal occurrences in each clause, which are updated if necessary; see Subsection 3.1.1.

Finally, for partial assignments two complementary structures are used:

- For search purposes, partial assignments are treated as stacks of assignments (moving down and up the search tree).
- Via an additional global vector of assignments we can check in constant time, whether a variable is assigned, and which value it has.

Local search typically works only with "total" assignments (i.e., with partial assignments φ with $\mathrm{var}(\varphi) = \mathrm{var}(F)$, where F is the input clause-set), while for the algorithms considered in this paper partial assignments are fundamental, and then the efficient implementation of the application of partial assignments is of utmost importance (needing additional data structures). Copying is perhaps the most fundamental enemy of efficiency, and the application of partial assignments is (in non-parallel computations) performed *in-place*; more on this and the two fundamental approaches, "eager" and "lazy", can be found in Subsection 4.2.

[3] The upcoming `OKlibrary` will give direct support for using these graph-theoretic abstractions.

2.4 Transformations

Let us close this section by some remarks on how to translate other problems into boolean CNFs. First there is the somewhat surprising fact that *boolean transformations are surprisingly efficient.* There are several important extensions of clauses, like

1. cardinality clauses, e.g., $v_1 + v_2 + v_3 \lesseqqgtr k$;
2. more generally pseudo-boolean clauses, allowing constant coefficients;
3. crisp CSP.

For all these cases, direct translation (avoiding sophistication) into boolean CNFs is an efficient way to deal with them (at this time), if a reasonable amount of "logical reasoning" is required by the problem. Boolean CNFs seem to be supported by superiorly efficient data structures — every deviation from this ideal is punished by a big loss in efficiency, which can be compensated only in special situations. But there is another important advantage by using a boolean translation: Not only do we get efficient data structures for free,

but the "atomisation" of information achieved by using boolean variables can be *inherently* more efficient for backtracking algorithms (with exponential speed-ups) than the original information representation.

This important point was raised in [39]: Chopping up a problem into boolean pieces in general increases the search space, but this richer space allows also for more efficient re-combinations.

3 Reductions: Poly-Time Methods

The purpose of this section is to introduce the main reductions used in SAT solving:

1. Unit clause propagation and generalisations are considered in Subsection 3.1.
2. Basic methods directly based on resolution are considered in Subsection 3.2.
3. Some basic comparisons between different notions of "local consistency" are given in Subsection 3.3.
4. Less common reductions are surveyed in Subsection 3.4.

A reduction here is simply a map $r : \mathcal{CLS} \to \mathcal{CLS}$ such that $r(F)$ is satisfiability-equivalent to F, and we consider here only polynomial-time computable r. Now one can study classes $\mathcal{C} \subseteq \mathcal{CLS}$ such that r is already sufficient to decide satisfiability for $F \in \mathcal{C}$, however this point of view in isolation is not very useful for SAT solving (at least not for practical SAT solving), as discussed in Subsection 3.5.

3.1 Generalised Unit-Clause Propagation

We define hierarchies $r_k, r'_k : \mathcal{CLS} \to \mathcal{CLS}$ of poly-time reductions for $k \in \mathbb{N}_0$ as follows:

1. $r_0 = r'_0$ detects the empty clause, and otherwise does nothing:

$$r_0(F) := \begin{cases} \{\bot\} & \text{if } \bot \in F \\ F & \text{otherwise} \end{cases}.$$

2. r_{k+1} reduces F to $r_{k+1}(\langle x \to 1\rangle * F)$ in case r_k yields an inconsistency for $\langle x \to 0\rangle * F$ for some literal x:

$$r_{k+1}(F) := \begin{cases} r_{k+1}(\langle x \to 1\rangle * F) & \text{for literals } x \text{ with } r_k(\langle x \to 0\rangle * F) = \{\bot\} \\ F & \text{otherwise} \end{cases}.$$

r'_{k+1} also notices when a satisfying assignment was found:

$$r'_{k+1}(F) := \begin{cases} r'_{k+1}(\langle x \to 1\rangle * F) & \text{for literals } x \text{ with } r'_k(\langle x \to 0\rangle * F) = \{\bot\} \\ \top & \text{for literals } x \text{ with } r'_k(\langle x \to 1\rangle * F) = \top \\ F & \text{otherwise} \end{cases}.$$

Main properties:

- Though the definition of r_k, r'_k is non-deterministic, these reductions yields unique results (are confluent).
- There always exists partial assignments φ, φ' such that $r_k(F) = \varphi * F$ resp. $r'_k(F) = \varphi' * F$; here φ is a forced (or "necessary") assignment.
- By applying $r_0, r_1, \ldots, r_{n(F)}$ (where $n(F) := |\mathrm{var}(F)|$ is the number of variables) until either an inconsistency is found or at the end we know that F is satisfiable, we obtain a SAT decision algorithm which *quasi-automatises tree resolution*, and which is the (real) essence of Stalmarck's solver. Obviously it is preferable to use the reductions r'_k, which for k large enough (at most $k = n(F)$) will also find a satisfying assignment if F is satisfiable. See [19] for a thorough treatment of the reductions r_k, r'_k.[4]
- In [26] the treatment of r_k, r'_k is extended to axiomatically given systems of problem instances with non-boolean variables (compare Subsection 2.2.3 in this article).

The fundamental open question is how efficiently r_k can be computed for general k:

1. r_1 is just unit-clause-propagation, and we will see in Subsection 3.1.1 that r_1 can be computed in linear time, that is, in $O(\ell(F))$ where $\ell(F) := \sum_{C \in F} |C|$ is the number of literal occurrences in F.
2. We obtain that r_k can be computed in time $O(n(F)^{2(k-1)} \cdot \ell(F))$ for $k \geq 1$.

[4] Using the hierarchy $G_k(\mathcal{U}, \mathcal{S})$ from [19] (with oracles $\mathcal{U} \subseteq \mathcal{USAT}$ for unsatisfiability and $\mathcal{S} \subseteq \mathcal{SAT}$ for satisfiability) we have $r'_k(F) = \{\bot\} \Leftrightarrow F \in G^0_k(\mathcal{U}_0, \mathcal{S}_0)$ and $r'_k(F) = \top \Leftrightarrow F \in G^1_k(\mathcal{U}_0, \mathcal{S}_0)$, where $\mathcal{U}_0$ is the basic oracle for unsatisfiability, just recognising the empty clause, and $\mathcal{S}_0$ is the basic oracle for satisfiability, just recognising the empty clause-set.

3. Thus already for r_2 in general we obtain only a cubic-time algorithm. Can we do better ? And what about general k ?

The reductions r'_k can be naturally strengthened via the use of (weak) autarky reduction (see Subsection 3.4.3; a similar early approach is [5]):

$$r^*_{k+1}(F) := \begin{cases} r^*_{k+1}(\langle x \to 1 \rangle * F) & \text{for literals } x \text{ with } r^*_k(\langle x \to 0 \rangle * F) = \{\bot\} \\ r^*_{k+1}(r^*_k(\langle x \to 1 \rangle * F)) & \text{for } x \text{ with } r^*_k(\langle x \to 1 \rangle * F) \subseteq F \\ F & \text{otherwise} \end{cases} .$$

Conflict-driven solvers only use r_1, while look-ahead solvers employ r_k for "$k \approx 2$": The `OKsolver-2002` (see [24]) uses exactly r^*_2 (apparently as the only solver at each node), while "modern" look-ahead solvers exclude "unpromising" variables from r_2 (thus go below $k = 2$), while employing r_3 for "promising" variables (see Subsection 5.2); the `march`-solvers also employ certain aspects of weak autarky reduction.

3.1.1 Unit-Clause Propagation

The special case r_1 is *unit-clause propagation* (UCP). UCP of central importance for backtracking solvers, and for efficiency reasons support for UCP needs to be integrated into the main data structure. The basic algorithm for UCP is the *linear time algorithm*, best understood as operating on the clause-literal graph (recall Subsection 2.3), which is represented by an adjacency list:

– a given unit clause $\{x\}$ is "propagated" by removing the literal occurrences of literal $\overline{x}$ (the graph representation yields quick access from a literal to its occurrences, and allows to remove edges efficiently), and it is checked whether this removal creates a new unit-clause (which by the graph representation is a constant-time operation);
– as soon as a new unit-clause is created, it is pushed on the buffer (typically a queue or a stack), used for the unit-clauses waiting to be propagated;
– a partial assignment with constant time access keeps track of the assignments, discarding multiple occurrences of the same unit clause, and detecting contradictory unit clauses.

Note that for "lazy UCP", which only needs to obtain the assignments resulting from the unit-clause propagation (used by conflict-driven solvers or in the look-ahead of look-ahead solvers), and not the resulting clause-set, we do not need to consider satisfied clauses. For faster UCP (achieving a better constant factor) the main problem is to make the propagation process more efficient, so that with less work we can detected (relevant) new unit-clauses. A first step is not to remove the occurrences of $\overline{x}$ but just to decrement counters for the clause-lengths, and if this counter reaches 1 then the clause is inspected for the new unit-clause. This is driven further by *watched literals*: We do not need to know the precise (current) length of all clauses, but we need only to be alerted if possibly we have less than 2 literals in a clause. So we can thin out the clause-literal graph by using only two literal-neighbours of a clause, and updating these neighbours (the watched literals) if one of them disappears. All these methods are only relevant for lazy UCP, which is also for look-ahead solvers of great importance due to the time spent during the look-ahead.

3.1.2 Failed Literals and Extensions

Reduction by r_2 is called "(full) failed literal reduction". It is not used by conflict-driven solvers, but essential for look-ahead solvers. Failed literal reduction relies on the efficient implementation of UCP, and, as already mentioned, the central question here is: How much can we do better than by just following the definition ?! (Better than just checking for all assignments to variables, whether UCP yields a conflict, and repeating this process if a reduction was found.) The current "front" of research (for look-ahead solver) considers weakenings and strengthenings of r_2 (trying only "promising" variables, and locally learning binary clauses encoding the inferred unit-clauses). See Subsection 5.2 for more information.

3.2 Resolution Based Reductions

Given two clauses C, D clashing in exactly one literal $x \in C \land \overline{x} \in D$, the **resolvent** is

$$\boldsymbol{C \diamond D} := (C \setminus \{x\}) \cup (D \setminus \{\overline{x}\}).$$

Given two clauses C, D and a clause R, the relation $\{C, D\} \models R$, that is, $C \land D$ logically implies R, is equivalent to the following:

- Either C and D are resolvable (clash in exactly one literal), and then $C \diamond D \subseteq R$,
- or C and D are not resolvable (clash in zero or at least two literals), and then we have $C \subseteq R$ or $D \subseteq R$.

Thus on the clause level, syntax and semantics coincide! Resolution calculi organise the iterated application of the resolution operation until either the empty clause has been derived, and thus the clause-set is unsatisfiable, or it is established that this is not possible (and thus the clause-set must be satisfiable). Resolution in its various forms, especially tree-like resolution, where the search process can be (relatively) efficiently inverted (starting with the goal), is the central tool for SAT. Via the correspondence between clauses and partial assignments, every backtracking solver is constructing a resolution refutation of its input (see Subsection 7.1). Additional resolution power (moving from tree resolution to full resolution) is gained by "clause learning", and is discussed further in Subsection 6.1 and Section 7. Via the following methods, the resolution operation can be involved in a more direct way:

Adding resolvents. Just adding arbitrary resolvents is highly inefficient (except of some special cases). So only short resolvents are added (of length at most 3), and this only during preprocessing.

DP-reductions. The **DP-operator** (also referred to as "variable elimination") is

$$\mathrm{DP}_v(F) := \{\, C \diamond D : C, D \in F \land C \cap \overline{D} = \{v\} \,\} \cup \{C \in F : v \notin \mathrm{var}(F)\}.$$

$\mathrm{DP}_v(F)$ is sat-equivalent to F, more precisely, $\mathrm{DP}_v(F)$ is equivalent to the quantified boolean formula $(\exists\, v \in \{0,1\} : F)$, and variable v is eliminated by applying DP_v. So by applying DP until all variables are removed we can decide SAT, but in general this is very inefficient (requiring exponential space). Thus DP is only applied (during preprocessing) in "good cases" (typically when size is not increased).

A general problem here (and elsewhere) regarding reductions is:

To remove or to add clauses ?
That is, simplifying the formula or adding inference power ?

Regarding resolvents, they are typically added.

3.3 Comparison with Local Consistency Notions for CSP's

UCP is the natural mechanism for extending a partial assignment by the obvious inferences. In the language of constraint satisfaction problems, UCP establishes node-consistency (while hyper-arc consistency for clause-sets is trivially fulfilled). More generally, for $k \geq 1$ call a clause-set F *r_k-reduced* if $r_k(F) = F$ holds (so r_1-reduced is the same as node-consistency). How is this consistency notion related to *strong k-consistency* for $k \geq 1$ and clause-sets F (i.e., for every partial assignment φ using strictly less than k variables and fulfilling $\bot \notin \varphi * F$ and for every variable v there is an extension φ' of φ with $\mathrm{var}(\varphi') = \mathrm{var}(\varphi) \cup \{v\}$, such that $\bot \notin \varphi' * F$)? Call a clause-set F *closed under k-bounded resolution* if for all resolvable $C, D \in F$ with $|C \diamond D| \leq k$ we have $C \diamond D \in F$. Now it is easy to see that $F \in \mathcal{CLS}$ is strongly k-consistent for $k \geq 1$ if and only if F is closed under $(k-1)$-bounded resolution. So the question is, how is being r_k-reduced related to being closed under k'-bounded resolution:

1. r_1 is sufficient to show unsatisfiability of all Horn clause-sets, while for every k there exists an unsatisfiable Horn clause-set which is closed under k-bounded resolution but $\bot \notin F$, simply due to the incapability of bounded resolution to handle large clauses.
2. Via small strengthenings of "bounded resolution" however, as discussed in [26], we obtain versions of "k-resolution" which properly generalise r_k for $k \geq 2$.

So in a sense by an adequate repair of the notion of strong $(k+1)$-consistency we obtain a consistency notion which is considerably stronger than being r_k-reduced, however the price is an explosion in memory consumption. One should note here the different contexts for "strong k-consistency" and "r_k-reduced":

- Algorithms for establishing strong k-consistency exploit that constraints as sets of (satisfying) tuples allow to remove arbitrary tuples (which have been found inconsistent), which is not possible with such simple "atomic constraints" as clauses.
- On the other hand, r_k-reduction exploits application of partial assignments by applying enforced assignments, which is supported due to the simplicity of clauses, while constraints typically only handle assignments which cover all their variables.

3.4 Other Reductions

3.4.1 Subsumption

Removing subsumed clauses is quite costly, and so mostly done only during preprocessing. See [47] for subsumption elimination also during search, which currently seems to be worth the effort only for harder problems like QBF. This is true for many somewhat more complicated algorithms:

SAT is too easy (currently) for them.

3.4.2 Equivalences

- Equivalences $a \leftrightarrow b$ often are detected (for conflict-driven solvers only during preprocessing), and substituted.
- In general, clauses which correspond to linear equations over $\mathbb{Z}_2$ are sometimes (partially) detected, and some elementary reasoning on them is performed; see [14] for an overview. Most recently however, these facilities seem to be getting removed from "practical SAT solving".

3.4.3 Autarkies

A partial assignment φ is an *autarky* for a clause-set F if every clause of F touched by φ is satisfied by φ:

1. The empty assignment is always an autarky.
2. Every satisfying assignment is an autarky.
3. Composition of two autarkies is again an autarky.
4. Autarkies can be applied satisfiability-equivalently, and thus we have *autarky reduction*.
5. A simplest case of autarky reduction is elimination of pure literals.
6. Since clause-*sets* contract multiple clauses there is also the concept of a *weak autarky* for a clause-set F, a partial assignment φ such that $\varphi * F \subseteq F$. Every autarky is a weak autarky, but not vice versa. Also application of weak autarkies yields satisfiable equivalent (sub-)clause-sets.

Autarkies emerged in a natural way from improved exponential upper bounds on SAT decision ([40,31,32,20]), while the accruing theory of autarkies ([22,25,23, 28,27]) focuses on polynomial-time SAT decision classes on the one hand (embedding matching theory, linear programming and combinatorial reasoning into the (generalised) satisfiability world), and on the other hand on the structure of *lean clause-sets* which are reduced w.r.t. autarky reduction. Via the notion of an *autarky system* we obtain also generalisations of the notion of minimally unsatisfiable clause-sets, parameterised by special notions of autarkies. In Subsection 3.1 we have already seen initial examples of the use of autarkies in SAT solvers; and see [33] for applications of the fundamental duality between autarkies and resolution. At this time the practical applications seem to be marginal, however I expect this to change within the next 5 years (perhaps especially regarding QBF); see [35] for a recent application.

3.4.4 Blocked Clauses

The concept of a blocked clause was introduced in [20], with a forerunner in [41], and allows to add to or delete from $F \in \mathcal{CLS}$ a special type of clause called "blocked":

- Clause C is **blocked** for F if there is some $v \in \text{var}(C)$ such that addition resp. removal of C does not change the outcome of applying DP_v (so one could speak of "inverted DP-reduction").
- Blocked clauses can be added / removed satisfiability-equivalently.
- Addition of blocked clauses containing possibly new variables covers Extended Resolution; so in principle this is very powerful, but we have no guidelines when to perform such an addition.
- Addition of blocked clauses without new variables still goes beyond resolution, as shown in [21], and could be interesting for SAT solvers (the obtained additional inferred assignments are applied directly in [41] for special cases).
- Elimination of blocked clauses was implemented (`lsat`; see [44]), and can help solving some special classes very quickly where all other solvers fail.

3.5 Poly-time Classes

Poly-time SAT-decidable classes can play a role for SAT solving as **target classes**:

The heuristics aims at bringing the clause-set closer to the class,
and finally the special algorithm is applied.

However, in practice poly-time classes play yet no role for SAT:

1. They do not occur (on their own!).
2. They do not provide good guidance for the heuristics.

The essential lesson to be learned here seems to me:

Algorithms are more important than classes!

Solvers are algorithm-driven, that is, algorithms are applied also "when they are not applicable", and they are only good, if they are better "than they should be". (And algorithms need a lot of attention and care; they have their own rights, and are not just "attachments" to classes.) For some examples, let us examine the 3 main Schaefer classes:

2-CNF. Unsatisfiable instances are handled by failed literal elimination, while satisfiable instances are handled by simple autarky reduction. So some look-ahead solvers solve them "by the way"; but it's not worth looking for them.

Horn. Unsatisfiable (renamable) Horn are handled by UCP; there have been many attempts to integrate also the satisfiable cases, but they all failed (in practice). Perhaps a main problem with this class is its brittleness (as clause-sets), while for example the closure under renaming makes it more complicated to deal with, and on the unsatisfiability side we still have a rather trivial class (solved by UCP).

Affine. This is the only case of some interest (and further potential), since **equivalences** do occur in special cases, and resolution cannot handle them efficiently. However, due to their special character, affine formulas (resp. their expressions as clause-sets) do not serve as a target class, but are handled by dedicated reasoning mechanisms (see Subsection 3.4.2 above), which could be understood as being handled by specialised "dynamic constraints".

The above statement "it's not worth looking for 2-CNFs" means

- Applying a special test for detecting the (narrow) class of 2-CNF seems to be rather useless.
- Heuristics aiming (just) at bringing down a clause-set to a 2-CNF are too crude.

However, for look-ahead solvers 2-CNFs are kind of basic:

Some algorithms used to solve 2-CNF (and Horn) are important — since these algorithms can solve much more than just 2-CNF (or Horn)!

I hope this illustrates the assertion "algorithms more important than classes".

4 DPLL in General

In this section now we outline the "modern DPLL scheme", with look-ahead solvers (see Section 5) and conflict driven solvers (see Section 6) as special cases. First a note on terminology: "DP, DLL, DPL, DPLL" — these four combinations have been used to describe backtracking algorithms with inference and learning:

1. "DP" is incorrect, since [7] only introduced DP-reduction (see Subsection 3.2) but not the splitting approach.
2. "DLL" refers to [6], the basic backtracking algorithm with unit-clause propagation, elimination of pure literals, and a simple max-occurrences heuristics.
3. "DPL, DPLL" acknowledge the influence of Putnam.

The following pattern seems reasonable (and not uncommon):

1. "DP" for DP-reduction (as it is standard now);
2. "DLL" for simple backtrackers;
3. "DPLL" for the combination of backtracking with resolution (including clause-learning).

4.1 Modern DPLL

A general scheme $\mathfrak{G}$ for DPLL algorithms is now presented, comprising look-ahead as well as conflict-driven solvers. The input is $F_0 \in \mathcal{CLS}$ (possibly after pre-processing). A global variable $\mathbb{L}_0$ contains the learned clauses, and we have $F_0 \models \mathbb{L}_0$ throughout. Thus learning as reflected by $\mathbb{L}_0$ is "global learning", that is, the learned clauses are always to be interpreted w.r.t. the original input F_0 (and not w.r.t. the respective residual clause-sets at each node). Initially $\mathbb{L}_0$ is empty. A parameter, the "history stack" $\mathcal{H}$, contains the information how to interprete $\mathbb{L}_0$ in the current situation, denoted by $\mathcal{H} * \mathbb{L}_0 \in \mathcal{CLS}$ (this might be just application of the partial assignment according to the current path, but it might also contain renamings, substitutions, etc.). Furthermore via $\mathcal{H}$ we can also perform "conflict analysis" (which is not further specified here; for a concrete example see Subsection 6.1.1), and relate a "residual conflict" to F_0. Besides the global variable $\mathbb{L}_0$ (which might be accessed from parallel processes or threads, and thus might need access-control), the procedure $\mathfrak{G}$ is a normal recursive function, with a clause-set F as first argument and $\mathcal{H}$ as second argument (following standard scope rules; initially F is F_0 and $\mathcal{H}$ is empty), and returning an element of $\{0, 1, *\}$, where "$*$" stands for "unknown". The history "stack" $\mathcal{H}$ actually needs to be readable as a whole for conflict analysis, and thus one should better speak of the "history list", but since we mention explicitely only the stack operations (mirroring the ups and downs of the current path) we stick to the notion of a "stack".

$\mathfrak{G}(F \in \mathcal{CLS}, \mathcal{H}) : \{0, 1, *\}$

0. Initialise the local history H as empty.
1. Reduction:
   (a) $F := r(F, \mathcal{H} * \mathbb{L}_0)$;
   (b) add the information about this reduction to H.
2. Analysis:
   (a) Success: If $F = \top$ then return 1.
   (b) Conflict learning and backtracking: If $\bot \in F$ then
      i. via $\mathcal{H}$ compute a set $\mathbb{L}$ of learned clauses;
      ii. $\mathbb{L}_0 := \mathbb{L}_0 \cup \mathbb{L}$;
      iii. $\mathcal{H}$.pop();
      iv. return 0.
   (c) Non-chronological backtracking: Otherwise, if appropriate then
      i. $\mathcal{H}$.pop();
      ii. return $*$.
3. Branching: Compute a finite set $\mathbb{B} \subseteq \mathcal{PASS}$ of partial assignments, and for all $\varphi \in \mathbb{B}$ do (possibly in parallel)
   (a) $\mathcal{H}$.push((H, φ));
   (b) $\delta_\varphi := \mathfrak{G}(\varphi * F, \mathcal{H})$;
   (c) if $\delta_\varphi = 1$ then $F := \top$;
   (d) if $\delta_\varphi = 0$ then $F := F \cup \{C_\varphi\}$.
   If these computations are not performed in parallel, then sort $\mathbb{B}$ appropriately before these computations, and break this loop (over $\varphi \in \mathbb{B}$) in case of $\delta_\varphi = 1$.
4. Goto Step 1.

Due to the given specifications, the returned result is always correct; we are not concerned here about establishing general rules for termination (which is not too complicated), nor are we concerned about completeness (equivalent to not performing a non-chronological backtrack at the root of the search tree) — these properties are easily established for the special cases we consider later. Explanations and remarks:

- The map $r : \mathcal{CLS} \times \mathcal{CLS} \to \mathcal{CLS}$ is any "reduction", where we just require that $r(F, \mathbb{L})$ is always satisfiability-equivalent to $F \cup \mathbb{L}$; the separation into two arguments enables special treatment of learned clauses.
- The purpose of the learning step is to enable the reduction r to circumvent the same conflict earlier in the future (when a similar situation arises).
- Non-chronological backtracking is performed if the current situation is better handled at a lower level (closer to the root) in the search tree; this includes the case of a (complete) restart as a special case (through repetition of this step).
- The elements of $\mathbb{B}$ are the "decision assignments":
  - For a look-ahead solver there is a variable v with $\mathbb{B} = \{\langle v \to 0\rangle, \langle v \to 1\rangle\}$, where v is the "branching variable". Thus if both branches returned "unsatisfiable", then the analysis step will confirm the current F as unsatisfiable.
  - For a conflict-driven solver there exists a variable v (again called the "branching variable)" and $\varepsilon \in \{0, 1\}$ with $\mathbb{B} = \{\langle v \to \varepsilon\rangle\}$, and thus here the iterative character of $\mathcal{G}$ is emphasised.
- Sorting of $\mathbb{B}$ shall take advantage of an early success (i.e., a satisfying assignment was found) in some branch: Imagine the situation where one branch is a hard unsatisfiable problem, while the other is an easy satisfiable problem — if not already performed in parallel, then we gain a large speed-up if we have put the satisfiable branch first.
- Note that the return value in Step 3b might be $*$, in which case just another iteration of the loop is performed.
- Pushing the item "(H, φ)" onto the history stack means that we can reconstruct how from the current $F_0 \cup \mathbb{L}_0$ we obtained the current F (through the successive reduction steps as stored on the (whole) stack) and which decisions were involved. The only point where $\mathcal{H}$ is used is Step 2(b)i, where we compute the learned clauses derived from the conflict (and since learned clauses are "global" we need to re-connect the current F to the global level); the loose concept of the history stack is just there to make the flow of informations more visible.
- At Step 3d a "local learning" step is performed (compare Subsection 7.2), that is, the learned clause C_φ is added to the residual clause-set F, and is not traced back to F_0. One could also apply conflict analysis here, but regarding the character of local learning it seems more appropriate to use the cheaper "full clause-learning" here.

There exist further global monitoring schemes:

- removal of "old" learned clauses;
- using some form of breadth-first search (typically at an early level);
- re-arranging the call order over an initial part of the search tree according to some statistical analysis (compare Subsection 5.3.3).

However yet these extensions have more the character of an "add-on", and time seems not ripe yet to formulate more general patterns, whence these schemes are not present in $\mathfrak{G}$. An important point here is that we have a recursive element of $\mathfrak{G}$, present in the branching step in case of $|\mathbb{B}| \geq 2$, and an iterative element by looping through Steps 1 to 3: Look-ahead solvers (see Section 5) only use this recursive (parallel(!)) aspect, while conflict driven solvers (see Section 6) focus on the iterative aspect (since there always $|\mathbb{B}| = 1$ is the case, the recursion can be eliminated altogether, as done in algorithm `cd` in Subsection 6).[5]

4.2 The Role of Partial Assignments and Resolution

There are two fundamental possibilities when applying partial assignments for branching in Step 3b:

eager. Really apply the assignment, so that at each node we only see the simplified instances;

lazy. Only record the assignment, and interprete the clauses as they are visited.

Since look-ahead solvers perform a lot of work at each node, they tend to be "eager" while conflict-driven solver are all "lazy" (if they perform non-chronological backtracking then the work would get lost anyway). Important:

> Application of partial assignments happens "in place", not by copying the instance, but by modification and undoing the modification.

Naturally, undoing the assignment(s) is easier for lazy data structures, but eager data structures pay off in case of heavy local workloads (as for the look-ahead).

DPLL solvers are based on a strong connection to tree resolution and strengthenings (see Subsection 7.1). I regard this as the backbone of SAT solving: Resolution is the "logic of partial assignments", for CSP and beyond, and can be based on a simple algebraic framework (see [26]). In this sense "SAT" and "CSP" can be seen as complementary: Where SAT emphasises the (global) operation of partial assignments on the problem instance, CSP puts more emphasise on exploiting (local) structure of constraints. The resolution connection explains also intelligent backtracking: By just computing the variables (really) used in the resolution refutation found, intelligent backtracking is possible (implemented in the `OKsolver`-2002; see Subsection 7.1).

5 Look-Ahead Solvers

In the history of look-ahead solvers, two lines can be distinguished:

1. `Posit` ([11], `Satz` ([34]), and `kcnfs` ([8])
2. Boehm-solver ([4]) and the `OKsolver`-2002 ([24])

while the `march`-solvers ([13, 15]) can be seen as combining those two lines.

[5] These algorithmic aspects should not be mixed up with the purely implementational aspects of simulating recursion in a look-ahead solver via an iterative procedure (using additional stacks). Actually, if the main data structures for the problem instance use in-place modifications (eagerly or lazily), and thus are not included in the recursion, then according to my experience the difference in resource consumption between the more elegant recursive approach and simulation of recursion is negligible.

The Boehm-solver introduced the special "two-dimensional linked-list" representation of clause-sets as a prototype for an eager data-structure ([18] discusses the same idea in a different context), which allows at each node to just see the residual clause-set (after application of the current partial assignment), while the `OKsolver`-2002, using this data structure, demonstrated at the SAT2002-competition that a generic solver with full failed-literal reduction (i.e., r_2), full look-ahead and autarky reduction can be quite efficient (see [45]). The line of `Posit`, `Satz` and `kcnfs` uses a simpler data-structure, more in direction of lazy data structures, where `Posit` uses partial r_2, while `Satz` and `kcnfs` use partial r_2 and partial r_3.

The general "world view" of look-ahead solvers could be summarised as follows:

- For hard problems (thus they can't be too big; say several thousand variables).
- Failed literal reduction and extensions, intelligent heuristics and special structure recognition at the centre.
- The heuristics considers both branches as independent and assumes the worst case. The choice of the first branch (the "direction heuristics", based on estimating the probability of satisfiability) is important on satisfiable instances.
- Eager data structures, with lazy aspects for the look-ahead.
- The aim is as much as possible reduction of problem complexity by inferred assignments (now and in the future).

5.1 The General Scheme for Look-Ahead Solvers

We present a simplified version of the general algorithm $\mathfrak{G}$ from Subsection 4.1, where now no conflict learning (and thus no history) is involved, and also no non-chronological backtracking (but see below for examples of restricted usage of global learning in look-ahead solvers to achieve "intelligent backtracking"); on the other side now more details are given about the heuristics for branching.

$\mathtt{la}(F \in \mathcal{CLS}) : \{0, 1\}$

1. Reduction: $F := r(F)$.
2. Analysis: If $F = \top$ then return 1, if $\bot \in F$ then return 0.
3. Branching:
   (a) For each variable $v \in \mathrm{var}(F)$ do:
      i. For each $\varepsilon \in \{0, 1\}$ do:
         A. Consider $F^v_\varepsilon := r'(\langle v \to \varepsilon \rangle * F)$,
         B. Compute a "distance vector" $\boldsymbol{d}^v_\varepsilon \in \mathbb{R}^m$, where component $\boldsymbol{d}^v_\varepsilon(i)$ for index $i \in \{1, \ldots, m\}$ measures the progress achieved from F to F^v_ε in "dimension i".
      ii. Summarise the distance vector $\boldsymbol{d}^v_\varepsilon \in \mathbb{R}^m$ by "distances" $d^v_\varepsilon \in \mathbb{R}_{>0}$.
      iii. Combine the two distance values d^v_0, d^v_1 into $\rho_v := \rho(d^v_0, d^v_1) \in \mathbb{R}_{>0}$.
   (b) Choose a branching variable v with minimal ρ_v.

(c) Return $\max(\mathtt{la}(\langle v \to 0 \rangle * F), \mathtt{la}(\langle v \to 1 \rangle * F))$, where in case of a non-parallel computation the first branch $\varepsilon \in \{0, 1\}$ is chosen such that F^v_ε appears more likely to be satisfiable than $F^v_{\overline{\varepsilon}}$ (and if the first branch returned 1 then the second branch is not considered).

Remarks

1. r' is the reduction used only for the look-ahead; typically $r \approx r_k$ and $r' \approx r_{k-1}$.
2. A distance vector $\boldsymbol{d}^v_\varphi \in \mathbb{R}^m$ measures the progress from F to F^v_ε in an m-dimensional way; a simple example would be to use the number of variables, the number of clauses and the number of literal occurrences (so here $m = 3$). Since there are m components, some of them could be zero (no progress) or even negative (deterioration) if the positive entries outweigh the non-positive entries. See Subsection 5.3.1 for further discussions.
3. See Subsection 5.3.2 for the discussion of the "projection" $\rho : \mathbb{R}_{>0} \times \mathbb{R}_{>0} \to \mathbb{R}_{>0}$.
4. In order to really present an "algorithm" (with well-defined semantics, and not just an "implementation"), the `OKsolver`-2002 performs the look-ahead step 3a fully for all variables, while all other look-ahead solver only perform a "partial look-ahead" on selected variables (this can be incorporated into the above scheme by using appropriate low distance values for variables which don't get selected).

5.2 Reductions: Failed Literals and Beyond

The most important reductions for look-ahead solvers is given by the range from r_1 to r_3, as presented in Subsection 3.1. The following methods are used to increase efficiency:

1. (Additional) lazy data structures are used, employing time stamps to avoid permanent re-initialisation.
2. Often only "promising" variables are considered for the failed literal reduction (thus weaker than r_2), while for "very promising variables" a double-look ahead is used (reaching r_3 here); see [15] for a recent study.
3. A main problem with r_k for $k \geq 2$ is the (apparent) necessity to run over the formula over and over again to determine whether one reduction triggered other reductions. The simplest thing (perhaps first used by the `OKsolver`-2002) is to realise that if from $x \to 1$ we obtain $y \to 1$ while $x \to 1$ does not yield a contradiction, then (later) $y \to 1$ on its own won't reach a contradiction neither (if nothing has changed meanwhile). This line of reasoning has been considerably strengthened by "tree-based look-ahead" as introduced in [13].
4. Strengthening of r_2 by "local learning": If unsuccessfully $x \to 0$ has been tested (i.e., it does not yield a contradiction), but at least $y \to 1$ was inferred, then the binary clause $(x \vee y)$ may be learned (locally).

What is the point of local learning: Isn't the clause $(x \vee y)$ already "contained" in the current formula, and we only get a shortcut? The point here is that $x \vee y$ is equivalent to $\neg x \rightarrow y$ as well as to $\neg y \rightarrow x$, and the first direction, "from $x \rightarrow 0$ infer $y \rightarrow 1$", is given by the current formula, but the second direction "from $y \rightarrow 0$ infer $x \rightarrow 1$" in general needs a higher level to be inferred; see Subsection 7.2 for more on local learning. All enforced assignments found (iteratively) by r_k strengthened with local learning are also found by r_{k+1}, and thus r_2 with local learning (discussed in Section 3.5 in [19], an experimental feature of the `OKsolver`-2002, and (partially) used by the `march`-solvers) can be seen as an approximation of r_3. An equivalent process to r_2 with local learning is "hyper binary resolution" ([2]).

Regarding autarkies, basic autarky testing was included in the `OKsolver`-2002, and further extended by the `march`-solvers, but yet it seems not of great importance.

5.3 Heuristics

Given a multi-dimensional "distance vector", the simplest possible way to pack it into one number is to use a linear combination; for further information see [30,29], while here we do not investigate this issue further, but assume that already a "distance" is given.

5.3.1 Distances

The first main task for the heuristics is:

Given the current F and the envisaged branch $v \rightarrow \varepsilon$, how "far" do we get when applying $v \rightarrow \varepsilon$ (and simplifications) ?

So for each variable v we get two positive real numbers (d_0^v, d_1^v) (the bigger the better). Motivated by the 3-SAT upper bound in [20], the `OKsolver`-2002 introduced as distance

the number of *new* clauses.

This might be surprising, since we are not reducing the problem size — but we want to maximise the future gains by the look-ahead reductions! Note that a partial assignment is an autarky iff it does not produce new clauses; this was the reason why autarky testing for branching assignments is included in the `OKsolver`-2002. This distance turned out to be far better than the earlier simple counts (which can be understood as approximations of the number of new clauses created), and has been taken over by the `march`-solvers. Now, since shorter clauses are better, we need clause weights. Despite many efforts, yet no convincing dynamic scheme exists. A reasonable heuristics gives weight $(\frac{1}{5})^{k-2}$ to new clauses of size $k \geq 2$. (For random 3-CNF an empirically optimal scheme is obtained around these values, and look-ahead solvers can be optimised quite well for general purpose SAT solving by just looking at random formulas.)

5.3.2 Projections

Assume now that for each variable the pair (d_0^v, d_1^v) of positive real numbers is given, and we want to "project" the pair to one value $\rho(d_0^v, d_1^v)$, so that the variable with minimal projection is best. For 2 distances, i.e., binary branching, it turns out, that the product $d_0^v \cdot d_1^v$ is good enough, that is, since we are here going for minimisation, the reciprocal value. For arbitrary branching width, in [30,29] a general theory on distances and projections has been developed (based on [20]), which shows that in general there is *exactly one projection*, namely the "τ-function", and that for width two the product is a good approximation (while in general, approximations are given by generalised harmonic means).

5.3.3 The First Branch

The most reasonable basic approximation for the probability of a clause-set F being satisfiable seems to consider F as a random clause-set in the constant-density model, with mixed densities for the different clause-sizes, and to compute the probability that a random assignment satisfies such a random F, which amounts to minimise

$$\sum_{C \in F} -\log(1 - 2^{-|C|}).$$

This was applied in the `OKsolver`-2002, and an alternative scheme of similar quality is to minimise $\sum_{C \in F} 2^{-|C|}$; for more information see [30]. Howsoever the approximation is obtained, the choice of the first branch is then to choose the branch which looks more likely to be satisfiable. [16] add an additional layer of control by introducing a monitoring depth m (for example $m = 15$), and when it comes to backtracking to this depth (where we have 2^m nodes minus the ones already decided), then the simple "chronological" backtracking order is interrupted, but search continues with another branch at this level according to the principles,

1. that the left branch is preferred over the right branch (because of the direction heuristics),
2. and that higher up the search tree the direction heuristics is more error-prone (since the problems are bigger).

Good results at the SAT2007-competition are demonstrated.

6 Conflict-Driven Solvers

In this section we give an overview on the main innovations of "conflict-driven" solvers, centred around the notion of "clause learning". The basic intuitions behind conflict-driven solvers seem to be as follows:

– for "simple" but big problems (up to millions of variables);
– optimised for problems from model checking and circuit verification;
– "fast and cheap" (light-weight), nowadays only lazy data structures are used;
– zero look-ahead for the heuristics, just unit-clause propagation reduction;
– the basic aim is: seek for conflicts, learn much.

Historically, one might distinguish 3 main phases:

1. Around 1996 learning was introduced to SAT by `Grasp` ([43]), motivated by previous work in the constraint satisfaction area, but adding the specific "SAT point of view", heavily exploiting clauses and their integration with the problem instance itself.
2. Around 2001 laziness and streamlined learning together with an associated heuristics was introduced by `Chaff` ([37]), emphasising the "fast and cheap" attitude. The success of this solver was the main breakthrough, and the related ideology of a "modern solver" started spinning.
3. Finally, `Minisat` started the "clean up" (for an introduction with algorithms and implementations see [10]).

6.1 Learning

For some initial theoretical analysis (in the framework of proof complexity) see [3]; in this article we focus more on the conceptual side.

6.1.1 The Basic Ideas

Assume a DPLL-solver reached a conflict, that is, for the current partial assignment φ we have $\bot \in \varphi * F$ (where φ collects all the assignments leading to the current node). The idea is to learn about the conflict so that we can early (!) avoid it at other places in the search tree (thus going beyond tree resolution):

More precisely, we want to learn a "conflict-clause" L
(adding it to the clause-set F)
such that $F \models L$ and $\varphi(L) = 0$.

The condition $\varphi(L) = 0$ is equivalent to $L \subseteq C_\varphi$, and so this part of the learning condition is perfectly clear. It is the first condition, which is equivalent to $\varphi_L * F$ being unsatisfiable, which has a wide scope (being a coNP-complete problem), and where all the variation lies. In Section 7, based on the general considerations from Subsection 2.2.3, approaches towards a general theory of learning are outlined, which might give better explanations of the fundamental ideas of "clause learning" and might open new and more powerful perspectives for the future, especially for the unification with look-ahead solvers. Here now I outline the "traditional" ideas underlying learning in the context of conflict-driven solvers.

First we specify the situation. Let $\mathrm{var}(\varphi) = \{v_1, \ldots, v_{n(\varphi)}\}$ be ordered according to the sequence of unit-clause-eliminations and decisions along the path, where $i_1 < \cdots < i_d$ are the indices of the decision variables. So $d \in \mathbb{N}_0$ is the current depth in the search tree, and, using $\varphi_i := \varphi \mid \{v_1, \ldots, v_i\}$ and $F_i := \varphi_i * F$ for $i \in \{0, \ldots, n(\varphi)\}$, we have that for $p \in \{1, \ldots, d\}$ the clause-set $F_{i_p - 1}$ is reduced w.r.t. r_1, so that decision variable $v_{i_p} \in \mathrm{var}(F_{i_p - 1})$ was needed for further progress, while for $p \in \{0, \ldots, d\}$ and $i_p < j < i_{p+1}$ (with $i_0 := -\infty$ and $i_{d+1} := +\infty$) we have $\{x_j\} \in F_{j-1}$, where $\mathrm{var}(x_j) = v_j$ and $\varphi(x_j) = 1$ (this just means that x_j was obtained by unit-clause-elimination). Furthermore assume

$\bot \in \varphi * F$, that is, we reached a conflict, and thus $n(\varphi) > i_d$; we also assume $d \geq 1$ (so that we are not already done). Now we have:

1. It's (nearly) completely senseless to learn L_φ (the full path), since the recursive traversal of the search tree will avoid this path anyway, except of schemes with restarts or non-systematic backjumping, where completeness is only guaranteed by clause-learning, and where actually this simplest scheme of learning just the full path is sufficient to establish completeness.
2. By definition there is a clause $L^0 \in F$ with $\varphi(L^0) = 0$. Now L^0 itself is also not very exciting (we know it already), but perhaps we can do something about it? We know that there is $i_d < j \leq n(\varphi)$ with $v_j \in \text{var}(L^0)$ (at least one implied variable from the current level must be involved).
3. Consider any inferred variable $v_i \in \text{var}(L^0)$, and thus $\overline{x_i} \in L^0$. When inferring v_i, an implication
$$\bigwedge_{j \in J} x_j \to x_i$$
has been established with $J \subseteq \{1, \ldots, i-1\}$. Thus we can replace $\overline{x_i}$ in L^0 and obtain $L^1 := (L^0 \setminus \{\overline{x_i}\}) \cup \{\overline{x_j} : j \in J\}$, using that the above implication is equivalent to $\neg x_i \to \bigvee_{j \in J} \neg x_j$.
4. This process of "conflict analysis", replacing inferred literals by their premises (using contraposition), can be repeated, obtaining $L^2, \ldots$, maximally until only decision variables are left, obtaining a "strongest" conflict-clause L^*.

That's all about the basics of (current) clause-learning, and all learning schemes essentially just vary in

- which inferred literals are replaced, and how far to go back with this process;
- what conflict-clauses then actually to "learn", i.e., to add to F (we might learn several of the L^i above, and potentially also some "side-clauses" involved in the derivation);
- and when learned clauses will be eliminated again (there might be a large number of learned clauses).

The point in adding clause L to F is to enable future tree pruning by more inferred variables (see Subsection 6.1.4) and to guide the "non-chronological backjumping" process (see Subsection 6.1.2). Performing "conflict analysis", that is, recovering the implications in Step 3, is quite simple for a conflict-driven solver: Since such a solver only performs unit-clause-propagation as reduction, every inference step, that is, every new inferred assignment, is witnessed by an existing clause, in the above case by the clause $C_i := \{\overline{x_j} : j \in J\} \cup \{x_i\} \in F$, and thus together with the assignment $x_i \to 1$ a pointer to C_i is stored. This doesn't cause much space overhead, and when doing the conflict analysis, one only has to look-up the inferred literals in the current envisaged learned clause L and to decide whether to perform the substitution process of Step 3 (which corresponds to a so-called "input resolution" step). We conclude this introduction into the learning process by some observations on clauses L^p obtained in the process of conflict analysis:

1. Every L^p contains some variable from level $d > 0$ (that is, there is $i \geq i_d$ with $v_i \in \mathrm{var}(L^p)$; this is not necessarily true for $d = 0$, since for this level there is no decision variable, but it is the initialisation of the process).
2. If L^p contains some variable from level q, then this is also true for all $p' \geq p$; for p' large enough we have $\mathrm{var}(L^{p'}) \cap \{v_i : i_q \leq i < i_{q+1}\} = \{v_{i_q}\}$ (with $\overline{x_{i_q}} \in L^{p'}$).

6.1.2 Dynamic Aspects of Clause-Learning

One aspect of learning seems most important to me, and it is here that CNF plays an important role: The learned clauses are fully integrated into the original problem, and this can happens here in a simple way, since the original problem is given as a CNF, and we learn *clauses*. It is this dynamic aspect which gives the power to clause-learning, and also makes it a much more complicated process than it first appears.

Consider the situation from the previous Subsection 6.1.1, and moreover we consider the point where the first clause is learned, that is, we are on the left-most branch of the branching tree and encounter the first conflict (so the current clause-database F_0 is the original input-clause-set[6]). The clearest approach is to consider a "purged" conflict-clause L^* (which only contains decision variables). To make the example simpler, let's assume that the depth d of the current leaf is 100. L^* necessarily contains the literal $\overline{x_{i_{100}}}$; in the best of all cases that's it (since conflict-driven solvers only use r_1-reductions, this would amount to a r_2-reduction for F_0), but in general many more of the literals $\overline{x_{i_1}}, \ldots, \overline{x_{i_{99}}}$ will appear in L^*. The worst case is that L^* contains all of them, and then L^* is essentially useless, but let's be optimistic and assume that $L^* = \{\overline{x_{i_1}}, \overline{x_{i_{50}}}, \overline{x_{i_{100}}}\}$.

We see now that at depth 50 of the current path, that is for $F_{i_{50}}$, we could have inferred the assignment $\langle x_{i_{100}} \rightarrow 1\rangle$ (in general, if the solver uses r_k-reduction, this amounts to a r_{k+1}-reduction). This is now the point where "non-chronological backtracking" sets in (actually taking an "eager approach"), and the whole tree starting with this node (which in our assumed case is yet just a path) is reworked, since already the node at level 50 is no longer reduced with respect to r_1, and decision variables $v_{i_{51}}, \ldots, v_{i_{99}}$ possibly could be turned into inferred variables (and furthermore, given the new situation the heuristics might decide differently).

Whatever the learned clause L is, all implied literals from level 100 should be eliminated (i.e., for $v_i \in \mathrm{var}(L)$ we have $i \leq i_{100}$), and then in any case the previous decision variable $v_{i_{100}}$ is now turned into an inferred variable, with the forced value the opposite of the previous value. Thus, while search in the "chronological" recursive approach would consider the second branch belonging to the decision level 100, the non-chronological approach goes back to decision

[6] With "current" we refer to the possible additions of learned clauses, and not to the "residual" clause-set obtained by applying the current partial assignment.

level 50 (99 in the worst case). We see now why the algorithms for conflict-driven solvers are better expressed as an iterative procedure (instead of the usual recursive presentation), where branching and backtracking is just managed by increasing resp. decreasing the decision level, and where the second branch is induced by the conflict clauses: It is not just a matter of convenience, but the clear tree structure (as used by look-ahead solvers) is blurred by doing "partial restarts" in the form of non-chronological backjumping (restarting at $F_{i_{50}}$ in the above example) and leaving the alternation of branches to the inference mechanism — so actually there is no "second branch", and in a sense we are always in the above situation, with just a left-most branch, only that F_0 grows over time (and in practice also shrinks, due to the removal of "inactive" learned clauses — this makes the whole process completely mysterious, and I will mostly ignore this aspect here). A final remark here: As already stated for the general case, the learned clause in the above case must include some variable from level 100 (since otherwise we would have found the conflict at an earlier level), and if we then would find (directly) a conflict at level 50, again the conflict clause must then use some variable from level 50; if however we backtrack to level 0, then there is no decision variable at this level, and thus here we might learn the empty clause and thus conclude that the original input is unsatisfiable.

6.1.3 The Iterative Solving Scheme

More precisely, the iterative ("conflict-driven") procedure $\mathtt{cd} : \mathcal{CLS} \to \{0,1\}$, a special case of the general procedure $\mathfrak{G}$ from Subsection 4.1, works as follows. We use $d \in \mathbb{Z}$ for the decision level (with $d = -1$ indicating an "impossible backtrack"). Instead of managing one global current partial assignment $\varphi \in \mathcal{PASS}$, which is expanded on branching and shrunk on backtracking, for the clarity of exposition we use $\psi_i \in \mathcal{PASS}$ for the initial partial assignment at level i, while $\psi'_i \in \mathcal{PASS}$ is the extended partial assignment with forced assignments added:

$\mathtt{cd}(F \in \mathcal{CLS}) : \{0,1\}$

0. Initialisation: $d := 0$, $\psi_d := \emptyset$. If $\bot \in F$, then return 0.
1. Reduction: Let ψ'_d be obtained by unit-clause-propagation on $\psi_d * F$, that is, $\psi'_d \supseteq \psi_d$ with $\psi'_d * F = r_1(\psi'_d * F) = r_1(\psi_d * F)$.
2. Analysis: Evaluate $\psi'_d * F$.
   (a) If $\psi'_d * F = \top$, then return 1.
   (b) If $\bot \in \psi'_d * F$ then backtrack:
      i. Compute a conflict-clause $L \notin F$ (with $\mathrm{var}(L) \subseteq \mathrm{var}(F)$) w.r.t. ψ'_d.
      ii. $F := F \cup \{L\}$.
      iii. While $d \geq 0$ and $\bot \in \psi'_d * F$ do $d := d - 1$.
      iv. If $d = -1$ then return 0.
      v. While $d \geq 1$ and $r_1(\psi'_{d-1} * F) \neq \psi'_{d-1} * F$ do $d := d - 1$.
      vi. Go to Step 1.
3. Branching:
   (a) Choose a branching variable $v \in \mathrm{var}(\psi'_d * F)$ and $\varepsilon \in \{0,1\}$.
   (b) $d := d + 1$, $\psi_d := \psi'_{d-1} \cup \langle v \to \varepsilon \rangle$.
   (c) Go to Step 1.

Explanations and remarks:

1. Procedure `cd` is correct and also always terminates (whatever the conflict-clause is — if only it is just a new clause).
2. The main choices are the choice of branching variable and first branch in Step 3a, which is discussed in Subsection 6.2, and the choice of the conflict-clause L, which is discussed in Subsection 6.1.4.
3. The conflict-clause L in Step 2(b)i can be chosen here according to the most loose semantics, namely any clause L with $L \subseteq C_{\psi'_d}$ and $F \models L$. It seems not being discussed in the literature, but in the `OKlibrary` we are experimenting with an "afterburner" for clause-learning, which uses additional reasoning power for strengthening the conflict-clause[7], and then in Step 2(b)iii possibly the "real backtrack" (the backtrack which is necessary) could be more than one step, but if learning is restricted to the conflict analysis from Subsection 6.1.1, then in Step 2(b)iii the depth d is decremented exactly once.
4. In the above formulation of `cd` we consistently used applications of partial assignments to the original input, and not to residual formulas, in order to emphasise the "lazy" aspect of handling the application of partial assignments.
5. Step 2(b)v is the non-chronological backtracking step:
   (a) Correctness and termination does not depend on this step (but only on the added conflict clauses): We could leave it out, or backtrack even further; backtracking to level $d = 0$ would be a full restart (while still keeping the learned clauses(!)).[8]
   (b) If $r_1(\psi'_{d-1} * F) \neq \psi'_{d-1} * F$, i.e., one level down there are further unit-clause eliminations possible, then we have also $r_1(\psi'_d * F) \neq \psi'_d * F$. So in general the decision levels after adding the conflict-clause can be divided into three connected parts: At the end we have the "contradicting levels" (only d in the standard situation), then come the "active levels" where further unit-clause propagations are possible, and Step 2(b)iii jumps to the beginning of this segment, and finally (i.e., at the beginning of the decision stack) we have the "unaffected levels" (this segment is empty iff we learned a clause L which after elimination of level-0-variables contains at most one literal).
   (c) Since we only learn one clause L (instead of several clauses) and only use r_1 (instead of for example r_2), the condition "$r_1(\psi'_{d-1} * F) \neq \psi'_{d-1} * F$" is equivalent to $|\psi'_{d-1} * L| = 1$, that is, all but exactly one literal in L is falsified by the "current" partial assignment of level $d - 1$. This is the condition as normally stated, but the condition in Step 2(b)v seems to be the real underlying reasoning (also allowing to use stronger means than r_1).

[7] This is more natural for "look-ahead solvers", since they have a stronger reasoning machinery anyway.

[8] However if this step is not performed fully then the "real backtrack" in Step 2(b)iii could involve more than one level even when sticking to the conflict analysis from Subsection 6.1.1, since earlier decision levels might have unprocessed unit-clause propagations.

6.1.4 Learning Schemes

Now we consider Step 2(b)i in algorithm cd from Subsection 6.1.3 in more detail; in Subsection 6.1.1 we have outlined the general idea, resulting in a non-deterministic sequence $L^0, \dots, L^*$ of conflict clauses (but with well-defined first and last element), and the question is now which L^p to learn.

As we have already explained, in order to make the backtracking system "aware" of the fact, that the decision $x_{i_d} \to 1$ was a failure, and hence $x_{i_d} \to 0$ is inferred, we need $\bot \in \psi_d * F$ after the learning step, i.e., all inferred literals from level d must have been eliminated in the learned clause L (such a conflict-clause then is called "asserting"). Just performing such elimination steps (that is, only eliminating inferred literals from level d) we obtain a well-defined conflict-clause L^+, which can be considered as the "weakest" conflict-clause, while at the other end of the spectrum we have the "strongest" conflict-clause L^*, where all possible elimination steps have been performed (and only decision variables are left). All existing learning schemes are situated between L^+ and L^*, where the choice of L^+ is called "1UIP" ([48]), while the choice of L^* has been given different names like "decision cut".[9] Unfortunately not much can really be said here, but let us consider the most fundamental decision, between L^+ and L^*. Above they have been called "weakest" and "strongest" according to the effort involved to obtain these clauses — now is this really true, that is, does the increased effort for computing L^* at least pay off in a smaller search tree? Due to the "sporadic" character of problem instances where conflict-driven solvers are successful (not on the most amenable instances for systematic studies, random formulas, where conflict-driven solvers perform badly), and due to the high sensitivity of solvers regarding the learning process (which is determinative for the heuristics), it seems that all empirical arguments are rather weak. And there are no theoretical results in any form. However, "in practice" L^+ (that is, the 1UIP-scheme) turns out to be the winner (and this for all learning schemes). It seems that despite all arguments for various schemes now the field of conflict-driven solvers converges on 1UIP.

As we have already mentioned, if some variable from level i is involved, then never level i can be emptied (because we only use the old inferences), and since the literal elimination process replaces literals by literals with smaller indices, we see that for the non-chronological backjump depth the clause L^+ is as good as L^* or anything between, so this step is not influenced by the learning scheme. In [9] it is argued that 1UIP has an advantage over other schemes because empirically more unit-clause propagations are enabled by these conflict-clauses (one could roughly say that 1UIP offers more "surface" for future conflicts (i.e., resolution steps), while going far back somehow narrows the choices). At other places it is furthermore argued that going "far back" conflicts with the idea of "locality", namely that statistics on usage of conflict clauses shall guide the branching heuristics.

[9] This is perhaps the right point to remark that I fail to see the point of the common "cut terminology", where a (fake) directed graph is constructed recording the events of literal inferences: The (in principal) very simple character of learning is obscured in this way, and if graph-theoretical notions shall be employed then one should use the appropriate notion of *directed hypergraphs* here.

6.2 Heuristic

Finally we investigate the heuristics for the choice of branching variable v and branching value ε in Step 3a of procedure cd from Subsection 6.1.3. It seems that regarding the heuristics, conflict-driven solvers still live in the "stone age" of simple literal counts, far behind look-ahead solvers — but they have one special weapon, based on the dynamic nature of the "clause database" due to the added conflict-clauses. However, compared to the situation for look-ahead solvers, where we have two independent branches and thus the total workload is minimised while as first branch one is chosen which could make a difference (i.e., where actually the independence breaks down, since in case of a satisfiable assignment found we simply abort), here now the notion of "branches" is broken open, and a more global situation has to be faced. The current guideline seems to be a greedy approach, seeking for as many "profitable" conflicts as possible.[10]

Regarding the branching value, there are two conflicting goals: Seeking for "good" conflicts, or trying to find a satisfying assignment. In [37] the greedy choice of searching for conflicts is also applied to the branching value, but apparently this wasn't very successful and the simple choice $\varepsilon := 0$ seems to be more popular and still the prevalent method (and only with SAT 2007 some discussions started regarding an improved choice). Perhaps due to their weak "statistical infrastructure", conflict-driven solvers don't have good measures at hand to estimate the probability of satisfiability, and thus do not employ a direction heuristics as discussed in Subsection 5.3.3 for look-ahead solvers. Furthermore it seems that on many instances coming from hardware verification actually truth value 0 is a reasonable choice due to the special encoding. And, as already mentioned, while for a look-ahead solver in principle the direction is clear (towards satisfiability), for a conflict-driven solver also a "fruitful conflict" might be attractive.

Now regarding the branching variable (for an overview on some techniques see [37]) the basic is just the (static) literal count for the input clause-set (preference for higher counts). This static count evolves into some *activity measurement* by dynamic updates:

- learning a clause containing the variable increases the activity;
- the activity decays over time.

Following the "locality idea", branching variables are chosen which have the highest activity. The motivation for such schemes might be summarised by

Where are many conflicts, there will be more.
And conflicts are good (since they cut off branches).

Various schemes about how much to increase and to decay have been proposed, but it seems to me that only the idea of variable activity regarding activity in conflict-clauses has fundamental virtues.

[10] Perhaps it is this analogy which drives proponents of conflict-driven solvers to call their approach "modern", while denying this qualification to "modern" look-ahead solvers: "Modern" in a sense of "modern (disaster) capitalism", and "old-fashioned" in the sense of "socialistic planning".

7 Towards a General Theory of Clause-Learning

In this final section I want to present some general ideas and methods which extend clause-learning, and can help putting it into a wider context.

7.1 From Branching Trees to Resolution Trees

The essential first step in understanding clause-learning is to understand the full translation of backtracking trees (possibly with local reductions like r_k) into resolution trees. A complete presentation in a general context is given in [26], but the principle is very simple:

1. Unfold the backtracking tree, replacing r_k-reductions by little sub-trees in (generalised) input-resolution-tree shape, finally obtaining a pure backtracking tree where nodes just represent splitting on variables.
2. At each leaf select a clause falsified by the partial assignment corresponding to the path to that leaf.
3. Starting from the leaves, perform resolution operations where branching variables now become resolution variables.
4. Cases, where actually one of the parent clauses of an envisaged resolution step does not contain the resolution variable, correspond to "intelligent backtracking": Only this branch is kept, while the other is discarded.

A "global learning step" corresponds to creating the tree from the leaves in a certain order and then allowing to link to the learned clause from later parts of the graph (which becomes a dag now). Learning "non-conflict clauses" corresponds to learning clauses from the "little sub-trees" corresponding to inference steps.

7.2 Local Versus Global Learning

Due to their lazy datastructures, conflict-driven solvers have problem seeing the current clause-set (that is, with the current partial assignment applied), and accordingly the clauses they learn are always "global", that is, all assumptions are carried out and the clause can be added to the original input clause-set. However, it might be worth learning clauses first "locally", and unfolding the assumptions (decisions) only when backtracking.

The simplest such learning scheme for look-ahead solvers has been already mentioned in Subsection 5.2: Recall the r_2-reduction from Subsection 3.1, and assume that when testing the assignment $x \to 1$ we derived $y \to 1$ (by unit-clause propagation) but we didn't reach a failed literal (and thus r_2 was unsuccessful in this case); actually any reduction which allows from assumptions $x \to 1$ to infer forced assignments $y \to 1$ can be used here. Though the reduction attempt failed, nevertheless we gained some inference information, and how can we use it? The derivation $x \leadsto y$ is just equivalent to the fact that the clause $\{\overline{x}, y\}$ follows from the current clause-set, and thus it can be learned here. Note that the clause $\{\overline{x}, y\}$ is a valid inference only "locally", that is, w.r.t. the current partial assignment (which is carried out in look-ahead solvers), while when backtracking then either the clause must be "unfolded", that is, some conflict analysis has to be performed to make the assumption about the previous decision variable explicit, or the locally learned clauses are just discarded when backtracking.

Let us again point out what is the use of locally learning the clause $\{\overline{x}, y\}$: For global learning, as described before, we mentioned the tree pruning effect, exploiting conflict analysis, which obviously cannot happen if we discard the clause upon backtracking. And the implication $x \to y$ is contained in the current formula anyway, so why learning $\{\overline{x}, y\}$? First, there is the aspect of a "short cut", but more importantly the clause $\{\overline{x}, y\}$ represents *both* implications $x \to y$ and $\neg y \to \neg x$, and while the first implication is found by the current reduction scheme like r_k, for the converse we need the next level of reduction r_{k+1} ! This effect is also active for global learning, and so learning of clauses L (as discussed in Subsection 6.1.1) which still contain inferred literals can yield inferences which are not obtained by the in a sense strongest learned clause L^* — again, the "forward" implications are all contained in L^*, but the "backward" implications need a higher level of reduction to come to light.

In the general DPLL-procedure $\mathfrak{G}$ from Subsection 4.1, local learning is the task of the reduction r: the learned clauses are discarded upon backtracking, and only via Step 2(b)i of $\mathfrak{G}$ (the formation of learned clauses) can information be saved from oblivion. $\mathfrak{G}$ doesn't allow intermediate steps, the gradual unfolding of learned information, and it seems to me that this should be an interesting possibility to explore; for now however we only consider "pure" local learning. We have already seen that r_k-reduction with local learning can be simulated by r_{k+1}-reduction. Asking about the strength as a proof system of branching trees with local learning in this sense is thus translated into the question about the strength of branching trees with reduction r_{k+1} at each node. If k is fixed, then such trees can save a polynomial factor over simple branching trees (that is, over tree resolution), which can be relevant for practice, but is less impressive from a proof-theoretical point of view: The reason about the reduced strength of local learning is that the learned clauses only gather information along the *path* from the root to the current node, while a global learned clause gathers information from its whole corresponding sub-*tree*. So aspects of global learning are definitely necessary to reach greater proof-theoretic strength.

However a (theoretical) possibility for a form of local learning which can even go beyond full resolution is that in Step 3d of $\mathfrak{G}$ we use a branching $\mathbb{B}$ such that $\{C_\varphi : \varphi \in \mathbb{B}\}$ is a hard unsatisfiable clause-set.[11] Yet such schemes were rarely considered, and they look rather difficult to make efficient; a possibly more accessible extension of resolution is considered in the following subsection.

7.3 Compressed Learning

In [26] resolution was generalised by using "oracles" $\bot \in \mathcal{U} \subseteq \mathcal{USAT}$ for unsatisfiability, where the only condition is that $\mathcal{U}$ is stable under application of

[11] The task then is to make $\mathbb{B}$ as "similar" to F as possible — the best case is to use $\mathbb{B} = \{\varphi_C : C \in F\}$!

partial assignments. Using $\mathcal{U} = \mathcal{USAT}$ trivialises everything, while $\mathcal{U} = \{\bot\}$ is exactly tree resolution. The abstract point of view of [26] is that from a problem instance P we "see" only whether for a partial assignment φ we have $\varphi * P \in \mathcal{U}$ or not, where in the positive case we learn C_φ, and the set of all learned clauses constitutes F which is the basis for some ordinary resolution process. For conflict-driven solvers, $\mathcal{U}$ would be just the set of clause-sets refutable by r_1 (that is, the set of clause-sets containing an unsatisfiable renamed Horn clause-set), while a reasonable stronger $\mathcal{U}$ here could be the set of clause-sets refuted by r_2 combined with some form of equivalence reasoning. Learning only clauses consisting solely of decision variables abstracts away from the inferences, and a strong $\mathcal{U}$ yields smaller branching trees as well as shorter learned clauses. One should remark here that in [26] a kind of simple "static" point of view is taken, and learning happens only at the leaves while at inner nodes just resolution steps happen, but obviously this can be made dynamic by taking the learned clauses into account for new learning steps (and by using non-chronological backtracking).

A big problem (quite literally) for the unification of look-ahead techniques with the conflict-driven approach is that the latter focuses on rather big instances, where the polynomial-time overhead of look-ahead solvers can actually result in weeks(!) of wasted run-time (on a single instance). As envisaged in the plans for the new `OKsolver` (which are contained in the `OKlibrary` themselves), the usage of an "after-burner" for learning, which only turns on the stronger inference machine for compressing clauses to be learned, could be a solution for this problem.

8 Conclusion

There is something more fundamental to "clause-learning" and "conflict-driven solvers" than, as suggested, "raw speed, super-efficient implementations and cleverly adapted heuristics". Instead, the old paradigm of backtracking-search has been made "reflective", reflecting the search meta-level onto the problem object-level. Perhaps only in the "purified context" of SAT (compared to CSP) this paradigm could have been evolved further, given our current (lack of) understanding.

However, also the traditional backtracking approach has seen substantial improvements through the look-ahead techniques and the related theory of backtracking heuristics. There is a certain agreement that SAT has reached in a certain way three "local optima" (for local search, look-ahead and conflict-driven), which actually seem to be three rather large plateaus. One opinion on this situation is that further progress with SAT lies mainly in considering applications, the "user interface", and integration with extensions. A (smaller) part of the SAT community however believes that we just started, and I hope that with this article some elements of the SAT-solvers to come have been outlined. If, metaphorically speaking, the learning-based approaches created a mirror-cabinet on the search process, the task is now to look behind the mirrors.

References

1. Ansótegui, C., Bonet, M.L., Levy, J., Manyà, F.: Mapping CSP into many-valued SAT. In: Marques-Silva and Sakallah [38], pp. 10–15, ISBN 978-3-540-72787-3
2. Bacchus, F., Winter, J.: Effective preprocessing with hyper-resolution and equality reduction. In: Giunchiglia and Tacchella [12], pp. 341–355, ISBN 3-540-20851-8
3. Beame, P., Kautz, H., Sabharwal, A.: Towards understanding and harnessing the potential of clause learning. Journal of Artificial Intelligence Research 22, 319–351 (2004)
4. Böhm, M.: Verteilte Lösung harter Probleme: Schneller Lastausgleich. Ph.D thesis, Universität Köln (1996)
5. Dalal, M., Etherington, D.W.: A hierarchy of tractable satisfiability problems. Information Processing Letters 44, 173–180 (1992)
6. Davis, M., Logemann, G., Loveland, D.: A machine program for theorem-proving. Communication of the ACM 5, 394–397 (1962)
7. Davis, M., Putnam, H.: A computing procedure for quantification theory. Journal of the ACM 7, 201–215 (1960)
8. Dequen, G., Dubois, O.: Kcnfs: An efficient solver for random k-SAT formulae. In: Giunchiglia and Tacchella [12], pp. 486–501, ISBN 3-540-20851-8
9. Dershowitz, N., Hanna, Z., Nadel, A.: Towards a better understanding of the functionality of a conflict-driven SAT solver. In: Marques-Silva and Sakallah [38], pp. 287–293, ISBN 978-3-540-72787-3
10. Eén, N., Sörensson, N.: An extensible SAT-solver. In: Giunchiglia and Tacchella [12], pp. 502–518, ISBN 3-540-20851-8
11. Freeman, J.W.: Improvements to propositional satisfiability search algorithms. PhD thesis, University of Pennsylvania (1995)
12. Giunchiglia, E., Tacchella, A. (eds.): SAT 2003. LNCS, vol. 2919. Springer, Heidelberg (2004)
13. Heule, M., Dufour, M., van Zwieten, J., van Maaren, H.: March_eq: Implementing additional reasoning into an efficient look-ahead SAT solver. In: Hoos and Mitchell [17], pp. 345–359, ISBN 3-540-27829-X
14. Heule, M., van Maaren, H.: Aligning CNF- and equivalence-reasoning. In: Hoos and Mitchell [17], pp. 145–156, ISBN 3-540-27829-X
15. Heule, M., van Maaren, H.: Effective incorporation of double look-ahead procedures. In: Marques-Silva and Sakallah [38], pp. 258–271, ISBN 978-3-540-72787-3
16. Heule, M.J.H., van Maaren, H.: Whose side are you on? Finding solutions in a biased search-tree (October 2007) (to appear)
17. Hoos, H.H., Mitchell, D.G. (eds.): SAT 2004. LNCS, vol. 3542. Springer, Heidelberg (2005)
18. Knuth, D.E.: Dancing links. Technical Report cs/0011047v1, arXiv (November 2000), `http://www.arxiv.org/abs/cs/0011047v1`
19. Kullmann, O.: Investigating a general hierarchy of polynomially decidable classes of CNF's based on short tree-like resolution proofs. Technical Report TR99-041, Electronic Colloquium on Computational Complexity (ECCC) (October 1999)
20. Kullmann, O.: New methods for 3-SAT decision and worst-case analysis. Theoretical Computer Science 223(1-2), 1–72 (1999)
21. Kullmann, O.: On a generalization of extended resolution. Discrete Applied Mathematics 96-97(1-3), 149–176 (1999)
22. Kullmann, O.: Investigations on autark assignments. Discrete Applied Mathematics 107, 99–137 (2000)

23. Kullmann, O.: On the use of autarkies for satisfiability decision. In: Kautz, H., Selman, B. (eds.) LICS 2001 Workshop on Theory and Applications of Satisfiability Testing (SAT 2001). Electronic Notes in Discrete Mathematics (ENDM), vol. 9. Elsevier Science, Amsterdam (2001)
24. Kullmann, O.: Investigating the behaviour of a SAT solver on random formulas. Technical Report CSR 23-2002, Swansea University, Computer Science Report Series (October 2002), `http://www-compsci.swan.ac.uk/reports/2002.html`
25. Kullmann, O.: Lean clause-sets: Generalizations of minimally unsatisfiable clause-sets. Discrete Applied Mathematics 130, 209–249 (2003)
26. Kullmann, O.: Upper and lower bounds on the complexity of generalised resolution and generalised constraint satisfaction problems. Annals of Mathematics and Artificial Intelligence 40(3-4), 303–352 (2004)
27. Kullmann, O.: Constraint satisfaction problems in clausal form: Autarkies and minimal unsatisfiability. Technical Report TR 07-055, Electronic Colloquium on Computational Complexity (ECCC) (June 2007)
28. Kullmann, O.: Polynomial time SAT decision for complementation-invariant clause-sets, and sign-non-singular matrices. In: Marques-Silva and Sakallah [38], pp. 314–327, ISBN 978-3-540-72787-3
29. Kullmann, O.: Fundaments of branching heuristics. In: Biere, A., van Maaren, H., Walsh, T. (eds.) Handbook of Satisfiability. IOS Press, Amsterdam (2008)
30. Kullmann, O.: Fundaments of branching heuristics: Theory and examples. Technical Report CSR 7-2008, Swansea University, Computer Science Report Series (April 2008),
`http://www.swan.ac.uk/compsci/research/reports/2008/index.html`
31. Kullmann, O., Luckhardt, H.: Deciding propositional tautologies: Algorithms and their complexity, 82 pages (preprint) (January 1997),
`http://cs.swan.ac.uk/~csoliver/`
32. Kullmann, O., Luckhardt, H.: Algorithms for SAT/TAUT decision based on various measures, 71 pages (preprint) (December 1998),
`http://cs.swan.ac.uk/~csoliver/`
33. Kullmann, O., Lynce, I., Marques-Silva, J.: Categorisation of clauses in conjunctive normal forms: Minimally unsatisfiable sub-clause-sets and the lean kernel. In: Biere, A., Gomes, C.P. (eds.) SAT 2006. LNCS, vol. 4121, pp. 22–35. Springer, Heidelberg (2006)
34. Li, C.M., Anbulagan: Heuristics based on unit propagation for satisfiability problems. In: Proceedings of 15th International Joint Conference on Artificial Intelligence (IJCAI 1997), pp. 366–371. Morgan Kaufmann Publishers, San Francisco (1997)
35. Liffiton, M., Sakallah, K.: Searching for autarkies to trim unsatisfiable clause sets. In: Kleine Büning, H., Zhao, X. (eds.) SAT 2008. LNCS, vol. 4996, pp. 182–195. Springer, Heidelberg (2008)
36. Lovász, L., Naor, M., Newman, I., Wigderson, A.: Search problems in the decision tree model. SIAM Journal on Discrete Mathematics 8(1), 119–132 (1995)
37. Mahajan, Y.S., Fu, Z., Malik, S.: Zchaff2004: An efficient SAT solver. In: Hoos and Mitchell [17], pp. 360–375, ISBN 3-540-27829-X
38. Marques-Silva, J., Sakallah, K.A. (eds.): SAT 2007. LNCS, vol. 4501. Springer, Heidelberg (2007)
39. Mitchell, D.G., Hwang, J.: 2-way vs. d-way branching for CSP. In: van Beek, P. (ed.) CP 2005. LNCS, vol. 3709, pp. 343–357. Springer, Heidelberg (2005)
40. Monien, B., Speckenmeyer, E.: Solving satisfiability in less than 2^n steps. Discrete Applied Mathematics 10, 287–295 (1985)

41. Purdom, P.W.: Solving satisfiability with less searching. IEEE Transactions on Pattern Analysis and Machine Intelligence 6(4), 510–513 (1984)
42. Segerlind, N.: The complexity of propositional proofs. The Bulletin of Symbolic Logic 13(4), 417–481 (2007)
43. Marques Silva, J.P., Sakallah, K.A.: GRASP: A search algorithm for propositional satisfiability. IEEE Transactions on Computers 48(5), 506–521 (1999)
44. Simon, L., Le Berre, D.: The essentials of the SAT 2003 competition. In: Giunchiglia, E., Tacchella, A. (eds.) SAT 2003. LNCS, vol. 2919, pp. 452–467. Springer, Heidelberg (2004)
45. Simon, L., Le Berre, D., Hirsch, E.A.: The SAT2002 competition. Annals of Mathematics and Artificial Intelligence 43, 307–342 (2005)
46. Urquhart, A.: The complexity of propositional proofs. The Bulletin of Symbolic Logic 1(4), 425–467 (1995)
47. Zhang, L.: On subsumption removal and on-the-fly CNF simplification. In: Bacchus, F., Walsh, T. (eds.) SAT 2005. LNCS, vol. 3569, pp. 482–489. Springer, Heidelberg (2005)
48. Zhang, L., Madigan, C.F., Moskewicz, M.H., Malik, S.: Efficient conflict driven learning in a boolean satisfiability solver. In: Proceedings of the International Conference on Computer Aided Design (ICCAD), pp. 279–285. IEEE Press, Los Alamitos (2001)

Author Index